Lakes of the Warm Belt

Frontispiece. Lake Maracaibo. North of the lake the Strait of Maracaibo extends into the Bay of Tablazo. The Guasare River joins the western bank of the Bay of Tablazo which is limited on its northern side by barrier bars and islands. Photo: Landsat, 18 January 1973

Lakes of the Warm Belt

COLETTE SERRUYA
Director
Israel Oceanographic &
Limnological Research
Haifa, Israel

UTSA POLLINGHER
Senior Research Biologist
Israel Oceanographic &
Limnological Research
Haifa, Israel

CAMBRIDGE UNIVERSITY PRESS
Cambridge
London New York New Rochelle
Melbourne Sydney

CAMBRIDGE UNIVERSITY PRESS
Cambridge, New York, Melbourne, Madrid, Cape Town, Singapore, São Paulo, Delhi

Cambridge University Press
The Edinburgh Building, Cambridge CB2 8RU, UK

Published in the United States of America by Cambridge University Press, New York

www.cambridge.org
Information on this title: www.cambridge.org/9780521105842

© Cambridge University Press 1983

This publication is in copyright. Subject to statutory exception
and to the provisions of relevant collective licensing agreements,
no reproduction of any part may take place without the written
permission of Cambridge University Press.

First published 1983
This digitally printed version 2009

A catalogue record for this publication is available from the British Library

Library of Congress Catalogue Card Number: 82-19857

ISBN 978-0-521-23357-6 hardback
ISBN 978-0-521-10584-2 paperback

Contents

Preface	ix
Introduction	1
Part I **The general features of the Warm Belt**	5
1 *Geodynamic history of the Warm Belt*	7
Introduction	7
Movement of continents and microcontinents	7
The Warm Belt in a new geodynamic perspective	9
2 *Climatological features of the Warm Belt*	12
The sun–earth relationship	12
The earth's atmosphere and heat balance	14
The global earth circulation	16
3 *Hydrology*	19
The water balance of the world	19
The water balance of the Warm Belt in comparison with the world's water balance	22
Part II **The aquatic ecosystem of the Warm Belt**	29
4 *South America*	31
Introduction	31
The Caribbean area	32
The Orinoco system	44
The Guiana area	50
Amazonia	55
The Andes	83
The Parana system and Rio de la Plata basin	96

5	*Central America*	104
	Introduction	104
	Lakes of Panama	105
	Lakes of Nicaragua	108
	Lakes of Guatemala	112
	Lakes of El Salvador	121
	Lakes of Mexico	124
	High mountain lakes of Central America	126
	Fishes of Central America	127
	West Indies	129
6	*Africa*	131
	Introduction	131
	The Nile basin	134
	The Upper Nile Valley	135
	The Albert Nile and Sudd Swamps	162
	The White Nile	166
	Lake Tana and the Blue Nile	168
	The Main Nile and Aswan High Dam	173
	Lakes of the Eastern Rift Valley and Ethiopian Rift Valley	192
	Eastern Rift Valley lakes	192
	Ethiopian lakes	204
	The Zaire basin	211
	The Zambezi basin	238
	The Niger basin	263
	Lake Chad	268
	The Volta basin	279
	Madagascar	284
7	*Middle East*	288
	Sinai peninsula	288
	The Jordan Valley	293
	The waters of Iraq	308
	The waters of Iran	312
8	*South-East Asia*	313
	Introduction	313
	India	314
	West Malaysia	349
	The Indo-Chinese peninsula	358
	Indonesia	372

	The Philippines	383
	Papua New Guinea	391
9	*Australia*	397
	Introduction	397
	Victoria	398
	New South Wales	403
	Queensland	406
	Northern Territory	408
	Western Australia	408
	South Australia	409
	The chemistry of Australian waters	413
	Biological features of inland waters of Australia	415
Part III	**The dynamic processes in lakes of the Warm Belt**	421
10	*Mechanisms and patterns of circulation*	423
	The main factors affecting hydromechanical events in tropical lakes	423
	Stratification and mixing	426
	A few types of lake and their circulation patterns	430
11	*Water chemistry*	436
12	*Biological diversity*	444
	Bacteria	444
	Phytoplankton	447
	Protozoa	465
	Rotifera	466
	Cladocera	471
	Copepoda	474
	Fish diversity in the Warm Belt	480
13	*Food webs: case studies*	490
	A detrital, aquatico-terrestrial food web: the várzea	490
	A plankton food web based on grazing: Lake Tanganyika	493
	The double food chain of Lake Kinneret	495
	The short food chain of a shallow equatorial lake: Lake George	497
	Appendix	499
	Bibliography	
	Index	

Preface

Limnology was born on the shores of Lake Geneva with the pioneer work of Forel (1892). This new branch of science developed mostly in Europe and North America, although scientific expeditions brought sporadic data from exotic countries. Centres of interest in limnological research varied with time. Limnology first focused on lake typology (1920–1930), then on energy-flow problems (1940–1950). From the 1960s, the acute problems of lake pollution caused limnological research to be directed toward the study of nutrient–algal production relations. These investigations led to practical applications; in particular, schemes of lake restoration were elaborated and gave spectacular results.

This remarkable development of limnology in less than 80 years concerned mostly cold and temperate lakes, as can be seen from the publications of the International Association of Theoretical and Applied Limnology. In the Congresses of Winnipeg (1974) and Copenhagen (1977), the publications devoted to the tropical areas represented less than 5% of the published material in comparison with 30% for North America and Europe. Only in the Congress of Kyoto (1980) was a special workshop held for the promotion of limnology in the developing countries which coincide with the tropical areas. This workshop was preceded by a more specialized meeting held at Nairobi in December 1979 on African limnology. Moreover, several pioneer books concerning various aspects of tropical limnology were published in the seventies: *The cichlid fishes of the Great Lakes of Africa* (Fryer & Iles, 1972); *The inland waters of tropical Africa* (Beadle, 1974); *Fish communities in tropical freshwaters* (Lowe-McConnell, 1975); and a series of monographs on the Nile, Lake Kinneret, Lake Chilwa and the Euphrates and Tigris.

This brief review indicates that, although there is a recent increase of interest for tropical limnology, the publications on the subject are not as abundant as required by the present development of the countries of the Warm Belt. Moreover, many relevant papers are scattered in numerous international and local journals and this dispersion does not favour a synthetical view of the problem specific to the tropics. An effort is needed to make the existing information on warm lakes accessible to a large public.

The present volume is a modest contribution in this direction. The first part of the volume describes the geological, climatological and hydrological features of the Warm Belt. The second and most extensive part of the book describes the main rivers, lakes and reservoirs of South America, Central America, Africa, South-East Asia and Australia. In the third and last part, the features described and analysed in the second part are utilized to define the specific physical, chemical and biological features of the lakes of the Warm Belt and to compare them with those of the corresponding temperate ecosystems.

We have consulted many sources of literature and have tried to include all the most important warm bodies of the Warm Belt, but we have probably missed valuable information and relevant data. We accept entire responsibility for these possible omissions. We do hope that, as it is, the book will be useful to all the scholars who are interested in warm limnology.

Acknowledgements

Our warmest thanks go to Mr O.H. Oren, from Israel Oceanographic & Limnological Research, for suggesting the idea of a book on warm limnology.

Because of the diversity of the subjects treated in this book, the discussions with and the reviews of various specialists were particularly valuable. We are especially indebted to Dr E. Holland, Landesanstalt für Umweltschutz Baden-Württemberg, Karlsruhe, with whom we discussed at length the problems of circulation patterns. We also wish to thank Dr Z. Ben Avraham, who brought to our attention the latest developments in tectonophysics. We are grateful to Dr Besch, LFU, Karlsruhe, who made available to us his own unpublished results of his studies in San Salvador, and to Dr J. Luna, Inter-Development Bank, Washington D.C., who sent us information concerning the fisheries of South and Central America.

Our special thanks are due to all the scientists who were kind enough to send the numerous reprints we requested.

We have consulted the libraries of many scientific institutions in Israel and abroad (the Institute of Limnology of the University of Constance, the Max Planck Institute of Limnology of Plön, and the Institute of Limnology of Uppsala), and gratefully acknowledge their valuable help. We wish to mention specially the library of the Israel Oceanographic & Limnological Research in Haifa, which has very rich material on the subject.

Introduction

The lakes of the Warm Belt of the earth are not characterized by a specific range of temperature. Many of them are 'warm lakes'; others are cold water bodies located in tropical or equatorial highlands. They all belong to an area which has small seasonal variations in temperature and which has been preserved from the Pleistocene glaciations. In the northern hemisphere, the Quaternary ice covered Greenland, much of Canada and the north-eastern part of the United States, Scandinavia, the Baltic and North Seas, Britain and vast areas of Germany, Poland, Russia and Japan. In the southern hemisphere, the glaciers covered Antarctica, the southern part of the Andean Cordillera, New Zealand and Tasmania. The Warm Belt is located in the area extending between these two domains of the Pleistocene glaciations, approximately between 30° N and 30° S.

The privileged climate of the Warm Belt is in fact much more ancient. It seems that, since the Permian, the Warm Belt has not experienced any marked cold period. The Early Mesozoic was drier and warmer than the Late Mesozoic; Australia and India were then more humid than America and Africa. During the Cenozoic, humidity seems to have been in general higher, and areas presently arid, such as central Australia, were then more humid. Conversely, areas presently very humid, such as South America, were then much drier. During the Pleistocene, alternation of dry and humid episodes accompanied the glacial periods. Humid phases occurred even in the Sahara, where a fauna of herbivores developed. Conversely, 'dead dunes' can be found below the present Soudanian savanna.

The dominance of warm climates without interruption since the end of the Palaeozoic has led to the idea that the present luxuriance of vegetation and animal life of the Warm Belt was the result of a slow evolution in a constant environment. The recent studies which have revealed the considerable variation of humidity in many regions of the Warm Belt have completely modified this concept. In dry periods flora and fauna have retreated to relatively wet sanctuaries from which they have dispersed again during the next humid phase. The concept of long evolution in a constant environment is now replaced by the idea of evolution resulting from contraction and expansion of humid areas.

Precambrian rocks dominate the Warm Belt which essentially belongs to the Gondwanalandian domain. The various regions of the present Warm Belt result from the division of the old continent of Gondwanaland into separate subcontinents along rift lines and their northward migration and drifting which started during the Mesozoic. The present distribution of fishes in the Warm Belt can be understood only in the light of this geodynamical history of the earth. In continental units which parted early from Gondwanaland, such as India and Madagascar, the primary freshwater fauna is essentially composed of ancient fishes such as Dipnoi and osteoglossids, since the modern primary freshwater fishes appeared in Gondwanaland after these units had drifted away.

In contrast to the fish fauna, the distribution of algae seems little affected by the history of the earth. The idea that there is less diversity among tropical algae in comparison with temperate phytoplankton is not supported by our analysis. So far, ideas about phytoplankton diversity depend mostly on sampling techniques. Authors who have worked on samples from which nanoplankton was not excluded have found that algal communities of the lakes of the Warm Belt were at least as diverse as those of the temperate lakes. Moreover, notions about 'cold stenotherm' species have to be revised since many of the so-called 'cold species' are abundant in warm waters.

Whereas most of the common algae of the Warm Belt are cosmopolitan species, several species are typically pantropical forms. The crustacean zooplankton of tropical freshwater is poor in number of species and of smaller size than its temperate counterpart. Whereas these features are not well understood, it is possible that the small size of the planktonic Crustacea has a favourable effect on the survival rate of juvenile fish.

There are striking overlaps between the industrial world and the temperate zone, and between the developing countries and the Warm Belt. This distinction governed the most recent evolution of 'temperate' and 'warm' lakes. The dense, industrial society which was established near the temperate lakes initiated the cultural eutrophication of these water bodies, and 'cold limnology' developed mainly as a consequence of this man-made trophic disturbance. The temperate lakes received a hypertrophied input of chemical energy (especially of nitrogen and phosphorus) which triggered the preferential hyperdevelopment of a relatively small number of species of net phytoplankton. In the large lakes of the Warm Belt, the energetic inputs have been, until now, more equilibrated; electromagnetic energy (radiation) and mechanical energy (wind) affect these water bodies as much as chemical inputs.

Instead of the simplified food chains of the polluted temperate lakes, the large lakes of the Warm Belt have a very complicated and diverse food web, where the exploitation of all possible niches replaces the nearly linear flow of matter occurring in polluted temperate lakes between chemical nutrients and algae. As far as small lakes and ponds of the Warm Belt are concerned, they have known sewage pollution much before the European and American lakes.

The present strategy applied in temperate lakes consists of preserving the water quality of lakes for water supply and recreation purposes. Considerable amounts of money are spent yearly to remove nutrients from watersheds and rivers of the temperate lakes. In Sweden, more than 70% of the domestic sewage undergoes tertiary treatment before reaching the lakes. In the Warm Belt, the management of lakes is oriented towards fisheries and food production. Efforts are directed at making lakes productive in order to increase fish yields, since freshwater fish is a major source of animal protein in tropical areas. With the exception of Eastern Africa, the Warm Belt has relatively few large lakes. As a result, riverine fish species dominate over lacustrine ones, especially in South America. The cichlids are lacustrine fish *par excellence* and they have formed remarkable endemic populations in East-African lakes. The increase in the number of man-made lakes will cause an increase and spread of cichlid species and a simultaneous decrease of riverine species, as already observed in Lake Volta. The study of the colonization of new reservoirs by cichlids will give us valuable information on the process of speciation in large lakes and on the formation of one of the most fascinating biotopes existing on earth.

PART I

The general features of the Warm Belt

1

Geodynamic history of the Warm Belt

Introduction

Rivers and lakes are relatively recent features on the face of the earth and it would seem that a succinct knowledge of the Plio–Pleistocene history of our planet would be an adequate introduction to the characteristics of the drainage system of the Warm Belt (WB). The geological events of these recent periods would indeed be sufficient to explain how the formation of tectonic depressions, preventing the rapid and direct drainage of freshwater to the oceans, made possible the accumulation of large masses of freshwater, especially in the equatorial zone. It would be more difficult to explain why Precambrian rocks, which directly affect the chemistry of rivers and lakes, are so abundant in the WB. It would be impossible to explain the distributions of plants and animals of the WB and their relations to other regions of the world.

It is worthwhile noting that 20 years ago, a discussion of the past of the WB would have been very hypothetical and fragmentary. The progress of geology has allowed the reconstruction of a strange and fascinating story.

Movement of continents and microcontinents

The idea that continents have not always been located where they presently are is not new. It was promoted by Taylor (1910), Baker (1911) and Wegener (1912) and revived by Du Toit (1937). Another fruitful idea was proposed by Dietz (1961) and Hess (1962) concerning the structure of the ocean floor. Convection currents in the earth's mantle determine giant convection cells, the mid-oceanic ridges and oceanic trenches corresponding, respectively, to the ascending and descending arms of the cells. The central part of the ridges is the site of formation of new crust which forms strips of sea floor on both sides of the crestal zone of the ridge, causing a permanent spreading of the ocean floor. A corresponding process of crust destruction occurs in the oceanic trenches.

The ideas of 'continental drift' and of 'ocean spreading' are the basic elements of the theory of plate tectonics: the external layer of the earth is composed of a small number of rigid plates moving upon a viscous mantle. Two plates may diverge, as in oceanic ridges. They also may converge with varying results according to the nature of the plates in contact. When an oceanic plate converges toward a continental plate, the denser oceanic material sinks under the lighter continental mass and is destroyed in the mantle (subduction). The convergence of two light and buoyant continental plates leads to a continental collision, generally associated with mountain formation and resulting in a thickening of the crust in the collision zone. Velocities of plate translation vary from 2 to 9 cm y^{-1}.

In this perspective, the ocean floor is distinct in nature and properties from the continental crust. This basic notion had to be revised with the identification, in the oceans, of numerous oceanic plateaux as fragments of continental crust surrounded by oceanic crust and therefore termed 'microcontinents' (Scrutton, 1976; Ben Avraham, 1981). They are specially abundant in the Indian Ocean and Western Pacific. They are generally small, rarely exceeding 500 km. Although they represent about 10% of the ocean floor, their origin and mode of formation are unknown. They do, however, raise the question of rifting with a renewed acuity. At any rate, we cannot any longer visualize the continental drift as the organized motions of a small number of large plates of oceanic or continental origin. The picture which emerges from the available information on microcontinents is much more complex. Ben Avraham (1981) hypothesizes that they represent the tails or leftovers of larger fragments. In the case of the oceanic plateaux of the Pacific Ocean, interpreted as microcontinents, he envisages that they derive from a 'Pacific continent' which broke and drifted away before the present cycle of plate motion, i.e. before Triassic times. It is after all not surprising that the pre-Triassic cycles of plate motion have left visible marks such as pieces of lost continents.

The microcontinents, moving with their oceanic plates, are bound to collide with and be accreted to near-by continents. This random accretion of bits and pieces of continents to large continental masses would account for the 'anomalous formations' described by geologists in coastal mountainous areas. These formations have completely distinct rock sequences and different faunas from the series in which they are included and have therefore been called 'allochthonous terraines'. New data indicate that many of these blocks became part of their present surroundings only recently. Ben Avraham, Nur, Jones & Cox (1981) hypothesized that the microcontinents observed in the present oceans are in fact the allochthonous terraines which will be found in a few million years in new coastal ridges.

It then appears that the external layer of the ocean is composed of fragments of different origin, different nature and different age. Similarly, mountain belts include numerous heterogeneous blocks of different origin which were incorporated by accretion. It is expected that many of the microcontinents and allochthonous terraines are made of pre-Palaeozoic and Palaeozoic rocks. This seems to be the case in many instances (e.g. Precambrian granite basement of the Seychelles Islands of the South China Sea). This emphasizes the fact that Precambrian rocks are not only the material of old cratonic areas but also a crust material which participated in various cycles of continental drift and is still involved in orogenic processes.

The Warm Belt in a new geodynamic perspective

There is evidence that plate motion has been active for some 2500 million years. Although attempts have been made to reconstruct the pre-Palaeozoic and Palaeozoic distribution of continents, clear and reliable information concerns essentially the most recent cycle of continental drift which started in the late Permian–early Mesozoic some 240 million years ago.

The available knowledge concerning the distribution of continents indicates that, in the Permian, the continents formed one single land mass (the 'Pangaea' imagined by Wegener) which was later divided into two continents corresponding to the 'Laurasia' and 'Gondwanaland' of Du Toit. The discovery of morainic glacial deposits (tillites) of Permian age in South America, South Africa, Madagascar, India, Australia and Antarctica, and of contemporary tropical deposits in Europe and North America led to the idea that, in the Upper Permian, Gondwanaland, composed of South America, South Africa, Madagascar, India, Australia and Antarctica, was located around the present South Pole whereas Laurasia, including North America, Greenland, Europe and Asia without India, was situated in the equatorial area of the earth. The post-Palaeozoic story of the continents seems, therefore, to be an equatorward movement of Gondwanaland. This latter land mass, however, did not move as a single block but was fragmented into smaller continents which moved separately, each at its own pace.

The present Warm Belt corresponds approximately to Gondwanaland and is an old polar continent which broke into several units which drifted in a divergent manner during the Mesozoic and Cenozoic times. *It is clear that the continental mass involved in this process had to be of Palaeozoic or pre-Palaeozoic age just like the microcontinents.*

This explains the overwhelming predominance of Precambrian formations in the WB. They outcrop or form the core of about two-thirds of the South-American continent and of nearly all Africa, Arabia and India. They also cover approximately 60% of Australia. These areas represent stable zones from the tectonic point of view.

The continents moving with their plates and the fragmentation of Gondwanaland in the Mesozoic implies that a deep continental rifting, concerning the whole lithosphere, took place at that period and led to the formation of a certain number of plates moving in a northward direction. Although the mechanism of rifting is of paramount importance in the understanding of evolution of continents, the theory of plate tectonics has not proposed any satisfactory model to explain it. The prevalent opinion that rifts develop along lines of weakness of the crust is not much of an explanation. The concept of microcontinents has so far not provided a better explanation of rifting than the classical theory. The question of knowing why a rift develops in a particular area is of crucial importance when hydrological problems of the WB are considered. The downfaulting related to the formation of the East-African Rift Valley has allowed a permanent storage of approximately 35 000 km^3 freshwater in old and deep lakes; the absence of rifts in South America explains its lack of large lakes in spite of the great humidity prevailing in this continent.

The fragmentation of Gondwanaland generated several continental units which moved away from one another. Flores (1970) assumed that India and Madagascar parted from Africa in the Lower Jurassic, 150 million years ago. McKenzie & Sclater (1973) estimated that South America started drifting from Africa in the late Cretaceous. Dietz & Holden (1970) hypothesized that Africa came into contact with Eurasia at the end of the Cretaceous, some 65 million years ago. According to McKenzie & Sclater (1973), India joined Asia only during the Eocene, 45 million years ago, whereas the most recent 'migratory' event of the WB seems to be the separation of Arabia from Africa which started 25 million years ago.

The collision of India–Africa with Eurasia caused the formation of the Alpine–Himalayan system whereas the Andes seem to be the product of collision of microcontinents with the western part of the South-American plate (Ben Avraham *et al.*, 1981). These recent high-altitude areas provide the WB with a new type of lakes, the tropical high mountain lakes, and play the role of land bridges between the temperate fauna and flora of both hemispheres.

Although the WB is essentially defined by certain types of climates, its geodynamic history has determined the nature of its rocks, the location of its deep lakes and, to a large extent, the distribution of its fauna.

2
Climatological features of the Warm Belt

There is evidence that, at certain periods in the past, climatological features were more uniform all over the earth. In the Tertiary Era, climatological belts developed and the present variation of meteorological features with latitude results from a specific earth–sun relationship which is the main factor to be considered to understand the climates of the Warm Belt. The second factor is the presence of a terrestrial atmosphere which operates a qualitative selection and noticeably reduces the amount of radiation which reaches the earth. The combined influence of the previous factors generates a third factor of second order, the atmospheric and oceanic circulation which redistributes heat and humidity on the earth.

The sun–earth relationship
General

It is common knowledge that the earth is moving counterclockwise around the sun along an elliptical orbit and at an average distance of 150 million km. As far as radiation flux is concerned, however, the most interesting fact is that the earth axis makes an angle of 66°30′ with the plane of the orbit and consequently an angle of 23°30′ with the plane of the Equator.

It follows that, at the solstices, one of the poles is tilted towards the sun and one of the hemispheres receives more radiation than the other. Conversely at the equinoxes, the earth is not tilted towards or away from the sun and the two hemispheres receive an equal amount of radiation.

The insolation of the Warm Belt

The absolute amount of radiation received by a given surface of the earth depends on the angle at which the sun's rays strike that surface and on the time of exposure to radiation.

If the axis of the earth was perpendicular to the plane of the orbit, the sun's rays would always strike the equatorial area at a right angle. The earth being a sphere, north and south of the Equator the angle of the sun's rays with the earth's surface would decrease gradually towards the Pole which would then never receive any radiation. This would in fact correspond to a permanent situation of equinox. In consequence, the radiation flux would be maximum and constant at the Equator all through the year and would decrease with latitude north and south of that line.

The inclination of the earth's axis from the plane of the orbit causes a certain redistribution of the sun's radiation: the equatorial area receives sun rays at a right angle only at the spring and autumn equinox periods and consequently has two periods of maximum insolation. In the summer solstice, the tilting of the North Pole towards the sun generates a greater insolation of the Northern Hemisphere and a deficit of insolation in the Southern Hemisphere. Then, the sun's rays strike the Tropic of Cancer at a right angle, whereas they strike the Equator at an angle of 60°30'. In the winter solstice the situation is identical for the equatorial zone but then the South Pole is tilted towards the sun and the sun's rays strike at a right angle the Tropic of Capricorn. Table 1 summarizes these different situations.

It follows that the equatorial zone shows the smallest variations of the angle at which the sun's rays strike the earth's surface (66°30'–90°). This explains that this area receives the maximal yearly insolation. The equatorial zone has two maxima and two minima but the quantitative difference between them is minor. Towards the tropics we note the predominance of one peak and one minimum. Moreover, the amplitude of variation of the angle increases (from 43° to 90° at the Tropics) and so does the amplitude of insolation.

Table 1 *The insolation in the Warm Belt during the year; angles at which the sun's rays strike the different areas of the Belt*

	Tropic of Cancer 23°30' N	Equator 0°	Tropic of Capricorn 23°30' S
Summer solstice	90°	66°30'	43°
Spring equinox	66°30'	90°	66°30'
Winter solstice	43°	66°30'	90°
Autumn equinox	66°30'	90°	66°30'

The earth's atmosphere and heat balance
General

The atmosphere, bound to earth by gravitational forces, has a total thickness of 80 km. Near the earth, the troposphere has an average thickness of 12 km. It is thinner at the Poles (8–10 km) than at the Equator (17 km). It contains air, water vapour and dust particles. Within the troposphere, temperature decreases at a constant rate of 6.4°C km^{-1}. At the top of the troposphere, the temperature is approximately $-57°C$. The stratosphere extends from the limit of the troposphere up to 50 km altitude. It has a very low density of matter and shows a constant increase of temperature up to 0°C. The ozone layer which surrounds the earth is found in this stratum. From 50 km to 80 km lies the mesosphere; in this layer the temperature drops again and reaches the lowest value found in the atmosphere ($-83°C$). Beyond the upper limit of the mesosphere, in the thermosphere, the temperature increases constantly. In this layer, gamma rays and X rays cause the ionization of atoms and therefore the lower thermosphere and upper mesosphere coincide with the ionosphere.

The weight of the air and water vapour column creates atmospheric pressure, the value of which decreases exponentially with altitude.

Only a tiny fraction of the sun's global radiation reaches the top of our atmosphere. Moreover, the gaseous envelope of the earth is a very efficient filter of the radiant energy which falls on the upper mesosphere.

In the lower thermosphere and upper mesosphere, the shortest-wavelength radiation (9% of total energy) is absorbed, creating the ionosphere and only the visible part of the spectrum (41% of total energy) and the infrared (50%) continue their trajectory earthwards.

In denser layers of the atmosphere, the short wavelengths of the spectrum are *reflected* in all directions by molecules and dust particles (diffuse reflection) causing a 6% loss of total energy. The long wavelengths penetrate further down but are partly *absorbed* by water vapour and dust, losing approximately 14% of the total energy. In a clear sky, as much as 80% of the total energy reaching the atmosphere may reach the earth's surface. Conversely, a heavy cloud cover in the troposphere reflects from 30 to 60% of the incoming radiation and absorbs another 5–20% allowing from 45 to 0% of the incoming radiation to reach the earth's surface.

It should be underlined that most of the reflected radiation is lost to space whereas the energy absorbed (by water vapour, carbon dioxide and the ground) is stored and acts as a secondary source of radiant energy. Active absorption in the stratosphere causes the increase of temperature observed in this layer. It is estimated that the total reflected energy (albedo) due to high-altitude scattering, clouds and the ground, amounts to 32% of the energy reaching the atmosphere and 68% is absorbed by clouds, atmospheric molecules and the ground.

The earth, being at a temperature above the absolute zero, emits electromagnetic radiations, the wavelengths of which depend on its surface temperature; the lower the temperature, the longer the wavelengths. Satellite studies of the global heat balance of the earth–atmosphere complex indicate that the total losses from the earth to space amount to 68% of the received energy and balance the total amount of energy absorbed by this complex.

The radiation balance of the Warm Belt

The satellites give us interesting information concerning the radiation balance at various latitudes. At the top of the atmosphere, in the polar region, about 130 kcal y^{-1} are received and only 18% of this amount reaches the ground because of the low angle of sun-rays and high albedo of the snow cover. The corresponding data for the equatorial region are 320 and 160 kcal y^{-1} indicating that 50% of the radiation received by the high atmosphere reaches the surface at that latitude.

If we now consider the net all-wave radiation (incoming radiation *minus* outgoing radiation) of the earth and of the earth–atmosphere complex, we see that the net balance of the earth is always positive (the earth surface gains more energy than it loses) and the net balance of the atmosphere is always negative. The global balanced budget previously described concerns the earth–atmosphere complex. Moreover the heat budget of the complex is balanced if we consider the radiation received and emitted by the whole earth and the whole atmosphere. The picture is very different if we look at the net all-wave radiation balance at various latitudes. There is a region of net radiation excess extending from 40 N to 30° S and a region of net radiation deficit north and south of this central area.

These considerations of energy balance define the Warm Belt with precision: *the Warm Belt corresponds to the portion of earth where the net all-wave radiation balance of the earth–atmosphere complex is positive.*

The existence of areas of net radiation surplus and deficit together with a balanced global budget requires a worldwide mechanism which transports heat from areas of surplus to areas of deficit. This is achieved through the fluids of the earth–atmosphere complex: the gaseous envelope and the oceans.

The global earth circulation
The heat transport system
Atmospheric heat transport

Globally, atmospheric pressure decreases with altitude. At a given altitude, however, atmospheric pressure is not equal at different latitudes mainly because of differences in temperature which modify the air density. This creates a gradient which generates an air motion from areas of high pressure (anticyclone) to areas of low pressure (cyclone), that is along the pressure gradient. The rotation of the earth introduces a significant perturbation known as Coriolis force. In the Northern Hemisphere (NH), the motion of fluids is deflected to the right of their trajectory as defined by pressure gradient. In the Southern Hemisphere (SH) a similar deflection exists towards the left of the fluid trajectories. There is no deflection at the Equator and it increases poleward.

It follows that, in the NH, the wind blowing towards a cyclonic area has a general anticlockwise direction and the wind blowing from an anticyclonic area has a clockwise direction. An opposite picture is observed in the SH.

In the equatorial area, the barometric pressure is constantly lower than the sea-level pressure of 1013 millibars. Conversely, at approximately $30°$ N and $30°$ S, the subtropical belts are areas of relatively high pressure. North and south of these belts extend wide low-pressure belts with minimum values around the latitude $60°$. The Poles, in contrast, have constantly high pressures. These pressure belts move north and south with the seasonal differences in temperature.

Atmospheric pressure and winds in the Warm Belt. In winter, in the NH, the subtropical belts of high pressure (SBHP) or Intertropical Convergence Belt (ITCB) is located about $30°$ N and centred in two oceanic cells over the Atlantic and Pacific Oceans. In summer, these cells move northwards and cover a much larger area as a result of low pressure centres developing on the adjacent continents. In the SH, the SBHP is more developed because of the greater extension of oceanic masses.

The SBHP represent permanent anticyclonic areas and consequently winds blow from these areas in different directions as imposed by the shape of the cells. This continuous movement of air away from the belts generates a vertical descending supply of air from the high atmosphere. Since descending air becomes increasingly dry, the SBHP are generally warm and dry areas. It is worthwhile noting that a large part of the SBHP are desert areas (Sahara, Arabia, Australia . . .).

Climatological features of the Warm Belt

As far as the Warm Belt is concerned, the most interesting feature is the general air motion from the SBHP towards the Equator. These winds known as 'trade winds' are deflected westwards in the northern hemisphere and eastwards in the southern hemisphere. The equatorial zone is the meeting zone of the north-east and south-east trades: therefore this is an area of winds of uncertain direction and calms. Similarly, but for a different reason, the SBHP, called also 'horse latitudes', have a high percentage of calm periods.

The unequal distribution of continents and oceans is strongly felt in the belt development of both hemispheres: whereas the SBHP are nearly continuous in the SH, they are interrupted and the general wind pattern disturbed in the vicinity of the great land masses of the NH. In summer, Asia and North America are the centres of pressure lows. Air is then compelled to move from the Indian Ocean and the Gulf of Mexico towards the centre of the cyclones. This movement of wet air is the summer monsoon bringing heavy precipitation over all South-eastern Asia and to a lesser extent over the central and eastern parts of North America. In winter, the continents are the centre of pressure highs and the air blows from the continents towards the ocean. This is the dry winter monsoon causing the long and dry winters of India.

Oceanic heat transport

A considerable amount of wind energy is transferred to the ocean surface. This input of mechanical energy sets in motion oceanic surface currents which are also affected by the Coriolis force. These currents, displacing surface water, generate deeper return currents.

Temperature differences are another cause of oceanic currents; the cold Arctic water tends to sink into deeper layers and thus causes an upward movement of warmer water.

In the Warm Belt, the north-east winds of the northern subtropical belt generate an oceanic current of south-west direction. In the Atlantic Ocean this current is deflected towards the north and north-east by the American Continent, creating the north subtropical gyre. In the SH the southern subtropical belt generates a similar current from south-east to north-west and a south subtropical gyre. This huge westward movement of water north and south of the Equator generates an eastwards equatorial countercurrent.

Schematically, the trade winds bring equatorial warm water on the western sides of the oceans whereas the equatorial return currents bring cold waters near the eastern coast of the oceans. It follows that a warm current moves along the coast of China, Siberia and the eastern coast of Australia in the Pacific and along the eastern coast of American Continent in the Atlantic whereas cold currents are found off the western coasts of America and Africa.

The humidity transport system

The amount of water vapour which can be held by the air is temperature-dependent: $1 m^3$ of warm air (30°C) can hold up to 30 g of water whereas at 0°C the same volume of air is saturated with only 5 g of water. It follows that air in cold areas can supply very small amounts of precipitation in comparison with air in warm regions.

In the equatorial zone, the warm and moist air has a relatively low density and rises up to the limit of the troposphere. During this ascending process, adiabatic cooling causes condensation and rain with a simultaneous release of latent heat. The rising air is replaced by the equatorward movements of the trade winds, loaded with water vapour evaporated from the oceans in the area of the SBHP.

This system generates the 'Hadley cells' with predominating evaporation in the SBHP, transport of humidity to the Equator, convection and massive precipitation in the equatorial area. The convergence of Hadley cells of both hemispheres towards the Equator causes daily precipitation and considerable condensation and heat release. It has been estimated by Sellers (1965) that the Hadley cells of both hemispheres transport yearly an order of magnitude of $40 000 km^3$ of water to the Equator. The cells of the SH convey larger amounts of water than those of the NH. This system generates an equatorial surplus of water with abundant runoff and a tropical evaporation excess.

3

Hydrology

The water balance of the world
General
From the beginning of the century, numerous attempts have been made at calculating the world's water balance (L'vovitch, 1975). In the present study, we have utilized the data of Baumgartner & Reichel (1975) and L'vovitch (1977).

The amount of 'active water' involved in the annual hydrological cycle is 496×10^3 km^3. As the earth's surface area is 510×10^6 km^2, this amount represents a layer of 973 mm over the globe (Table 2).

The oceans play a major role in the world's water exchange: 80% of all the exchanges occur over the oceans. Three times more precipitation (P) falls on the oceans than on the continents and the evaporation (E) from the oceans is six times greater than that from the continents. In absolute terms, the water balance ($P-E$) of the continents is positive ($+39.7 \times 10^3$ km^3) whereas that of the ocean is negative (-39.7×10^3 km^3).

The amount of precipitation is nearly equal in both *hemispheres* but there is a net excess of E over P in the Southern Hemisphere (SH) (Fig. 1A). It follows that the water balance of the Northern Hemisphere (NH) is positive ($+73 \times 10^3$ km^3) and the water balance of the SH is negative (-73×10^3 km^3). This unequal distribution of E generates an oceanic current from the NH to the SH and an atmospheric transport in the opposite direction.

The absolute amounts of P and E of the *continents* are larger in the NH than in the SH and in such a way that the runoff ($R = P - E$) is greater in the NH than in the SH (Fig. 1B). The situation is reversed if we compare the *oceans* of both hemispheres: the deficit of the water balance is five times greater in the SH than in the NH.

Table 2 *Water balance of the world*
(P = precipitation; E = evaporation; R = runoff; yearly values)

	Area (10^6 km^2)	P (10^3 km^3)	P (mm)	E (10^3 km^3)	E (mm)	R (10^3 km^3)	R (mm)
Northern Hemisphere							
Land	100.3	68.0	678	43.6	435	24.4	243
Ocean	154.6	179.4	1160	185.3	1198	−5.9	−38
Total	254.9	247.4	970	228.9	897	18.5	73
Southern Hemisphere							
Land	48.6	43.1	888	27.8	572	15.3	316
Ocean	206.5	205.6	996	239.4	1160	−33.8	−164
Total	255.1	248.7	975	267.2	1048	−18.5	−73
Earth							
Land	148.9	111.1	746	71.4	480	39.7	266
Ocean	361.1	385.0	1066	424.7	1176	−39.7	−110
Total	510.0	496.1	973	496.1	973	–	–

Hydrology

Water depth is the ratio between the volume of water falling on or evaporating from a given area and this surface area; it is generally expressed in mm. As far as P and E are concerned, water depths are higher for oceans than for continents. The water depths of continental P and E are only 70 and 41%, respectively, of the water depths of oceanic areas. Continents of the NH have smaller water depths of P and E than those of the SH. Conversely, oceans of the NH have higher P and E water-depth values than the oceans of the SH.

The components of the hydrosphere

The different components of the hydrosphere are presented in Table 3. Freshwater represents 2% of the hydrosphere. The lakes of the world contribute only 0.02% to the water reserve of the earth and no more than 1% of the total 'freshwater'. In fact, the term 'freshwater' is improper since L'vovitch (1979) estimates that the volume of the earth's lakes (280 000 km³) can be broken down into 150 000 km³ natural freshwater lakes with an outlet, 5000 km³ freshwater storage reservoirs and 125 000 km³ salt lakes.

For short-term purposes, it is justified to consider glacier water as stored in an inactive form. The volume of active water is then 4380×10^3 km³ or

Fig. 1. Precipitation and evaporation at different latitudes. Each sign represents the yearly amount of precipitation (dots) and evaporation (crosses) corresponding to 5° latitude. The vertical dotted lines delimit approximately the Warm Belt. A = the whole globe (land and ocean); B = land only; dotted areas = areas where evaporation exceeds precipitation

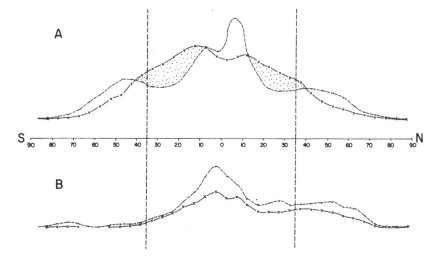

0.3% of the hydrosphere and 15% of the freshwater reserve. Since an amount of 496×10^3 km^3 is involved in the annual hydrological cycle, the overall turnover of active water on earth is about 8.8 years.

The water balance of the Warm Belt in comparison with the world's water balance

General

Whereas the Warm Belt (WB) occupies only 57% of the earth's surface, the annual P and E represent, respectively, 66 and 75% of the total amount of water annually circulated in the world's hydrological cycle.

As in the case of the world's water balance, the WB is characterized by a net predominance of oceanic over continental water exchanges. The continental P and E of the WB are only 31% and 17%, respectively, of the oceanic P and E (Table 4).

The amount of water annually evaporated from the WB is larger than the amount of precipitation received by this region. The water balance of the continental areas of the WB $(P-E)$ is positive ($+25629$ km^3). The strongly negative water balance of the WB oceans (-70772 km^3), however, causes an overall WB deficit of -45143 km^3 which represents as much as 10% of the total amount of water of the hydrological cycle. By comparing Tables 2 and 4 it is easy to see that the oceanic WB deficit is nearly twice the world oceanic deficit. This emphasizes the major role played by the oceans in the world's atmospheric circulation.

In contrast with the world's water balance, the continental masses of the

Table 3 *The components of the hydrosphere (according to L'vovitch, 1979)*

	Volume ($\times 10^3$ km^3)	Per cent total volume	Per cent freshwater	Turnover rate (years)
World ocean	1 370 323	93.941 8		3000
Deep groundwater	60 000	4.113 2		5000
'Freshwater'				
'Active' groundwater	4000	0.274	14.094	330
Glaciers	24 000	1.645	84.566	8000
Lakes	280	0.019	0.987	7
Soil moisture	85	0.005 8	0.299	1
Atmospheric vapour	14	0.000 96	0.049	0.027
Rivers	1.2	0.000 08	0.004	0.031
Total freshwater	28 380.2	1.92		
Grand total	1 458 703	99.999 84	99.999	2800

Table 4 Precipitation, evaporation and runoff of the Warm Belt; yearly averages

	Area (10³ km²)			Precipitation (km³)			Evaporation (km³)			Runoff (km³)		
Absolute values	Land	Ocean	Total	Land	Ocean	Total	Land	Ocean	Total	Land	Ocean	Total
WBN	44 348	101 475	145 823	39 929	131 533	171 462	26 290	147 327	173 617	13 639	−15 794	−2 155
WBS	32 147	113 694	145 841	38 331	118 774	157 105	26 341	173 752	200 093	11 990	−54 978	−42 988
WB	76 495	215 169	291 664	78 260	250 307	328 567	52 631	321 079	373 710	25 629	−70 772	−45 143
Water depths					(mm)			(mm)			(mm)	
WBN				900	1296	1176	593	1452	1191	308	−156	−15
WBS				1192	1045	1077	819	1528	1372	373	−483	−295
WB				1023	1163	1127	701	1492	1281	335	−329	−154
WB as % globe	51	60	57	70	65	66	75	76	75	65		−178

WBN = Warm Belt north; WBS = Warm Belt south; WB = Total Warm Belt

NH and SH of the WB have nearly equal amounts of P and E. The oceans of the northern WB receive more precipitation and lose less water through E than the oceans of the southern WB (Table 4). This is caused by the previously mentioned atmospheric water movement from the SH to the NH.

Table 4 shows that the WB has a total water deficit of 154 mm mostly due to the considerable oceanic evaporation losses. The deficit of the WB water balance is unequally distributed between the two hemispheres: the NH has a deficit of 15 mm whereas the deficit of the SH is as high as 295 mm. This results from high values of water depth of P and low values of water depth of E in the northern WB in comparison with the southern WB.

In Table 5 we have compared the total yearly runoff of the Warm Belt with the yearly discharge of the main rivers of the WB, utilizing as source of information the data of the Hydrological Decade and the data of the Water Encyclopaedia. Whereas the total runoff of the Warm Belt is 25.6×10^3 km^3, the discharge of the main rivers of the equatorial, tropical and subtropical belts is 13.4×10^3 km^3. Considering that the groundwater runoff of these areas varies between 33 and 50% of total runoff (L'vovitch, 1979) and assuming an average value of 35% we estimate the groundwater runoff of the WB to be approximately 9.0×10^3 km^3 per year. The additional 3.0×10^3 km^3 correspond to the small rivers not taken into account in Table 5.

The water balance of the different continents of the Warm Belt

Table 6 shows that Africa is the continent with the largest WB area (39%), followed by Asia (23%) and South America (21%). South America contributes the highest amount of precipitation (34%), however, in comparison with Africa and Asia (26% each). From Table 6 it is clear that Africa is the 'driest' continent of the Warm Belt with the highest evaporation (33%) and lowest runoff (13%). This is even more obvious from the water-depth data: Africa has a smaller yearly amount of precipitation (697 mm) than the dry Australia (775 mm). In contrast, South America is the wettest continent of the WB. This is also well illustrated by the yield per area of the large rivers of the WB (Table 5).

Lakes of the Warm Belt

In Table 7 are listed the main lakes and reservoirs of the WB. The volume of water stored in these water bodies is $\sim 40\,000$ km^3, that is 14% of the total volume of the lakes of the world. This is a remarkably small volume and minor percentage for an area which receives 70% of the world's continental precipitations.

Hydrology

The uneven distribution of the lakes in the Warm Belt is striking. Approximately 95% of the volume of the lakes of the WB is located in Africa, the driest continent of the WB, whereas only 4% is found in South America, the wettest zone of the WB, and < 1% in South-East Asia and Australia (Table 8).

This paradoxical situation has a geological explanation. The African Rift Valley is the only large-scale tectonic event which has generated

Table 5 *Discharge, drainage area and specific discharge of the main rivers of the Warm Belt; yearly values*

River	Discharge (km^3)	Drainage area (10^3 km^2)	Specific discharge (mm)	Source
Colorado	18	279.5	64.4	HID
Rio Grande	3	352.2	8.5	HID
Mississippi	547	3222	169.8	WE
Amazon	6570	5778	1137.0	WE
	5700	4686	125.9	HID
Orinoco	547	881	620.9	WE
Parana	475	2305	206.0	WE
Tocantins	321	907	353.9	WE
Sao Francisco	88	673	130.8	WE
Uruguay	120	233	515.0	WE
Niger	192	1114	172.3	WE
Nile	88	4015	21.9	WE
	84	3100	27.1	HID
Zaire (Congo)	1248	4015	310.8	WE
	1314	3700	355.1	HID
Zambezi	221	1295	170.7	WE
Chari	38	600	63.3	HID
Senegal	25	290	108.7	HID
Tigris	38	194	195.9	HID
Euphrates	16	274	58.4	HID
Indus	173	927	186.2	WE
Ganges + Brahmaputra – Meghna	1250	1760	710.2	HID
Godavari	110	299	367.8	HID
Krishna	61	308	198.0	WE
	47	251	187.2	HID
Irrawadi	484	430	1125.6	WE
Salween	47	280	167.9	HID
Hwango-ho	103	673	153.0	HID
Mekong	430	660	652.0	HID
Total (approx.)	13 350	32 600	410.0	

HID = Hydrological International Decade; WE = Water Encyclopaedia.

Table 6 *Precipitation, evaporation and runoff of the Warm Belt area of the different continents*

	Area (10³ km²)(%)	Precipitation (km³) (%)	Evaporation (km³) (%)	Runoff (km³) (%)	Precipitation (mm)	Evaporation (mm)	Runoff (mm)
North America							
0–35° N	4648(6)	4688(6)	3418(6)	1270(5)	1009	735	273
South America							
0–15° N	3140	6745	3738	3007	2148	1190	958
0–35° S	13039	19582	12554	7025	1502	963	539
Total	16179(21)	26327(34)	16292(31)	10032(39)	1627	1007	620
Africa							
0–35° N	19740	10908	9099	1809	553	461	92
0–35° S	9803	9686	8110	1576	988	827	161
Total	29543(39)	20594(26)	17209(33)	3385(13)	697	582	115
Asia							
0–35° N	16820	17588	10035	7553	1043	597	449
0–15° S	1043	2661	1359	1302	2551	1303	1248
Total	17863(23)	20249(26)	11394(35)	1134	638	496	
Australia							
0–35° S	8262(11)	6402(8)	4318(8)	2084(8)	775	523	252
Total	76495(100)	78260(100)	52631(100)	25626(100)	1023	688	335

Hydrology

topographic conditions necessary for the storage of large amounts of water. Amazonia, which is relatively flat, harbours only small-scale river-lakes unless we consider the whole basin as a vast fluvio-lacustrine system. The situation would have been very different in the Pliocene since immense lakes covered Central Amazonia and the central part of the Zaire basin. Finally, in the Warm Belt, the glaciers which have caused the formation of

Table 7 *Volume of the main lakes and reservoirs in the various continents*

Water body	Volume ($\times 10^6$ m^3)	Water body	Volume ($\times 10^6$ m^3)
SOUTH AMERICA		AFRICA	
Cienagas	4000	Awasa	1300
Maracaibo	300 000	Shala	37 000
Valencia	6300	Langano	3800
Tota	1920	Abiata	1600
Poopó	200 000	Ziway	1000
Guri	18 000	Chilwa	2200
Brokopondo	30 000	Kivu	583 000
Titicaca	893 000	Tanganyika	23 100 000
CENTRAL AMERICA		Bangweulu	11 200
Gatun	5480	Mweru	36 000
Madden	810	Mwadingusha	1000
Nicaragua	80 000	Nzilo	5000
Managua	10 000	Tumba	2000
Peten Itza	4000	Maji Ndombe	11 500
Izabal	8300	Chad	72 000
Guija	740	Kariba	157 000
Atitlan	13 000	Malawi	8 400 000
Patzcuaro	1000	Cabora Bassa	62 000
Chapala	11 000	Kainji	14 000
AFRICA		Volta	165 000
Victoria	2 700 000	Internal delta	1000
Kyoga	16 200	SOUTH-EAST ASIA	
Edward	78 000	Bhakra	80 000
George	600	Kinjar	240
Bunyoni	800	Stanley Reservoir	2700
Albert	140 000	Chilka	2000
Tana	28 000	Pulicat	1000
Nasser	157 000	Great Lake of Kampuchea	80 000
Mariut, Edku, Burullus,		Bum Borapet	600
Manzala	3200	Songkla	1000
Qarun	800	Lanao	21 000
Eyasi	1200	Mainit	18 000
Kitangiri	4800	Laguna de Bay	2000
Manyara, Magadi, Natron	2000	Wisdom	21 000
Rudolf	360 000	Dakataua	3300
Shamo	2200	AUSTRALIA	
Abaya	8200	Corangamite	10 000

hundreds of thousands of lakes in the temperate zone have affected only limited areas in high mountains, and Lake Titicaca, the only large lake of high altitude, is not of glacial origin.

Table 8 *Volume of lakes and reservoirs in various continents*

Continent	Volume of the main lakes and reservoirs	
	(10^3 km^3)	(%)
South America	1.6	4.2
Africa	36.2	95.3
Asia	0.2	0.5
Total	38.0	100

PART II

The aquatic ecosystem of the Warm Belt

4
South America

INTRODUCTION

The South-American continent, having its maximum width at 4° S, is the greatest continental mass of humid tropical climate on the earth. The climatological and geographical implications of such a position are, however, strongly diversified by the Andean Cordillera on the western part of the continent which can then be divided into two large units: the non-Andean East and the Andean West (Sick, 1969).

The Andean West corresponds to one of the youngest orogenic belts of the earth and is presently characterized by an intense tectonic activity underlined by a high frequency of earthquakes and vertical movements especially in the coastal area. In the geological past, extensive transgressions and regressions have occurred in the Andean province, modifying endlessly the distribution of flora and fauna, leading to the disappearance of certain species, and enhancing the development of the most adaptive plants and animals. In the recent geological past, the Andes and their prolongation in North America have been a favoured biological bridge allowing the Arctic and Antarctic species to cross the Warm Belt. Presently, they offer one of the most varied ecosystems on earth: narrow coastal plains of Colombia and Ecuador with luxuriant rain forest, contrasting with the dry coast of Peru and Chile where agriculture is only possible in the valleys, mountain forest with fern-trees, grass and herb pastures of the '*tierra fria*' above 3500 m altitude, valleys and intra-cordilleran basins with humid savanna, and the dry altiplano of Bolivia where agriculture is practised above 4000 m.

Moving eastwards, one leaves orogenic turbulence and penetrates into one of the most stable and ancient continental shields of the planet. The last orogenesis occurred in the late Precambrian and early Cambrian, and transgressions were of limited extent. Such tectonic stability combined with humid tropical conditions explain the development of ecosystems of 'maximal biotic maturity' (Schwabe, 1968). In sharp contrast with the changing – in space and time – Andean biotopes, the Amazonian ecosystems have achieved highly sophisticated development based, not on simple competition for external nutrients, but on the optimal utilization of nutrients stored in living biomass by a complex community. The rain forest of Amazonia–Orinoco, covering an area of 4.5×10^6 km^2, is the best example of such mature ecosystems that it seems appropriate to describe it as *rich in structure*. The incredibly numerous species of plants and animals are optimally integrated to utilize the limited nutrient resources of the crystalline substratum.

In return, these ecosystems protect the natural substrate from demineralization by minimizing the mineral losses; it is therefore clear that their destabilization generates a very damaging chain of reactions including soil erosion, laterization and irreversible disappearance of species. Unfortunately, the increasing rate of interference by man in South-American ecology does not seem to take this warning into account. Extensive cattle breeding has deeply modified the savanna grass cover and African grass species have been introduced in the newly cleared woodland area (Sternberg, 1968). Whereas the demographic density is below one inhabitant km^{-2} in Amazonia, urban centres are overpopulated and becoming rapidly industrialized. Pollution of freshwater is already a severe problem in various areas such as the Pernambuco area where freshwater biota have been considerably decreased, the Magdalena basin in Colombia, and the copper-mining areas of Peru. Modification of the hydrographic systems by the building of dams and impoundment of very large volumes of water for the production of hydroelectric power is well under way: the Brokopondo Reservoir on the Surinam was flooded in 1964 and since then many others have been built or planned. One of the biggest projects, the Itaipu Dam, on the Parana River will flood 600 km^2 of Paraguayan forest and 800 km^2 of Brazilian farmland.

In the following study of South-American freshwaters, we have adopted the hydrographic subdivisions utilized by Ziesler & Ardizzone (1979) in 'The inland waters of South America'.

THE CARIBBEAN AREA

This area corresponds approximately to the watershed of Rio Magdalena in Colombia, and Lake Maracaibo and Lake Valencia in North Venezuela.

The Caribbean area

The watershed of Rio Magdalena

The watershed of Rio Magdalena (Fig. 2) covers an area of 257 000 km² between 2 and 10° N and 73 and 77° W. The Colombian Andes are divided into three cordilleras separated by two parallel valleys where the Magdalena and its main tributary, the Cauca, run northwards until El Banco. At this point, the River divides into two branches which flow at a 90° angle from its upper course. The Rio Cauca joins the southern branch of the Magdalena at the latitude of El Banco. The waters of the Magdalena and its tributaries originate in the Andes at an altitude of 2000–4000 m.

The Magdalena River

The Magdalena has a total length of 1316 km and its main tributaries are the Cauca, the Sogamoso, the Cesar and the San Jorge which cover approximately 50% of the watershed.

In the Magdalena watershed, the meteorological cycle includes a wet season from April to November (80% of annual precipitation) and a dry winter. In consequence, the discharge of the river is bimodal with low yields

Fig. 2. The watershed of Rio Magdalena, Lake Maracaibo and Lake Valencia

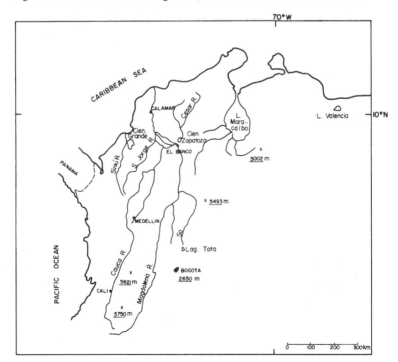

in January–February (approximately 4000 m³ s⁻¹). The river discharge increases from March onwards and reaches peaks of 10 000 m³ s⁻¹ in November–December, these values representing the average yields for the period 1965–1973 (Ducharme, 1975). The long-term yearly average discharge of the river is 6800 m³ s⁻¹ at its entry into the Caribbean Sea. The average annual fluctuation of water level is 4.40 m but in specific years variations up to 7.3 m have been reported. It follows that the lower parts of the valley are periodically flooded (~1.3 million ha are flooded for from three to six months a year). When the river recedes, an area of 320 000 ha remains under water: these are the 'river-lakes' or '*ciénagas*', located in the flood plain of the river.

Turbidity is generally high and prevents the development of submerged vegetation. Floating meadows composed of *Eichhornia crassipes* and *Pistia stratiotes* may cover the '*canos*' (the narrow and dry channels connecting the *ciénagas* with the river), and the quiet areas of the river banks.

About 80% of the Colombian population is concentrated in the Magdalena watershed leading to a demographic density of 54 inhabitants km⁻² which is relatively high when compared to 0.24 km⁻² in Amazonia.

It is not surprising then, that problems of water pollution appeared especially in the lower course of the river. The salt content of the Magdalena and the Cauca, very low in the upper course (conductivity 31.87 μmhos cm⁻¹) increases downstream where conductivity reaches 225 μmhos cm⁻¹. Similar increases in phosphates, nitrates and silicates were observed. These are mainly caused by the drainage of the *ciénagas* into the river and the confluence of polluted tributaries such as Rio Bogota. The downstream increase in salt content of the Magdalena is mainly related to agricultural development. High concentrations of copper (up to 0.68 mg l⁻¹ in the Magdalena and up to 0.45 mg l⁻¹ in the Cauca), probably originating from algicides utilized in agriculture, have been reported by Ducharme (1975). Industrial development also affects the quality of various tributaries of the Magdalena and the Cauca. According to Hernandez Devia (1976) six areas have severe pollution problems. The Bogota area drains the nearly untreated domestic sewage of the three million people of Bogota town and the sewage of about 200 industries (chemicals, leather, plastics, food industry) into the Bogota River. The Medellin area drains the sewage of 1 200 000 inhabitants and 180 industries (paints, textiles, soap, fertilizers, chemicals, food industry, paper industry, metal industry) into the Medellin River (an affluent of the Cauca). In the Cali area, on the upper Cauca, the sewage water of one million inhabitants together with the sewage of 100 factories (mainly paper industry) represents a dangerous polluting load for

The Caribbean area

fish reproduction. In the Baranquilla area, 670 000 inhabitants and 100 factories (food industry, textiles, chemicals and metal industry) drain their untreated sewage into the river. In the Barrancabermeja, the oil refineries introduce a special type of pollution due to leakage or inadequate treatment of the refinery effluents. The Cartagena Bay is mainly threatened by refinery effluents and potential spilling of tankers. Intensive man-made deforestation is the major cause of turbidity which has considerably altered the fisheries of the river and of the *ciénagas*. The rather isolated fish fauna, derived mostly from the upper Orinoco, includes 150 species with 65 characoids and gymnotoids, 65 catfishes, 10 cyprinodonts and cichlids and various species of marine origin (Lowe-McConnell, 1975).

The river-lakes or 'ciénagas'

The main *ciénagas*, the rivers to which they are connected and their average surface area are as follows:

'*Ciénaga*' (River)	Area (km^2)
Zajatoza (Magdalena)	340
Ayapel (San Jorge)	128
Grande (Sinù)	118
Guajaro (Magdalena)	115
Zarate (Magdalena)	80
La Raya (Cauca)	37

In certain cases, the '*ciénagas*' are permanently connected to the main river by a canal. In other cases, the canal dries up with river subsidence.

These river-lakes are shallow (2–7 m) and their bottom is covered with a thick layer of fine mud. Their temperature fluctuates from 24 to 32°C. The labile thermal gradient which is established during the afternoons of quiet days is rapidly destroyed by wind and precipitations and complete homothermy prevails in the morning. Like many other shallow, tropical water bodies, the *ciénagas* have a diurnal thermal cycle and no marked seasonal variations.

High turbidity is a general feature of *ciénagas* and is mainly caused by the river sediments at the flood period and by resuspension of bottom sediments at the low-water period. Turbidity limits productivity to the upper 30–50 cm. The highest values were found in the Guajaro (0.02–0.57 g O_2 m^{-3} h^{-1}). In contrast, the floating vegetation, mainly composed of *Eichhornia crassipes*, grows very rapidly in periods of high water level and takes up the large amounts of nutrients, especially phosphorus, dissolved in the *ciénaga* water during the flooding process. With increasing water level, large areas are flooded and nutrients are released from flooded vegetation and from cattle manure.

The dominance of floating vegetation over phytoplankton explains the relatively low concentrations of oxygen in various *ciénagas* (e.g. 4–6 mg l^{-1} oxygen in upper layers of the *ciénaga* of Ayapel in October 1974). Moreover, the lower layers of the deepest *ciénagas* are devoid of oxygen in the afternoons but in the morning the profiles are again homogeneous.

An experimental fish sampling, carried out by Kapestky, Escobar, Arias & Zarate (1978) in ten *ciénagas* of the Magdalena flood plain, revealed that the fish biomass varies from 0.2 to 251 kg ha^{-1} with an average of 68 kg ha^{-1}, a relatively low value when compared with the fish biomass of similar water bodies in Niger (196–270 kg ha^{-1}) or in Zambia (337 kg ha^{-1}). The small characid *Roeboides dayi* was the fish most frequently found. The annual catch of the Magdalena watershed is estimated at 65 000 tons, 38 000 tons of which come from the river and 27 000 from the *ciénagas*.

Lakes and reservoirs

Lake Tota. Lake Tota is located in the eastern Cordillera at an altitude of 3015 m above mean sea level (MSL) and has a surface area of 56 km^2; its maximal and average depth are, respectively, 67 and 34 m, and its volume is 1920×10^6 m^3.

The climate of the Lake Tota watershed includes a wet season from April to November and a dry period from December to March. The annual precipitation amounts to 779 mm. The air temperature fluctuates from 0 to 20°C with an average of 10.7°C. The average discharge of the lake is 1.3 m^3 s^{-1}.

Lake Tota is a tropical lake of high altitude with a polymictic regime and is never stratified. The light penetration is excellent with a range of Secchi-disc measurements of 9–12 m (Ducharme, 1975).

The salt content is very low (conductivity 75 μmhos cm^{-1}). Although the lake has been considered as oligotrophic in the past, there is presently an increase in nutrient concentration (10 ppm NO_3, 0.09 ppm PO_4) caused by the local agriculture. The use of pesticides and fungicides explains the relatively high levels of copper (0.10–0.40 ppm) of the lake water.

The oxygenation of the lake is excellent and caused by the mixing action of the wind. In the quiet bays of the lake, submerged macrophytes develop (e.g. Canadian pondweed, *Elodea canadensis*) but there is no floating vegetation. The primary productivity of the phytoplankton per unit volume is very low (0.04 mg O_2 l^{-1} h^{-1} at 1 m depth) but the trophic zone extends nearly to the lake bottom (productivity of 0.01 mg O_2 l^{-1} h^{-1} at 40 m). It follows that the primary productivity per unit surface area is far higher than in *ciénagas*.

Reservoir El Prado. This man-made reservoir has been built for hydroelectric purposes on the Prado River, which is a left-hand tributary of the Magdalena. The reservoir is 10 km long, 1.5 km wide and 70 m deep at the dam wall. The filling of the reservoir was not preceded by any clearance of the vegetation.

The reservoir is located at 355 m above MSL and its superficial temperature oscillates between 27 and 31 °C. It is stratified most of the year, although the difference in temperature between epi- and hypolimnion is rarely greater than 1 °C. The thermocline located at 3–4 m in the dry period (January–February) deepens to 10 m in August.

The hypolimnion is devoid of oxygen but, during the rainy season, cold and well-oxygenated waters penetrate into the hypolimnion, and then concentrations of oxygen up to 4 mg l^{-1} may be observed locally below 20 m. This oxygen is rapidly consumed by the decomposing vegetation (Ducharme, 1975).

The primary productivity is high per unit volume (0.32 mg O_2 l^{-1} h^{-1}) but, as in the *ciénagas*, the trophogenic zone is limited to 2 m by turbidity. In May, at the beginning of the wet season, floating plants (mainly *Pistia stratiotes*) and phytoplankton blooms develop simultaneously.

Lake Maracaibo

Location and regional history

Lake Maracaibo (see Fig. 2) is located in western Venezuela at 71°02′–72°08′ W, 9°02′–10°57′ N and belongs to a tectonically unstable area. The Maracaibo basin was the site of marine sedimentation during most of the Mesozoic period and during the Palaeocene. Then, a series of marine regressions and transgressions caused the alternation of marine brackish, freshwater and terrestrial deposits. During the Oligocene, the early phases of uplifting of the Andes led to the emersion of the Maracaibo basin. A long-lasting subsidence process characterizes the area and more than 3000 m of post-Eocene sediments underlie the lake. The resulting entrapment of organic matter explains why the Maracaibo basin is one of the world's most productive oil fields (Redfield, 1958).

At present, Lake Maracaibo is of lagunar type with seasonal connections to the marine water of the Gulf of Venezuela. In its northern part,

Lake Maracaibo basin is located in a semi-arid area with yearly average rainfall of 500 mm; conversely, the upper basin receives an annual average amount of 2500 mm precipitation. The boundary between the two regimes depends on the latitudinal movement of the equatorial rainfall belt. The system generates a dry season in spring–summer and a wet one in winter. The rainfall intensity is variable.

General description

Lake Maracaibo covers an area of $12\,000\,km^2$ with a maximum length (N–S axis) of 160 km and a maximum width of 120 km. This relatively shallow water body (maximum depth 35 m and mean depth 25 m) drains an area of $90\,000\,km^2$ and has a volume of $300 \times 10^9\,m^3$. The watershed area : lake volume ratio is 0.3, intermediary between subalpine lakes (0.10) and lakes of semi-arid areas (0.69) (Serruya, 1980).

The lake receives numerous tributaries, the most important being the Catatumbo River, originating from the Cordillera Oriental. The lake flows into the Caribbean Sea through the Maracaibo Strait with an interannual average volume of $13 \times 10^9\,m^3\,y^{-1}$, determining a water-residence time of approximately seven years.

The hydrographic system

Redfield (1958) thinks that, in recent times, Lake Maracaibo stood above sea level and was drained by a deep valley located on the western side of the Strait of Maracaibo. The post-glacial rise in sea level drowned the valley and allowed the present connection with the sea.

The Strait of Maracaibo is a channel 40 km long and 10–20 km deep, extending north of the lake and flowing into Tablazo Bay, which is a shallow embayment about 2 m deep. The movement of sea water to the bay is limited by barrier bars and barrier islands and occurs through four natural channels. The main channel, running between the islands of San Carlos and Zapara, is 3 km wide and has a natural depth of 4 m. In order to facilitate navigation, especially of oil tankers, this channel was dredged three times: down to 5.5 m in 1938–39, to 10.6 m during the period 1953–56 and to 13.6 m in 1957–62 (Boscan, Capote & Farias, 1973).

Tidal action forces sea water into Tablazo Bay where it mixes with outflowing freshwater. The mixture, heavier than lake water, sinks and flows southward along the bottom to form a nearly meromictic saline hypolimnion in the lake. As a result of a strong, wind-driven current, the light freshwater of the epilimnion circulates anticlockwise which provides an excellent mixing of the upper water mass. The distribution of chlorides reflects this hydrodynamic pattern: low and homogeneous concentrations in the epilimnion and high concentrations increasing with depth in the hypolimnion. Moreover, the cyclonic circulation of epilimnic water results in a concave shape of the hypolimnion in such a way that the isochloride curves protrude upward in the central part of the lake.

The amount of sea water admitted into the Tablazo Bay depends on two antagonistic processes: the tidal current favouring the bayward movement of sea water (Redfield, 1961) and the discharge of freshwater from the lake to the sea pushing the salty water seaward. Since the second factor varies more widely than the first, the salinity of the lake water depends finally on the precipitation over the watershed. Measurements by Redfield (Redfield & Doe, 1965) in March 1954 (dry season) clearly showed the saline stratification in the lake with chloride concentrations of 750 and 2000 ppm in the epilimnion and hypolimnion, respectively, and high chloride concentrations in the Tablazo Bay (> 18 000 ppm).

Similar measurements in November 1955 (wet season) indicated a complete mixing in the lake with homogeneous chloride concentrations < 600 ppm, compared to nearly 10 000 ppm in the bay. The system works in such a way that salty water is admitted into the lake mostly during dry seasons. The long-term changes in salinity of the lake are affected considerably by the cumulative effect of several consecutive dry or wet years.

A tentative yearly balance has been established by Friedman, Norton, Carter & Redfield (1956) from meteorological and hydrological data and compared with the deuterium balance of the lake. The gains indicated a runoff of $32.2 \times 10^9 \, m^3 \, y^{-1}$, rainfall of $22.0 \times 10^9 \, m^3 \, y^{-1}$ and a sea water intrusion of $0.7 \times 10^9 \, m^3 \, y^{-1}$ which corresponded to a total of $54.9 \times 10^9 \, m^3 \, y^{-1}$ and a heavy water value of $0.811 \times 10^7 \, m^3 \, y^{-1}$. The losses indicated an evaporation of $32.9 \times 10^3 \, m^3 \, y^{-1}$ and a seaward outflow of $22.0 \times 10^9 \, m^3 \, y^{-1}$ which corresponded to a total of $54.0 \times 10^9 \, m^3 \, y^{-1}$ and an amount of heavy water of $0.805 \times 10^7 \, m^3 \, y^{-1}$. The adjustment of the deuterium balance would require an 18% reduction of evaporation with a compensating increase of losses to the sea.

The biological features of Lake Maracaibo

The variations of salinity with depth and time, limit the lake invertebrate fauna to blue crabs and teredos. Fish are abundant, however, and commercial catches include 112 species, although only 20 contribute most of the landings from the lake itself. The average annual catch ranges from 10 000 to 17 000 tons. In spite of its variable salinity, Lake Maracaibo has been described by Gessner (1953) as the lake which may well be the richest in plankton of all the waters on earth. Lake Maracaibo is characterized by heavy phytoplankton blooms, dominated by the blue-green alga *Anabaena circinalis* but also including species of *Microcystis* and *Lyngbia* and the diatom *Aulacoidiscus*.

The nutrient balance

The concentration of total phosphorus in Lake Maracaibo is five times higher than that in the open sea (1.5 and 0.3 μg atomic P l^{-1}, respectively). Redfield (1958) concludes that the phosphorus of the lake is mainly derived from the drainage area and that the high algal productivity of the lake comes from both the high retention time of this element in the lake and its rapid recycling. In the deep layers of the lake, the high concentration of total phosphorus (up to 6 μg atomic P l^{-1}) is due to the sedimentation of organic matter in the saline hypolimnion; in these heavy waters the lack of circulation does not allow the oxygen renewal and organic matter accumulates nearly undecomposed.

The high productivity of the upper water of Lake Maracaibo, together with the special hydromechanic structure of the deep layers preventing rapid oxidation of organic matter, are factors favouring the formation of hydrocarbons (Redfield, 1958). Lake Maracaibo with its shallow sill and low tidal action is considered by Redfield to be a model environment for the development of oil fields.

Evolution of water quality

Boscan *et al.* (1973) report that in the past the Lake Maracaibo upper waters were utilized for domestic and agricultural purposes. Its chloride content was 600–750 mg l^{-1} during the period 1937–48, increased to 1022–1118 mg l^{-1} in 1948–49 and returned to 700 mg l^{-1} in 1954. From 1957 onwards, a continuous increase in salinity took place and chloride concentration reached 2400 mg l^{-1} in 1967. A subsequent decrease was observed in latter years and in 1972 the chloride content was 1300 mg l^{-1}. Redfield (1965) showed that in 1957–58 the increase in salinity coincided with relatively dry years. Boscan *et al.* (1973) note that the period of rapid increase of salinity (1957–62) coincided with the widening and dredging of the Barrier of Maracaibo.

It is clear that both mechanisms have a part in the evolution of the salinity of the lake. Besides the saline contamination, however, other types of pollution endanger the water quality. Large amounts of raw sewage are dumped into the lake and in certain areas of low turbulence, decrease in oxygen accompanied by distressing odours have been reported (Boscan *et al.*, 1973). The dredging process in the Barrier area recirculates large amounts of fine particles and organic matter into the lake water and diminishes its oxygen content. In addition to all the previously mentioned sources of organic load, numerous oil spills are occurring (1500 cases for the year 1970). The oil spills have a lethal effect on zooplankton, especially copepods, and on many fish species.

Lake Valencia (Lake Tacarigua)

Lake Valencia (see Fig. 2) is located in northern Venezuela at 67°35'–52' W, 10°05'–16' N and at an altitude of 405 m above MSL. Its present area is 350 km² (maximum width 16 km, maximum length 29 km). It drains an area of 3140 km² with, according to Apman (1973), a demographic density of 350 inhabitants km^{-2} (data of 1970). The watershed is nearly completely deforested. The major inflowing river is the Aragua River. The lake is presently endorheic. Its mean depth is 18 m and maximum depth 39 m. The regional precipitation is 1070 mm y^{-1} (Bockh, 1956).

Lake Valencia occupies an east–west tectonic depression between the northern Coastal Range and the Serrania del Interior in the south. The depression is bordered by longitudinal faults. The sediments in the graben include remnants of *Megatherion*, indicating that the graben was formed before the end of the Tertiary, probably during the late Pliocene. Four different stages have been distinguished in the evolution of Lake Valencia (Peeters, 1968, 1971). Lake Valencia I was established in the late Pliocene to early Pleistocene under a relatively humid climate. It was deeper than the present lake and had probably become semi-arid. This is clearly indicated by the increase of undecomposed feldspath found in the sandy sediments of this period when compared with similar sediments of the previous period. Peeters believes that this reduction of humidity was responsible for stage II of Lake Valencia, which was much more limited in depth and extension than Lake Valencia I. Recent studies of palaeolimnology seem to indicate that the lake basin was dry from at least 13 000 to 11 000 years before present (BP) (Frey, 1979; Bradbury *et al.*, 1981). During the Holocene, humidity rose again and as a result of weathering of feldspath its concentration dropped rapidly in the sediments of this period. This new modification of the climate was accompanied by an increase in the lake level which is clearly indicated in the landscape by lacustrine terrasses located far above the present lake level (+427 m). This Holocene lake is known as Lake Valencia III. The lake level was still around +427 m at the period of the Spanish conquest. During the last centuries, the level has dropped by more than 20 m. It seems that human activity is the main reason for this hydrological disturbance in spite of the present humid tropical climate. The intensive deforestation, the repeated burning of vegetation and the tapping of fluvial and underground sources of water have been suggested (Bockh, 1956, 1968) as the main reasons for the lowering of the lake level.

These long-lasting, man-made modifications in the watershed have also led to a considerable increase in salinity. In 1950, Bonazzi, compiling the existing data on salinity from 1830 to 1950, showed that the total solids of the lake increased from 500 to 980 mg l^{-1}. Lewis & Weibezahn (1976) report that, in 1971, the Instituto Nacional de Obras Sanitarias found a content of 1309 mg l^{-1} dissolved solids. With the exception of Rio Limon, all the affluent streams have high salt content electrical conductivity 500–800 μmhos cm^{-1}). They are however less loaded than the lake (conductivity 1830 μmhos cm^{-1}) and therefore contribute to the dilution of the lake water. The situation is different as far as nutrients are concerned: the tributaries, very loaded in organic matter from wastewater origin (some of the streams carry only sewage water), are the main sources of nutrients of the lake. Measurements made in July 1974 by Lewis & Weibezahn (1976) indicate considerable phosphorus loading values from these affluents (0.76–1.12 mg m^{-2} per lake per day).

The direct effect of this high nutrient load is enhanced by the hydromechanical conditions resulting from level lowering. Given the present maximum depth of 39 m and the amount of wind energy received by the lake, the lake has become nearly epilimnic and a stable thermocline is found only at 30 m. When the effect of wind and/or cooling is strong enough, the whole epilimnion (that is nearly the whole lake) is mixed, and sinking organic matter is mineralized and recycled. In periods of calm weather, the water of the polluted tributaries, less saline and lighter than the lake water, remains in the photic zone and the incoming nutrients are immediately utilized by the planktonic algae. This chemical eutrophication, accelerated by the alteration of physical factors, accounts for the high productivity values of net production measurement by Lewis & Weibezahn in 1976 (2148 mg C m^{-2} d^{-1}). Blue-green species dominate the algal population with 55% of the total biomass; species like *Oscillatoria* and *Lyngbia*, representing 50% of the blue-green biomass, are probably not actively consumed by zooplankton and decompose in the water. Among the tributaries of Lake Valencia, Rio Limon is an exception in the sense that its watershed is partly forested and its water relatively free of sewage pollution. Its main primary producers include an algal mat on the sides of the stream and the epipelic red alga *Hildenbrandia*. The algal mat, composed of blue-greens (*Oscillatoria* and *Lyngbia*) and of diatoms (*Surirella, Melosira, Amphipleura, Navicula*), has a daily net production per m^2 500 times greater than that of the planktonic algae of the stream and 100 times greater than the epipelic *Hildenbrandia*. Since the diatom–blue-green mat has a wider spatial extension than the *Hildenbrandia* mat, the former algae contribute most of the autotrophic input and cover the respiratory requirements of the total community (Lewis & Weibezahn, 1976).

Kiefer (1956), Brehm (1956) and Hauer (1956) gave lists of the main zooplankton components of Lake Valencia. In 1968, Gessner published quantitative results which showed that Rotifera dominated the zooplankton at nearly all depths, followed in abundance by the Copepoda, whereas the Cladocera group was relatively poor.

Sanders (1980), working on rotifers collected from Lake Valencia, demonstrated a clear diel pattern in egg deposition for *Keratella americana* and diel variations in egg deposition for *Brachionus havanaensis*. The author did not offer any explanation of this interesting observation.

Infante (1981) studied the diet of the larval stages of *Notodiaptomus venezolanus, Mesocyclops crassus* and *M. brasilianus*. The nauplii of the three species start consuming algae at stage 3, mainly *Nitzschia* sp. and *Oocystis* sp. The diet of copepodites includes *Cyclotella meneghiniana, Cosmarium* sp. and *Tetraedron* sp. In general, *C. meneghiniana, Nitzschia* sp. and *Oocystis* sp. are quantitatively the main algal species the copepodites feed on, whereas *Lyngbya* is ingested but not digested.

Grospietsch (1975) reports 39 species and subspecies of benthic *Thecamoeba* belonging to 10 genera. Most species are the same as those found in European lakes. Comparison of the relative distribution of the different genera in the sediments of Lake Valencia and Lake Balatòn, however, shows that the genera *Centropyxis, Cyphoderia* and *Lesquerensia* are more common in Lake Valencia, whereas *Difflugia* is 30% less common in Lake Valencia than in Lake Balatòn. Most of the rare planktonic Rhizopoda are imported from other biotopes.

The floating meadows offer ideal conditions of development to many Testacea. Forty species belonging to six genera have been identified. The genus *Arcella* is abundant, especially with *A. rota*, endemic to the southern hemisphere and common in the Amazonian region. The genus *Difflugia* is mostly represented by *D. oviformis* and *D. tuberculata* which have small and fine shells. Large numbers of *Lesquerensia* sp. can be found in the littoral area. These forms are also common in the sediment of European lakes.

The natural and man-made modifications which have occurred during the last 10 years have considerably modified the biomass and composition of the flora and fauna of the lake.

Infante (1978a,b) reported that the numbers of phytoplankton cells were about 1000 times higher than the values found by Gessner in 1968. Moreover, comparing the quantitative results of Gessner (1968) with his measurements of zooplankton populations in 1977, Infante concluded (1978a,b) that in nine years the zooplankton abundance had almost tripled. The most spectacular increase was that of the rotifers which increased up to eight times at certain depths. *Keratella americana* was the leading species in both surveys. The genus *Brachionus*, however, had shown the greatest rate of increase with dominance of *B. calyciflorus* and *B. havanaensis*. The Cladocera group which was not abundant in 1968 had decreased and, in the Copepoda, there was a replacement of *Notodiaptomus venezolanus* by *Thermocyclops hyalinus*.

In conclusion, the historical process of lake-level lowering enhances the effects of chemical eutrophication, since increasing amounts of nutrients flow into a diminishing volume of water; moreover, the salinization of the lake causes the nutrient-rich inflowing waters to remain in the euphotic zone allowing an optimal utilization of nutrients. The dominance of the blue-green algae and the increase of heterotrophic bacteria are the main biological characteristics. The very significant and rapid increases in phytoplankton and zooplankton biomass serve as indicators of the dynamism of the eutrophication process.

The combined effect of level lowering and increasing eutrophication led recently to an extensive mortality of fish and zooplankton (Infante *et al.*, 1979). A calm period, favouring accumulation of reduced substances and particularly H_2S in lower layers, was followed by a stormy period from 29 November to 6 December 1977. The sudden appearance of H_2S in surface layers killed a large number of fish, the genus *Rhamdia* being particularly affected. Moreover, 90% of the zooplankton was destroyed. Such occurrences, which were not reported in the past, are expected to take place more frequently in the future. The feeding ecology of one of the most commercial fish of Lake Valencia, the cichlid *Petenia Kraussii*, an omnivorous fish feeding on bottom material and submerged plants, has been studied by Infante (1981).

THE ORINOCO SYSTEM

This area corresponds to the watershed of the Orinoco River and coastal basins in Venezuela and Colombia (Fig. 3).

The Orinoco River

The Orinoco River has its source in the Parima Mountains, at an altitude of 900 m, near the Brazilian border. It follows first a SW–NE course, then flows to the north-west and north along the Colombian border. After the confluence of the Azure River, it follows an easterly course to the Atlantic. Its total length has been estimated between 1200 and 1700 miles (1930–2785 km) by Gresswell & Huxley (1965) and is reported to be 2148 km by Ziesler & Ardizzone (1979). According to 'The water encyclopaedia' (Todd, 1970), its yearly average yield is 547×10^9 m^3 for a drainage area of 881 000 km^2. Marcinek (1964) reports a yearly yield of 940×10^9 m^3 for a drainage area of 1 086 000 km^2. The annual yield per area is then 0.62 m^3 km^{-2} in the first case and 0.87 m^3 km^{-2} in the second.

The thinly populated basin of the Orinoco extends over savanna and tropical rain forests. Navigation is possible up to 1600 km from its delta. In its upper course, the Orinoco River is connected to the Amazon system through the Casiquiare River and the Negro River. A priest, named Acuna, was the first, in 1680, to make this hydrological link known to European geographers who did not believe him. In May 1800, Humboldt established definitively the existence of the Orinoco–Casiquiare waterway and during expeditions carried out from 1958 to 1962, Vareschi investigated the area in detail. He found that here, as in many other areas of the Orinoco and Amazon basins, the difference in altitude between the waterdivide lines and the rivers is minor; at low waters, the waterdivide line functions normally but it is submerged at high water and the direction of the flow is then variable and depends on meteorological conditions. Temporary connections between small tributaries of the Orinoco and tributaries of the Guainia River, itself an affluent of the Rio Negro, can be explained in this way; however, the major connection occurs through the Casiquiare River. In the Orinoco riverbed, before the bifurcation, sandbanks visible at low waters, divide the river current into two distinct jets: the northern jet, representing 65% of the total discharge continues with the Orinoco whereas the southern jet (25% of the discharge) is diverted towards the Casiquiare and the Amazonian watershed (Fig. 3).

Edwards & Thornes (1970) studied the chemistry of the Upper Orinoco and Casiquiare Rivers. The waters are acid (4.2–6.8 pH) and have a remarkably low electric conductivity (8.5–33 μmhos cm^{-1}). Their high concentration of SiO$_2$ (3.75 mg l^{-1}) reflects the large extension of Precambrian granites and gneiss of the watershed. Moreover, the waters belong to the sodium-chloride type and not to the calcium-carbonate type as do most of the inland waters. Expressed in $mg\,l^{-1}$, the average concentrations are $0.56\,Ca^{2+}$, $0.18\,Mg^{2+}$, $0.77\,K^+$, $2.6\,Na^+$ and $4.53\,Cl^+$.

On its way to the Atlantic, the Orinoco receives many tributaries, such as the Guaviare, the Meta, the Arauca and the Apure Rivers on the left hand and the Ventuari, the Caura and Caroni Rivers on the right hand.

At Ciudad Bolivar, 420 km from the delta, the water level is minimal in March and maximal in August–September. The annual fluctuation of the water level may reach 15 m. At high level, the electrolytic content of the water is very low but the concentration of suspended matter increases. pH ranges from 6 to 7 and temperatures from 27 to 29°C. At low waters, blooms of green algae develop.

The Orinoco delta starts 160 km from the sea and covers an area of $22\,000\,km^2$ (Gessner, 1965). Near the city of Barrancas, the Orinoco divides into numerous ramifications: the Manamo Grande, the Pedernales, the Capure, the Macareo, the Maruiso, the Rio Grande *etc.*

The delta has two main branches: in the north (Pedernales), the brackish water penetrates far inside the delta at low levels of the river whereas in the south (Rio Grande) mixing between sea water and river water takes place outside the delta.

Fig. 3. The watershed of the Orinoco River

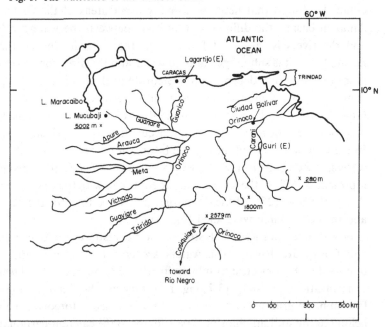

The Orinoco system

In the Pedernales, a salinity of 7.8‰ allows the development of mangrove forest (*Rhizophora mangle* and *Avicennia nitida*); red algae dominate with *Bostrychia pilulifera* and *Catenella impudica*. In a ramification of the Pedernales, at Tucupita, freshwater algal species are found (*Pediastrum* sp., *Staurastrum* sp., *Surirella*, *Melosira*) as well as rotifers and *Diaptomus*. Luminescent Elateridae are common on the trees.

The delta of the Orinoco comprises many lagunas of various sizes with floating vegetation (*Eichhornia*) and 'walls' of *Montrichardia arborescens*. The plankton is very rich in species and number of organisms and does not show massive development of blue-green algae as in the river-lakes of Amazonia. These lagunas have also a rich aquatic fauna of fish and aquatic snakes (*Anaconda*). The lagunas of Manami and Manamito, near Tucupita are rich in phytoplankton (*Surirella* sp., *Melosira* sp., *Eudorina elegans*, *Staurastrum*, other Desmidiales) and zooplankton (*Brachionus*, *Keratella*, *Diaptomus*). *Macrobrachium jelskii* is found in Manamito. At high water, the planktonic biomass is flushed into the river and supplies an excellent fish food.

Lakes and reservoirs
Lake Mucubaji

Lake Mucubaji (see Fig. 3), located at 3550 m altitude in the watershed of the Apure River in the Andes, is one of the few natural lakes of this region. This small water body (0.26 km^2) has a maximum depth of 15.5 m and mean depth of 5.6 m. With a conductivity of 12.6 μmhos cm^{-1} and a sulphate concentration of 0.2 mg l^{-1}, the Mucubaji waters are typical of a Precambrian rock watershed. Lake temperature varies from 12.5°C in the upper layers to 12°C at the deepest point. In spite of this small thermal difference, a stable thermocline exists at 8–9 m and the resulting hypolimnion is depleted of oxygen.

The population of algae includes mostly Chlorophyta, representing 24% of the biomass and dominated by *Oocystis*, and Pyrrhophyta, contributing 75% of the biomass and consisting of *Peridinium* spp. The importance of *Peridinium* in Mucubaji is in agreement with Löfler's observation (1972) that *Peridinium* is commonly observed in high mountain lakes of South America. Gessner (1955) interpreted the algal assemblage of Lake Mucubaji as the result of a gradual impoverishment of the algal assemblages of lakes of lower altitudes. It is interesting to note that Lake Kinneret, a subtropical lake, is similarly dominated by *Peridinium* and Chlorophyta (Pollingher, 1978).

The total absolute biomass is relatively high because of the large volume of *Peridinium* cells. Samples from June 1974, examined by Lewis & Weibezahn (1976), show a total biomass of $1.8 \times 10^6\,\mu g\,ml^{-1}$ that is $1.8\,g\,m^{-3}$. However, the productivity measurements indicate a low net carbon fixation (168 mg C $m^{-2}\,d^{-1}$) which is probably a combined result of low nutrient level and low temperature. Similarly, the littoral vegetation is very poor. The respiration rates have also been found to be very low (<3 mg C $m^{-3}\,h^{-1}$). This strengthens the hypothesis of a very stable thermocline, a necessary condition to produce an oxygen depletion in waters having so low a metabolic rate.

Lagartijo Reservoir

The Lagartijo Reservoir was formed as a result of the damming of Lagartijo River, a tributary of the Tuy River which does not belong to the Orinoco watershed but flows directly into the Caribbean Sea. The reservoir contributes to the water supply of Caracas and its watershed is partly protected. It is located at 190 m above MSL, has an area of 4.5 km², a mean depth of 17.5 m and a maximum depth of 54.8 m. It drains a watershed of 298 km².

The water temperature varies from 27°C in October at the end of the rainy season to 30°C in May. The reservoir is stratified. In July, Lewis & Weibezahn (1976) observed a clear stratification with an epilimnion at 30°C and a hypolimnion at 24°C. In this reservoir, the thermal stratification seems to be permanent in spite of the cooling occurring from October to January. The stratification is kept stable by the inflow of the Lagartijo River: the river water, cooler than the hypolimnion water at its entry into the lake, flows on the lake bottom as a density current.

The abundant runoff in this region of high precipitation (2 m y^{-1}) allows a considerable inflow of cold water into the lake during the rainy season. The oxygenated river water prevents the deoxygenation of the hypolimnion in the wet season in spite of absence of mixing. Conversely, during the dry season, the hypolimnion is devoid of oxygen.

Lagartijo Reservoir has a conductivity of 185 μmhos cm^{-1} in the epilimnion; the hypolimnion has lower conductivity values because of the diluting effect of the river water. In contrast with Lake Mucubaji, the water of Lagartijo Reservoir belongs to the bicarbonate type and has a relatively high buffer capacity (95 mg l^{-1} alkalinity). Expressed in ppm, the concentrations of a few elements of nutrients are as follows: 18 Ca^{2+}, 10 Mg^{2+}, 8 Cl^-, 22 SiO_2, 0.01 PO_4–P and 0.06 NO_3–N. The concentrations of nutrients in the epilimnion are monitored by an eventual partial mixing of the epilimnion with the upper layers of the hypolimnion. This is in fact

The Orinoco system

the only mechanism which supplies nutrients to the euphotic zone since the watershed load is directly channelled into the hypolimnion.

This explains why primary productivity is strictly dependent on turbulence. In July 1974, Lewis & Weibezahn (1976) determined a net primary production of 1252 mg C m^{-2} d^{-1}. Mean respiration in the upper layers was found to be 229 mg C m^3 d^{-1}. Per unit area of the euphotic zone, respiration was nearly equal to net production.

The phytoplankton examined by Lewis & Weibezahn (1976) is dominated by the Cyanophyta with *Anabaenopsis raciborskii*, *Rhabdoderma sigmoidea*, *Lynbgya limnetica* and *Chroococcus* sp. as dominant algae. The Cyanophyta represent 89% of the total biomass. The Chlorophyta are represented by numerous species, in particular *Scenedesmus ecornis* and *Oocystis* but contribute only 2% to the total biomass. Euglenophyta contribute 4% and the Cryptophyceae *Rhodomonas minuta* and *Cryptomonas erosa*, 5%. The total biomass is 2.7 g m^{-3}. It is interesting to note that, in his detailed plankton survey of 1966–68, Reyes (1972) does not mention *Lyngbya* which in 1974 represented 15% of the total algal biomass.

Guanapito Reservoir

The Guanapito Reservoir is located in the watershed of a tributary of the Apure River. It has an area of 10 km^2 and a maximum depth of 40 m at the spillway. Its watershed receives an average amount of rainfall of 1 m y^{-1}.

The reservoir was studied by Lewis & Weibezahn (1976) in July 1974. An upper thermocline was observed between 6 and 7 m. The hypolimnion was devoid of oxygen. A second thermocline, observed at 15 m, can be destroyed only by heavy winds. The upper layers had a temperature of 27°C and the bottom water was at 25°C.

The Guanapito waters have a relatively high conductivity (240 μmhos cm^{-1}), a high alkalinity (115 mg l^{-1}) and a calcium concentration of 24.5 mg l^{-1}.

The primary productivity was lower than in Lagartijo Reservoir: 496 mg carbon m^{-2} d^{-1}. This low productivity is due to nutrient depletion generated by a high algal standing crop (1.7 g m^{-3}) under conditions of low turbulence.

The phytoplankton is dominated by the Bacillariophyceae with *Cyclotella stelligera* and *Synedra acus* as the main species. The Bacillariophyceae contribute 46% of the biomass. The Cyanophyta, which represent 27% of the biomass, are mainly composed of *Lyngbya limnetica*, *Anabaenopsis raciborskii* and *Chroococcus minutus*. The Chlorophyta (14% of the biomass) are dominated by *Oocystis* sp. whereas *Tetraedron minutum* and *Cryptomonas* sp. contribute 4% to the biomass.

Guri Reservoir

The Guri Reservoir (see Fig. 3) is the largest freshwater body in Venezuela. It is part of a development plan of eastern Venezuela and results from the construction of a dam on the Caroni River in 1968. At present, the reservoir has an area of 800 km² and a volume of 18×10^9 m³. Its mean and maximum depths are, respectively, 22 and 31 m. It drains an area of 95 000 km² near the reservoir. The vegetation in the watershed varies from savannas with small trees to deciduous forests and evergreen forests. The area is hardly inhabited.

The reservoir was surveyed by Lewis & Weibezahn (1976) in July 1974. The surface water was at 29°C; a first thermal discontinuity at 5 m represented the limit of the daily wind mixing. A second thermocline was noted at 20 m, and the hypolimnic temperature varied from 27 to 26.5°C. Lewis & Weibezahn assumed that no mixing took place below 20 m during the dry season.

The electrical conductivity of Guri water is as low as 9 μmhos cm^{-1}, the alkalinity 3.2 mg l^{-1} and Ca^{2+} level is lower than 1 mg l^{-1}. This nearly distilled water is acid (pH 6.5) and coloured dark by humic substances originating from the forests of the upper valley. The resulting inhibition of light penetration considerably limits primary production. Lewis & Weibezahn (1976) measured a rate of 65 mg C m^{-2} d^{-1}.

The phytobiomass does not exceed 0.45 g m^{-3}. It is dominated by the Pyrrhophyta, mainly *Peridinium* sp., which form 84% of the biomass. The Cryptophyceae *Rhodomonas minuta* and *Cryptomonas* sp. represent 7% of the biomass. The Chlorophyta (6% of the biomass), represented by numerous species, are dominated by *Scenedesmus quadricauda* and *Coelastrum* sp. The Cyanophyta, dominated by *Rhabdoderma sigmoidea* fa *minor*, contribute only 2% to the total biomass. The massive dominance of Pyrrhophyta in these nutrient-poor waters is in good agreement with the interpretation given to the blooms of *Peridinium cinctum* in Lake Kinneret by Serruya, Gophen & Pollingher (1980).

THE GUIANA AREA

The 'Guiana area' corresponds to the northern watershed of the oriental region of the Plateau of Guiana. This is a Precambrian mountainous area at about 1000 m altitude plunging northward to swampy coastal areas (Fig. 4).

The Guiana area

The drainage system is composed of numerous and relatively short rivers such as the Essequibo River (970 km long), the Mazaruni River (560 km), the Courantyne River (720 km), the Suriname River (400 km) and the Litani–Lawa–Maroni River (720 km). The yearly precipitations range from 2000 to 3000 mm and the resulting abundance of water is well expressed in the Amerindian word *Guiana*, 'the land of waters'.

In order to develop its aluminium industry, Surinam needed additional electric power. The Suriname River was dammed, approximately 160 km from the Atlantic Ocean for hydroelectric purposes. The dam was completed in February 1964 and the flooded area is now known as the Brokopondo Reservoir. An area of 12 550 km² is drained by the dam and the average outflow discharge is 270 m³ s⁻¹ (period 1966–70). The surface area of the lake is 1500 km², the shoreline 2000 km and maximal depth 38 m. Its capacity is 180 MW.

In 1963, the Netherlands Foundation for Scientific Research in Surinam and the Netherlands Antilles undertook observations on the Suriname River in its natural conditions and in the reservoir during the filling stage.

The Suriname River, flowing on Precambrian rocks, has a very reduced solid load. It is very poor in dissolved minerals and has a low pH. The water

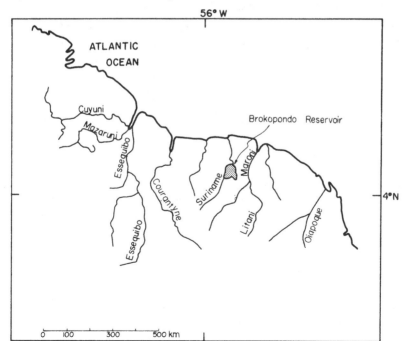

Fig. 4. The Guiana area and Brokopondo Reservoir

is turbid and brown. This does not result, as in Amazonia, from the presence of humic acids, but from iron and silica in suspension. In its natural state, the river flowed through tropical rain forests with the trees reaching the banks of the river. Leentvaar (1966) reports that Podostemaceae were the only water plants in the river; they were restricted to small pools, swamps and portions of the river banks and served as shelter for insects and the freshwater shrimp *Macrobrachium*; stones were covered with freshwater sponges and the freshwater snails *Pomacea* and *Doryssa* were found. A plankton sampling, carried out in late 1963 at Afobaka station, indicated a dominance of diatoms and desmids and a relative scarcity of zooplankton, reflecting oligotrophic conditions.

During the dry season, 'river plankton' used to develop with the diatoms *Eunotia asterionelloides, Rhizosolenia eriensis, R. longisete* and *Surirella* sp. and the Desmidiales *Teilinga granulata, Gonatozygon monotaenium, Cosmarium* sp., *Staurastrum* sp., *Staurodesmus mamillatus, Cosmocladium* sp., *Euastrum* sp., *Closterium* sp., *Micrasterias braziliensis* and *M. arcuatus*. Chlorophyta were less common: *Pediastrum* spp. were rare but *Eudorina elegans* was common at periods when physical conditions were changing rapidly in the river. Cyanophyta were scarce with *Merismopedia convoluta* and *Oscillatoria* sp. as the most common representatives. Zooplankton was observed only during the dry season and comprised the rotifers *Polyarthra* sp., *Keratella americana* and *Conochiloides coenobasis* and the crustaceans *Cyclops* sp., *Bosminopsis* sp. and *Moina* sp. (van der Heide, 1976).

In May, with the onset of the rainy season, this community was being washed downstream and replaced by a new assemblage of *Arcella, Heliozoa, Rhipidodendron* and Fungi Imperfecti but at the peak of the rainy season, in July-August, the river water was replaced by pure rain water devoid of algae. This periodic flushing helped to maintain oligotrophic conditions in the river.

Before the dam was closed, populations were transported to new villages, nearly 10 000 animals were rescued and released around the lake but the lake area was not cleared of trees. As the trees reached 20–30 m in height, in about 70% of the full reservoir the tops of the trees remain above the water surface (van Donselaar, 1968) since in 1967 the reservoir reached a depth of 30.5 m. During the filling process turbulence decreased, not only as a result of the passage from river to lake conditions, but also because of the physical obstacles opposed by the dead trees to water movements. The resulting stagnation generated the thermal stratification of the reservoir: instead of the uniform temperature of 30°C of the river water before the closing of the dam, the water gradually warmed up in the upper layers

The Guiana area

(32°C in 1964 and 34–35°C in 1966) and became colder near the bottom (26°C). There was a tendency towards a general warming up from top to bottom however, and in 1970 the bottom temperature was 27.8°C. The high oxygen consumption of the dead forest turned the hypolimnion (below 5 m) into an anoxic water body rich in H_2S. This caused a massive killing of bottom-dwelling fish such as the stingray and the catfish.

Van der Heide (1978), reporting a detailed survey carried out in May 1964, noted that temperature fluctuated widely in the upper 3–4 m but, below this layer, temperature decreased gradually with depth without any marked thermocline. Oxygen was present only in the uppermost layer. At dawn, oxygen even disappeared from this layer. The complete absence of oxygen in deep layers and the time-constant vertical gradient of conductivity indicate that the superficial temperature gradient is quite sufficient to maintain a durable thermal stability in the absence of a clear deeper thermocline. Vertical mixing is then reduced to the upper 3–4 m and, when hypolimnic waters loaded with decaying organic matter are recirculated, oxygen of the upper layer is rapidly consumed.

The concentration of oxygen in the deep layers increases after each winter turnover, however, indicating a progressive diminution of the organic load. Leentvaar (1973a) mentioned the temporary presence of oxygen in deep layers during the anoxic period from 1968 onwards; this was probably due to the increase of turbulence resulting from the disappearance of the drowned trees by slow decomposition.

In the period following the closing of the dam, the pH varied between 5 and 6. Conductivity of surface water was around $25\,\mu$mhos cm^{-1} but reached $35\,\mu$mhos cm^{-1} during dry periods. Conductivity increased more rapidly near the bottom where it reached $70\,\mu$mhos cm^{-1}.

In early 1964, at Afobaka station, the river plankton and the diatom *Eunotia asterionelloides* disappeared whereas *Eudorina elegans* and Euglenophyta developed. This transitional phase was followed by a burst of unicellular flagellates. Thick algal mats of *Spirogyra* and *Mougeotia* covered the dead trees and harboured large populations of snails, crustaceans and insects. The zooplankton community increased; it included crustaceans and additional species of rotifers such as *Filinia longiseta, Asplanchna* sp. and *Ascomorpha saltans*. The rotifer *Sinantherina spinosa* and the turbellarian flatworm *Catenulla lemnae* appeared near the anoxic zone of the lake (van der Heide, 1976).

The most dramatic modification, however, was the expansion of water plants. During the first three years, mostly free-floating species developed: the water hyacinth (*Eichhornia crassipes*), floating ferns (*Ceratopteris pteridoides* and *C. deltoidea*), and duckweeds (*Lemna valdiviana* and *Spirodela natans*) (van Donselaar, 1968). Sometimes these aquatics formed floating islands which were colonized by plants of the forest flora. The remarkable development of the water hyacinth was favoured by the sudden drop in turbulence. The plant spread from four quiet areas in the northern part of the river where it was present before inundation, firstly along the banks, but from July 1964 onwards it also colonized the dead drowned forest. In November 1964 the water hyacinth covered 5000 ha, in June 1965, 17 900 ha and in April 1966, 41 200 ha, i.e. 40% of the reservoir surface (van Donselaar, 1968). In July 1964, a control campaign was undertaken. In 1965 and 1966, systematic air sprayings of the auxin 2,4-D were carried out and allowed a costly but efficient control of the weed. *Ceratopteris pteridoides*, which spread from 1200 ha in November 1964 to 11 700 ha in April 1966, especially in Sara Creek, retreated noticeably after the control campaign.

With time, factors hampering turbulence, such as the presence of dead trees or the development of water plants, have disappeared. Wind-driven mixing, again possible, allows a better reoxygenation of the reservoir. These new conditions slowly modify the plankton assemblage. In his 1968 survey, Leentvaar (1973*b*, 1979) found that the plankton of the flowing water of the upper river was poor. In contrast, brown detritus with spiculae of sponges, iron bacteria and flagellates was abundant. In the upper part of the lake, where water became stagnant, the plankton was less brown and zooplankton more abundant. Further into the lake, *Volvox, Eudorina* and *Cosmarium* sp. dominated. In the central part of the reservoir abundant *Cosmarium* sp. and *Eudorina elegans* were accompanied by large populations of *Cyclops* and *Ceriodaphnia cornuta*. In the dam area the plankton was clearly dominated by desmids and ostracods. In 1968 the lake plankton was dominated by desmids, crustaceans, rotifers, and Volvocales (*Volvox* and *Eudorina*). The abundance of desmids and the diminution of flagellates indicated that the lake water was returning to more oligotrophic conditions.

Since prior to inundation little was known about the fishes, it is difficult to determine the modifications introduced by stagnation. Nijssen (1969) reports that many species disappeared as a result of the oxygen deficiency in deep layers, destruction of niches and the physical obstacle of the dam which prevents the upstream migration of certain species. In comparison with the upper reaches of the river the fish fauna of the reservoir is poor. The most common species are *Hoplias malabaricus, Cichla ocellaris, Serrasalmus rhombens, Acestrorhyncus falcatus, A. microlepus, Leporinus freiderici, Curimatus schomburgki, C. spilurus, Moenkhausia* and *Creatochanes* spp. In 1964, the river bed below the dam fell almost dry and was fragmented into numerous isolated pools rich in plankton, supersaturated in dissolved oxygen and having a high conductivity because of evaporation. Since 1964 was a relatively dry year, sea water penetrated into this stretch of the river, increasing the chloride concentration of the water. In May 1964, the chloride level exceeded $10 g l^{-1}$ at Domburg.

Another hydroelectric scheme is planned on the upper Mazaruni River about 290 km from the coast. Its planned capacity is 1000 MW expandable to 3000 MW in the future.

AMAZONIA

This area corresponds to the watershed of the Amazon River. With an area of 6.9×10^6 km² (Lyra, Oliveira & de Meno, 1976), it is the largest drainage area of the world. 'Amazonia' covers more than one-third of the South-American continent (Gresswell & Huxley, 1965) and conveys to the Atlantic Ocean one-fifth of the freshwater of the world.

The Amazon Valley has developed in a very old graben between two Precambrian massifs: the Guiana and Brazil shields (Fig. 5).

Geology and geography

In the pre-Andean period, the Amazonian region was drained westward into the Pacific Ocean. In the Pliocene, the rising of the Andes blocked the western outflow of this drainage system. An immense lake covered the lowlands and up to 300 m of fluvio-lacustrine sediments were deposited. In eastern Amazonia, the freshwater formations, the Barreiras series, originated from the erosion of the Precambrian shields of Guiana and Brazil. In eastern Amazonia, the lacustrine deposits are known as Pebas formations and made of Andean material. Marine and brackish horizons are interbedded in the Pebas formations, which points to minor marine transgressions from the Caribbean Sea. Later, Pleistocene epeirogenic movements in the Andean area tilted the continent eastward and the lake drained into the Atlantic Ocean. New rivers carved their beds into the soft freshwater sediments. The general rise of ocean level in post-glacial times drowned part of the former valley and initiated a new sedimentation cycle in lower Amazonia.

In its middle and lower course, the wide alluvial plain, exceeding 100 km in certain areas, is bordered by thin strips of sedimentary rocks of Palaeozoic age (schists, sandstones, clays, limestones and gypsite). North

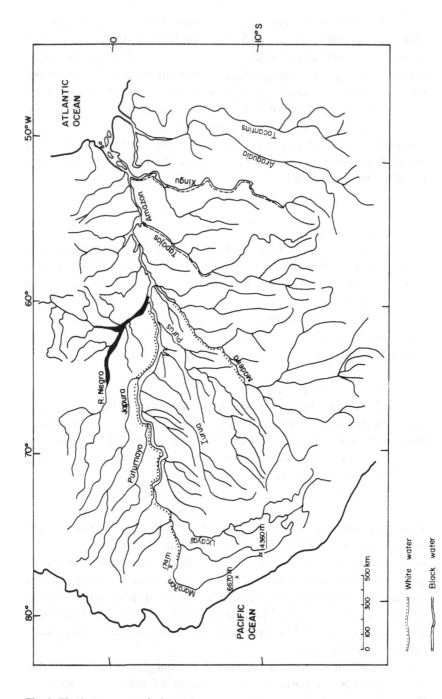

Fig. 5. The Amazon watershed

Amazonia

and south of these narrow belts, extend the crystalline formations of the Precambrian shields which cover large areas of the watershed. The Upper Amazon (Marañon) flows on the sedimentary and volcanic series of the Andes.

Amazonia, extending from 5° N to 19° S, has a typical 'rainy tropics' climate. Temperature is high and uniform. At Manares, the yearly average is 27.4°C with a March minimum of 26.9°C and an October maximum of 28.3°C (Heintzelmann & Highsmith, 1973). Humidity is high and rainfall ranges from 1800 to 3500 mm y^{-1}. Its monthly distribution varies with latitude and altitude. At Juarete (0°18′ N, 68°44′ W, 122 m alt.), monthly precipitation varies from 200 to 300 mm; however, a small peak (400 mm) is observed in April, May and June. At Boa Vista (2°48′ N, 60°42′ W, 90 m alt.) precipitations are less than 100 mm per month most of the year but a clear peak is observed from May to August with monthly averages of 300–400 mm. At Taperinha (2°30′ S, 54°20′ W, 20 m alt.) the wet season extends from January to June with monthly amounts of 200–350 mm followed by a dry season with precipitation not exceeding 100 mm per month. It then follows that from a typical and uniform equatorial regime without dry seasons one passes rapidly to a boreal tropical or to an austral tropical regime, with a clear wet season as one moves north or south from the Equator.

It is easy to understand that the lowlands with uniform rainfall or limited dry season – and these features apply to nearly the whole non-andic Amazon basin – are covered with the dense rain forest (the hylaea of von Humbolt). The humid savanna develops only on the north and south borders of the basin where the dry season is more accentuated.

Hydrology

The River Marañon (the upper course of the Amazon) has its source in the Cordillera Huay Huash in Lake Lauricocha at about 5000 m altitude. The Marañon flows in a south–north direction over 600 km at high altitude, then leaves the Andes by heading north-east in such a way that, 900 m from its source, the Marañon is at an altitude of 174 m above MSL. Approximately 1000 km further, at Tabatinga, the altitude of the river is only 81 m above MSL. From Tabatinga the river is called Solimoes, whereas the name Amazon should be restricted to its middle and lower course. In fact the term Amazon is often used to designate the whole river. The Amazon has a total length of 6770 km but only 14% of the river system is located at an altitude exceeding 200 m. This results in extremely low gradients of the order of magnitude of 1.5%.

The Amazon receives numerous affluents. Their upper course, generally carved in rocky formations, has a 'normal river bed' but, when they enter the Amazon alluvial plain, the river bed widens considerably and the current velocity drops significantly. This causes an immediate sedimentation of the solid load with formation of islands on both sides of the major channels and swampy areas with reduced circulation. This sedimentation zone, which reaches 100 km long in the Tapajos Valley (Sioli, 1975a), cleans the river water from suspended matter. Clear and slowly moving waters enter the mouthbay of the river which is often wider than the sedimentation zone and has been given the name of 'river-lake'. At certain periods, the water is stagnant or may even, as in the lower Amazon, flow upstream under the effect of the ocean tides which are felt up to 870 km in the Amazon River (Guilcher, 1965). It is then clear that, in the 'river-lakes', there is no bottom erosion but sedimentation of fine organic ooze 3–4 m thick (Sioli, 1975b). This typical Amazonian morphology is observed on the last 2500 km of the Amazon. The mouthbay of the Rio Negro is especially famous. In its sedimentation zone, known as the Anavilhanas Archipelago, a special type of forest has developed, the 'Igapo forest' which is partly flooded at the high-water period. The mouthbay of Rio Negro is a deep lake reaching a depth of 100 m.

All these tributaries and their lower mouthbays join the immense alluvial plain or flood plain of the Amazon, called the várzea. The width of the várzea varies from 20 to 100 km; it is limited by the slopes of the 'terra firme'. The várzea has a very fertile soil, renewed every year by additional sedimentation of Andean material (Gibbs, 1967). The várzea is a complex system of islands, lakes connected to the river by channels called *furos*, sidearms (*paranas*) of the river and sandbanks. Moreover, the small gradient does not allow a high current velocity: 0.5 to 1.0 m s^{-1} in the dry season and 1.0 to 2.0 m s^{-1} in the rainy period. As a consequence, a complicated network of meanders and anostamoses has developed.

In spite of the previously mentioned physical and hydraulic features which characterize a very lazy fluvio-lacustrine waterway, the Amazon River conveys enormous amounts of water and suspended matter. It discharges at a rate of 200 000 m^3 s^{-1} into the Atlantic Ocean, that is an annual volume of 6.5×10^{12} m^3 water, and an annual amount of 1×10^9 tons of suspended solids. Guilcher (1965) reports that the maximal floods of the Amazon River may reach 280 000 m^3 s^{-1} or 1×10^9 m^3 d^{-1}. If one pumped out Lake Geneva at such a rate, the lake would dry up in less than four days.

The Amazon River has numerous affluents: 20 of them are more than

100 km long. The most important on the northern left bank are the Pastaza, Tigre, Napo, Putumayo-Ica, Caquetá, Japurá, Negro, Trombetas, Paru and Jari, and on the southern right bank, Huallaga, Ucayali, Javari, Jandiatuba, Jutai, Jurná, Tefé, Purus, Madeira, Parana, Tapajos, and Xingu. It is interesting to note that certain tributaries of the Amazon are among the biggest rivers of the world: for example the Rio Negro has an annual yield similar to that of the Zaire River, although its drainage area is four times smaller. The contribution of the southern tributaries to the total yield of the Amazon is dominant; however, the regime of the Amazon is affected by the flood periods of northern and southern affluents. The Madeira, the main southern tributary, has its maximum in March and minimum in September, and this regime also characterizes most of the other southern rivers. Rio Negro, which is the main northern tributary, has a June–July maximum and a December minimum. At Obidos, after the confluence of Rio Negro and Rio Madeira, the Amazon has one single maximum in May–June between the maxima of the southern and northern tributaries. In spite of receiving water from both hemispheres, the Amazon has a clear flood period, although it always maintains a very high yield. Guilcher (1965) mentions $72\,500\,m^3\,s^{-1}$ as a minimum rarely reached.

At the period of high water, the Solimoes River level rises by 20 m, the mouthbay of Rio Negro by 12 m, and the mouthbay of the Tapajos by 7 m. At maximal extension, the várzea covers more than $50\,000\,km^2$.

Substrate vegetation and water chemistry

The different geological and pedological formations outcropping in the Amazonian watershed determine the type of vegetation cover and affect the composition of the water (Klinge & Ohle, 1964). The granites and gneiss of the Precambrian massifs weather into brown loam soils. The water draining these soils is poor in minerals, especially calcium, sulphate and bicarbonate, has a low pH (4.0–6.6) (Sioli, 1975*b*) and is relatively richer in sodium and potassium. The Barreiras series, which were formed from erosion material of the Archean massifs, have similar chemical features. However, whereas the weathering of the crystalline rocks is a primary process which ends up with soluble compounds, the leaching of the Barreiras series releases only minor amounts of soluble minerals. The weathering processes of these formations took place during the Tertiary and the erosion, transport and redeposition processes, together with thorough washing over long periods by tropical rains, have left only the most insoluble components (Sioli, 1968*a*).

The Barreiras series and the crystalline massifs are covered with dense rain forest with high species diversity, which prevents erosion of the soil and utilizes rapidly the degradation products of organic matter. This forest typically lacks the species of commercial timber (*Swietania, Bertholetia, Hevea*) (Fittkau, Junk et al., 1975).

These various features explain that the water draining these areas is devoid of suspended matter and extremely poor in minerals, nutrients and organic matter. They are in fact among the purest natural waters. Sioli (1968a) has described this type of water as 'clear water'. The Tapajos River is a classical example of this type of water.

In various areas of the Precambrian massifs and of the Barreiras series, such as the Cururú Valley on the upper Rio Negro, large outcrops of white sands are covered with caatinga vegetation, an open forest including species able to withstand semi-arid conditions such as cactus and mimosa. The drainage water of these areas is chemically very pure and devoid of suspended matter but, in contrast with the clear water, has a brown, coffee colour which has given the Rio Negro its name. Klinge (1967) has shown that these white sands represent tropical podzols with a characteristic lower illuvial horizon. The incompletely oxidized organic matter forms acid humus, and brown soluble humic substances are released into the drainage water. These 'blackwaters' are even more acid than the clear water and are richer in iron and aluminium. They have very low concentrations of Ca^{2+}, Na^+ and K^+, probably as a result of cation exchange by humic substances. Sioli (1968a), thinks that humus retains these cations and releases H^+ ions which cause the acidification of the waters.

The water draining the sedimentary palaeozoic series has a higher salt content and a higher pH. This chemical composition allows the development of snails which host *Schistosoma mansoni*, the agent of the disease bilharzia, which is not found in other types of water.

The most common type of water in Amazonia is a third type named 'the white waters'. These are turbid waters, very rich in suspended solids originating mainly from the Andes. Gibbs (1967) showed that 82% of the suspended solids discharged by the Amazon was supplied by the mountainous part of the watershed representing only 12% of the total area. These waters have a neutral pH.

The Amazonian paradox

The luxuriant vegetation of the Amazonian forest and extreme species diversity of its fauna suggest an abundance of nutrient sources and fertility. As early as 1903, however, Katzer discovered the mineral poorness of the Amazonian waters which was fully confirmed by later studies (Sioli,

Amazonia

1950, 1954, 1955; Schmidt, 1972b). Furch (1976) has shown that, with the exception of copper, zinc, chromium and cadmium, the concentration of ions in Amazonian waters is one to two orders of magnitude lower than the world average for freshwater. These waters which have been described as 'slightly polluted, distilled waters' drain various types of soils which have themselves a lower than world-average content of alkali and alkali-earth metals. This is clearly shown in Table 9, where data collected by Furch & Klinge (1978) are compared to the average composition of soils according to Bowen (1966).

The Amazon mineral soils are especially poor in calcium, magnesium, potassium and sodium. Moreover, the work of Furch & Klinge (1978) indicates that there is a decline in most elements from values near to world standards in extra-Amazonian soils to values which are two to three orders of magnitude lower in soils of central Amazonia. A similar difference, mainly for calcium and magnesium is observed in the composition of leaves, stems and roots of Amazonian terrestrial vegetation when compared to extra-Amazonian plants. There is a significant decline in calcium, potassium, magnesium, barium and strontium in the living material from extra-Amazonian regions towards central Amazonia. The humid tropical climate in itself is not the cause of these low concentrations in the Amazonian vegetation, since Nye (1958) found much higher levels in leaves

Table 9 *Chemical composition of Amazonian soil in comparison with average soil composition*

Soil type	Na^+ (ppm)	K^+ (ppm)	Mg^{2+} (ppm)	Ca^{2+} (ppm)	Sr^+ (ppm)	Source
Mineral soils (0.30–50 cm depth)						
Tropical rain forest, tall Amazon caatinga	–	252	380	122	–	Klinge, Medina & Herrera, 1977
Season evergreen forest, Manaus	10	20	5	5	–	Klinge, 1976a, b
Clay fraction, upper soil						
Guiana shield	600	700	280	700	90	Irion, 1976
Palaeozoic rocks	1125	750	290	375	135	Irion, 1976
Tertiary rocks	160	225	100	350	75	Irion, 1976
Várzea sediment	280	230	640	5400	–	Howard-Williams & Junk, 1977
Average soil	6300	14 000	5000	13 700	300	Bowen, 1966

and stems of Ghanaian trees growing in a similar humid tropical climate. As a consequence, the litter of the Amazonian forest is also much poorer in most elements than average. If we compare the concentration ratios of extra-Amazonian and central Amazonian soil, water and vegetation (Table 10), it is clear that although the ratios are very high for mineral soil and water they are much smaller for living material compared to soil or water, indicating the ability of the vegetation to concentrate elements from soil and water. It is, nevertheless, difficult to explain why the Amazonian vegetation is doing so well with six times less calcium than extra-Amazonian plants (Howard-Williams & Junk, 1977).

It quickly appeared to Amazonian scientists that 'the closed circulation of the nutrients within the living biomass' (Sioli, 1979) was the solution to the paradox. This is achieved by countless strategies which all converge to prevent nutrient leakage out of the system.

For example, the rain water and the groundwater of the Amazonian forest have very similar composition and both are very poor in nutrients but the water dripping from the leaves along the tree is rich in nutrients and major elements (Sioli, 1979); the excretions of the numerous tree dwellers are washed from the leaves and rapidly taken up by epiphytes on the tree stems or by the tree roots. The rapid uptake by the root system is made possible by the unusual development of the root mass in the Amazon forest. In the caatinga forest of Southern Venezuela, Klinge & Herrera (1978) have measured an average of 132 tons ha^{-1} dry matter of root mass which represents about three times the average value for a temperate forest and is the highest root mass ever reported for tropical forests. Moreover, this exuberant root system is very superficial: Klinge & Herrera also reported that 86% of the root mass is located in the upper 30 cm of the soil

Table 10 *Relative concentrations of various elements in several compartments of tropical forest in Panama and Brazil, expressed as a ratio of Panamanian:Brazilian values (from Furch & Klinge, 1978)*

	Na^+	K^+	Mg^{2+}	Ca^{2+}	Sr^{2+}	Ba^{2+}
Mineral soil	46.2	2.1	104	1171	–	–
Natural water	105	28.0	239	1724	196	–
Primary vegetation						
Leaves	0.1	3.8	1.1	6.0	1.7	23
Stem	0.2	7.4	2.4	8.5	0.9	52
Leaf fall	0.9	1.4	1.6	6.5	–	–
Leaf litter	0.8	8.2	1.7	11.0	–	–

Amazonia

and 70% have a diameter less than 6 mm. In the rain forest of Brazil, Klinge (1973, 1976b) reports root-mass values of 40–60 tons ha^{-1}.

This abundant and superficial root system is in permanent contact with a rapidly decomposing litter. Klinge & Herrera (1978) observed that, below the fresh litter layer a few centimetres thick, the leaves were already decomposed and covered by a network of fine roots. The enormous area of contact between litter and roots is of major importance in the conservation of nutrients within the living system (Herrera, Merida, Stark & Jordan, 1978). Stark (1969) showed that fungi were instrumental in nutrient transfer and Herrera, Merida *et al.* (1978) observed that fungi fixed on roots were directly pumping labelled phosphorus from leaf to root.

This incessant flux of nutrients from the litter to the roots can only be achieved by a rapid decomposition of this litter. The studies of Klinge & Rodrigues (1968) on terra firme rain forest near Manaus show that the litter accumulates during the dry season, when trees lose their leaves. Later studies by Klinge (1977) on terra firme forest, várzea forest and *igapo* forest show a higher leaf fall in the rainy season. The average annual fine litter fall was found to vary between 8 and 10 tons ha^{-1} dry matter. In comparison with other tropical forests, the Amazon forests retain much lower amounts of nutrients, via litter to the soils, except for nitrogen.

In contrast with terrestrial vegetation, the Amazonian aquatic macrophytes have a 'normal' composition. They have higher values of total ash than the terrestrial plants and are richer in calcium, potassium and sodium. The concentrations of the major elements in the aquatic plants are also much higher than the concentrations of these elements in the water in which the plants grow. This is especially true for potassium and phosphorus (Howard-Williams & Junk, 1976). According to these authors, 1 ha of aquatic plants and 1 ha of water in Lago do Castanho contain, respectively, 356 kg and 45 kg potassium. This emphasizes the role of the nutrient reservoir of the macrophytes. Moreover, the macrophyte production being higher than the algal production, the absolute amount of nutrients stored in aquatic plants is considerable. It returns to the food chain through leaching, bacterial decomposition and direct grazing. Experiments by Howard-Williams & Junk (1976) on decomposition rates of plants sampled and exposed in the Amazonian floating meadows have shown a very fast rate of decomposition: 50% of the original material was decomposed in the first month whereas the times for 50% decomposition of similar material in England varied from 200 to 425 days (Mason & Bryant, 1975).

These decomposition studies have also shown that nutrient levels of the litter end up with uniform values and do not depend on the original concentrations in the plants. It follows that, whereas herbivorous animals are very selective in plant grazing, detritivores have a much wider food substrate at their disposal. The rapidity of decomposition of aquatic plants and the ecological advantage of the detritivores are essential to the understanding of the Amazonian food chain.

Aquatic biotopes of Amazonia

From the point of view of chemical composition, nutrient and suspended material content of the water, the Amazon and its tributaries of right-bank and Andean origin contrast sharply with the other tributaries draining the Precambrian rocks. The Amazon River is a 3000 km long tongue of white water, rich in suspended matter with an electrical conductivity of $60\,\mu$mhos cm^{-1} and a pH of 6.8, whereas numerous tributaries carry a sediment-free water having a conductivity of $10\,\mu$mhos cm^{-1} and a pH of 4.5.

In addition to the ecological differentiations caused by such variable chemical features, the considerable fluctuations of the Amazon water level divide the landscape into two basic entities: the never-inundated terra firme and the periodically flooded várzea (Fig. 6). The combined effect of these two factors creates a large variety of biotopes with marked seasonal variations.

Fig. 6. The várzea forest at high water. Left shore of the Lower Amazon between Santarém and Monte Alegre (by courtesy of Prof. H. Sioli, Max Planck Institute, Plön, West Germany)

Blackwaters

Gessner (1959) thought he could infer the presence of blackwaters from the absence of Gramineae. The extensive work of Junk (1970) has shown that these plants can thrive in blackwaters as long as these are periodically mixed with white waters. In blackwaters, Gramineae are therefore limited to the lower courses of the rivers. Lago Janauari, in the mouth area of Rio Negro, has large populations of *Paspalum repens* and *Echinochloa polystachya*. Upstream, in pure blackwaters, these populations do not exist. *Eichhornia crassipes*, very sensitive to low pH, is also absent from blackwaters.

Similarly, the phytoplankton populations of blackwater are not abundant but present a great diversity (Schmidt, 1976). In the waters of Rio Negro, Uherkovich (1976) found 204 taxa including: 7 Cyanophyta, 6 Euglenophyta, 5 Pyrrhophyta, 3 Volvocales, 54 Chlorococcales, 86 Conjugatophyceae, 6 Cryptophyta, and 33 Bacillariophyta. The population was dominated by *Melosira granulata* var. *angustissima* and *Tabellaria fenestrata* var. *asterionelloides* and included the common following species: *Dictyosphaerium pulchellum, Closterium kuetzingii, Kirchneriella lunaris, Staurastrum quadricauda, Sphaerocystis schroeteri*, and *Diatoma elongata* var. *minor*. In a later survey of Rio Negro, Uherkovich & Rai (1979) studied the vertical distribution of algae and found that, in the upper layers, small Cryptophyceae formed more than 80% of the algal cells.

Productivity measurements by Schmidt (1973) indicated that rates of net carbon fixation at optimal depth ranged from 0.012 to 0.27 g C m^{-3} d^{-1} and net productivity varied from 0.011 to 0.18 g C m^{-2} d^{-1} for a trophogenic zone of ~ 2 m. Both poor nutrient supply and unfavourable light conditions were suggested by Schmidt to explain these very low values of productivity. Fisher (1979), working on the lower Rio Negro, found very small algal biomass (1 g wet mass (wm) m^{-3}) mostly composed of Chrysophyta and Chlorophyta and low rates of productivity (0.19 g C m^{-2} d^{-1}). Although the C:N:P atomic ratio of these waters (245:45:1) seemed to indicate a phosphorus limitation, the addition of this element did not enhance the carbon uptake. The system is then more limited by light than by nutrients.

The low primary productivity does not seem to affect the zooplankton. Unexpectedly, in the blackwater Lago Taruma-Mirim, Brandorff (1978) found an abundant crustacean plankton. The Cladocera are dominated by the bosminids with *Bosmina hagmanni, Bosminopsis deitersi, Bosminopsis* sp., and *B. negrensis* being discovered and described by Brandorff (1976*a*). The other Cladocera include *Diaphanosoma fluviatile, Holopedium amazonicum, Ceriodaphnia cornuta, Moina minuta* and *Moina* sp. The cyclopids are represented only by *Oithona amazonica* but the calanoids include *Aspinus acicularis, Dactylodiaptomus pearsei, Notodiaptomus* sp., *Rhacodiaptomus calatus, R. retroplexus, Diaptomus coronatus* and *D. negrensis*. In terms of biomass, the cyclopids dominate the crustacean plankton especially at low water. We note the significant increase of diaptomids at high water. Brandorff (1976*b*) found an average number of 237 individuals l^{-1} at high water and 113 at low water. Although the blackwater zooplankton is of smaller size than the white-water animals, Brandorff did not find significant differences between the areal zooplankton biomass of Lago Taruna-Mirim and Lago Castanho, a white-water lake. He concludes that it is probable that the secondary production of the crustacean plankton (in blackwater) is not so low as often imagined.

H. Rai (1978) found a very low chlorophyll concentration ($1-14\,\mu g\,l^{-1}$), indicating limited algal populations. Surprisingly he also found that large populations of zooplankton may develop with subsurface peaks of 60–120 individuals l^{-1}. At certain periods, these peaks correspond with subsurface peaks of chlorophyll and phaeo-pigments. Rai interprets this peak as a layer of detritus directly eaten by zooplankton or rapidly mineralized and channelled towards algal production and zooplankton.

At high waters, the number of bacteria increases although the rate of glucose uptake is very low ($V_{max} = 0.09\,\mu g$ glucose $l^{-1}\,h^{-1}$). At low waters, the number of bacteria diminishes but the rate of glucose uptake (V_{max}) jumps to $9.3\,\mu g$ glucose $l^{-1}\,h^{-1}$ (Rai, 1979*a,b*). In a later paper, Rai & Hill (1980) showed that, in the rainy season, the Rio Negro supplies organic matter to Lake Tupe, mainly allochthonous amino acids which are easily degraded by heterotrophic bacteria which are not active towards glucose. In the dry season, the decay of a bloom of diatoms and *Peridinium* releases sugars which are rapidly taken up by a small but active population of saprophytic bacteria. The succession of organic substrates of different nature and origin causes the succession of bacterial populations which have different substrate affinity. Since the saprophytic bacteria represent less than 1% of the total number of bacteria, it is clear that glucose is far from being the universal substrate for measurement of heterotrophy, especially in lakes receiving large amounts of allochthonous organic matter.

A blackwater lake: Lake Tupe. Lake Tupe is located on the left bank of the Rio Negro; it is 3 km long and 0.3 km wide and has an area of 68 ha. This is a 'river-lake' whose shoreline has many indentations. Its depth varies from a few metres in years of low level to ~ 15 m. The banks of the lake are covered with '*igapo*'. The lake has no floating meadows.

The Secchi-disc transparency varies from 1.4 to 2.7 m; the maximal value is caused by heavy rainfall. Conversely, at high levels, the entry of the dark waters of Rio Negro diminishes the transparency.

Lake Tupe is generally stratified. The most stable stratification occurs in September with a top layer above 30°C and the bottom layer at 25.5°C whereas in January, at the end of the low water, the thermal difference is only 1.5°C. During stable stratification the hypolimnion is devoid of oxygen. Turnover may occur at low water (Reiss, 1977a), but since the lake is protected from the wind it does not occur frequently. From mid-June to mid-November, hydrogen sulphide is produced in the hypolimnion.

Lake Tupe water has a pH of 4.5–5.1 and a mean conductivity of 6.8 μmhos cm^{-1} at the surface and 7.8 at the bottom. Na$^+$, K$^+$ and Mg^{2+} have concentrations lower than 1 mg l^{-1} and Ca^{2+} reaches 1 mg l^{-1}. Schmidt (1970) reports that the waters of the Rio Negro are rich in bacteria specialized in using the humic matter. Schmidt also mentions that, on standard culture plates, bacterial colonies originating from várzea water are yellow or red but on culture plates with water from the Rio Negro, 20–50% of the colonies are blue and represent the typical microflora of the blackwaters.

Reiss (1977a) studied the macrobenthic fauna of Lake Tupe. In the profundal zone, a mean abundance of 721 individuals m^{-2} was observed, with a minimum of 66 m^{-2} in August at decreasing water level and a maximum of 1705 m^{-2} in April at rising water level. At high water level, the benthos decreases rapidly, probably because of the presence of H$_2$S. The benthic fauna is limited to Chaoboridae, Ostracoda, Acari and a few Nematoda. The Ostracoda are the most abundant animals, their population is minimal at decreasing water levels and maximal at rising water. Although the total number of individuals in Lake Tupe is not markedly different from that of the várzea lakes, the biomass in the profundal zone of Lake Tupe is much smaller because of the small size of the animals. With a biomass of 0.136 g m^{-2} Lake Tupe has the lowest mean total benthic biomass of the lakes of central Amazonia. The Ostracoda feed on detritus and are a food source to the Acari and the larvae of *Chaoborus* which probably feed also on zooplankton.

In the littoral zone, the same seasonal pattern is observed with maximal number of animals at rising water, decreasing number at high water and minimal number at low water. During this latter period, the population is dominated by Chironomidae. At rising water, Ostracoda and Chaoboridae are more abundant. In absolute terms, the littoral populations are larger than the profundal ones (mean biomass of 0.31 g wm m^{-2}). Besides the more favourable chemical conditions prevailing in the littoral, the presence on the sediments of a thin layer of algae consisting of the filamentous bluegreens *Hapalosiphon* and *Lyngbya*, together with diatoms, represents an appreciable food source to the phytophagous animals.

Clear waters

The good-light climate of clear waters allows a relatively high productivity in spite of their low nutrient concentrations. In their study of Rio Tapajos, Schmidt & Uherkovich (1973) found chlorophyll concentrations ranging from 61 to 162 mg m^{-2} and net productivity values varying between 0.44 and 2.41 g C m^{-2} d^{-1}. In a more detailed work on algae, Uherkovich (1976) reports that as in blackwaters, *Melosira granulata* and *M. granulata* var. *angustissima* dominate but, unlike blackwaters, *Tabellaria fenestrata* is very rare. In contrast, *Microcystis aeruginosa* fa. *aeruginosa* and *Oscillatoria limosa* are common. The Chlorococcales *Pediastrum duplex* and *Treubaria crassispina*, and the Conjugatophyceae *Staurastrum quadrinotatum, Mougeotia* sp., *Hyalotheca dissilieus* and *Staurastrum hystrix* are characteristic. In the clear waters of Tapajos River, Uherkovich found 154 taxa: 13 Cyanophyta, 2 Euglenophyta, 2 Pyrrhophyta, 3 Chrysophyta, 1 Volvocales, 40 Chlorococcales, 70 Conjugatophyceae, and 22 Bacillariophyta. Koste (1974), working on the plankton of the clear water of Rio Tapajos and Lago Raroni, found *Bosminopsis, Ceriodaphnia, Bosmina* and copepods as the most frequent crustacean zooplankton. He observed 73 species of pelagic and benthic rotifers.

Biotopes of the várzea

The main river Solimoes–Amazon (Fig. 7). At the high-water period it is very common to see masses of floating vegetation drifting with the current (Junk, 1970). Since at the low-water period they are very rare, it is clear that, at high waters, the river moving vegetation is supplied from other biotopes by a purely mechanical effect. Depending on local conditions, the drifting plants not only remain alive but grow due to satisfactory nutrient conditions. Robust species such as *Eichhornia, Reussia* and *Pistia* resist these difficult conditions well; however most of the drifting vegetation is damaged and dries off. The succession of species in the river depends on the growth conditions in their original biotope. The species which are able to float in isolated waters are the first to appear in the river with the water level rise. In 1967, during Junk's survey, *Paspalum repens* belonging to this type appeared in March. *Echinochloa polystachya*, rooted in the bottom of lakes, resisted longer and appeared only in April. In May, these two species represented 90% of the drifting vegetation. The timetable of the drifting process depends on the timing and intensity of the Amazon floods.

In contrast, phytoplankton is neither abundant nor active and primary productivity is extremely low: $0.063 \, g \, C \, m^{-2} \, d^{-1}$ (Fisher, 1979). As in blackwaters, light limitation is responsible for the low carbon uptake, however in this case, light penetration is limited by the turbidity caused by suspended material. Moreover, as high turbulence limits the time algae can remain in the euphotic zone and the high respiration rates consume more carbon than is produced by photosynthesis, net production cannot occur over long periods.

In contrast with the blackwaters of Rio Negro, the Amazon waters are characterized by very large seasonal fluctuations of algae and bacteria. During the low-water period, the number of algae does not exceed 3000 cells ml^{-1} and the number of bacteria is also minimal (Schmidt, 1970). In October, algae peak with 15 000 cells ml^{-1} and bacteria with 500 000 ml^{-1}. The low productivity, strong current and high respiration do not allow the development of an autochthonous algal population; at high level, the

Fig. 7. The várzea of Lower Solimoes near the junction with Rio Negro (by courtesy of Prof. H. Sioli, Max Planck Institute, Plön, West Germany)

connection between the river and the surrounding lakes allows a constant import of algae and bacteria. When the Amazon decreases and this connection ceases, so does the supply of allochthonous organisms. The small absolute biomass of the organisms is due to the dilution of lake water by river water.

The white waters of the Amazon appear then as a collection of the different biotopes which develop in the more quiescent water bodies of the várzea.

The biotope of the banks and sedimentation zones. The diminution of the water currents of the tributaries near their junction with the Amazon River causes immediate sedimentation. At low water, large banks of soft sediments emerge from the water. They are rapidly colonized by *Echinochloa polystachya, Paspalum repens, Hymenachne amplexicaulis, P. fasciculatum* and *Salix humboldtiana* (Junk, 1970). When the water rises, the vegetation grows at an accelerated rate and forms vast floating meadows. From this biotope, large masses of *P. repens* are torn and drift towards the Amazon.

The várzea lakes

These lakes receive their main water supply from the Amazon (Fig. 8); to a much lesser extent they receive water of different quality from the terra firme. When the Amazon recedes, the lake water then flows toward the Amazon causing a significant flushing of the lakes. This special hydrology is also characterized by an enormous variation in water level (up to 10 m). Because of their very nature, the várzea lakes have generally a high development of shoreline favourable to the colonization by higher rooted plants and by floating meadows (Figs 9 and 10). Their chemistry, temperature, oxygen regime and biological features are affected by their depth, exposure to wind, turbidity and connection with the Amazon. Lakes like Lago do Castanho, which are connected all the year round to the main river, are filled with the rising waters and nearly emptied at the low-water period. In such lakes, the very high flushing rates considerably affect chemical and biological features (e.g. massive migration of fish with lowering waters).

Lake Redondo. Lake Redondo is located in the várzea of the Solimoes, 25 km south-west of Manaus. In contrast with the majority of várzea lakes it is round and has a surprisingly low amount of shoreline development (1.04). It is surrounded by river sediments and receives only river water. It

Amazonia

is, however, not permanently connected with the main river and has a mean depth of 2 m and maximum depth of 3.5 m during the low-water period. In the high-water period (June–August) it reaches a maximum depth of 4.5 m. The waters are turbid, especially at the low-water period, because of sediment resuspension by wind action (Secchi disc 0.5 m). The maximum difference of temperature observed between surface and bottom was 1.8°C. This small thermal difference allows polymixis and the presence of oxygen at all depths except at the high-water period. The water has a neutral pH and an unusually low electrolyte content during the dry season, probably because of bottom seepage.

Fig. 8. 'Furo' (natural channel) connecting the Lower Amazon to a várzea lake through the várzea forest. Between Santarém and Monte Alegre at middle to low water (by courtesy of Prof. H. Sioli, Max Planck Institute, Plön, West Germany)

Marlier (1967) studied in detail the food web of Lake Redondo. He found very low values of primary productivity (96 mg C m^{-3} d^{-1} or 140 mg C m^{-2} d^{-1}). The marginal floating meadows, formed mostly of *Paspalum repens* and *Panicum*, seem to be a more important source of organic matter. Marlier estimates their daily average production as 6300 mg C m^{-2} or 50 tons dry mass (dm) ha^{-1} y^{-1}. Junk (1970) thinks this value is too high and estimates that the production surplus does not exceed 3–5 tons dm ha^{-1} y^{-1}. Both authors agree on the order of magnitude of the meadow biomass

Fig. 9. Floating vegetation in a várzea lake near Obidos, Brazil, at high water (by courtesy of Prof. H. Sioli, Max Planck Institute, Plön, West Germany)

Amazonia

(9.1 tons dm ha^{-1}, Marlier, 1967; and 6–8 tons dm ha^{-1}, Junk, 1970). It is not clear whether the meadows contribute much assimilable organic matter to the zooplankton, which forms a biomass of 3.22 g wm m^{-2} and is composed of Cladocera (*Diaphanosoma, Moina, Ceriodaphnia, Bosmina* and *Daphnia*), Copepoda (Calanoidea, Cyclopoidea, Harpacticoidea) and Rotifera. The meadows harbour a very diversified fauna of invertebrates

Fig. 10. *Victoria amazonica* in a várzea lake near Alenquer at middle water (by courtesy of Prof. H. Sioli, Max Planck Institute, Plön, West Germany)

such as snails and insects, representing an average biomass of 40 g fresh mass (fm) m^{-2}. The benthic fauna includes Campsurid nymphs, Chironomid larvae, Hydracari, *Chaeoborus* larvae, clams and snails, nematodes and copepods, representing a dry mass biomass of 0.245 g m^{-2}.

Four species of prawns were collected by Marlier: *Macrobrachium amazonicum, M. jelskii, Palaemonetes carteri* and *Emnyrhynchus burchelli*. The *Macrobrachium* species examined for feeding habits were found to live on grass leaves and roots only. Forty-nine species of fish have been identified, among them 18 species were found to be obligate herbivores (eight species) or predominantly herbivorous (ten species). Among the obligatorily herbivorous fish, three species had a large-spectrum vegetal diet with phytoplankton, epiphytic diatoms, plant debris, leaves and roots, and five species were specialized feeders living on grass seeds or water grass or fruits or filamentous algae. The predominantly herbivorous fish had a mixed diet of algae, higher plants and zooplankton. The carnivorous fish included 18 obligatorily and two predominantly carnivorous species. The specialized carnivores were found to feed on fish (five species), insects (two species) and zooplankton (three species).

In such an ecosystem, the floating meadows represent a semi-allochthonous organic matter, since the plants partly draw their nutrients from the soil, but they constitute an excellent food source for small invertebrates and fish. In contrast, phytoplankton represents a minor source of organic matter. The plant-eaters (primary consumers) include a large number of invertebrates (Oligochaeta, decapod prawns, phyllopods, planktonic copepods and Cladocera, insects, snails) and fish. A smaller group of invertebrates feed on detritus. The secondary and tertiary consumers feeding on zooplankton, insects and molluscs including carnivorous insects, *Chaoborus*, ostracods and fish.

In the várzea lakes of the Redondo type, we again find the rain forest paradox: a shallow water body, poor in electrolytes and nutrients, supports an abundant vegetation biomass (mainly higher plants and epiphytic algae) which is able to maintain a diversified and elaborate animal population; this is made possible by a very rapid and efficient process of mineralization of organic matter and periodic injections of the Amazon nutrients.

Lake Janauaca. Lake Janauaca is a large and shallow lake (2–3 m at low water level) located 60 km south-west of Manaus. The Amazon water penetrates into the lake in late November–December and, at the maximum of the flood, the lake reaches 10–12 m in depth. Lake Janauaca may have up to 80% of its volume yearly filled by and flushed to the main river. The lake water belongs to the 'white decanted type' with a Secchi-disc transparency of 0.60–1.1 m. The penetration of Amazon water generates a turbid area of mixed river–lake water.

Lake Janauaca has very low levels of inorganic nutrients but a high concentration of chlorophyll a (52 μg l^{-1}) and high rates of primary productivity (2.2 g C m^{-2} d^{-1}) (Fisher, 1979). Fisher measured a phytobiomass of 15–24 g wm m^{-3}, dominated by Cyano- and Chlorophyta. The high productivity and low nutrient levels indicate a rapid recycling of nutrients (1–3 h as turnover time for nitrogen and phosphorus). This lake also has a dense population of zooplankton (1000–3000 individuals l^{-1} or 13–15 g wm m^{-3}). Fisher estimated that the grazing pressure of such a population requires a phytoplankton turnover of 100–300% per day.

In late November, the inflow of Amazon nutrient-rich water causes a temporary decrease in primary productivity due to increasing turbidity. The suspended matter is rapidly sedimented, however, and the nutrient flux causes a bloom of blue-green algae and diatoms with a significant increase in biomass.

In a transect representing the phyto- and zoobiomass in the river water, the river–lake mixed zone and the lake water, Fisher showed that the zoobiomass, very low in the river, increases in the mixed zone, culminating in a predominance of blue-greens, and decreases in the lake where green algae are dominant. This maximum of the phytobiomass in the mixed zone emphasizes the fertilizing role of the Amazon water. Lake Janauaca, less turbid and more affected by the Amazon nutrients than Lake Redondo, has a rich planktonic life which supplies an abundant 'inoculum' to the Amazon.

Fisher & Parsley (1979) noted that the nitrate and phosphate stripping and accumulation of blue-green algae probably represent a very important biological process, since they occur on a very large scale. These authors estimated that, between Tefe and Santarém, the surface of potentially floodable lakes is about 15 000–30 000 km^2. For a depth of 10 m at flood time, a volume of $0.15–0.30 \times 10^{12}$ m^3 water is stored in the lakes, which represents 5–10% of the Amazon discharge during the floods. Fixation of the nutrients into organic matter in the lakes prevents their loss to the ocean; the current and turbidity of the Amazon would not allow such high photoassimilation. Furthermore, it seems that blue-green algae do not only accumulate the dissolved inorganic nitrogen, but are also capable of fixation of atmospheric nitrogen.

The abundance of zooplankton noted by Fisher in Lake Janauaca was also reported by Brandorff & De Andrade (1978) in Lake Jacaretinga, another várzea lake near Manaus. Before the Amazon flood, the zooplankton population, dominated by *Diaphanosoma sarsi*, *Daphnia gessneri, Notodiaptomus amazonicus, N. coniferoides* and *Thermocyclops minutus*, includes approximately 200 individuals l^{-1}. In March, just before the influx of the Amazon white water, the population decreases to a few individuals per litre, in late March–April it increases again and is then dominated by *N. amazonicus*. In late April, the zooplankton population almost disappears.

The short-lived increase in zooplankton is accompanied by a remarkable increase in animal size and egg-production: for example, in *N. amazonicus* the average egg number per female increases from 3.4 before the flood to 24.3 after the flood, which clearly indicates the improvement in trophic conditions. The poor trophic conditions of zooplankton at the low-level period, when the lake is isolated from the Amazon, have been confirmed by the enrichment experiments of Zaret, Devol & Santos (1979). Whereas addition of phosphorus to Jacaretinga water did not enhance algal production, the addition of nitrogen only or nitrogen + phosphorus caused a considerable biomass increase. It seems then that nitrogen limits algal growth at low-level time and consequently zooplankton reproduction. The disappearance of the zooplankton following the explosion of productivity is tentatively explained by Brandorff & De Andrade by the predation pressure of carnivorous fish which then enter the lake to spawn.

Lago do Castanho. Lago do Castanho also belongs to the group of lakes connected to the Amazon by the Parana of Janauaca. At low water, Lago do Castanho is almost dry; at high water, it is 11–12 m deep and has an approximate surface area of 2 km². At its southern end, the lake is bordered by terra firme covered by *igapo*, the flooded forest. Part of the lake shore is occupied by floating meadows, mostly composed of *Paspalum repens* on the flood-plain side (Junk, 1970) but *Oryza perennis, Leersia hexandra* and *Cyperus* sp. are also found. *Eichhornia crassipes, Reussia rotundifolia, Pistia stratioides, Neptunia oleracea, Salvinia auriculata* and *Azolla* sp. are frequently encountered with varying abundance. *Utricularia* sometimes covers large areas.

The lake water level, depending on the Amazon flood, starts rising in November and remains high until mid-June and then decreases suddenly. The seasonal fluctuations of the water level depend on the Amazon regime. Schmidt (1972a) notes that, for example, the minimal level in 1968 was 5 m higher than in the previous and the following years.

At the onset of the rainy season, the Solimoes water generally has a high

Amazonia

concentration of suspended matter (up to 164.5 mg l^{-1}) which drops down to 10 mg l^{-1} at low water (Schmidt, 1972b). When the river water penetrates into the lake its mineral suspended material adds to the organic load (vegetal detritus and plankton) of the lake water. Therefore, in October, the total suspended matter of the lake may be as high as 260 mg l^{-1}. However, the mineral load of the river is rapidly sedimented and, at maximal water level, the Secchi-disc transparency reaches 2 m. Conversely, the lowering of the lake water is accompanied by a decrease in transparency caused by sediment stirring and wave action on newly uncovered soft deposits. It follows that, at that period, significant amounts of suspended matter are returned to the river with the riverward flow.

At low water, the water average temperature fluctuates from 28.7°C in early morning to 30.7°C at noon. Homothermal morning profiles develop then into afternoon stratification, the thermal difference along this 2 m deep profile not exceeding 2°C. In the high range of temperature, however, this small difference is enough to determine a temporary stratification. At high water, an upper layer of 4–6 m is affected by daily warming. At 11 m, the water remains at 27°C. Between the upper and bottom layers, no clear metalimnion develops and thermal variation is progressive with depth. Oxygen being absent below 5 m, it is clear that mixing seldom occurs below this depth at high water, with the exception of short periods of 'friagem' (air temperature drop) or heavy rainfall. Schmidt (1973) noted that, as a result of heavy rainfall, cool runoff from the terra firme flows into the hypolimnion, increasing temporarily its total salt content and alkalinity and making the stratification more stable. Schmidt describes Lago do Castanho as polymictic at low water and oligomictic at high water.

For levels below 3 m, the lake is oxygenated down to the bottom. Even in such a shallow water body however, oversaturation due to photosynthesis occurs at the surface, and undersaturation occurs near the bottom, especially on calm days. At high water, the complete disappearance of oxygen in lower layers is followed by the appearance of hydrogen sulphide from February until September–October. Eventual mixing due to friagem causes massive fish killing, even if, according to Geisler (1969), Amazonian fish are particularly tolerant to low oxygen conditions.

In the lake, the Solimoes waters, with their conductivity of 45–84 μmhos cm^{-1}, are diluted by rainwater falling directly on the lake surface and conductivity varies from 20 to 50 μmhos cm^{-1}. The lake water belongs to the calcium-bicarbonate type although it has a relatively high sodium content at low levels. The alkalinity is low (31.7–3.7 mg l^{-1} HCO$_3^-$) and pH ranges from 6 to 7. Free CO$_2$ is abundant (5–30 mg l^{-1}) and continually renewed by decomposition of organic matter; it may however decrease significantly in the upper layers as a result of photosynthesis. At low water levels, peaks of total phosphorus, total iron and dissolved silica are caused by sediment resuspension. In contrast, at high water levels, there is a distinct increase in dissolved PO_4-P (30 mg l^{-1}) and dissolved bivalent iron (4 mg l^{-1}), which are released from the sediment under the low redox conditions prevailing in the hypolimnion; nitrate is of course absent from the hypolimnion.

Primary productivity has been studied by Schmidt (1973) and phytoplankton assemblages by Uherkovich & Schmidt (1974). In samples taken with a 10 μm net, 209 algal taxa were found including 19 Cyanophyta, 58 Euglenophyta, 1 Pyrrhophyta, 108 Chlorophyta (3 Volvocales, 54 Chloroccales, 51 Desmidiales), 9 Chrysophyta and 14 Bacillariophyta. The maximal number of species was found from December to February and from July to September and the minimal number in October–November. The phytoplankton is dominated by an association of *Melosira granulata* var. *angustissima*, *Sphaerocystis schroeteri* and *Closterium*. From time to time are observed mass developments of Cyanophyta (*Anabaena hassalii*, *Aphanizomenon flos-aquae*, *Oscillatoria limosa*) and flagellated Chlorophyta (*Eudorina elegans*, *Pandorina morum*, *Volvox weissmanni*). Euglenophyta and Conjugatophyceae are always present with tropical and subtropical forms. This algal population includes endemic species, among them five species of Euglenophyta, one species of Chlorophyta and many species of Conjugatophyceae.

At low water, primary productivity takes place only in the upper 60–70 cm because of turbidity, but very high values of productivity have been measured (2.15 g C m^{-3} d^{-1}) (Schmidt, 1973). At high water level, the optical conditions improve and the productivity curve shows a maximum at 0.5–1 m depth but the dilution of phytoplankton concentrations by inflowing water decreases the productivity in the optimal depth which may be as low as 0.32 h C m^{-3} d^{-1}. This period of rising water is also the period when the lowest areal production values are obtained (0.3 g C m^{-2} d^{-1} for a chlorophyll concentration of 0.04–0.08 mg m^{-2}). At high water in the year 1969, a mean value of 1 g C m^{-2} d^{-1} was found.

Schmidt, on the basis of his experimental data of 1967–68, calculated an annual net production of 298 g C m^{-2}, that is an average net production of 0.8 g C m^{-2} d^{-1} and an annual gross production of 358 g C m^{-2} or nearly 1 g C m^{-2} d^{-1}.

Brandorff (1978) studied the crustacean plankton of Lago do Castanho. The Cladocera are represented by a large number of species. The daphnids include *Daphnia gessneri, Ceriodaphnia cornuta, C. reticulata, Moina minuta, M. reticulata,* and *Moina* sp. whereas the sidids and bosminids are, respectively, represented by *Diaphanosoma fluviatile* and *D. sarsi* and by *Bosmina tubicen* and *Bosminopsis deitersi*. These species of Cyclopoids are known: *Mesocyclops leuckarti, Thermocyclops minutus* and *Oithona amazonica*. The calanoids include *Argyrodiaptomus azevedoi, Dactylodiaptomus pearsei, Notodiaptomus amazonicus, N. coniferoides* and *Diaptomus ohlei*.

In terms of biomass, the crustacean zooplankton is more abundant at low water (846 individuals l^{-1}) than during the flood (154 l^{-1}). In both cases, it is largely dominated by the cyclopoids (65% of total biomass at high water and 80% at low water).

In relatively deep várzea lakes, the large fluctuations of water levels and the thermal stratification and consequent hypolimnic anoxia which establishes at high water considerably affect the lake-bottom fauna. At rising water, the fauna of the anoxic bottom waters is limited to the chironomids *Chironomus gigas* and *Ablabesmyia* sp. At this period, the decrease of oxygen renewal due to the extension of floating meadows in shallow areas also limits the development of the littoral benthic fauna. The animals have adapted to this variable environment by migrating into the floating meadows which, at rising water, become a very populated biotope, especially rich in chironomid larvae (Reiss, 1977b). These also colonize all floating particles. It is interesting to note that under the prevailing conditions, floating biotopes are the most stable environments.

The fishes of Amazonia
Species

The freshwater fishes of South America belong to relatively few basic groups but active processes of adaptive radiation within these groups have led to the largest number of species known in any zoogeographical unit in the world (Lowe-McConnell, 1975). In Amazonia alone, more than 2000 species of fishes have so far been described, i.e. one order of magnitude more than the fish species known in Central Europe (Geisler, Knoppel & Sioli, 1975). Approximately 85% of the fishes belong to the Ostariophysi with the characoids and siluroids as dominant groups. This fauna occupies all possible niches of the very diversified habitats of the Amazonian plain.

The systematic study of Amazonian fishes started in the 1930s and a detailed bibliography is given by Lowe-McConnell (1975). Recently, during the Alpha Helix phase IV expedition, samples were collected in the lake and in the main river and studied by Fink & Fink (1979). What follows is a brief description of their findings.

The most primitive fish group encountered during the expedition, the Potamotrygonidae, belongs to the Chondrichthyes, the cartilaginous fishes including sharks, rays and skates.

Freshwater stingrays belonging to the genus *Potamotrygon* were commonly found; they live on the bottom and feed on invertebrates. The lungfish *Lepidosiren paradoxa* is a member of another ancient group of bony fishes belonging to the subclass Sarcopterygii. Its ancestors were particularly abundant from the Devonian to the Triassic period. It now dwells near the shores of swampy areas. Another old group of bony fishes, the Osteoglossidae, is represented by three species: the common food-fish *Arapaima gigas*, a piscivorous species mostly found in river-lakes which may reach 4 m and 200 kg; *Osteoglossum bicirrhosum*; and *O. ferreirai*. *Osteoglossum* species are mouth-brooders and valuable commercial fish. The clupeid *Ilisha amazonica* is essentially piscivorous.

The Ostariophysi are massively represented by 12 families of characoids, four families of gymnotoids and 14 families of siluroids. It seems that, in South America, the Ostariophysi occupied relatively 'free' ecosystems and underwent explosive radiation with a resultant adaptation to an extraordinary variety of niches. Among these, the characoids have reached a very large range of trophic specializations. One of the most famous and common fish of this group is the genus *Serrasalmus* (piranha).

Hoplias malabaricus is another member of this group. It is a common piscivorous fish which contributes a large fraction of the fish biomass in South American freshwater. When oxygen decreases in the water, this fish which is not an air-breather lies in shallow water under the surface. In similar conditions, *Hoplerythrinus*, which is very closely related to *H. malabaricus* and an air-breather, lives on atmospheric air. *Prochilodus* is a mud-eating characoid important as a food-fish. The gymnotoids or electric fishes form the second largest group of Ostariophysi. *Electrophorus electricus* emits powerful electric discharges. It is an air-breather as is *Gymnodus carapo*. The electric fishes generally move and feed at night and hide during the day.

The catfishes (Siluriformes) are the third group of Ostariophysi. They are generally bottom-dwellers. With the exception of the filter-feeder *Hypophthalmus*, the South-American catfish belong to the 'night fish fauna'. The level of endemism is very high in this group. The family Pimelodidae includes 250 species of catfish, many of them of large size, such as *Brachyplatystoma*, *Hemisorbium* and *Pseudoplatystoma*. These fish are piscivorous but may adapt to other diets. *Hypophthalmus* is a very important commercial fish, living in large schools in pelagic water and

Amazonia

feeding on plankton. The sucker-mouth catfishes form the largest catfish family, probably because this feeding mechanism enables them to colonize a wide variety of habitats. Finally, the cichlids are bottom-feeders, such as *Geophagus*, or piscivorous, such as the genus *Cichla*.

Main biotopes

The water chemistry considerably affects the distribution of fishes in Amazonia. In his comparative study of fish fauna of Lake Redondo, a white-water lake, and Lake Preto de Eva, a blackwater lake, Marlier (1967, 1968) found 47 species in the former water body and 49 in the latter but only six species were common to both water bodies.

In white waters, where no submerged flora, phytoplankton, zooplankton and lamellibranchs can develop, the most common fishes are siluroids and the characoid *Colossoma*, which feed on allochthonous material, especially nuts from rubber trees. The floating meadows which develop in white waters harbour the characoids *Hemigrammus* and *Hyphessobrycon* and cichlids such as *Cichlosoma severum*, *C. festivum*, *Gymnotus carapo* and *Synbranchus marmoratus*. Other predatory species include characoids (*Hoplias malabaricus*, *Serrasalmus* sp.) gymnotoids, siluroids and cichlids (Junk, 1973a,b).

In the clear-water rivers (Xingu, Tapajos), the catfish *Hypophthalmus edentatus* is abundant, since the phytoplankton blooms of these waters represent an excellent food source for this fish.

The blackwaters have the poorest fish populations; the fish which survive the low pH are often killed by upwelling of anoxic water from deep layers.

In most lakes however, the fish communities are not stable; at low water the fish migrate towards the river and at high water a different population invades the lake. Moreover, many fish predators feed on these populations. This makes it very difficult to have significant samples of these populations in a given biotope, since a vast number might have been just brought in by the floods or flushed away by the water fall.

Fish feeding and the role of fish in the food chain

Knöppel (1970) has studied the gut contents of the fish populations found on the three main types of water by poison sampling. The fish were sampled in the blackwater Igarape Taruma, in the clear-water Igarape Barro branco, both located on the terra firme of the lower Rio Negro, and Lago Calado, a typical várzea lake in the Solimoes flood plain. The study is based on a collection of 3200 fish including 53 species (22 characoids, six gymnotoids, seven siluroids, 17 percoids mainly cichlids and one cyprinodont). Of 1121 stomachs examined, 4% were full, 21% three-quarters full, 46% half full, 21% one-quarter full, and only 6% were empty.

These percentages seem to indicate the abundance of fish food. Knöppel mentions the following food items:

> insect larvae: present in 27 species, main food in nine species and reaching 20% of total food in five species;
>
> detritus present in 19 species;
>
> algae found at low frequency and exclusively in fishes of Lake Calado;
>
> plant remains in more than 25 species;
>
> crustacea in 21 species;
>
> fishes (whole fishes, scales or flesh) were found in 20 species;
>
> fishes are the main food of *Cichla ocellaris* and are also eaten by large characoids and cichlids;
>
> ants were found in 13 species; other terrestrial insects (termites, beetles, flies and midges) were also observed.

Insect larvae, plant remains and terrestrial insects form the main diet of the fish, however it is generally a mixed diet and few species are specialized. Moreover, fishes do not seem to be restricted to one particular spatial niche since the same fish was found to feed at the same time on benthic and surface food. A great variety of species is frequently observed. All these features emphasize the fact that the Amazonian waters, in spite of their poverty in nutrients and plankton, are rich enough in food to support a large and varied population of fishes which are specialized neither for food source nor for territory.

The striking abundance of aquatic insect larvae in the fish stomachs and their small density in the creeks indicate that the 'abundance of fish food' is based more on a rapid turnover of the food items than on their large biomass at a given moment. Moreover, the stable conditions prevailing in this biotope allow both a stable supply of food and a more-or-less constant and high density of fishes.

The Amazonian food chain is essentially based on allochthonous food sources, especially in clear and blackwaters. The rapid turnover of plant material and insects sustain a large biomass of fishes and invertebrates. These, in turn, because they have slow life-turnovers, mobilize a large fraction of the nutrients and continuously mineralize a small portion of the material they ingest. The resulting nutrients allow the growth of the limited planktonic population.

In such a food chain where detritus is directly transformed into large-animal biomass and where food exerts no drastic pressure, the fish population is regulated by piscivorous species which are abundant in the Amazon; more than 40 species of characoids are piscivorous, and additional piscivorous species are also found in cichlids, siluroids and gymnotoids (Lowe-McConnell, 1975).

Large amounts of nutrients being stored in animal biomass, the migration of the animals or their removal by fishing considerably affects the nutrient potential of a given area. For example, Fittkau (1970, 1973a) reports that the migration of fish from the Amazon and várzea lakes into the affluent rivers at the spawning period supplies food to caimans, turtles and other piscivorous animals. These predators, through their excreta, supply the nutrients necessary to plankton growth, the favourite food of the new-born fishes. It follows that when the caiman population was destroyed by man-hunting, the survival of young fishes diminished accordingly.

Another experimental sampling was carried out by Saul (1975) in Northern Ecuador in the watershed of Rio Napo, a tributary of the Amazon, at 340 m altitude. Two streams, one lake and swamp were investigated. In the streams, characoids were heavily dominant (64–89%): *Aphyocharax alburnus* living in sheltered, shallow waters and feeding on insects; *Astyanax abramis* found in swift-flowing creeks and rivers and feeding on plants and insects; *Bryconamericus beta* living in swift and in sluggish waters of river systems, feeding on bottom detritus and insects; *B. breviceps* and *Creagrutus beni*, both found in swift water; and *Odontostilbe fa madeirae* abundant along shorelines and feeding on benthic algae. In the streams, locariids and cichlids represented no more than 8–26% of the fish populations. In the swamps, 90% of the fish belonged to the three groups characoids, gymnotoids and cyprinodontids; most characoids were erythrinids (*Hoplias, Hoplerythrinus, Erythrinus*). In contrast, 62% of the lake fish were cichlids (*Aequidens tetramerus, A. vittatus* and *Crenicichla lepidota*).

THE ANDES
General

With the exception of Lake Titicaca, information concerning the limnology of water bodies of the Andean region is very scanty. A recent investigation by Hegewald, Aldave & Hakulit (1976) has provided new data on chemistry and phycology of Peruvian waters.

The coastal plain of Peru is located between 0 and 1000 m altitude and is 10–100 km wide. The waters of the semi-arid and warm northern part of the coastal plain are dominated by green algae, especially *Pediastrum simplex*. In the central and southern sections, the water chemistry is very variable depending on the presence or absence of local mineral deposits: e.g., the fountain water of Casa Grande (Province of Trujillo) has a high salinity (527 mg l^{-1}) and is rich in Ca^{2+} and SO_4^{2-}; its high content of NO_3^- and Na^+ originates from the vicinity of local deposits of sodium nitrate. Lake Huacachina, some 250 km south of Lima, is famous for its population of *Spirulina platensis*, whereas the tropical blue-green alga *Anabaenopsis magna* was observed in a brackish pond near Viru, south of Trujillo.

In the Sierra, between 2500 and 4500 m, the water bodies have a low salt content (144 mg l^{-1} at Huanuco waters), neutral pH and relatively high concentrations of Ca^{2+} (20 mg l^{-1}), SO_4^{2-} (15 mg l^{-1}) and HCO_3^- (101 mg l^{-1}). Huanuco pond was characterized by a bloom of *Melosira granulata* and the presence of *Scenedesmus intermedius* and *Treubaria triappendiculata*. *S. intermedius*, known in Peru only from this pond, present cell-wall ornaments which differ from European and African specimens by a 'rosette structure' (Hegewald, Aldave & Schnepf, 1978). Lago Junin is located north-west of Lima at 4100 m altitude; its temperature in July 1973 was 11°C. The lake shallows are covered with *Scirpus totora* and other aquatic plants. The water chemistry has the characteristics presented in Table 11.

Lakes Titicaca and Poopó
General

Lakes Titicaca and Poopó are located at 3500–4000 m altitude, in a NW–SE basin 1000 km long and 200 km wide in the Peruvian–Bolivian altiplano which originated from the erosion of mountains formed in Cretaceous times into a vast peneplain (Fig. 11). Miocene vertical faulting of

Table 11 *Chemical features of Lago Junin*

Lago Junin	Results of 27.7.1973	
Total salt content	(mg l^{-1})	258
Conductivity	(μmhos cm^{-1})	352
pH		7.3
K^+	(mg l^{-1})	5.1
Na^+	(mg l^{-1})	17.8
Mg^{2+}	(mg l^{-1})	19.3
Ca^{2+}	(mg l^{-1})	26.6
Cl^-	(mg l^{-1})	25.5
SO_4^{2-}	(mg l^{-1})	46.9
HCO_3^-	(mg l^{-1})	125.7

The Andes

the peneplain created the present depression and drainage system. During and after the Pliocene, the whole area was uplifted to freezing altitude. The milder climate of a Pleistocene interglacial caused the formation of successive lacustrine systems: Lake Ballivian which included the present Titicaca and Poopó and Lake Minchin which extended over the basin of the present Lake Poopó. During the Pleistocene, Lake Ballivian reached a level of 100 m above the present level and remained for a certain time at 85 m. Newell (1949) founded this conclusion on the remnants of two terraces discovered at these levels. During the drier post-glacial period, Lakes Titicaca and Poopó have gradually shrunk to their present size (Gilson, 1964).

The Titicaca depression is an endorheic basin extending from 71° to 67° W and 14°25′ to 22°50′ S and covering an area of 19 000 km². On its western and eastern sides, the basin is bordered by high cordilleras reaching

Fig. 11. The hydrographic system of Lake Titicaca and Lake Poopó

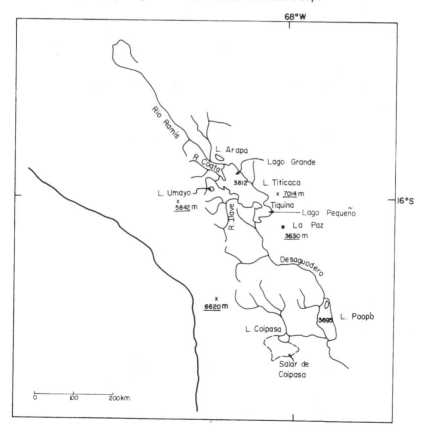

6500 m altitude. The general altitude of the basin decreases from north to south; similarly, the annual precipitation, amounting to 1800 mm in the north drops to 300 mm near the 19th parallel. The waters of the Peruvian highlands drain first into the Titicaca. The outlet of the Titicaca, the Desaguadero flows into Lake Poopó, which has no effluent.

Lake Titicaca is composed of two basins: el Lago Grande or Lago Mayor and el Lago Pequeño. Both basins are connected by the Tiquina Canal. The water level of the lake fluctuates between +3806 and 3810 m (period 1918–76), data from Carmouze, Arce & Quintanilla, 1977). The annual variation of water level rarely exceeds 1 m but long-term fluctuations of 2–4 m were observed, the most noticeable was the continuous drop in water level from 1933 to 1944. Gilson (1964) notes that the flat and swampy areas of the north-western and northern margins of the present lake indicate that Lake Umayo and Laguna Arapa were within the lake in recent times. Raised beaches of the old Lake Ballivian are additional evidence of the greater extension of the lake in the past.

The main morphometric characteristics of the lake are included in Table 12.

Hydrological balance
Lake Titicaca (Fig. 12). Various authors have tried to establish a hydrological balances of Lake Titicaca (Monheim, 1956; Kessler, 1970;

Table 12 *Morphometric characteristics of Lake Titicaca (from Carmouze et al., 1977)*

Maximum length	175 km
Maximum width	67 km
Mean length Lago Mayor	45 km
Shoreline	1125 km
Lake area	8372 km^2 (100%)
Lago Mayor area	6489 km^2 (77.5%)
Lago Pequeño area	1331 km^2 (15.9%)
Puno Bay	552 km^2 (6.6%)
Mean depth Lago Mayor	134 m
Mean depth Lago Pequeño	10 m
Mean depth Puno Bay	14 m
Maximum depth of lake	281 m
Total volume	893.0 × 10^9 m^3 (100%)
Lago Mayor volume	870.7 × 10^9 m^3 (97.5%)
Lago Pequeño volume	14.3 × 10^9 m^3 (1.6%)
Puno Bay volume	8.0 × 10^9 m^3 (0.9%)

The Andes

Richerson, Widmer & Kittel, 1977). We here report a new hydrological balance by Carmouze *et al.* (1977) who have calculated the average flow of the lake over a long-term period (1956–73) and have thus obtained results which have a more general value.

The 58 000 km² watershed of Lake Titicaca drains into the lake through five main rivers – Ramis, Coata, Have, Huancane and Suchez – which supply an annual amount of 6.75×10^9 m³ water (average of 1956–73 data). The average meteorological input is 7.5×10^9 m³ y^{-1}. Assuming that the lake has no sublacustrine spring, its annual water income is 14.25×10^9 m³.

Only a minor amount of water leaves the lake through the Desaguadero, its natural outlet (0.67×10^9 m³ y^{-1}). On the basis of chloride balances, the authors have estimated the losses by infiltration to be 0.60–0.63×10^9 m³ y^{-1}. The losses by evaporation, estimated by difference, amount to 12.95×10^9 m³ y^{-1} (see Fig. 13A).

The residence time of water is 68 years for Lago Mayor and five years for Lago Pequeño. This corresponds to a yearly renewal of 1.5% of the lake water by the annual income (1.45% for Lago Mayor and 19.7% for Lago Pequeño).

In conclusion, Lake Titicaca, with its overwhelming predominance of losses through evaporation (91% of annual income) is nearly a closed lake. However, a considerable difference exists between Lago Mayor and Lago

Fig. 12. Lake Titicaca, Bolivia (by courtesy of Dr J. Luna, BID, Washington D.C., USA)

88   South America

Pequeño: the latter evacuates 36.5% of its yearly income in comparison with only 11.6% for the former.

Lake Poopó. Lake Poopó is located in Bolivia at 3695 m altitude. With an area of 2600 km² and a maximum depth varying between 3 and 5 m, its

Fig. 13. The hydrological balance of Lake Titicaca (A) and Lake Poopó (B)

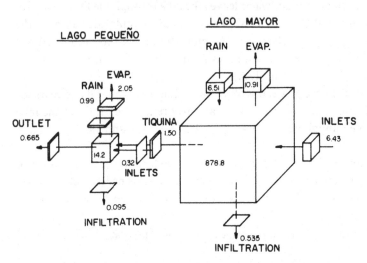

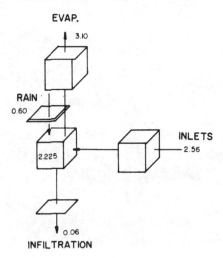

The Andes

approximate volume is 2.22×10^9 m^3. The lake has no outlet but it outflows southward into the Salar de Caipasa. The data concerning this lake are much less precise than for Lake Titicaca and Carmouze *et al.* (1977) have presented the following estimation:

Annual water income through natural inlets	2.56×10^9 m^3
Annual water income through precipitation	0.60×10^9 m^3
	3.16×10^9 m^3
Losses through infiltration	0.06×10^9 m^3
Losses through evaporation	3.10×10^9 m^3
	3.16×10^9 m^3

The lake receives 81% of its water income through natural inlets but evaporation accounts for 98% of the losses (see Fig. 13B). The residence time of water is only 8.5 months.

Thermal regime

The Titicaca basin has a tropical subalpine climate with a short and rainy summer from December to March and a long winter. Gilson (1964) reports the annual average of the monthly mean temperatures (1938–50) to be 8.4°C at Puno (6.2°C in June and 9.9°C in November). Local winds are generally not strong enough to affect significantly the mixing of the lake water but allow a very high rate of evaporation from the lake surface.

The Percy Sladen Expedition made extensive water measurements from May to September 1937. In May, the temperatures of the surface and the epilimnion were, respectively, 13.7°C and 12.8°C; a distinct thermocline was observed between 50 and 70 m and, in the hypolimnion, the temperature varied from 11.8°C to 10.8°C. In mid-July the surface temperature was only 12.4°C and the thermocline had deepened to 90–100 m; in September, the first 2 m had a temperature exceeding 12°C but the rest of the profile remained between 11.6 and 11.0°C. Later measurements by Schindler (1955) on January 1954 and by Richerson, Widner, Kittel & Landa (1975) from January to December 1973 confirmed that the lake is stratified in summer to autumn with a thermocline between 40 and 70 m in March and 70 and 100 m in June. By mid-winter, Richerson *et al.* (1975) found the lake isothermal at 11.1–11.2°C at least down to 100 m. Low oxygen concentrations below this depth seem to indicate an incomplete mixing for the 1973 winter. However, since these authors measured 5 mg l^{-1} oxygen at 200 m depth on 2 May 1973, that is in late summer after a long stratification period, it is likely that a better mixing had occurred during the 1972 winter and renewed the oxygen reserve. The stratification reappeared in late September to October and developed during the next summer. During this 1973 survey, Secchi-disc measurements ranged from 4.5–6 m in February–March, during the phytoplankton maximum and flood period, to 10.5 m in August.

Richerson et al. (1977) have estimated that in June, just before overturn, the hypolimnic oxygen deficit is about 250 g m^{-2} which corresponds roughly to the oxidation of 90 g C m^{-2}. This amount represents approximately two months of ^{14}C primary production. Since the lake is stratified during eight months, the authors concluded that only 20% of net annual production was mineralized below the thermocline.

Richerson et al. (1975) calculated the heat budget of Lake Titicaca. From the February maximum heat storage to the July minimum, the heat loss was only 18 000 cal cm^{-2}. In comparison with temperate lakes, where values of 50 000 cal cm^{-2} are common, the heat loss of Lake Titicaca is very low. It is due to the small seasonal differences of temperature prevailing on the altiplano. Heat inputs and outputs are nearly equilibrated and the lake is only weakly stratified. In contrast with the temperate climate where water temperatures 'lag behind the season' (Kittel & Richerson, 1978), the Titicaca epilimnic temperatures were above median air temperature for every month of 1973. The lake is then a heat source for the environment. The small vertical differences of water temperature in this low range of temperature should cause a permanent instability and polymixis. This probably occurs in shallow lakes; Lake Titicaca is too deep and/or too protected from wind to be similarly affected. Because the annual fluctuations of heat storage are small however, the year-to-year variation of atmospheric parameters, and especially air temperature, can strongly affect water mixing. Kittel & Richerson (1978) emphasized the irregularity of Lake Titicaca circulation and noted that the winter 1974 circulation affected a greater volume of water than the 1973 incomplete mixing.

Water chemistry

The salt concentration of the Titicaca inlets is generally much lower than that of the lake water. This is unexpected in an area of high rainfall and low temperature. However, the predominance of evaporation (91% of annual water income and water losses) over other types of water losses explains this gradual concentration of salts.

The Andes

A mixed-water sample composed of water from various depths at a deep station was taken by the Percy Sladen Expedition on 16 July 1937 in Lago Mayor and its analysis supplied the results of Table 13.

Gilson (1964) reports also that Lago Pequeño is 10% more concentrated than Lago Mayor and that Lago Poopó has a chloride concentration of 5 g l^{-1}. Around Lake Poopó, the land is impregnated with salt and completely unsuitable for agriculture and human settlement.

A sample taken by Löffler (1960) on 16 April 1954 in Puno Bay gave very similar values. Moreover, Richerson et al. (1975) report that their own 1973 analysis of the lake water gave results which were very close to Gilson's values.

In December 1976, Reyssac & Dao (1977) took new samples which were analysed by Carmouze; the results are presented in Table 14. Here again, the results obtained for Lago Mayor are very near to the values reported by previous authors, indicating a present stability in the water chemistry and consequently in the hydrological balance of the lake. Similarly, as in 1937, the Lago Pequeño is approximately 10% more concentrated than Lago

Table 13 *Chemical analysis of Titicaca water. Percy Sladen Expedition, July 1937 (results in mg l^{-1})*

Na^+	167.7	Fe	0.04	Total salts:
K^+	14.9	Al^{3+}	0.4	788.1
Li^+	0.9	Cl^-	249.5	
Ca^{2+}	65.4	SO_4^{2-}	246.5	
Mg^{2+}	34.5	SiO_3	8.6	

Table 14 *Chemical analysis of Lake Titicaca water (Reyssac & Dao, 1977)*

	Lago Mayor	Lago Pequeño	Lago Poopó
pH	7.9–8.4	8.2–8.7	8.9
Alkalinity	2.12 me l^{-1}	2.0 me l^{-1}	4.65 me l^{-1}
Na^+	177 mg l^{-1}	195 mg l^{-1}	12.65 g l^{-1}
K^+	15.6 mg l^{-1}	16.7 mg l^{-1}	0.55 g l^{-1}
Ca^{2+}	66.0 mg l^{-1}	60.8 mg l^{-1}	0.96 g l^{-1}
Mg^{2+}	32.8 mg l^{-1}	35.7 mg l^{-1}	0.39 g l^{-1}
Cl^-	248 mg l^{-1}	259 mg l^{-1}	17.64 g l^{-1}
SO_4^{2-}	244 mg l^{-1}	274 mg l^{-1}	8.26 g l^{-1}
SiO_2	0.77 mg l^{-1}	0.17 mg l^{-1}	0.90 g l^{-1}

Mayor, with the exception of calcium. Tutin (1940) explained this anomaly by the uptake of the abundant Charophyta of Lago Pequeño. Conversely, Lago Poopó seems to have undergone a completely different evolution: the chloride concentration is more than three times the 1937 value reported by Gilson (1964).

Biological features

Phytoplankton. Tutin (1940) published one of the first lists of phytoplankton of Lake Titicaca. He emphasizes the fact that the 30 samples taken in various places down to 50 m showed a great uniformity which he explains by the limited number of species; since the survey was carried out in winter, a good mixing of the upper layers is another possible explanation.

In Lago Mayor, the six following species were present in nearly all the samples: *Ankistrodesmus longissimus; Botryococcus braunii; Dictyosphaerium ehrenbergianum; Staurastrum paradoxum; Ulothrix subtilissima;* and *Peridinium* sp. *Dictyosphaerium* and *Botryococcus* were the most abundant species but the whole algal community was rather thin with an average of 3000 organisms (cells or colonies) per litre without variation with depth. It should be noted that nanoplankton was not examined.

Lago Pequeño has a similar assemblage with *Mougeotia* and *Oedogonium* as additional species and complete lack of *Dictyosphaerium*.

In Lago Poopó, the algal community is limited to *B. braunii, Glenodinium* sp., *Rhabdoderma salina* and *Lyngbya aestuarii*.

It is interesting to note the ubiquity of *B. braunii* which was found in lagunas (68 mg l^{-1} Cl^-), in Lake Titicaca (250 mg l^{-1} Cl^-) and in Lake Poopó (> 5000 mg l^{-1} Cl^-).

Phytoplankton was investigated in 1973 by Richerson *et al.* (1977). In this case, the algae were sampled with a water sampler and consequently small algae were not excluded. The species assemblage was very different from that of 1937: in particular the 1937-abundant *Botryococcus braunii* was absent in 1973 and the 1937-rare *Stephanodiscus astraea* was extremely abundant in the winter of 1973. It is difficult to conclude that there was a change in the algal population, however, since the sampling techniques were very different.

In autumn–winter, *S. astraea* was especially abundant and accompanied by the Bacillariophyta *Cyclotella stelligera, C. meneghiniana,* the Chlorophyta *Mougeotia viridis, Selenastrum minutum, Closterium* sp., *Oocystis borgei, Staurastrum gracile* and the Cyanophyta *Gloeocapsa punctata*. In spring–summer, the algal population was dominated by the Cyanophyta *Anabaena sphaerica, Lyngbya vacuolifera* (in spring only) and by the chlorophyte *Ulothrix subtilissima*. This period corresponded to the lowest value of the index of diversity; in contrast, this index was constantly high from April to October.

Reyssac & Dao (1977) sampled net-phytoplankton in Lake Titicaca in December 1976. These authors found a predominance of Chlorophyta ($7-21 \times 10^6$ cells l^{-1}) and a relatively small number of diatoms ($1.5-2.0 \times 10^6$ cells l^{-1}). This rarity of diatoms, generally abundant in salty waters, had already been noted by Tutin (1940). The blue-green algae were rare and dinoflagellates were more abundant in Lago Pequeño than in Lago Mayor. Lago Pequeño was especially rich in Desmidiales (*Cosmarium, Staurastrum, Closterium*), which, according to Reyssac & Dao, was due to the relatively low calcium concentration. Lago Poopó was rich in diatoms, especially *Navicula* (11×10^9 cells l^{-1}). Chlorophyta were rare and Cyanophyta absent.

Widmer, Kittel & Richerson (1975) measured the primary productivity of Lago Mayor from January to December 1973. In winter, conditions were excellent for light penetration and the euphotic zone extended down to 15–20 m. A year-round light inhibition was observed in the upper layers because of the high intensity of radiation (especially ultraviolet radiation) at high altitude in tropical areas. The productivity maximum was therefore found at 5 m. The isothermy which probably caused a vertical dispersion of the algae was followed by a spring minimum of productivity (around 1000 mg C m^{-2} d^{-1}) and biomass (13 mg C m^{-3}). This minimum occurred simultaneously with the disappearance of *Coscinodiscus* sp. from the algal assemblage. In early summer, the productivity and biomass rose suddenly (2800 mg C m^{-2} d^{-1} and 28 mg C m^{-3}, respectively). In summer–autumn, productivity remained between 1100 and 1600 mg C $m^{-2}d^{-1}$ whereas biomass reached 30–39 mg C m^{-3}. The seasonal average values of productivity and biomass are shown in Table 15.

Table 15 *Primary productivity and algal biomass in Lake Titicaca, 1973. Averages calculated from Widmer et al. (1975)*

	Primary productivity (mg C m^{-2} d^{-1})	Biomass (mg C m^{-3})
Summer	1188	28.17
Autumn	1409	27.95
Winter	1609	26.78
Spring	1008	14.32

Zooplankton. The zooplankton, sampled in 1937 by the Percy Sladen Expedition and studied by Harding (1955a,b) was sparse and included the following species: *Boeckella titicacae* (the dominant copepod), *Microcyclops leptotus, B. occidentalis, Eucyclops neumanni, Paracyclops finitimus* and *Mesocyclops annulatus* among the copepods and *Ceriodaphnia quadrangula, Bosmina coregoni* and *Daphnia pulex* among the Cladocera, this latter group being much less abundant than the copepods. Moreover, P. Beauchamp (1939) described nine species of rotifers.

In April 1954, Thomasson (1956) found numerous *Keratella quadrata* and an abundant population of copepods in the Puno Bay.

Ueno (1967) who, in April 1961, examined the zooplankton at two stations in Lago Pequeño and three stations in Lago Mayor gave the following list:

Cladocera: *Daphnia pulex, Ceriodaphnia quadrangula, Bosmina* cf. *hagmanni, Macrothrix palearis, Camptocerus aloniceps, Alona cambouei, Pleuroxus aduncus, Alonella excisa clathratula, Dunhevedia odontoplax, Chydorus sphaericus, C. eurynotus;* and
Copepoda: *Boeckella titicacae, B. occidentalis, Eucyclops neumani, Microcyclops* sp.

Widmer *et al.* (1975) who made a quantitative study of the lake zooplankton during the whole year 1973 found that *B. titicacae* was always massively dominant and *Microcyclops leptotus* subdominant; the other species rarely exceeded 10% of the biomass. The total biomass was nearly constant throughout the year, a feature already reported by Burgis (1969) in Lake George and by Gophen (1972) in Lake Kinneret in spite of the greater seasonality of the latter. The absolute values of biomass varied from 5 to 20 g m^{-2}.

The flora of the lake. Tutin (1940) recorded only eight species of higher plants. In sheltered and silty areas, the bulrush *Scirpus tatora* is especially abundant. On more exposed and clayey shores, the *Zannichellia palustris – Potamogeton strictus* association develops. At about 1 m depth, *Myriophyllum elatinoides* and *Elodea potamogeton* begin to occur; they are found down to 8–10 m. The chara (*Kenoyeri howe*) meadows follow down to 12 m and then the moss *Scianorum* and its epiphyte *Nostoc sphaericum* cover gentle slopes down to 29 m. No plants with floating leaves were observed.

The fauna. The Percy Sladen Expedition found an abundant population of molluscs, especially the genera *Taphius* (Planorbidae) and *Littoridina*

(Hydrobiidae). Haas (1955) described 24 species including five new genera and 14 new species, seven of which belonged to the genus *Littoridina*; most of these are endemic to Lake Titicaca.

The Crustacea are dominated by the genus *Hyalella* (Amphipoda) which has supplied a large number of species occupying specialized niches.

The fish population is almost composed of only one genus *Orestias* (Cyprinids) with 20 species, two sub-species and three forms; 14 species are endemic to Lake Titicaca basin (Gilson, 1964).

In 1941–42, the lake trout *Salvelinus namaycush* and the rainbow trout *Salmo gairdneri* were introduced into the lake. The trout, especially the latter species, grew well in the lake and provided good landings from 1948 to 1953. Since then, the trout fishing has gradually deteriorated. The lack of food does not seem to have caused the decline of the trout, since scale-reading indicated that specimens of 2.5 kg were in their third or fourth year. The overfishing by Indian fishermen is a more plausible reason strengthened by the observation that a reduction of fishing activities in the 1960s was followed by an improvement of the trout catch.

The introduction of the trout had a secondary negative effect: a sporozoan parasite associated with the trout spread to the *Orestias* population and contributed to the disappearance of certain forms (Villwock, 1972). This emphasizes the danger of uncontrolled stocking with foreign species.

Reflections on the flora and fauna of Lake Titicaca. The existence of cold high mountain lakes inside the Warm Belt raises the question of the origin of their flora and fauna. It seems, *a priori*, that the surrounding Warm Belt represents a severe barrier to cold, stenothermic animals. In the case of Lake Titicaca, two possibilities are to be considered. Lake Titicaca is the only tertiary lake among the tropical high mountain lakes (Löffler, 1968). As such, it might harbour Pliocene–Pleistocene relicts as is the case for Lake Kinneret (Serruya, 1978a). Although we know that the old Lake Ballivian had a higher level than the present Titicaca and probably a lower electrolyte content, we know nothing about its flora and fauna. According to Villwock (1963) and Lowe-McConnell (1975), for example, the ancestors of the cyprinodont genus *Orestias* 'are thought to have been present before the mountains were formed and to have risen with the mountain chain'.

The second possibility of colonization of such lakes consists in passive dispersal by birds: Löffler (1968) notes that birds can carry stages of aquatic animals on themselves or even in their digestive tracts. Not only are birds abundant in the lake vicinity (Andean gull, coot, moorhen, cormorant, waders, ducks, tyrant-birds, humming birds, flamingo and flightless grebe) but many of the waterfowl migrate along the Andes. Löffler (1968) indicates that 21 species and subspecies of wildfowl are characteristic of high mountain lakes of the tropical Andes in comparison with only three species of lake-dwelling birds in East-African high mountains.

Most of the phytoplankton of Lake Titicaca is composed of cosmopolitan species. This is not surprising because planktonic algae represent an easily transportable material. Many macrophytic species are also cosmopolitan; the bulrush *Scirpus totora*, reported by Löffler (1964, 1968) to be endemic to Lake Titicaca has also been found in Lake Junin, north-east of Lima, by Hegewald *et al.* (1976). Similarly to algae and for the same reasons, certain groups of animals are more or less exclusively cosmopolitan (Protozoa, rotifers, tardigrades). Other groups such as the copepods include both cosmopolitan and endemic species whereas the molluscs are dominated by endemic species.

It is interesting to note that the temperate and subantarctic parts of the South-American continent seem to be the centres of origin of many species of Lake Titicaca. Among the non-malacostracan crustaceans, 20% of known species also occur abundantly in the southern part of the continent, whereas only 2% of the species show a northern distribution. The case of the genus *Boeckella* is a good example: many species belonging to this genus exist in the temperate and subantarctic zones of the continent. The progression of the genus towards northern and high altitude Andean areas is accompanied by an impoverishment in species and increasing modifications of the fifth legs of the males.

The temperate and subantarctic zones of the South-American continent represented a permanent reservoir for cold flora and fauna. The extension of the Andes from 50° S to 12° N and the presence of numerous migratory birds offered both a continuous pathway and efficient agents for the dispersal of organisms. This, however, did not prevent the occurrence of intralacustrine speciation and speciation in different lakes. Villwock (1962) hypothesized that the various species and subspecies of *Orestias* originated from an ancestor living in Lake Ballivian: this old stock had been split into various littoral groups which had developed into different species.

THE PARANA SYSTEM AND RIO DE LA PLATA BASIN

With an area of 3.2×10^6 km^2, the Rio de la Plata basin is the second largest watershed of South America extending over Brazil, Bolivia,

The Parana system

Argentina, Paraguay and Uruguay. The basin includes the most densely populated areas of South America (Sao Paulo, Brasilia, Buenos Aires, Montevideo) and areas of very low demographic intensity (upper valley of Paraguay).

Two main river systems form the Rio de la Plata basin: the Parana–Paraguay system, which covers 84% of the total watershed and supplies 75% of the total discharge (23 000 m^3 s^{-1}); and the Uruguay system. The Rio de la Plata basin is periodically connected with the Amazon system in the Matto Grosso area: during the flood period, the Guapore, a tributary of the Amazon, and the Jauru, a tributary of the Paraguay, join their head-waters at approximately 15°S and 59°W (Bonetto, Dioni & Pignalberi, 1969). This explains the biological similarities existing between the Amazonia and the Paraguay–Parana basin. In addition, common biological features are shared by the Upper Parana, Uruguay and Sao-Francisco basins.

Geology and geography

In the north-eastern part of the basin (see Fig. 14), the Upper Parana flows in highland areas composed of Precambrian rocks partially covered with younger sediments and Mesozoic basalts. The river flows between steep banks and its course is interrupted by numerous falls and rapids. The weathering of the Precambrian formations generate infertile soils rich in aluminium and iron oxides and where kaolinite is the dominant clay-mineral. Similarly, the Paraguay rivers drain Precambrian rocks and sedimentary formations. In contrast, the Middle and Lower Parana drain the soft and silty soils of the Pampa plains. In this flat landscape, the river bed widens into a vast flood plain. In its middle course, the Parana has a steep left bank and a flat area on the right side which is periodically flooded. In the lower course of the Parana the situation is just opposite with a steep right bank and a left-side flood plain. Unexpectedly, the river-channel gets narrower downstream: from 5 km at Corrientes, the river section drops to 2 km at Rosario. Between these two locations, the width of the flood plain grows from 14 to 57 km. The Parana flows into a 300 km long delta covering an area of 15 000 km^2.

A general north–south climatic gradient divides the basin into tropical areas in the uplands of the Paraguay and Upper Parana in the north and a subtropical, temperate to warm southern zone with dry winters and rainy summers. The Andes mountains in the east and the Atlantic Ocean in the west modify this gradient locally.

At the same latitude, desert areas with < 100 mm annual rainfall cover

98 South America

the eastern part of the basin, whereas annual precipitations exceed 2400 mm in south-eastern Brazil. The overall north-to-south decrease in precipitation on the South-American continent, however, is clearly indicated by comparing the specific yields of the Amazon and the Rio de la Plata: 1 m² of watershed of these two rivers yields 1 m³ y⁻¹ and 0.22 m³ y⁻¹, respectively.

The vegetation of the basin is characterized by the contrast between the Pampa landscape and the gallery-forest forming a continuous strip along the river banks. Within the flood plain the vegetation varies with the proximity of the water. *Salix humboldtiana* dominates on the main banks of the river. On the islands, an association of *Sapium haematospermum*, *Cathormion polynatum*, *Erythrina crista-galli* and *Nectandra falcifolia* is

Fig. 14. The watershed of the Parana–Rio de la Plata system

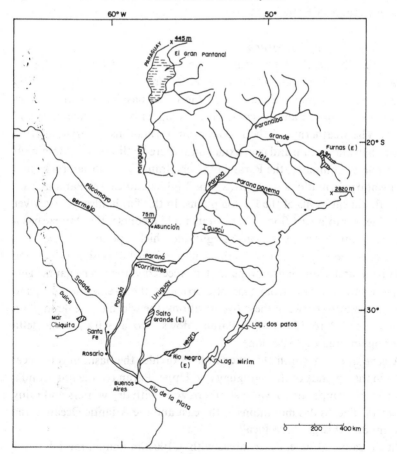

The Parana system

observed. On low grounds, *Panicum prionitis* and *Solanum malacoxylon* replaces the previous association. Near the open water, *Ludwigia peploides* and *Polygonum punctatum* predominate. In pools and ponds, rooted plants with floating leaves (*Nymphoides indica, Nymphea amazonica*) and submerged rooted plants (*Myriophyllum brasiliense, Elodea densa, Ceratophyllum demersum*) compete for light. Under certain conditions, floating plants (*Eichhornia crassipes, Salvinia herzogii, Azolla caroliniana*) eliminate other aquatic vegetation.

Swamps and wetlands play an important role in the Rio de la Plata basin. At the head-waters of the Paraguay, the Gran Pantanal, extending over 80 000 km², is the most important complex of wetlands of the basin. During the rainy season, numerous temporary shallow lakes retain the flood water and thence have a major regulating effect on the discharge of the Paraguay water. Bonetto (1975) emphasizes the danger of certain drainage projects which, if carried out, would cause disastrous floods in the lower valley.

Water chemistry and hydrological regime

The Middle Parana water has a low content of electrolytes and its conductivity varies from 60 μmhos cm^{-1} in the rainy season to 200 μmhos cm^{-1} in dry periods; it belongs to the sodium bicarbonate type and is rich in silica and iron. Its usually high content of CO_2 and poor buffer capacity account for a frequently low pH. Sodium is more abundant than calcium, whereas sulphates and potassium are especially low. Depetris (1976) has shown that high discharge produces a calcium–hydrogen-carbonate type of water and low discharge a sodium-chloride type. He interprets this observation as an indication of the role of saline environments as sources of dissolved load at low water.

The Parana River has its floods in January–March and its low water in July–August. In normal years, the water level rises ~ 3 m and overflows the flood plain. The floods modify the concentration of suspended solids and soluble elements and affect also the flora and fauna by flushing vast amounts of plankton and macrophytes and enhancing the development of various animal species. In contrast, during low-water periods, many organisms die; the resulting losses have been estimated at 500 kg ha^{-1}. About 50% of these losses concern species of economic importance; it is estimated that they exceed the yearly fish catch in the basin (12 000 tons).

The water bodies of the Middle Parana Valley

Bonetto *et al.* (1969) distinguished four types of water bodies in the Middle Parana Valley (ponds, ox-bows and meander scrolls, swamps and wetlands) and describes their main limnological features.

The thermal regime of such water bodies depends on their size and depth, exposure to wind, and density of aquatic vegetation. In large water bodies poor in macrophytes, warm weather causes thermal stratification during the day, mixing at sunset and inverse stratification at night. The presence of dense aquatic vegetation accentuates the diurnal increase of temperature of the upper layers by a few degrees centigrade.

The optimal conditions are considerably affected by the turbidity following the floods. Later on, each water body develops its own optical climate according to its size, plankton density and exposure to wind. In general, shallow water bodies, protected from the wind and partially covered with floating plants, have a high transparency and a dense submerged vegetation. In shallow water bodies, exposed to wind, the turbidity due to constant mixing of sediments limits the development of submerged vegetation.

In water bodies rich in macrophytes, the phytoplankton is usually poor and primary productivity very low (50 mg C m^{-2} d^{-1}). In open water, the spring and summer phytoplankton may reach high biomass values (chlorophyll content 25 mg m^{-3}) and Bonetto *et al.* (1969) underlines the role of nanoplankton in such water but relatively modest values of primary productivity have been measured (1000 mg C m^{-2} d^{-1}). In contrast, primary productivity of macrophytes seems very high. The spring growth of vegetation decays during the following winter. The resulting massive bottom sedimentation of organic matter orientates the food chain towards a detritus pathway with a rich fauna of detritus-eaters such as molluscs, other invertebrates and mud-eating fishes.

Zooplankton is especially abundant in spring and summer in water bodies with scarce aquatic vegetation where biomass values as high as 1200 individuals l^{-1} have been recorded. Rotifers represent the leading group, calanoid copepods are abundant during the warm season, while Cladocera form a much smaller group.

The floating vegetation harbours a very diverse population of Acarina, Amphipoda, Mollusca, etc., whereas the benthic fauna has a large biomass and a small number of species, a typical feature of a nutrient-rich environment. The microfauna is dominated by *Thecamoebina* and flagellates; Nematoda, Ostracoda, Oligochaeta and chironomid larvae constitute the mesofauna. The low content of calcium and the low pH values do not seem to limit the development of molluscs although their shells are much thinner than in other environments: unionids with *Diplodon* sp. are

The Parana system

particularly abundant but *Pisidium* and *Naiades* are also represented. Trichodactylinid crabs and snails (*Ampullaria insularum*) are frequent members of this benthic macrofauna.

The very abundant fish population (100 kg ha^{-1}) is largely dominated by the mud-eating species *Prochilodus platensis* which represents 62% of the biomass. Two typical 'river species', *Hoplias malabaricus* and *Symbranchus marmoratus*, account for another 10% of the biomass. The remaining 28% includes *Pimelodus clarias, Schizodon fasciatum, Leporinus obtusidens, Plecostomus* sp., *Loricaria* sp. and *Serrasalmus* sp. Lowe-McConnell (1975) notes that there is a noticeable decrease in species and biomass in the Upper Parana and Uruguay Rivers when compared to the Paraguay–Parana system. The striking predominance of the detritus-eater *Prochilodus platensis* emphasizes one of the major features of the South-American freshwater food chain: the overwhelming predominance of 'detritus pathways' over the 'grazing pathways' (phytoplankton → zooplankton → fish).

Experimental fish samplings were carried out in 1972 by Cordiviola de Yuan & Pignalberi (1981) at Corrientes and Santa Fe in temporary pools on islands located in the flood plain of the Middle Parana River. In the Corrientes area, the dominant species in terms of biomass were all characoids: *Triportheus paranensis, Moenkhausia dichroura, Acestrorhynchus falcatus* and *Prochilodus platensis*. In Santa Fe, *Prochilodus platensis* was generally dominant. The specific diversity was higher in the Sante Fe area with 97 species than at Corrientes with 44 species. In general, the animals were of small to medium size including adults of small species or juveniles of larger growing species. Species occurring in all the ponds were mostly carnivorous species such as *Lycengraulis simulator, Acestrorhynchus falcatus, Hoplias malabaricus*, but also included the mud-feeders *Apareidon affinis, Prochilodus platensis* and the omnivore *Astyanax fasciatus*.

Economic development

In the highlands of the Rio de la Plata, water and steep slopes have been used for generation of hydroelectric power. On the Parana River, Brazil has already built several dams: the Jupia (1400 MW), the Ilha Solteira (3200 MW) and the Tiete (1200 MW). In 1973, Brazil and Paraguay agreed to build a 10 700 MW dam at Itaipu on their common border. In 1973 also, Paraguay and Argentina decided to build a multipurpose dam on the Parana River at Yacireta-Apipe (4050 MW). On the Uruguay River, the common Argentinian–Uruguayan dam of Salto Grande will supply 1260 MW. Several other projects are under consideration.

Navigation is the second most important use of these rivers. The total length of the rivers of the basin is 8300 km (Bonetto, 1975) but only 6250 km are navigable (Cano, 1978) and no more than 4600 km are actually navigated.

Besides wide use of the basin water for domestic, municipal and industrial use, especially in Argentina and Brazil, irrigation of rice fields is common in Argentina. Many large cities such as Rosario, Buenos Aires, Sao Paulo, Columbia and others discharge their raw sewage directly into the rivers of the basin.

The lakes of Brasilia

The lakes around Brasilia are of small dimensions but their association with the large city confers upon them a special role. The eutrophication of one of them, Lake Paranoa, led to the Lago do Paranoa Restoration Project which also included Lago Santa Maria, Lago Descoberto and another six natural lakes (Cronberg, 1977).

Lago Paranoa

This artificial lake, 25 m deep, east of Brasilia, was created in 1960 by damming the Paranoa River. It was then a desmid lake. In 1969, the lake was sewage-polluted and dominated by blue-green algae. The dominant alga seemed to be the heterocyst-forming *Anabaenopsis raciborskii* (24 mg wm l^{-1}), but this has been given different names because of a certain polymorphism probably related to the variable nutrient concentration of the water. A survey, carried out a few days after sewage dumping, showed a 30% increase of *Anabaenopsis* and a simultaneous increase in the number of heterocysts and of the cell size. When experimentally diluted with nutrient-poor water, the heterocysts disappeared and the algal biomass declined. The presence of the fish *Tilapia nilotica* in the experiments increased the algal growth whereas the condition factor of the fish decreased. It is clear that the algae were able to use the fish excretions but the fish fed poorly on the blue-greens. The results were opposite with zooplankton: rotifers present in the experiment grazed the algae.

Lago Santa Maria

This artificial lake has an annual mean biomass of 0.4 mg l^{-1} and a mixed population of *Peridinium, Dinobryon sociale* and *Staurastrum* sp. It is the most oligotrophic lake in the region.

The Parana system

Lago Descoberto

Lago Descoberto is intermediate between Lake Paranoa and Lake Santa Maria. Its biomass oscillates around 5 mg l^{-1}; it has the same algal species as Lake Santa Maria but these are accompanied by *Kirchneriella, Ankistrodesmus* and *Scenedesmus*.

Most of the small natural lakes were extremely rich in desmids and *Peridinium*. Lake Feia, north-east of Brasilia, deserves special mention since it harbours the agents of schistosomiasis.

5
Central America

INTRODUCTION

Central America has been partially emergent since the Palaeozoic. In the early Triassic it reached its maximal terrestrial area and extended from the North-American to the South-American continent. In the mid-Cretaceous, the connections with both continents were interrupted by the Mexican geosyncline in the north and the Venezuelan–Peruvian geosyncline in the south. During most of the Cenozoic, marine connections existed in the north, in the Tehuantepec region, and in the south. In the Pliocene, Central America became a terrestrial bridge between the northern and southern continents (Maldonado–Koerdell, 1964). The work of Patterson & Pascual (1963) on South-American mammals indicate that the isolation of South America ended only in the late Pliocene to early Pleistocene, when South America came in contact with Central America in the Panama region.

Tectonically speaking, Central America is a very complex area. It is bordered on its western side by the Middle American Trench, a subduction zone prolongating northward into the Peruvian–Chilean Trench. In these areas, the oceanic crust of the Pacific plate sinks and is destroyed below the South-American continental plate. This activity is accompanied by volcanic eruptions and earthquakes. The volcanic and seismic axis is parallel to the western coast.

The Mexican Plateau is an uplifted region (the *Mesa del Norte* and *Mesa Central*) bordered by steep escarpments (*Sierra Madre Oriental* and *Occidental*) and composed of tertiary volcanic material in the west and marine sedimentary formations in the east. Central America is formed of central ridges of Palaeozoic to Cenozoic rocks uplifted by numerous tectonic episodes and Quaternary volcanic cones exceeding 4000 m.

Mesozoic and Cenozoic marine and continental sediments outcrop mainly on the Atlantic side. The western and eastern coastal lowlands are narrow except in Honduras and Nicaragua. The mountainous ridges are interrupted by three main transversal passes: the Honduras pass along the Ulua, the pass of the Lake Nicaragua basin and the pass of Panama.

Central America is located between 7°12′N and 18°3′N. A humid tropical climate is typical of the swampy coastal lowlands where mosquitoes, flies and parasitic insects are abundant. With altitude, the climate becomes drier and cooler and the highest demographic concentrations are observed between 1000 and 2000 m.

The rainy season extends from May to November; the Atlantic coast receives about twice as much rain as the Pacific side. The total yearly rainfall varies from 500 mm to >5000 mm (Carlson, 1952).

The streams which efficiently drain the highlands meander in the lowlands into swamp and marshes. The main stream is the San Juan in Nicaragua which drains Lake Nicaragua, the largest lake of Central America, with a surface area of 8264 km^2.

LAKES OF PANAMA
Gatun and Madden Lakes

Both lakes are man-made reservoirs, integrated into the operation of the Panama Canal; they were studied in detail by Gliwicz (1976). The main morphometric and chemical features of the lakes are shown in Table 16.

Gatun Lake

The lower course of Rio Chagres was dammed and the valley flooded in 1914. Trees were not cleared before flooding and are covered with periphyton: *Oscillatoria, Lyngbya* and *Spirogyra* down to 5 m, then desmids and diatoms down to 12–15 m and, below, colonies of Bryozoa. The periphyton has a limited biomass but a high productivity. The shoreline is bordered by a belt of chara (*Kenoyeri howe*), very poor in invertebrate macrofauna.

Rio Chagres, draining Lake Madden, is the main affluent of Lake Gatun, but most of its waters do not reach Lake Gatun and flow directly into the Panama Canal to the Pacific Ocean.

Lake Gatun undergoes frequent vertical mixing. A labile thermal stratification may develop with a difference in temperature of 0.8°C between surface and 25 m depth (29.7°C and 28.9°C, respectively). The polymixis allows a good oxygenation of bottom water which may explain the low content of organic matter of the sediments.

Madden Lake

Madden Lake resulted from the damming of the Upper Rio Chagres; the flooding was completed in 1935. Here the bottom only is covered with the stumps. The low water transparency explains the small amounts of periphyton and the lack of higher vegetation.

Madden Lake is nearly always stratified. Even during periods of maximal mixing, a difference in temperature of $1.5°C$ exists between the epi- and hypolimnion; this difference rises to $5°C$ in periods of calm weather, due to the river water which flows into the lake at its northern end and flows through the lake, on its bottom, as a density current. That is why, in spite of a nearly permanent stratification, no oxygen depletion of deep waters is observed but only a metalimnic minimum.

Transparency is poor, probably because the large increase in density at

Table 16 *Main characteristics of Gatun and Madden Lakes (from Gliwicz, 1976)*

	Gatun Lake		Madden Lake	
	Max.	Min.	Max.	Min.
Altitude (m)	26 (Jan)	25 (Jun)	76 (Nov–Jan)	71 (Mar)
Surface (km^2)	431	422	50	37
Mean depth (m)	12.7	12.4	16.2	12.7
Maximum depth (m)	29	28	48	43
Shoreline (km)	1750		250	
Development of shoreline	24		10	
Volume (10^6 m^3)	5480	5260	810	470
Watershed area: lake volume ratio (m^2 m^{-3})	0.42	0.44	1.27	2.18
Dissolved O$_2$ (mg l^{-1}) 1 m	7.73	7.03	8.40	7.69
11 m	7.78	6.52	6.88	0.52
25 m	7.70	1.04	7.06	0.28
Ca^{2+} (mg l^{-1})	10.2	9.0	–	
Mg^{2+} (mg l^{-1})	3.8	3.5	–	
Cl$^-$ (mg l^{-1})	5.0		–	
SO$_4^{2-}$ (mg l^{-1})	6.0		–	
SiO$_2$ (mg l^{-1})	16.2	12.1	–	
NO$_3$-N (mg l^{-1})	0.06	0.04	–	
Alkalinity (mgl^{-1} CaCO$_3$)	90		–	
pH	7.2	7.1	7.2	7.0

the thermocline prevents sinking of organic material. An additional reason lies in the very high value of the watershed area : lake volume ratio.

Phytoplankton

Gliwicz (1975, 1976) sampled the total plankton community, including nanoplankton, and gives a detailed list of algae of both lakes. Gatun Lake is dominated by an assemblage of desmids (*Cosmarium moniliforme, Staurastrum* sp. and *Staurodesmus clepsydra*), diatoms (*Melosira* sp. and *Fragilaria crotonensis*) and numerous small blue-green algae. Madden Lake has a year-round population of filamentous blue-greens (*Oscillatoria limnetica, O. geminata, O. redeckei, Lyngbya limnetica* and *Microcystis aeruginosa*) and dinoflagellates (*Ceratium hirundinella* and *Peridinium palatinum*). Chlorophyll is twice as high in Madden Lake (5 mg m^{-3}) as in Gatun Lake (2.5 mg m^{-3}). In Gatun Lake, nanoplankton accounts for at least 40% of the chlorophyll. Although the contribution of nanoplankton is more difficult to estimate in Madden Lake, it seems that here too it accounts for a large fraction of the chlorophyll content.

Gliwicz (1976) measured the primary productivity of both lakes. He showed that incident radiation and temperature are never limiting factors. Conversely, the data indicate that, in both lakes, maximal primary productivity (P_{max}) values are recorded after deep and continuous mixing. However, this can also be related to the inflow of nutrient-rich river water which occurs simultaneously with the mixing process. The algal biomass itself obviously affects the level of primary productivity but with a relatively low correlation coefficient. Phaeopigments were also considered as an indicator of nutrient cycling but no significant correlation was found between phaeopigments and P_{max}. In contrast, it was noted that, in Lake Gatun, the higher the proportion of nanoplankton in the algal biomass, the higher the productivity. The most significant correlation was found between zooplankton grazing and P_{max}. In their grazing experiments on nanoplanktonic algae of Gatun Lake, Weers & Zaret (1975) distinguished three groups of algae: one group, which included *Cyclotella stelligera, Oocystis lacustris, Scenedesmus bijuga, Chroomonas* and others, was actively grazed and the concentration of these algae in the experimental medium was very low; a second group of small blue-greens was apparently not grazed since their number in the medium remained unchanged; and a third group, including *Ankistrodesmus* sp., not only was not affected by grazing but the number of algal cells increased in the presence of grazers. This work demonstrates the determining role of selective grazing in the specific composition of algal communities.

It appears, then, that in the tropical environment, where temperature and light are not limiting, the ecosystem is regulated by the interaction of biotic factors. In particular, zooplankton grazing regulates primary productivity through various pathways: firstly, zooplankton excretion is a continuous source of nutrients; secondly, zooplankton grazing maintains the algal biomass at a low level and thus limits the stocking of nutrients in biomass. In certain cases however, the pressure of zooplankton grazing on nanoplankton may enhance the development of inedible algae which stock nutrients for long periods of time. Such an evolution reduces the food sources of grazers and consequently their population: in Madden Lake, the January–February dominance of nanoplankton is accompanied by a peak of filter feeders but the overgrazing of nanoplankton induces the appearance of filamentous blue-greens and a sudden and prolonged drop in filter feeders. A similar situation has been described in Lake Kinneret (Serruya *et al.*, 1980) where the winter peak of zooplankton exhausts the nanoplankton and favours the development of *Peridinium cinctum fa westii*.

LAKES OF NICARAGUA
Lake Nicaragua and Lake Managua

Lake Nicaragua and Lake Managua are located in the Nicaraguan depression, a late Tertiary to Quaternary graben (Swain, 1964). It is likely that Lake Nicaragua and Lake Managua formed one single lake which was later separated into two basins. At present Lake Managua drains into Lake Nicaragua, about 7 m below it, by the 26 km Tipitapa River. The Quaternary volcanic eruptions, distributed on a nearly continuous line on the western coast of Guatemala, El Salvador and Nicaragua, also occurred along the southern, north-western and northern part of Lake Nicaragua and surround Lake Managua. The recent volcanic material also formed most of the 310 islands of Lake Nicaragua. There are several active volcanoes in or near the lakes: the Volcanic Ometepe Island in Lake Nicaragua emits dust and ash nearly all the time.

An old theory (Hayes, 1899) claimed that the depression occupied by the lakes was formerly an arm of the Pacific Ocean cut off from the sea by volcanic material. According to this view, the lake basin drained originally into the Pacific Ocean but additional eruptions reversed the hydrographic system which then drained into the Caribbean Sea via the San Juan River. This theory explained the presence in Lake Nicaragua of sharks, which were believed to be of Pacific origin. More recent work (Zoppis & Del Guidice, 1958) emphasized the role of faulting and subsidence in the formation of the lakes and assumed that the graben was never invaded by the sea. This is in good agreement with the results of Swain (1966) who did not find 'undoubted marine sediments'.

Earthquakes are also very frequent in the lake area. Managua, the capital of Nicaragua located on the western shore of Lake Managua, was almost destroyed by the quake of 1931 (Carlson, 1952).

Lake Nicaragua is the largest lake to be found in America between Lake Titicaca and the North-American Great Lakes. The general morphometric features of the lakes are shown in Table 17.

Lake Nicaragua is located at 34 m above MSL so that the lake bottom lies at 26 m below sea level. Lake Managua, located at 40 m above MSL, is entirely above sea level.

The water chemistry of the lakes is summarized in Table 18. Lake

Table 17 *Morphometric features of Lakes Nicaragua and Managua*

	Lake Nicaragua	Lake Managua
Length (km)	170	60
Width (km)	75	25
Surface (km^2)	7700[a]*8864[b]	1295
Maximum depth (m)	60	30
Mean depth (m)	15	10
Volume (10^9 m^3)	80	10
Watershed area (km^2)	40 000	–

[a] value generally quoted
[b] value of the Instituto de Formento Nacional, Managua, 1975

Table 18 *Chemical features of Lakes Nicaragua and Managua*

	Lake Nicaragua			Lake Managua
	(a)	(b(i))	(b(ii))	(a)
pH	7.0	–	–	8.7
TDS (mg l^{-1})	151	–	–	747
Ca^{2+} (mg l^{-1})	19	14.7	13.2	9.4
Mg^{2+} (mg l^{-1})	3.5	5.4	7.2	22.1
Na$^+$ (mg l^{-1})	17.7	–	21.0	230.8
K$^+$ (mg l^{-1})	3.9	–	4.6	35.9
HCO$_3$+CO$_3$ (mg l^{-1})	82.4	65	60	500
Cl$^-$ (mg l^{-1})	15.9	21	18	132.9
SO$_4{}^{2-}$ (mg l^{-1})	9.1	22	20	30.3

(a) Unpublished data from Yen lin, quoted in Cole (1963)
(b) Swain (1966); (i) surface water; (ii) bottom water

Nicaragua has a low salt content with sodium chloride and calcium sulphate as dominant salts. There is no significant difference in composition between surface and bottom waters. This chemical homogeneity is due to the absence of stratification and good winter mixing caused by easterly winds (Swain, 1966). Lake Managua has a much higher salt content, dominated by sodium chloride and sodium hydrogen carbonate. The lake level varies up to 2.5 m between the wet and the dry season; at very low levels, the Rio Tipitapa does not flow and Lake Managua becomes a closed lake with evaporation as sole water loss. Winds frequently cause the resuspension of bottom sediments and Lake Managua is more turbid than Lake Nicaragua (0.5–2.0 m Secchi-disc).

Biological features (Swain, 1966)

The planktonic algae of both lakes seem dominated by diatoms. In Lake Nicaragua, the following genera have been found: *Melosira, Fragilaria, Synedra, Navicula, Pinnularia, Cymbella, Epithemia, Rhopalodia* and *Surirella. Elaeophyton coorongiana* is an abundant benthic alga in both lakes (Swain & Gilby, 1964). This alga is found in brackish coastal lakes of Australia and Africa.

The copepods are all limnetic animals belonging to the genera *Mesocyclops, Cyclops, Macrocyclops, Diaptomus, Paracyclops, Eucyclops* and *Cletoamptus*. Cladocera are abundant but have not been studied in detail. Ostracoda of freshwater type such as *Physocypria granadae, Heterocypris nicaraguensis, Potamocypris islagradensis nicaraguensis* are dominant, but estuarine species have also been found (*Pericythere marginata, Darwinula managuensis, Metacypris ometepensis*) (Swain & Gilby, 1964). The nematodes show an interesting mixture of freshwater, brackish and marine species (Hartman, 1959).

Lake Nicaragua is remarkable for its marine fishes adapted to a very dilute water and lacking in the more saline Lake Managua. This marine fauna consists of two elasmobranchs (a 2 m long shark and a saw-fish reaching 300 kg) and the Atlantic tarpon. These fishes probably reached the lake via the San Juan River since they have phylogenetic affinities with the Atlantic fauna; in particular, Lake Nicaragua sharks are closely related to the Caribbean shark *Carcharhinus leucas*. The same three marine animals are also found in Lake Izabal in Guatemala.

The freshwater fish fauna includes Siluridae (*Rhamdia managuensis, R. nicaraguensis, R. barbata*), Characidae (*Astyanax nasutus, A. aenus, Brycon dentex, Bramocharax bransford, B. elongatus*), Poeciliidae (*Paragambusia nicaraguensis, Poecilia sphenops*), Atherinidae (*Melaniris guatemalensis*),

Nicaragua

Cichlidae (*Cichlasoma managuense, C. lovii, C. granadense, C. citrinellum, C. dorsatum, C. erythraeum, C. labiatum, C. lobochilus, C. centrarchus, C. rostratum, C. nicaraguense, Neetroplus nematopsis, Hrotipapia multispinosa*) and Lepisosteidae (*Lepisosteus tropicus*).

The assumed marine origin of the basins and the presence of marine organisms in the lakes led Swain (1966) to investigate the sediments of the lakes and look for an eventual transition from marine to freshwater conditions.

The inorganic fraction of the sediments is mainly composed of Quaternary volcanic ash and clays where quartz, plagioclase feldspaths, montmorillonite and volcanic glass are the main constituents. The sediments are rich in diatom frustules and faecal pellets. Organic nitrogen, carbohydrates and amino-acid levels are similar to those of eutrophic lake sediments whereas the bitumen content is low and near to that of oligotrophic lakes. No indication of a past marine phase was found in the examined material. A definite answer would require long cores in the deepest part of the lake.

Lake Managua has a flat bottom composed of volcanic lava and conglomerates. The combined effect of rocky shores, very strong wind-generated turbulence and little light penetration prevents the development of aquatic plants. The water temperature oscillates from 26°C in the morning to 30°C at mid-day. Prior to 1922 the sewage of Managua, then a small town, was disposed of in septic tanks, the lake was clear and people used to bathe in the town beaches. In 1922, an urban sewage collector was built and the town sewage has been dumped into the lake since then. The lake became rapidly eutrophic and ceased to fill its recreational functions (Lin, 1961).

The ichthyological fauna of Lake Managua includes the following species: the Characidae *Astyanax aeneus* and *A. nasutus*, the Clupeidae *Dorosoma chavesi*, the Poeciliidae *Paragambusia nicaraguensis* and *Mollienesia sphenops*, the Cichlidae *Cichlosoma citrinellum, C. dorsatum, C. erythraeum, C. labiatum, C. lobochilus, C. centrarchus, C. rostratum, C. longimanus, C. nigritum, C. nicaraguense, C. balteatum* and the sardine *Melaniris sardina* among the non-predators. The predatory species include *Lepidosteus tropicus*, the Siluridae *Rhamdia managuensis, R. nicaraguensis, R. barbata*, the Characidae *Brycon dentex, Bramocharax elongatus, Roeboides guatemalensis*, the Cichlidae *Cichlosoma managuense* and *C. dovii* and *Gobiomorus dormitor*.

Experimental fishing with nylon nets (mesh 15–50 mm), performed by a FAO mission, showed that *Dorosoma chavesi* is the most abundant fish of the lake (48% of the catch), followed by the cichlids (26%) and the predatory characids (19%).

LAKES OF GUATEMALA

Guatemala is particularly well provided with lakes. Whereas the lakes of the northern lowland province of Peten are of karstic origin, several lakes of the southern highland originated from volcanic activity.

The Peten lake district

The Peten lake district lies in karstic dolomitic limestone of Cenozoic age; the whole area is endorheic.

The area is characterized by uniformly high temperature (monthly means 20.8°C in December and 27.1°C in May) and heavy seasonal rainfall (1600 mm y^{-1}). Here, as in Africa, the 1960s were unusually rainy and the 1970s unusually dry (1213 mm in 1974).

The sediments of some lakes have been studied by Cowgill & Hutchinson (1966), Cowgill *et al.* (1966) and Deevey, Vaughan & Deevey (1977). More general limnological studies were carried out by Brezonik & Fox (1974), Deevey, Brenner, Flannery & Yezdani (1980) and Deevey, Deevey & Brenner (1980).

The general features of the Peten lakes are presented in Table 19. All these lakes, except Petenxil and Sal Peten, have a stable stratification underlined by the absence of oxygen and presence of sulphide in deep waters. As usual in the tropics, a few degrees difference between surface and bottom waters confers a certain stability to the system, especially if it is protected from winds or too small for winds to have any efficient mixing effect. In Yaxha and Sachab lakes, Deevey, Brenner *et al.* (1980) observed that, in spite of persistent stratification, occasional mixing takes place as a result of nocturnal cooling and sinking of surface water. Deevey, Brenner *et al.* (1980) found that the bottom of Lakes Yaxha and Sachab is sealed with a thick clay layer which originates from erosion due to deforestation and which isolates the lake from groundwater; the observed nonevaporative water losses which prevent these closed lakes from becoming saline occur in the marginal zones.

The water chemistry of the lakes of Peten reflects the lithology of the basin: hydrogen carbonate, sulphate, calcium and magnesium are the predominant ions (Table 20). The salt concentration of Sal Peten is one order of magnitude higher than that of other lakes; it is also richer in zinc, copper and strontium. This lake is probably fed by groundwater flowing

Guatemala

through gypsum formations. Nutrient concentrations were generally low with the exception of ammonia accumulation in the hypolimnion of several lakes. Most of these lakes, however, receive sewage and agricultural drainage and are dominated by blue-green algae, especially *Lyngbya* sp., *Oscillatoria* sp., *Raphidiopsis* sp., and *Microcystis aeruginosa*. In addition to the blue-greens, Peten Itza, Yaxha and Sachab also have an abundant flora of Chrysophyta, Chlorophyta and *Peridinium* sp.

The zooplankton is mainly composed of copepods (*Diaptomus dorsalis, Mesocyclops inversus, M. edax* and *Tropocyclops prasinus mexicanus*), ostracods (*Cypria petenensis*) and Cladocera (*Eubosmina tubicen*). The small biomass and low diversity of the zooplankton are attributed by Deevey, Deevey & Brenner (1980) to the predation pressure of planktivorous fish: the clupeid *Dorosoma petenense*, the characid *Astyanax fasciatus*, and the atherinid *Melaniris* sp.

The sedimentation rates in Lakes Yaxha and Sachab have been studied by Deevey *et al.* (1977). The authors showed that sediment-trap results were invalidated by horizontal transport and they calculated a deposition rate of $1 cm^{-3} y^{-1}$ in both lakes from microfossil abundance in core sediments and phytoplankton carbon production.

The lakes of southern Guatemala
Lake Izabal

Lake Izabal is located in the lowlands of southern Guatemala on the Atlantic side at 88°58′ W – 89°25′ W and 15°24′ N – 15°38′ N.

Its watershed has a total area of 6862 km² and penetrates 205 km from the Caribbean coast to the highlands of Guatemala. The drainage basin

Table 19 *General features of the Peten lakes*

Lake	Area (km²)	Altitude (m)	Maximum depth (m)	Oxygen (mg l⁻¹)	
				Surface	Bottom
1. Eckixil	2	119	21	7.8	0
2. Juleque	0.03	110	25.5	6.75	0
3. Macanche	2	160	57.5	8.1	0
4. Paxcamen	0.09	110	30	6.7	0
5. Peten Itza	99.6	110	60	7.7	2.3
6. Petenxil	0.52	114	4.2	7.2	8.0
7. Sal Peten	2.4	104	5.6	7.8	7.8
8. Comixten	0.09	110	14	6.3	2.2
9. Yaxha	7.4	–	27	6.5	0
10. Sachab	3.9	–	13	7.0	3.0

Table 20 *Chemical composition of the Peten Lakes*

	Conductivity (µmhos cm⁻¹)	pH	Alkalinity (mg l⁻¹)	SO_4^{2-} (mg l⁻¹)	Cl^- (mg l⁻¹)	Na^+ (mg l⁻¹)	K^+ (mg l⁻¹)	Mg^{2+} (mg l⁻¹)	Ca^{2+} (mg l⁻¹)	SiO_2 (mg l⁻¹)
1. Eckixil	192	8.1	85	7	6.3	4.3	2.0	3.0	31	18
	294			3					63	
2. Juleque	545	7.5	67	138	7.2	3.8	4.9	4.0	75	16
	720	7.0	230						140	
3. Macanche	700	8.6	145	130	29	15.5	7.3	50.0	28	47
4. Paxcamen	347	7.1	74	43	6.7	2.7	2.0	2.0	38	19
	508	6.5	235	80				3.0	56	
5. Peten Itza	485	8.3	76	110	10.3	7.1	2.8	25.5	39	32
6. Petenxil	260	7.7	79	36	4.4	3.5	0.9	2.1	40	18
7. Sal Peten	4100	7.8	60	3000	111.0	95.0	22.5	300.0	750	62
8. Comixtun	650	7.1	281	58	1.0	1.4	1.1	19.0	105	15
9. Yaxha				9.6	11.2	10.0	6.2	3.4	32	
10. Sachab				12.0	14.4	14.2	8.6	5.6	246	

Lakes 1–8: Data from Brezonik & Fox (1974)
Lakes 9 & 10: Data from Deevey, Brenner et al. (1980)

Guatemala

culminates at 3015 m altitude above sea level at only 50 km from the lake shore. The geological formations include Permian limestones, Tertiary detrital beds, metamorphic Precambrian strata and igneous rocks. Rio Polochic, which enters the lake in its western corner by a vast delta, is the main inlet of Lake Izabal, since it drains 76% of the watershed.

The trade winds which reach the Polochic upper valley without finding serious obstacles cause orographic rains, ranging from 2000 to 4000 mm y^{-1} with an average of 2992 mm y^{-1} (Brinson & Nordlie, 1975).

The discharge of the six main rivers, measured monthly during the hydrological cycle 1971–72, was found to be $13\,290 \times 10^6$ m^3 y^{-1} for a total area of 6862 km^2 or an annual runoff of 1937 mm which represents 65% of the precipitation.

Lake Izabal drains into the Gulf of Honduras, in the Caribbean Sea by the Rio Dulce, 42 km long. Half-way to the sea, the river broadens into a shallow water body, El Golfete, 7.5 m deep.

The main morphometric features of Lake Izabal and its watershed are shown in Table 21. It is interesting to note that as the lake level is only at 1 m above sea level and the maximum depth is 16 m, most of the lake volume is below sea level.

The thermal structure of the lake has been studied by Brinson & Nordlie (1975) during a whole year. The upper 10–12 m were generally isothermal with the exception of an upper 2–3 m layer temporarily warming during the day and cooling again at night. During the wet season, a cold bottom layer caused by the density current of the Rio Polochic water was observed at a distance of 30 km east from the delta area. The absolute temperature varied from 25.5°C to 30.4°C.

The lake water belongs to the calcium–hydrogen-carbonate type and has a pH generally in the range 6–7. Since the lake is not stratified, the oxygen never disappears in open water. Whereas the conductivity of Rio Polochic varies generally from 170 to 225 μmhos cm^{-1} and exceptionally reaches 260 μmhos cm^{-1}, the lake water may reach a conductivity of 465 μmhos

Table 21 *Main morphometric features of Lake Izabal*

Altitude	+0.9 m to a few metres	Volume	8300 10^6 m^3
Area	717 km^2	Watershed area 6862 km^2	
Max. depth	16.8 m		
Mean depth	11.6 m		
Max. length	70 km	Watershed area: lake volume	
Max. width	20 km	ratio 0.83 m^2 m^{-3}	

cm^{-1}, indicating a source of brackish water on the bottom as far as 12 km from San Felipe, the location where Rio Dulce leaves the lake.

Other observations confirm this hypothesis: red mangroves and barnacles are common on the western shore of the lake and along the Rio Dulce and El Golfete as well as many other organisms with marine affinities, for example the bull shark (*Carcharhinus leucas*) found in the mangroves of Rio Dulce and Lake Izabal (Thorson, Cowan & Watson, 1966; Brinson, Brinson & Lugo, 1974). Moreover, the lake fish fauna is very poor in primary freshwater species and mostly composed of euryhaline forms. These features suggest an upstream movement of sea water which was studied by Brinson *et al.* (1974). These authors carried out conductivity and temperature profiles from the sea to the lake in March, May, June and October 1972. In March (dry season), conductivities higher than 10 000 μmhos cm^{-1} were measured from the sea to approximately the central part of El Golfete. On vertical profiles, a distinct chemocline, at 5 m, separated a deep salty water mass from an upper freshwater layer. The salty water was also slightly warmer than the upper layer. From the dynamic point of view, the upper layer flowed downstream with a strong current on a quiet, salty lower layer.

Later in the dry season, when the outflow of Lake Izabal diminished, the direction of the water flow was reversed under the influence of winds and tides. Water from the Gulf of Honduras flowed into Lake Izabal and conductivity increased in the San Felipe area. That is why conductivity values of 5000 μmhos cm^{-1} were recorded in June at the outlet of the lake where, in other seasons, conductivity was low. In October, the rains caused a significant increase in the discharge of Lake Izabal which flushed the salty water of the Rio Dulce and El Golfete which then had a uniform conductivity of 170 μmhos cm^{-1}.

It is therefore established that during the dry season the combined effect of tides, gravity and wind allows an upstream current of sea water into Rio Dulce and Lake Izabal. During the wet season, the pressure of freshwater reverses the direction of the current and flushes out the salty water. Previous measurements by Brooks (1970) and Nordlie (1970) performed during the wet season failed to show the first phase of the process.

The studies of the lake plankton in August 1969, March 1970 and from January to November 1972 indicated a great variability of the populations. In August 1969, there was an algal bloom with abundant *Lyngbya* sp. and zooplankton was rather inconspicuous and mainly composed of copepods; in March 1970, copepods remained at the same level but were accompanied by Cladocera and rotifers and phytoplankton was extremely low. In 1972, two peaks were recorded in March and in August–September.

Brinson & Nordlie (1975) give the following list of planktonic species. *Melosira granulata* and *Synedra ulna* are the most common diatoms. The blue-greens include *Anacystis cyanea*, *Anabaena flos-aquae* and *Lyngbya* sp. The green algae are mainly represented by *Staurastrum pingue*, *Cosmarium* sp., *Pediastrum simplex*, *Eudorina* sp. and *Coelastrum* and the Pyrrhophyta by *Ceratium hirundinella*. The Cladocera comprise *Bosmina longirostris*, *Eubosmina tubicen*, *Moina micrura*, *Ceriodaphnia cornuta* and *Diaphanosoma fluviatile*. The Copepoda are composed of *Diaptomus dorsalis*, *Pseudodiaptomus culebiensis*, *Mesocyclops edax* and *Thermocyclops inversus*. Among many other species, *Brachionus falcatus* is an abundant component of the Rotifera.

A free-living phycomycete seems to be a frequent member of the plankton community. Its occurrence, abundance and periodic 'blooms' are a peculiar feature of Lake Izabal.

Algal gross primary productivity reaches its maximum in March–April with 2600 mg C m^{-2} d^{-1} and its minimal values in February with 430 mg C m^{-2} d^{-1} and in September with 640 mg C m^{-2} d^{-1}.

The benthic fauna of the lake is characterized by its paucity in Chaoborinae and Tendipedinae (< 100 individuals m^{-2}) and its abundance (up to 1500 individuals m^{-2}) in mud-dwelling Crustaceans (Tanaidaceae) typical of marine environments.

Similarly, the fish fauna is mainly composed of euryhaline marine species such as the snook and the jack (*Oligoplites paloma*). The details of the food web are unknown.

Lake Atitlan

Lake Atitlan (Fig. 15) is located at 1555 m altitude in the southern highlands of Guatemala at 91°07′–91°17′W – 14°45′N. It has a surface area of 130 km² and a maximal depth of 324 m.

The lake was studied by Meek (1908), Juday (1915), Deevey (1955, 1957) and Brezonik & Fox (1974).

Lake Atitlan lies in a depression the drainage of which has been blocked by volcanic material. The lake has no visible outlet: it is likely that bottom infiltrations feed one spring located below the lake on its southern side. The lake shores are steep and water very transparent (6–13.7 m Secchi-disc).

The lake heat budget is 20 000 cal cm^{-2}, very close to the value obtained for Lake Titicaca (19 000 cal cm^{-2}); such values are intermediate between temperate lakes (52 000 cal cm^{-2} for Lake Michigan) and equatorial lakes (3400 cal cm^{-2} for Lake Ranu Klindugan). The lake is stratified but there is enough thermal instability to prevent meromictism. There is at least one turnover a year.

The Atitlan waters are relatively rich in sodium; hydrogen carbonate and sulphate are the dominant anions (Table 22). Nutrient concentrations are extremely low, which may explain the absence of blue-green algae in this lake.

Atitlan is an oligotrophic lake and during the 1969 survey of Brezonik & Fox (1974) the algal cell number was as low as 72 ml^{-1}; blue-green algae were absent. Besides a few Chlorophyta, and a small number of *Synedra* spp., the algal assemblage was dominated by *Peridinium* spp.

In his 1950 survey, Deevey observed a total lack of diaptomid copepods in Atitlan. The zooplankton was dominated by cyclopids and *Daphnia* spp. The known fish population of Lake Atitlan before 1940 included *Profundulus guatemalensis, Mollenosia sphenops, Cichlasoma nigrofasciatum* and *Cyprinus carpio*. In or after 1940, the following species were introduced: *Poecilestes pleurospilus, Astyanax fasciatus, Cichlasoma motaguense, C. spilurum, C. godmanni, C. macracanthus, C. guttulatum* and *Micropterus salmoides*.

The indigenous crab *Potamocarcinus guatemalensis* is abundantly fished all the year round.

Until 1962, the fish catch was high, especially the *Astyanax* catch. In the following years, *Astyanax*, the cichlids, poecilids and cyprinodonts declined rapidly (Lin, 1963).

Fig. 15. Lake Atitlan, Guatemala (by courtesy of Dr J. Luna, BID, Washington D.C., USA)

Table 22 Physico-chemical features of lakes of South Guatemala (from Brezonik & Fox, 1974)

	Conductivity (μmhos cm^{-1})	pH	Alkalinity (mg l^{-1})	SO$_4^{2-}$ (mg l^{-1})	Cl$^-$ (mg l^{-1})	Na$^+$ (mg l^{-1})	K$^+$ (mg l^{-1})	Mg^{2+} (mg l^{-1})	Ca^{2+} (mg l^{-1})	SiO$_2$ (mg l^{-1})
Amatitlan	830	7.7	152	18	163.0	120.0	14.5	9.4	32	84
Atitlan	420	8.3	150	33	19.0	42.0	8.0	15.0	13	81
Calderas	232	7.9	115	–	2.1	5.8	8.0	11.7	16	26
Encantada	102	7.7	51	4.0	1.0	3.8	3.0	3.6	10	40
Tortugas	137	7.7	60	7.0	1.6	9.0	2.4	3.0	10	55
Izabal	210	7.9	70	30.0	7.0	4.7	1.0	6.5	25	10

Lake Amatitlan

Lake Amatitlan is located at 1189 m altitude in the southern highlands of Guatemala, south of Guatemala City. The lake is formed of two basins connected by a narrow canal. The western basin has a surface area of 8 km² and a maximal depth of 33 m.

Rio Lobos is the main inlet of the lake, which is drained by Rio Michatoya towards the Pacific through the lava material surrounding the lake.

The lake is stratified but circulates in winter at a temperature of 19–20°C. The transparency does not exceed 3.5 m, probably because of the silt carried into the lake by Rio Lobos.

Lake Amatitlan water has a relatively high salt content (conductivity 830 μmhos cm^{-1}). It belongs to the sodium–chloride type (Table 22). This unusual composition results from the presence of hot and saline springs along the southern shore. One of these springs contains 2864.6 mg l^{-1} total solids, 1220 mg l^{-1} of which are chloride (Deevey, 1957). The lake water is relatively rich in iron. The nutrient levels, especially PO_4–P, are higher than in Lake Atitlan.

Closterium aciculare (Chlorophyta) and *Peridinium* sp. (Pyrrhophyta) dominate the phytoplankton, but the Chlorophyta *Eudorina elegans* and *Staurastrum* sp., the Bacillariophyta *Melosira granulata* and *Synedra ulna* and the blue-green *Microcystis aeruginosa* are also present (Brezonik & Fox, 1974).

Productivity measurements by Deevey (1957) yielded values of daily photosynthetic carbon fixation of 2000 mg C m^{-2}.

The zooplankton is dominated by copepods (80%) whereas Cladocera do not exceed 20% of the population.

The bottom fauna is affected by the very low oxygen content of the bottom water; below the littoral zone it is restricted to the chironomids, especially *Tendipes* larvae. Below 20 m, *Chaoborus* larvae dominate and, below 25 m, the sediments are azoic. The mean benthic standing crop is therefore very low (39 kg wm ha^{-1}) (Deevey, 1955). Since the productivity is relatively high (1920 mg C m^{-2} d^{-1}) (Deevey, 1955) the question arises whether unused organic matter accumulates in the sediments. Deevey writes that the sediments are 'brown, not black and appear to be mainly composed of fine mineral particles'. It seems consequently that most of the organic matter produced in upper layers is utilized in the water and never reaches the sediments.

The fish fauna is abundant and dominated by cichlids. Data of 1956 (from commercial catch and from experimental sampling tests) revealed that the fish population of the shallow zone consisted of *Cichlasoma guttulatum* (52% by mass), poecilids (22%), *C. nigrofasciatum* (19%), *C. motaguense* (5%) and *Astyanax* (2%). In 1959 the lake was stocked with *C. managuense* and *C. macracanthus*, and in 1960 with *C. managuense* which became so abundant that in 1961 it represented 60% of the commercial catch (38 tons y^{-1}) and contributed another 25 tons y^{-1} to sport fishing (Lin, 1963).

LAKES OF EL SALVADOR
Lake Ilopango

With a surface area of 70 km², Lake Ilopango (Fig. 16) is the largest lake of El Salvador. The volcanic origin of the lake is underlined by its circular shape, its great depth (248 m) and its steep shores.

The lake was studied by Utermöhl (1958). The dry season, extending from November to April, is accompanied by strong storms which generate a more-or-less thorough mixing of the lake. In January 1953, Utermöhl observed a nearly complete homothermal profile at 25.6°C. In February, a weak stratification was re-established. The shallow thermocline deepened

Fig. 16. Lake Ilopango, El Salvador. Floating vegetation (by courtesy of Dr Besch, LFU, Karlsruhe, West Germany)

during the spring–summer period and reached 40 m in June. Then the epilimnion was at a temperature of about 27°C and the hypolimnion at 25.7°C.

The lake receives many tributaries and consequently a relatively high nutrient load. This is in good agreement with the existence of water blooms in this lake and the brown-green colour of its water.

It is therefore not surprising that the lengthy stratification causes a progressive decline of oxygen in the hypolimnion followed by a complete lack of oxygen in July. It should be however emphasized that the lake does not receive sewage water even from the nearby city of San Salvador, since the wastewater of the town is drained toward the Tempa River. The lake water constitutes in principle a good water source for the city. The water is however, rich in dissolved salts (1000 mg l^{-1}) and has a high level of boron (8–10 mg l^{-1}) and arsenic (0.8–1.0 mg l^{-1}).

In May 1980, samples of Lake Ilopango water, examined by Dr Hickel (personal communication) contained *Aphanocapsa, Chroococcus, Sphaerocystis, Eutetramorus, Oocystis, Closterium* and *Cyclotella*. The maximum development of algae occurs in summer during the stratification period; then the Secchi-disc visibility drops down to 2.1 m compared to 13–15 m in January.

The benthic fauna is mainly composed of chironomid larvae which are numerous in February–March (up to 5200 individuals m^{-2}) and decrease significantly in July.

Lake Coatepeque

Lake Coatepeque is a crater lake of 24.5 km² and 120 m maximum depth. The lake has few tributaries and is more oligotrophic than lake Ilopango. This is indicated by its blue colour and the permanence of oxygen in the hypolimnion, even during stratification. Utermöhl (1958) reported 37–59% oxygen saturation in the hypolimnion after 7 months stagnation. Moreover, the oxygen measurements of Utermöhl (1958) carried out in 1953 were identical to those of Juday (1915), which indicates that no deterioration of water quality had taken place between these two dates.

No algal bloom has been observed in Lake Coatepeque, and Secchi-disc values are much higher than in Lake Ilopango: 19 m in March and 8.5 m in September.

Blue-green algae are dominant with *Aphanocapsa, Chroococcus* and *Microcystis*, but the green algae *Botryococcus* and *Oocystis* are also found. *Ceratium hirudinella*, absent from Ilopango, is one of the components of the netplankton of Coatepeque.

The benthic fauna is dominated by chironomid larvae.

Lake Guija

Lake Guija is located on the boundary between Guatemala and El Salvador at 426 m altitude. Its surface area is 45 km², its volume 739×10^6 m³, and its mean and maximal depth are 16.5 and 26 m, respectively. Numerous volcanic craters near the lake are extremely recent.

The lake water has a very low salt content (123.6 mg l^{-1}). Hydrogen carbonates, chloride and calcium concentrations are, respectively, 90.5, 4.4 and 14.5 mg l^{-1} (Deevey, 1957).

The lake is stratified in summer. In October 1950, at the end of the stratification period, the epilimnion was at 27°C, the thermocline was located between 15 and 20 m and the hypolimnion was at 24.5°C. Alkalinity and phosphorus increase with depth: the former parameter increases from 90.5 mg l^{-1} HCO_3^- in surface to 110 mg l^{-1} at 25 m and the latter from 45 to 140 mg m^{-3}.

On 12th June 1980, Besch (personal communication), examined the zooplankton and reported the following list: *Brachionus patulus, Keratella cochlearis, Filinia longiseta, Polyarthra vulgaris* (rotifers); *Ceriodaphnia affinis, Bosmina longirostris* (Cladocera); and *Mesocyclops leuckarti, Thermocyclops crassus* and *Mastigodiaptomus albuquerquensi* (cyclopids).

Chaoborus is especially abundant in the bottom fauna which reaches 13.6 kg wm ha^{-1}.

In 1961, an experimental fish sampling was carried out and indicated that the relatively poor fish population is dominated by *Calcichthys guatemalensis* (FAO, 1963). In the 1960s the lake was stocked with *Cichlasoma managuense*.

Reservoir 5th of November

The 5th of November Reservoir is located near the border with Honduras on the River Tempa; it was built in 1954. Its surface area is 20 km² and its maximal depth 42 m near the dam.

Because of poor soil conservation measures, the floods wash considerable amounts of silt and clay into the lake. As a result in 1975 the operational volume of the reservoir was only 100×10^6 m³, i.e. 50% of its initial value in 1954.

The conductivity is about 350 µmhos cm^{-1}. During a short survey in 1973 the phytoplankton was dominated by *Oscillatoria* sp., *Tribonema* sp. and *Anabaena* sp.; rocks were covered with *Lyngbya* sp., while *Oscillatoria* sp. and *Spirulina* sp. were the most common epiphytic algae. Floating masses of *Eichhornia* sp. were frequently found in the lake and *Nayas* sp. was the dominant aquatic plant in the river.

Chironomid larvae, abundant in the benthos, is a good food source to the catfish *Arius taylorii* which feeds also on the numerous ostracods associated with the *Eichhornia* meadows (Rubio, 1975).

LAKES OF MEXICO
The lakes of the Lerma system
Lake Patzcuaro

Lake Patzcuaro is located in Mexico at 101°33′–101°42′ W and 19°32′–19°41′ N at 2035 m altitude (see Fig. 17). De Buen (1944a,b) reported a surface area of 111 km², whereas Ziesler & Ardizzone (1979) assigned the lake a surface area of 70 km². Deevey (1957) mentions a maximum depth of 15 m in the northern part of the lake. The lake is not stratified although superficial warming occurs. In February, the temperature of deep waters varies from 13 to 15.5°C and, in July, the deep waters have a temperature of 21 °C. Oxygen is always present. The lake is generally turbid because of suspended volcanic silt.

Although the lake has no outlet it has a low salt content (428 mg l^{-1}). It is a carbonate lake with Na$^+$ + K$^+$ (1975 mg l^{-1}) higher than Ca^{2+} (3.2 mg l^{-1}) and Mg^{2+} (2.5 mg l^{-1}). Chloride and sulphate are low (21.3 and 0.2 mg l^{-1}, respectively). The phytoplankton of Lake Patzcuaro is dominated by diatoms. Osorio Tafall (1944) gave a list of plants and animals of the lake together with their trophic role.

The bottom fauna of Lake Patzcuaro is dominated by chironomid larvae (mainly *Tanypus*), *Chaoborus* and leeches; Deevey (1957) reports a mean mass of 1.77 g m^{-2} or 18 kg ha^{-1}.

The fish fauna includes characins of the family Goodeidae, and

Fig. 17. The lakes of Mexico

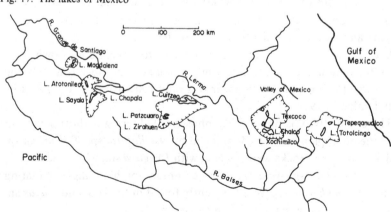

Mexico

Atherinidae (genus *Chirostoma*). The natural fauna has been considerably modified by the introduction of the black bass (*Micropterus salmoides*).

Lake Cuitzeo

Lake Cuitzeo is located in Mexico at 100°58′ W and 19°54′ N and has an area of 220 km². As Lake Patzcuaro, it has no outlet.

Lake Chapala

Lake Chapala is located in Mexico at 102°41′–103°25′ W and 20°06′–20°21′ N at an altitude of 1500 m in a tectonic trench. It is the remnant of a series of tertiary lakes which have been captured and formed the present Lerma drainage system. The Lerma River is the main inlet and carries an annual amount of 2.0×10^9 m³ water to the lake which has a surface area of 1109 km², an average depth of 8 m, a volume varying from 6×10^9 m³ to 11×10^9 m³ and is the largest lake in Mexico. The lake is drained by the Rio Grande de Santiago.

The lake circulates at all seasons and is well oxygenated. The deep water is at 20–21 °C in February and 24 °C in July (De Buen, 1945; Deevey, 1957).

Lake Chapala has a low salt content (267 mg l^{-1}) and, as Lake Patzcuaro, is a carbonate lake with $Na^+ + K^+$ (110 mg l^{-1}), higher than Ca^{2+} and Mg^{2+} (2.0 and 1.1 mg l^{-1}, respectively). Chloride and sulphate are low (17 and 12.4 mg l^{-1}, respectively).

The bottom fauna, particularly meagre (0.47 g m^{-2}), is dominated by Tubificids (Deevey, 1957).

The fish catch is 1000 tons y^{-1} and mainly composed of the small 'charral' *Chirostoma* spp.

Eichhornia crassipes has invaded different areas of the lake, hampering navigation and fishing activities.

Lake Zirahuen

The deepest lake of the Lerma system has been studied by De Buen (1943). It seems that the lake is stratified, probably monomictic. The water is very transparent since the phytoplankton, dominated by green algae and especially desmids, is not abundant.

The Valley of Mexico lakes

The 'Valley of Mexico' is a graben limited by faults and volcanoes. Tertiary lava covered the Cretaceous sediments; the Valley was then drained to the south. Basaltic material, located south of the Valley, dammed the river in late Pliocene and the water converged to the central part of the basin.

The closed basin of Mexico lies at an altitude of 2236 m and is surrounded by mountains of 3000–6000 m altitude. In 1900, it was drained artificially. The hydrographic system consisted then of a north–south series of lakes which joined into one large lake during the rainy season and separated again during the dry season, with the Aztec Lake Texcoco as the lowest and most saline. The development of Mexico City and the necessity of evacuating its growing amount of sewage led to the complete drainage of the lakes.

Bradbury (1971) studied the palaeolimnology of ancient Lake Texcoco in cores from lacustrine sediments beneath Mexico City, that is on the margin of the old lake. Alternations of planktonic and benthic epiphytic assemblages reflect the changes of water levels. The water level was especially high 100 000 years ago, whereas the dominance of the marsh flora during the last 10 000 years indicates a relatively dry climate or tectonic modification of the drainage system.

Lake Texcoco is intimately linked with Aztec civilization. In the fourteenth century, the Aztecs, migrants from the North, settled on a small island near the western shore of Lake Texcoco and, in 1325, founded the town of Tenochtitlan which became the capital of the Aztec Empire (West & Angelli, 1976).

HIGH MOUNTAIN LAKES OF CENTRAL AMERICA

High mountain lakes of Central America were described by Löffler (1972). The principal areas above 3000 m are the Sierra de los Cuchumantanes in north-western Guatemala (1500 km^2), part of the Cordillera Talamanca in Costa Rica (400 km^2) and a few isolated volcanoes such as Tajumulco (4210 m altitude), the highest peak of Central America in western Guatemala. In contrast with the high mountains of Mexico and Colombia, the highlands of Central America do not reach the snow line. Pleistocene glaciation has, however, occurred in the Cordillera Talamanca. The high volcanoes do not show any sign of glacial forms since they reached their present altitude after the last glacial period. Most of the high mountain lakes are associated with volcanism and are relatively recent. Notable exceptions are the Altos of Cuchumantanes, a mountain of Mesozoic limestones. The Cordillera Talamanca consists of volcanic and plutonic material and some 30 water bodies of glacial origin are known.

In March 1979, one of these was visited by Gocke (1981). The Laguna Grande de Chirripo is located at 9°21′ N and 3520 m altitude in a valley of glacial origin. The Laguna has a surface area of 5.43 ha, a maximal and mean depth of 22 and 8 m, respectively, and a volume of $0.45 \times 10^6 \, m^3$.

Although, during the survey, the air temperature varied daily from 4 to 14°C, the water temperature was always between 10 and 11°C. The lake was thoroughly mixed at night. The water with a conductivity of 14 μmhos cm^{-1} and alkalinity of 0.15 meq l^{-1}, has the composition of rain water.

Peridinium sp. dominates the phytoplankton but has a low productivity: 95 mg C m^{-2} d^{-1} net productivity and 155 mg C m^{-2} d^{-1} gross productivity. The zooplankton is represented by pigmented copepods and the benthos consists mainly of amphipods and chironomid larva. In the bordering areas of Mexico and Colombia, high-altitude lakes lie in young volcanic areas or in plutonic rock mountains. It follows that most of these lakes have very low salt content with the exception of the lakes located in calcareous areas. They are characterized by an algal assemblage dominated by *Peridinium* (*P. willei*, *P. volzii*) and including *Botryococcus braunii* and *Dinobryon*. There is a great contrast between the uniformity of the algal composition and the selective distribution of benthic and planktonic crustacea. The distribution of the harpacticoids is a good example: genera of holarctic origin (*Moraria*, *Bryocamptus*) do not exist south of the northern limit of the tertiary marine transgression. Similarly, the South-American subgenera *Attheyella*, *Delachauxiella* are not found north of this barrier. The planktonic copepod *Mesocyclops lepotus*, abundant in the Andean lakes, is absent from Central America. The North-American calanoid *Leptodiaptomus* is abundant in Guatemala but does not cross the transgression barrier. Cladocera seem to have been less affected by palaeogeographic barriers, although the North-American *Daphnia middendorffiana* and *D. peruviana* do not exist in Central America. Since Central America was reunited to the continents in the Pliocene, the slow dispersal rate of the crustaceans in high-altitude lakes is surprising. We shall see that fish distribution shows a nearly identical pattern.

FISHES OF CENTRAL AMERICA

Myers (1966) and Miller (1966) commented on the geographical distribution and origin of the ichthyological fauna of Middle America.

A total of 445 species of fish is known from Central America. This includes 104 species of primary freshwater fishes, 165 secondary freshwater species and 187 peripheral forms.

All the primary fishes are ostariophysans (characoids, gymnotoids, cyprinoids and catfishes). They belong to 18 families, the most important of which are the Characidae (42 species) and two catfish families, the Pimelodidae (23 species) and the Loricariidae (14 species). The distribution of these animals is remarkable: two-thirds of the families and 75% of the species are concentrated in Panama and Costa Rica.

What best characterizes the area between the Costa Rican–Nicaraguan border and the Isthmus of Tehuantepec is the poverty of the primary freshwater fish fauna (5 families and 27 species only). In this area, the fish population is heavily dominated by secondary and peripheral forms. The scarcity of cypriniforms in this area is puzzling: in North America, where the Cyprinidae appeared only in the early Miocene, they now constitute a prominent element of the fauna; similarly they became very prosperous in Africa where they penetrated in the late Tertiary. Myers thinks that the rarity of cypriniforms in middle Central America is due to the fact that they are still very new to the area where they were preceded by the secondary Poecilidae and Cichlidae.

Fig. 18 shows the distribution in Central America of primary and

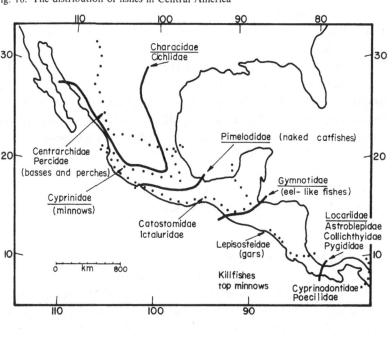

Fig. 18. The distribution of fishes in Central America

secondary fishes of northern and southern origin. Among the North-American families, the Poecilidae, Cyprinodontidae and to a lesser extent the Lepisosteidae show the most successful colonization. Among the South-American families, the Cichlidae and Characidae have spread all over Central America and Southern Mexico. The large number of Cichlidae species (~ 100 species) and the presence of endemic species all related to the common South-American genus *Cichlasoma* suggest that the Cichlidae colonized the area before the Characidae. The large extension of the Poecilidae and their evolution into predators of the small species and fingerlings indicate that they were established in Central America prior to the Cichlidae and developed without much competition.

Myers concluded that the Ostariophysan fishes of the primary freshwater type are new to Central America and in the process of colonizing the area. They probably did not exist in Central America before the Pliocene whereas secondary fishes such as the poecilids invaded Central America in the early Tertiary, maybe the Oligocene, and cichlids colonized the area in the Miocene.

These conclusions are in agreement with both geological facts and fish physiology. The marine transgressions which covered the Tehuantepec area during various periods of the Tertiary were certainly not an absolute obstacle for secondary Poecilidae as they were for Cyprinidae. Similarly, the southern Panamanian–Colombian gap, which persisted from the Mesozoic to the Pliocene, was a serious barrier to South-American Ostariophysan fishes which could spread northward only after the Pliocene bridge was established.

WEST INDIES

The West Indies or Antilles form an archipelago spreading from Florida and Yucatan to Venezuela. The archipelago is composed of numerous islands located between $10°$ and $27° N$ and $59°$ and $85° W$: the Bahamas, the Greater Antilles (Cuba, Jamaica, Santo Domingo, Puerto Rico) and the Lesser Antilles. The high temperatures and daily effects of trade winds cause high evaporation and a drastic decrease in water levels during the dry season.

The West Indies have relatively few natural lakes. Cuba, for example, has only several lagoons and artificial reservoirs. The largest lagoon is the Laguna Arivuanabo, a shallow eutrophic water body of 72 km^2. The largest lake of the West Indies is located in Santo Domingo. Lake Enriquillo lies at an altitude of 48 m in the central-southern part of the

Dominican Republic, has a surface area of 281 km², maximum and mean depths of 30 and 6 m, respectively, and a volume of 1.3×10^9 m³. It is a salt lake with a total salinity of 72‰.

Harrison & Rankin (1976), studying the freshwater invertebrate fauna of the eastern Lesser Antilles, showed that much of the fauna seems to be derived from an older stock occupying South America, Africa and Madagascar: for example, the planorbid genus *Bronyhalaria*, several ephemeropteran families, and the caddis genus *Lytonema* have such a distribution. These observations are consistent with a theory of drifting continents including an ancient separation of Africa and Madagascar from South America and the Caribbean area and a more recent drifting of the Antillean plate eastward of the mainland.

6
Africa

INTRODUCTION

The Warm Belt of Africa is mainly composed of Precambrian formations. Most of the continent has been emergent since late Precambrian times and has known no orogenesis since then. The basic hydrographic network, mainly of east–west direction, is very ancient; Balek (1977) reports the opinion that it was elaborated during the Gondwanaland phase of Africa. In any case, this hydrographic network is much older than most of the Great Lakes, Lake Chad excepted.

From Miocene onwards, uplifting and downfaulting have produced a morphology of basins and ridges; the old drainage system found new ways to the seas through leaks in the ridges and closed basins were formed (e.g. Lake Chad). From the Zambezi to Northern Israel (6000 km), a crustal extension along one of the major fault lines of the earth led to the drifting apart of the continental masses on each side of the fault and the resulting formation of the Syro-African Rift Valley. In eastern Africa, the Rift divides into two branches, the Western and the Eastern Rift Valleys. These are trenches lying some 1000 m below the top of steep escarpments and bearing signs of intense volcanic activity. The Rift Valleys were ideal traps for freshwater and most of the East-African lakes are located in these depressions with the exception of Lake Victoria situated in the uplifted area between the two Rift Valleys.

Under the present climatic conditions prevailing in 'tropical' Africa, surprisingly few areas have abundant enough precipitation to maintain evergreen forest. This is the case of the Congo basin and the Gulf of Guinea but, for example, eastern Africa is rather dry.

Moreover, although the Pleistocene extension of glaciers was limited to high mountains, the Pleistocene–Holocene period was far from having uniform and homogeneous climatic conditions; recent research has demonstrated that lake levels underwent considerable fluctuations during this period. The knowledge of these events is fundamental to our understanding of the preservation of the terrestrial flora and the evolution of the lacustrine fauna.

As Livingstone (1975) points out, we have known for a long time, from the position of moraines on the high mountains of equatorial Africa, that, during certain periods of the Quaternary, the climatic belts were much lower than at present. During the last 20 years, analysis of pollen and chemistry of lacustrine sediments, accompanied by radiocarbon dating, has significantly enriched our knowledge concerning climatic variations of the late Pleistocene and Holocene. From previous studies and their own pollen analysis in eastern Africa, Livingstone & Kendall (1969) and Livingstone (1975) concluded that a major climatic change took place 12 000–10 000 years ago. In tropical Africa, humidity increased and in equatorial high mountains, the mean annual temperature rose by $\sim 6°C$. These warm, humid conditions lasted ~ 3000 years. From 7000 to 5000 years before present (B.P.), slightly drier conditions prevailed and a brief increase of moisture around 5000 years B.P. was followed by a drier than present climate until 3000 years B.P. Moreover, in all the cores which reached late Pleistocene sediments, the pollen analysis indicated arid conditions.

Williams, Bishop, Dakin & Gillespie (1977), studying Lake Besaka in the central Ethiopian Rift and lakes of the Afar, concluded that in central Afar levels 'were high from around 10 000–11 000 years B.P. to around 8500 years B.P. and again from 7500 to 4000 years B.P.'. Gasse & Street (1978), also studying the Ethiopian lakes, especially Lakes Shalo, Abhe and Asal, traced a very arid phase between 17 000 and 10 000 years B.P., followed by a humid phase which culminated 9400 years B.P. The levels decreased between 6000 and 4000 years B.P. but a new high-level period took place between 2700 and 1000 years B.P. Similar results had been published by Grove & Goudie (1971) for the Galla lakes, located in the southern part of the Ethiopian Rift. In particular, these authors presented evidence indicating that around 9200 years B.P. the four Galla lakes were united in one large water body. Hecky & Degens (1973) and Hecky (1978) found that in the late Pleistocene, the Lake Kivu water level was 300 m below its present position. Then, the lake had no outflow. At ~ 9500 years B.P., the water level was 100 m above the present level and the lake had an active outlet. Around 4000 years B.P., the level dropped again and the lake again became a closed basin but, since 1200 years B.P., the level has risen and the Ruzizi River drains the lake into Lake Tanganyika.

All these results are in good agreement with previous results reviewed

Introduction

and compared by Butzer, Isaac, Richardson & Washbourn-Kaman (1972). We have redrawn their Fig. 4 and added some of the newly published data on other lakes (Fig. 19). It is remarkable that distant lakes show high levels in the early Holocene and arid conditions in the late Pleistocene.

The accumulating evidence has invalidated the simple correlation between glacial episodes in high latitudes and fluvial periods (with high lake levels) in the tropics. It is at least not true for the late Pleistocene period. Theories explaining the distribution of animals and based on such a correlation have to be revised. For instance, mountain species of birds and butterflies having clear affinities and found both in western Africa (Mount Cameroon) and in high mountains of East Africa, were assumed by Moreau (1966) to have crossed the continent during the last Ice Age when an hypothetical general lowering of climatic belts would have made it possible. The arid conditions which then prevailed deny this possibility and no satisfactory explanation is available.

This lengthy arid period also raises the question of sanctuaries which allowed the moist, forest species to survive such unfavourable conditions. Finally, the fluctuations of lake levels had obviously a considerable impact on evolution and preservation of freshwater fauna in various lakes.

The factors which originated such modifications are still uncertain. Their spatial extension and their temporal simultaneity strongly suggest that these were of climatic nature. The debate is then limited to whether the modification of the vegetation shown by pollen analysis (increase in forest species) and the increase of lake water levels which occurred in the early Holocene, were caused by a lowering of evaporation rates due to a drop of temperature or by an increase in precipitation. From their study of the

Fig. 19. Study of sediment cores. Black areas indicate high water levels, striped areas intermediate levels and dotted areas low water levels. Redrawn from Beadle (1974)

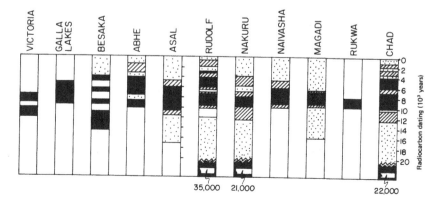

calcareous deposits of the early Holocene in Ethiopian lakes, Gasse & Street (1978) concluded that there was an increased input of $CaCO_3$ into the lakes resulting from leaching of soils in the watersheds which would imply increased rainfall.

With this general background and unsolved problems in mind we shall study the lakes of the main hydrological basins of Africa.

THE NILE BASIN

The Nile system seems to be composed of portions of rivers of different origin and age (Fig. 20). Berry & Whiteman (1968) think that the valley of the Blue Nile and its continuation to the sea date back to the Eocene. Recent reviews of the history of the Nile are found in Beadle (1974) and Rzoska (1976, 1978). At present, the Nile, flowing through 35° of latitude, is the only tropical river reaching the Mediterranean Sea.

The hydrographic map of the Nile basin shows in a dramatic way that the

Fig. 20. The Nile system: A = Nile tributaries and watershed; B = quantitative network. Within figure, V = Lake Victoria; G = Lake George; E = Lake Edward; A = Lake Albert; K = Lake Kyoga

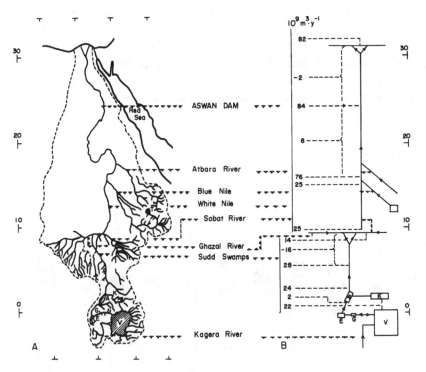

The Nile Basin

water which reaches Egypt originates very far from the Lower Valley. From the Sudd Swamps down to the sea (~ 4000 km), the Nile does not receive any significant tributary from its western watershed and, on the eastern bank, a similar situation is observed from Berber to the sea (2800 km). In other words, the equatorial plateau of Lake Victoria and the mountains of Ethiopia are the only water sources of the Nile and 44% of the Nile watershed (3×10^6 km^2) does not contribute water to the river (Rzoska, 1978).

Although the Nile, with its 6695 km (length of the main river from the head-waters of the Kagera River) is the longest river of the world, its yearly discharge is only 84×10^9 m^3, less than 2% of the Amazon's yearly discharge. The yields per area of the Nile and the Amazon, 27 and 1215 m^3 km^{-2} y^{-1}, respectively, highlight the aridity of the Nile basin.

The present hydrographic system of the Nile is represented in Fig. 20 with the mean annual discharge of the river at different stations according to data of Balek (1977). Since the Sobat River drains essentially the Ethiopian plateau, this latter mountainous area contributes 84% of the total discharge of the Nile in comparison with 16% only for the African lake plateau. The relatively small contribution of the White Nile is essentially due to a huge 16×10^9 m^3 y^{-1} loss through evaporation and evapotranspiration in the Sudd Swamps.

The seasonal regime of the White Nile, rather uniform because of the buffering effect of Lake Victoria and Lake Albert, contrasts with the floody characteristics of the Blue Nile. This latter river, originating from Lake Tana and draining the highlands of Ethiopia, has a flood peak from July to October reaching 5950 m^3 s^{-1} (average 1912–62, Balek, 1977) and low discharges in the first half of the year with minima in April (138 m^3 s^{-2}, same period, same source).

The Upper Nile Valley

The Upper Nile Valley corresponds mainly to the sources of the river, i.e. the East African lakes plateau. Lake Victoria (1134 m altitude) occupies a large part of the plateau and its watershed is bordered on its eastern and western side by highlands 1000 m higher than the lake basin; these highlands form the escarpments of the Rift Valleys on both sides of the Victoria basin.

The hydrography of the sources of the Nile presents peculiar features reflecting the recent tectonic activity which took place in the area. The morphology of the Kafu–Nkusi, Katonga–Mpanga and Kagera Rivers and their tributaries suggests a westward flow to Lake Edward and Lake

Albert. Such a westward drainage existed in pre-Pleistocene times.

Pleistocene uplift of the western part of the basin reversed the direction of the rivers which then flowed eastward. The formation of the Western Rift depression determined a new, low base level which caused a renewed but so far modest westward flow from the ill-defined swampy waterdivide.

The Kagera River, the main inlet of Lake Victoria, is considered to be the true source of the Nile. The main outlet, the Victoria Nile, flows into Lake Kyoga which drains into Lake Albert. Since Lake Albert lies at 620 m altitude in the western Rift, the waters of Lake Kyoga reach Lake Albert through a series of falls (Atura Falls, Murchison Falls). Lake Albert also

Fig. 21. The lakes of the Western, Eastern and Ethiopian Rift Valleys

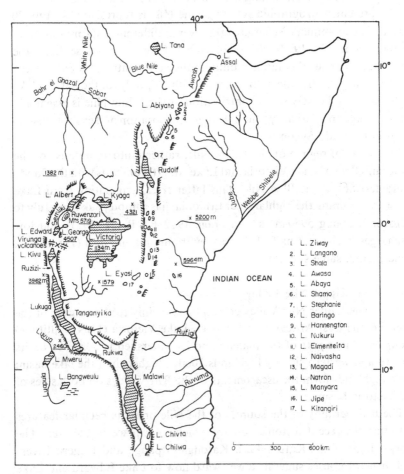

drains the southern part of the western Rift through the Semliki River, the main outlet of Lake Edward and Lake George.

Lake Victoria
General geography

Lake Victoria (Figs. 20 and 21) is located in a shallow depression between the two uplifted ridges which border the Western and Eastern Rift Valleys. The drainage area of the lake itself extends over 184 000 km^2. The basin is mainly composed of Precambrian formations including metamorphosed sediments, intrusive and igneous rocks, some of them at least two billion years old (Haughton, 1963). Three outcrops of Palaeozoic Karroo formations with *Glossopteris* fossils are known in southern Uganda. Miocene volcanic material covers vast areas of the eastern watershed which is bordered in its northern part by the Elgon Volcano. The western watershed has been less affected by vulcanism although the Virunga volcanic complex considerably modified the local hydrology.

The Victoria basin is crossed by the Equator at the level of Entebbe in the northern part of the lake. The rainfall pattern is bimodal with peaks in March–May and October–December. A permanent low-pressure centre over Lake Victoria allows both active evaporation and abundant direct precipitation. The easterly trades are the dominant winds with net southerly and south-easterly components. In absolute values, the eastern shore of the lake receives less precipitation (760 mm at Musoma) than the western shore (2000 mm at Bukoba). Air temperatures vary from 16 to 27°C. Similar to conditions which prevailed around 1875, lower temperature, higher rainfall and subsequent lower evaporation have been observed since 1960, causing a 2 m increase in the Lake Victoria level. The vegetation is of savanna type; the dominant tree species are *Acacia, Albizzia, Butyrospermum* while *Cymbopogon, Hyparrhenia* and *Londetia* are among the common grass species.

Swamps represent a frequent landscape in the Lake Victoria basin. They are especially developed in the northern Kagera Valley. Thompson (1976) described the vegetation of the northern and southern shores, respectively, as 'savanna-like, derived from forest' and 'savanna woodland', the former being more modified by agriculture. In certain cases, the swamps border the woodland; in others, 'a swamp forest' represents an intermediate zone between both types of vegetation. The swamps are dominated by *Cyperus papyrus* with *Miscanthidium violaceum, Sphagnum* spp. and *Nymphaea caerulea*. Near the shores, the papyrus swamp is replaced by a *Phoenix reclinata*-dominated swamp forest. The valleys of the western watershed are generally swampy and papyrus-choked. Wave action or variations of level may cause portions of the papyrus swamp to separate from the main swamp and create 'floating islands' which considerably hamper navigation. Papyrus dominates many swamps in the watershed of the Kagera River. Above 2000 m it is replaced by *Typha australis, Phragmites australis, Juncus* spp. and *Cyperus denudatus*.

Origin and history of Lake Victoria

As previously mentioned the general opinion is that, from the Miocene until at least mid-Pleistocene, the Victoria basin was drained by a westward-flowing hydrographic system. The Pleistocene uplift of the western region changed the direction of the rivers, causing the drowning of the lower valleys of the new system. Lake Kyoga originated from the drowning of the Kafu River and Lake Victoria is probably the result of the junction of the lower drowned portions of the Kagera and Katonga Rivers. If this hypothesis is correct, the lake is then not older than mid-Pleistocene.

Three horizontal, post-tectonic beaches have been identified along the northern and southern shores at 18.2, 12–14 and 3.4 m above the present lake level. It is clear that a water level 20 m higher than the present could not be held with the outlet of the lake in its present position. Kendall (1969) assumes that these high levels correspond to a different situation of the outlet and that the drops in water level correspond to various stages in the erosion of the Nile gorge.

From two ^{14}C-dated sediment cores from Pilkington Bay, on the northern coast of the lake and one in open water off Entebbe, Kendall (1969) gave a detailed account of the last 15 000 years. From 14 500 to 12 000 B.P., the Victoria water was more concentrated than now and carbonate minerals precipitated in shallow waters: it is probable that, during this period, the lake had no outlet. An outlet functioned from 12 000 to 10 000 B.P. and, after a short interruption around 10 000 B.P., continued until the present to drain the lake into the Nile. Pollen analysis indicates that forest vegetation developed on the northern shore of the lake after 12 000 B.P., declined around 10 000 B.P. and reached its maximal development around 9000 B.P. Between 7000 and 6000 B.P. evergreen was replaced by semi-deciduous forest which remained abundant until 3000 B.P.

The agreement between chemical events in the lake and the composition of the vegetation in the lake watershed clearly indicates that they were caused by large-scale climatic changes. Kendall then concluded that 'the northern Victoria basin was dry from about 14 500 to 12 000 B.P., moderately wet from 12 000 to 10 500 B.P., moderately dry from 10 500 to

The Nile Basin

9500, wet from 9500 to 6500 and slightly drier or with a more seasonal rainfall after 6500 B.P.'. Kendall considers that the three raised beaches have been formed during the last 12 000 years; however, only the age of the lower one could be estimated (3720 years B.P.).

It is clear that, whereas Lake Victoria originated from mid-Pleistocene tectonic movements, climatic fluctuations also played an important role in its development.

Morphometric and hydrologic features of the lake

The main morphometric features of Lake Victoria are presented in Table 23.

On the western and to a lesser extent northern and southern sides of the lake, the bottom slope is very gentle and the wide belt of shallow waters includes numerous islands with a total area of 1500 km². The eastern coast is significantly steeper. The coastline is very irregular, especially in the northern and southern areas where bays and gulfs are numerous.

The water budget was first established by Hurst & Phillips (1931) and Hurst (1952). It indicates that rainfall is approximately equal to evaporation and the inflow from streams (about 10% of the rainfall) balances the Victoria Nile outflow. A more recent budget by Krishnamurthy & Ibrahim (1973) gives the following values in 10^9 m³ y⁻¹:

Inflow from streams	18 ± 5%	Victoria Nile	23.4 ± 5%
Rain over lake	100 ± 10%	Evaporation	100 ± 10%

From 1954, the lake has been dammed and the Victoria Nile now leaves Lake Victoria through the Owen Falls.

Table 23 *Morphometric parameters of Lake Victoria*

Altitude	+1134 m		
Length	300 km NS	Max. depth	79 m
Width	280 km EW	Mean depth	40 m
Shoreline	3440 km	Surface area	68 000 km²
Development of shoreline	3.7	Volume	2700 × 10⁹ m³
		Watershed area (lake area not included)	184 000 km²
		Watershed area : lake volume ratio	0.068 m² m⁻³

Physical and chemical limnology

Talling (1966), reflecting Newell's opinion (1960) that open water was stratified throughout the year into three distinct layers, concluded that the offshore area is probably monomictic with a full circulation from May to August. It seems that the evaporation, enhanced by the south-east trade winds, is the main cause of cooling and mixing. The stratification period includes three main phases: 1) from September to December there is a superficial warming but isotherms are loose; 2) in December–May the thermocline deepens; and 3) from June to August the thermal gradient disappears.

A weak thermal gradient and small differences in temperature between surface and bottom waters (24.8–23.8°C on a 60 m-deep profile) characterize Lake Victoria in comparison with lakes of higher latitudes. These small vertical variations are sufficient, however, to prevent mixing of deep and superficial water, clearly shown by the concomitant 'chemical stratification'. During the first phase of stratification (September–December, 1960), Talling observed a loose gradient of oxygen with 7 ppm at the surface and 2 ppm at 60 m. From January to June, the concentration of oxygen, which was still 6 ppm at 40 m, decreased with depth and was < 1 ppm at 60 m. In July–August, the concentrations were more homogeneous with 7 ppm at the surface and 5 ppm at 60 m. Nitrate, phosphate and silicate distribution is also governed by stratification, with vertical homogeneity during the isothermal period and accumulation in lower layers the rest of the year.

During a vivid controversy on the existence of internal seiches, Fish (1957), who described one occurrence of seiche, opposed Newell (1960) who claimed that no clear upwellings had been observed. Kitaka (1971) described a cyclonic upwelling in the southern offshore waters of the lake. In contrast with an internal seiche upwelling which is due to the wind-induced accumulation of surface water on one side of the lake and the rise of deep cold water on the opposite side, a cyclonic upwelling occurred in a limited area of low atmospheric pressure. On a first north–south cruise (8–9 February 1969), Kitaka observed regular flat isotherms with 23°C on the bottom, 24.8–25°C at the surface and maximal thermal gradient around 50 m. On a second north–south cruise (20–23 March 1969), a violent northerly storm developed. About 60 km from the southern shore Kitaka observed that the isotherms formed a local conic structure as a result of 'a cyclonic upwelling'.

The chemistry of Lake Victoria waters was studied by Graham (1929),

Carter (1955), Lind & Visser (1963), Livingstone (1963), Talling & Talling (1965), Visser (1974) and Visser & Villeneuve (1975). Table 24 shows the composition of the water of Lake Victoria and Lake Nabugabo, a small lake on the north-western side of Lake Victoria, separated from the main lake some 4000 years B.P.

The chemical composition of Lake Victoria is characterized by its low salt content and the dominance of sodium among the cations. Moreover, the calcium level is much lower than the average value for African waters (6.5 ppm). Nitrogen and phosphorus are also notably low, probably because of the retention of these elements by the swamps choking most of the rivers (Visser & Villeneuve, 1975).

Biological features

Phytoplankton. The description of the phytoplankton of Lake Victoria is based on the survey carried out by Talling (1966) from October 1960 to August 1961. Inshore waters had an abundant and spatially variable algal population; in contrast offshore phytoplankton had a more homogeneous distribution. During the year, the general following succession was observed: *Melosira nyassensis* var. *victoriae* and *Stephanodiscus astraea* dominated during the isothermal period (August–September); *Anabaena flos-aquae* peaked in December whereas *Nitzschia acicularis* started developing in January.

In general, the Victoria phytoplankton is characterized by a great variety of species, a nearly year-round dominance of small Cyanophyta (*Aphanocapsa* spp.) and relatively small seasonal variations in biomass. Concentrations of chlorophyll in the epilimnion varied only between 1.2 and 5.5 mg m^{-3}. In other words, the peaks of diatoms and blue-green algae previously mentioned are superimposed on a constantly abundant population of small *Chroococcales* and green algae. Talling's work indicates that the August maximum of *Melosira* comes both from a resuspension of bottom cells caused by physical mixing and by a horizontal transport of *Melosira* cells from inshore areas. Conversely, *Anabaena flos-aquae* and *Anabaenopsis tanganikae* decline whenever there is a partial or complete mixing, probably because a large portion of the biomass is driven below the photosynthetic zone.

In such a climate, where solar radiation and seasonal variations of water temperature are small, vertical mixing plays a dominant and complex role: with increasing turbulence, cells are maintained in suspension, nutrients are recirculated and illumination per cell is reduced. A very similar

Table 24 Chemical composition of Lake Victoria and Lake Nabugado (Results from Visser & Villeneuve, 1975)

	Conductivity (µmhos cm⁻¹)	pH	Na⁺ (mg l⁻¹)	K⁺ (mg l⁻¹)	Ca²⁺ (mg l⁻¹)	Mg²⁺ (mg l⁻¹)	Mn²⁺ (mg l⁻¹)	Fe³⁺ (mg l⁻¹)	Al³⁺ (mg l⁻¹)	Cl⁻ (mg l⁻¹)	SO₄²⁻ (mg l⁻¹)	NH₄⁻ (mg l⁻¹)	NO₂⁻ (mg l⁻¹)	NO₃⁻ (mg l⁻¹)	PO₄³⁻ (mg l⁻¹)	SiO₂ (mg l⁻¹)
Lake Victoria	96.8	7.8	9.0	4.1	3.9	2.7	0	2.3	0.1	3.0	2.5	0.1	0	0.1	0.14	4.7
Lake Nabugado	25.0	4.7	2.0	2.6	0.5	0.1	0	0.1	0	0.8	1.7	0.1	–	0.3	–	–

The Nile Basin

situation has been described for Lake Kinneret (Serruya *et al.*, 1980).

A not too different picture was obtained by Akiyama, Kajunnolo & Olsen (1977) in their 1973-74 survey of the Mwanza Gulf in the southern part of Lake Victoria. *Melosira* and *Anabaena* were the dominant species as in offshore water; however, in contrast to Talling's offshore observations, both genera peaked during the same period (November-March), whereas *Nitzschia* reached a maximum during the June-August period. The algae were much more abundant in the Mwanza Gulf than in offshore water: for *Anabaena* the respective peak numbers were 6×10^8 cells l^{-1} and 1×10^6 cells l^{-1}.

In their study of Mwanza Gulf's zooplankton, Akiyama *et al.* (1977) found Copepoda (calanoids and cyclopids) to be the dominant zooplankton. Their average number ranged from 100 to 400 individuals l^{-1}. Cladocera were present all the time with 10 individuals l^{-1} in June-July and up to 30 in November-December. *Bosmina longirostris*, *Moina dubia*, *Ceriodaphnia cornuta*, *Alona* sp., *Diaphanosoma* sp., *Chydorus sphaericus* and *Daphnia* sp. dominated the population. Rotifera, with concentrations of 30 individuals per litre, was mainly represented by *Brachionus angularis* var. *bidens*, *B. falcatus*, *B. forficula*, *B. diversicornis*, *B. quadridentatus*, *Keratella*, *Tetramastix*, *Filinia*, *Conochilides* and *Asplanchna* spp.

The number of individuals is much lower in open water since Green (1971) found no more than 1.4 Cladocera individuals per litre (mean value) at a deeper station (27 m depth). The general picture described by Akiyama *et al.* is in good agreement (dominance of copepods, main species found) with a nine-station survey carried out by Rzoska (1957) in April 1956.

Benthic fauna. Mothersill, Freitag & Barnes (1980) investigated the benthic fauna of north-western Lake Victoria. Sixteen macroinvertebrate taxa were found; the gastropods *Melania tuberculata* and *Bellamya* sp., the insects *Chaoborus* sp. and chironomids, the pelecypod *Corbiculina* sp. and an unidentified oligochaete were the dominant taxa. Frequent associations of *M. tuberculata* and *Bellamya* sp. were observed.

Fish. The riverine origin of Lake Victoria implies the adaptation of riverfish to lake conditions. Presently, the lake is isolated by a series of falls: the Owen Falls which, since 1954 has controlled the Victoria outflow, converted 'the partial barrier (for fish) of the Ripon Falls, now under water, into a total barrier' (Worthington, 1972). In any case, the Murchison Falls form an impassable barrier to fish ascending from Lake Albert. It follows that the Victoria fish fauna lacks certain nilotic species such as *Polypterus, Hydrocinus*, citharinids and *Lates*. However, fossil specimens of this genus, found in Miocene beds east of Lake Victoria, testify to its existence in the Victoria basin in pre-Rift times. It is possible that a former nilotic fauna was destroyed in the Pleistocene and replaced by the present one which is heavily dominated by cichlids (>170 species).

In the present state of knowledge, Lowe-McConnell (1975) reported a total number of 208 fish, but this is an underestimated value since many cichlids are still unknown. Certain species move into rivers to spawn such as *Labeo victorianus* and *Schilbe mystus*; others, like most of the cichlids, spawn in the lake.

The overwhelming success of the cichlid genus *Haplochromis* in Lake Victoria is the main ichthyological feature of this lake. From the genus *Haplochromis* stemmed some 150 endemic species which fill all possible niches in the lake (Greenwood, 1974). The adaptive radiations concern main feeding specializations leading to modifications of the mouth and digestive tract.

About 30% of the *Haplochromis* feed on insects and detritus; others feed on molluscs, but even on this single food source two distinct specializations are observed: extraction of the mollusc from the shell and shell crushing. Four species are algal grazers, one feeds on macrophyte leaves and 30% of the species are piscivorous. These latter species are not bound to a special habitat (Lowe-McConnell, 1975). In addition to the *Haplochromis* population, there are four endemic monotypic genera (*Macropleurodus bicolor, Hoplotilapia retrodens, Platytaeniodus degeni* and *Paralabidochronis victoria*), derivatives of *Haplochromis* species and two endemic *Tilapia* species (*T. esculenta* and *T. variabilis*) which feed on phytoplankton.

The mechanism of speciation is still a subject of debate: isolation of fishes in different water masses (allopatric speciation) or isolation within a lake (sympatric speciation). Kendall (1969) describes the example of Lake Nabugado, isolated from Lake Victoria less than 4000 B.P. by a sand bar, where five of six species of *Haplochromis* are endemic to the small lake. This example indicates that the divergence process may be extremely rapid as far as cichlids are concerned. Worthington (1954) placed emphasis on the lack of competition that *Haplochromis* enjoyed in Lake Victoria. He claimed that the absence of predator fish in Lake Victoria was a factor favourable to speciation. However, in 1957, Lake Kyoga was stocked with the Nile perch, *Lates niloticus*, which rapidly colonized the lake; from 1961 to 1968, the total catch of fish other than *Lates* increased (Worthington, 1972). *L. niloticus* was introduced into Lake Victoria about 1960 but did not acclimatize as quickly as in Lake Kyoga. In contrast, Okemwa (1981)

The Nile Basin

reports that, in Nyanza Gulf, *Haplochromis* dominated the catch until 1979 when the *L. niloticus* catch surpassed *Haplochromis* landings. Okemwa relates the decline of *Haplochromis* to predation by *Bagrus docmae, L. niloticus, Clarias mossambicus* and man. In 1951–53, four non-indigenous *Tilapia* species were introduced in Victoria: *T. zillii, T. nilotica, T. leucostecta* and *T. rendalli. T. zillii* competed with the endemic *T. variabilis* for nursery grounds and caused a rapid decrease in this latter species catches.

There are 38 non-cichlid species of fish in Lake Victoria, 16 of which are endemic. A study by Corbet (1961) indicates that insects contribute a great deal to the diet of these fish, especially for species of *Alestes, Barbus* and *Schilbe*. Molluscs represent the main food of *Protopterus*.

Lake Kyoga

Lake Kyoga is a dendritic lake receiving the Victoria Nile and the Kafu River and draining to Lake Albert. It has the same origin as Lake Victoria. It is located in Uganda at $0°36'–2°0'$ N and $32°20'–34°30'$ E (see Fig. 21). Its main morphometric features are presented in Table 25.

The chemical composition of Lake Kyoga water is shown in Table 26. The large differences in concentrations observed between these two sources are probably related to the pronounced horizontal gradients observed by Evans (1962) and depending on how much the measured station is affected by the Nile inflow.

Although the Victoria Nile brings into Lake Kyoga an abundant inoculum of *Melosira granulata, Lyngbya* is the dominant alga. Lake Kyoga has a less diverse phytoplankton than Lake Victoria. It includes *Melosira nyassensis* var. *victoriae*, the green algae *Coelastrum reticulatum, C. cambricum, C. microporum, Sorastrum spinulosum, Tetraedron trigonum, Staurastrum leptocladum* and the blue-greens *Lyngbya circumcreta* and *Aphanocapsa elachista*.

The Cladocera, studied by Green (1971) in 1962, are represented by *Ceriodaphnia cornuta, Moina micrura* (which is the quantitatively domin-

Table 25 *Main morphometric features of Lake Kyoga*

Altitude	1100 m		
Lake area	2700 km^2	Max. depth	8 m
With swamps	6300 km^2	Mean depth	6 m
Watershed area	75 000 km^2		

Table 26 Chemical composition of Lake Kyoga

	(a)	(b)		(a)	(b)		(a)	(a)
Conductivity (μmhos cm^{-1})	320		Ca^{2+} (mg l^{-1})	11.4	21.7	NH_4^- (mg l^{-1})	0.2	SiO_2 (mg l^{-1}) 15
pH	7.7	7.6–9.0	Mg^{2+} (mg l^{-1})	2.0	13.8	NO_2^- (mg l^{-1})	0	
Na$^+$ (mg l^{-1})	6.9	10.8	Cl$^-$ (mg l^{-1})	1.6	12.0	NO_3^- (mg l^{-1})	0.2	
K$^+$ (mg l^{-1})	7.2	9.6	SO_4^{2-} (mg l^{-1})	3.9	31.0	PO_4^{3-} (mg l^{-1})	0.52	

(a) from Visser & Villeneuve (1975)
(b) from EAFRO (1954)

ant species), *Diaphanosoma* spp. and *Chydorus barroisi*. The genus *Daphnia*, found abundantly in Lake Victoria, is completely lacking. At the seven stations sampled by Green, the total number of Cladocera ranged from 962 to 8567 individuals m^{-2}.

The total annual fish catch was estimated to be 17 700 tons by Fryer & Iles (1972), 23 000 tons by Worthington (1972) and 48 900 tons by Welcomme (1972).

Lake Edward
General

Lake Edward is located between 29°15′ and 29°55′ E and its southern shore is at 0°45′ S (see Fig. 21). The lake lies in the Western Rift Valley at 920 m altitude. The western shore of the lake is in contact with the western escarpment of the Rift Valley. Alluvial plains from lacustrine origin occupy the other sides. The oriental plain between the lake and the eastern escarpment of the graben is drained by the Ishasha River. The southern watershed is limited by the Virunga volcanoes, the northern part of which is drained by the Rwindi and Rutshuru Rivers. In the north, the Ruwenzori, over 5000 m altitude, extend on the eastern side of the Semliki River which drains Lake Edward into Lake Albert.

Verbeke (1957) describes the temperature as particularly stable with an annual average of 22.5°C. The average of monthly maxima varies from 26.3°C in January to 30°C in September and the average monthly minima range from 15.5°C in July to 17.8°C in September. Two dry periods, December–January and June–July, are observed. Rainfall is variable from year to year. The maximal values (>1500 mm) were recorded on the western side of the lake.

The wind regime is very regular: south-westerly, land wind during the night and strong, north-easterly lake breeze during the day blowing in the lake axis.

The tropical Ituri forest of Central Africa crosses the western graben and the Semliki River and reaches the slopes of the Ruwenzori.

The typical vegetation of the alluvial plains is savanna with *Temeda trianda, Acacia, Phoenix reclinata* and *Euphorbia dawei*. The lower slopes of the steep western shores of the lake have a similar vegetation with higher frequency of arborescent plants.

Origin of the lake

The fine-grained early Pleistocene sediments are rich in fossil fish, reptiles and molluscs. We note that the fossil molluscs, very similar to those found in the Kaiso beds of the Lake Albert basin, are now extinct. Moreover, these formations include fossils of *Crocodilus niloticus* and of the fish *Lates* and *Hydrocinus* found nowadays in Lake Albert and the Nile but absent from Lake Edward and George. These sediments do not lie horizontally but have been tilted by tectonic movements. It is clear then that in the communicating Lakes Edward and Albert 'the Kaiso mollusc fauna' had developed and was destroyed by tectonic disturbances and vulcanism.

The mesolithic sediments of Ishango, at the mouth of the Semliki River have yielded remains of *Lates niloticus* and *Barbus bynni*. These sediments being covered with volcanic ash, it is likely that there is a connection between mesolithic volcanic activity and the extinction of the fishes.

It follows that the present Lake Edward is relatively recent and its fauna results from a late recolonization.

Morphometric features

The main morphometric characteristics of Lake Edward are presented in Table 27. The lake bottom is strongly asymmetric: it slopes down gently on its western side and the deepest point (117 m) is located at the foot of the 2500 m-high escarpment.

Physical and chemical limnology

The monthly average temperature of the surface water varies from 25°C in August when the whole lake is homothermal and mixing occurs (Verbeke, 1957) to 27.2°C in May–June and in December. At 50 m, the temperature rarely departs from 25°C. A second mixing, probably partial, occurs in February.

The transparency ranges from 1.9 to 3.0 m. The chemical composition of the lake is represented in Table 28. The total salinity is 720 mg l^{-1}. The high buffer capacity of the lake water results in very stable pH. In the upper

Table 27 *Main morphometric parameters of Lake Edward*

Altitude	914 m		
Length	80 km	Max. depth	117 m
Width	40 km	Mean depth	34 m
		Surface area	2325 km²
		Volume	78×10^9 m³

Table 28 *Chemical composition of Lake Edward*

	pH	Na$^+$ (mg l^{-1})	K$^+$ (mg l^{-1})	Ca^{2+} (mg l^{-1})	Mg^{2+} (mg l^{-1})	Cl$^-$ (mg l^{-1})	SO$_4^{2-}$ (mg l^{-1})	HCO$_3^-$	CO$_3^{2-}$	SiO$_2$ (mg l^{-1})
Verbeke (1957)	8.9–9.2	112	79	9.7	44.5	27	30–35	580	57 mg l^{-1}	2–4
Talling & Talling (1965)	8.5–9.3	110	90	12.4	47.8	36	31	9.85 meq l^{-1}		6.5

layers, the oxygen content is 6–9 mg l^{-1}. The lake water is devoid of oxygen from 40 to 115 m with the exception of the mixing period when oxygen reaches the bottom but is rapidly consumed by the highly organic sediments.

Biological features

The phytoplankton of Lake Edward was investigated by the Damas Mission in 1935–36. In fact, the long list established by Damas (1937) and quoted by Van Meel (1954) concerns mostly the littoral flora. This list includes 33 species of Cyanophyta, 185 species of diatoms and 63 species of Chlorophyta. Van Meel stresses that the diatoms are dominated by *Nitzschia*, especially *N. fonticola, N. lancettula,* and by *Surirella engleri, Stephanodiscus damasi* and *Coscinodiscus rudolfi*.

The plankton develops after the May–June turnover and the zooplankton peaks from August to October. It is then mainly formed of Cladocera; a second peak of Cladocera, less conspicuous, occurs in January–February after the winter mixing. Conversely, the copepods dominate during the stratified period. The main species of Cladocera are: *Daphnia longispina, Moina dubia, Ceriodaphnia rigaudi* and *C. bicuspidata*. The copepods are represented by *Cyclops hyalinus* and *C. leuckarti*.

The littoral biotopes are rich and varied. In calm bays, *Pistia stratiotes* harbours an abundant fauna of Chironomidae and Culicidae, Ephemeroptera, and Ostracoda. At a depth of ~0.5 m, *Ceratophyllum* spp. are frequent and shelter Ostracoda, Copepoda and Ephemeroptera; *Najas* sp. found down to 0.75 m depth are rich in Ostracoda, Copepoda, Decapoda, Acarina and Diptera; the areas covered with *Potamogeton* have an abundant fauna of Hemiptera, Chironomidae, Copepoda and Acarina.

The bottom of the lake is covered with a fine-grained, organic mud with the exception of a shallow belt near the shores and near the estuaries of rivers, where sediments are sandy and compact. The mud contains 20–33% organic matter and 51–68% silica because of their abundance of diatom frustules.

The sands are inhabited only by a few Mollusca, Ostracoda and chironomid larvae. In contrast, the mud substrates have an abundant fauna of Mollusca (*Melanoides, Sphaerium* and *Byssanodonta*), Oligochaeta, Ostracoda and larvae of Chironomidae and Chaoboridae. These latter two groups represent the bulk of animal benthic biomass, but their distribution is limited from 0 to 40 m depth because of lack of oxygen in deeper layers. Verbeke (1957) gives the following figures:

Chaoborus: 1250 larvae m^{-2} (5.6 kg dm ha^{-1} y^{-1})

The Nile Basin

Chironomids:

0– 7 m 1850 larvae m^{-2} (111 kg dm ha^{-1} y^{-1})
7–15 m 200 larvae m^{-2} (12 kg dm ha^{-1} y^{-1})
15–40 m 500 larvae m^{-2} (30 kg dm ha^{-1} y^{-1})

The *Chaoborus* larvae are exceptionally rich in protein (67% dm) and fat (13% dm) in comparison with respective values of 46% and 7% for chironomid larvae. *Chaoborus* larvae are therefore an excellent potential food for fishes, but at present are not consumed, whereas chironomid larvae are abundantly preyed upon.

From this detailed study of the benthic fauna, Verbeke concludes that certain invertebrates such as Mollusca, Trichoptera and various groups of worms are represented by a limited number of species.

The fish fauna presents very similar features of 'incomplete recolonization'. The non-cichlid fish include 30 species. Entire families are missing and various genera present in Lake Albert and the Nile are lacking (*Polypterus, Hydrocynus, Distochodus* and *Citharinus*). The cichlids are more numerous than previously thought however: in Lake Edward and Lake George, some 100 species are known at present and many of them are endemic (Beadle, 1974). This underlines once more the extraordinary capability of cichlids for adaptation and speciation. As already mentioned, predator fishes such as *Lates* are missing.

Lake George

Lake George is located at 0°0′ and 30°10′ E, at 913 m altitude, in Western Uganda (see Fig. 21). The western and eastern shores of the lake are bordered by papyrus swamps whereas eroded savanna grassland forms the western and southern sides of the lake. The Ruwenzori mountains rise on its western side to 4000 m above the lake level. Part of the precipitation falls as snow, which diminishes the wet-season runoff and, by melting, increases the dry-season flow.

Lake George has been intensely investigated by an IBP team and a non-exhaustive list of studies includes Burgis (1969, 1970, 1971, 1973, 1974, 1978), Burgis *et al.* (1973), Burgis & Dunn (1978), Burgis & Walker (1971), Dunn (1973, 1975), Dunn *et al.* (1969), Ganf (1972, 1974*a,b,c,d*, 1975) Ganf & Blazka (1974), Ganf & Horne (1975), Ganf & Milburn (1971), Ganf & Viner (1973), Gwahaba (1973, 1975, 1978), Haworth (1977), Horne & Viner (1971), McGowan (1974), Moriarty (1973), Moriarty & Moriarty (1973), Moriarty *et al.* (1973), Viner (1969*a*, 1972*a,b,c,d,e*, 1977*a*), Viner & Smith (1973).

Lake George has a surface area of 250 km² and a mean depth of

2.4 ± 0.1 m. The shoreline is not indented much but has a large extension to the north-west, the shallow Hamakungua bay, generally covered with *Pistia stratiotes*. Three large islands, located on the western side of the lake, disrupt the monotonous bathymetry.

The lake receives inflows from the northern side and empties into Lake Edward by the shallow Kazinga Channel. Lake Edward and Lake George are almost at the same level; the rock sill of the Semliki River which drains Lake Edward serves as a regulator of the water level of both Lake Edward and Lake George and keeps it nearly constant.

During the mid-Pleistocene, a larger lake covered the Lake George and Lake Edward area, but arid conditions which developed in the late Pleistocene about 100 000 years ago restricted the lakes to the topographically depressed areas and led to the formation of two separate lakes. The sediments of Lake George consist of a 2.5 m-thick layer of soft, organic mud, overlying a much more compact and older lacustrine formation. The ^{14}C-dating of the mud bottom showed that the deposition of the recent layer started 3600 ± 90 years ago (Viner & Smith, 1973); the underlying formation includes volcanic ashes deposited in freshwater. Since, according to Bishop (1969), the volcanic activity in the Lake George area came to an end 8000 years ago, it is probable that this eruptive phase was followed by a complete drying of the lake which lasted about 4000 years. The present lake came into existence only 3600 years ago. The study of the recent sediments shows a progressively increasing fertility with time. In the upper 70 cm, however, the deposition rate of organic matter became constant; this is understood by Viner & Smith to show a steady state of primary productivity controlling the deposition rate.

The water balance of Lake George is shown in Table 29. The mean retention time is 4.3 months.

The dominant wind is from the east. The wind velocity is maximal during the wet seasons in February–March and September–October. On a daily basis, the wind shows a maximum in mid-afternoon and a minimum at night. The protected situation of the lake explains the generally low wind velocity.

Table 29 *Annual water budget of Lake George (Viner & Smith, 1973)*

	10^6 m^3		10^6 m^3
Inflow	1948	Evaporation	456
Direct rainfall	205	Outlet	1697

The lake surface receives a mean annual solar energy of 1970 J cm^{-2} d^{-1} ($\pm 13\%$). All the morphological features of the lake and lake surroundings, its equatorial location and hydrological regime concur to give Lake George a remarkable stability which concerns winds, solar input, temperature, inflow, outflow and water levels. As underlined by Burgis (1978) the short-term events have a greater importance on biological phenomena than seasonal variations.

As far as the diurnal cycle of temperature is concerned, the lake is homothermal at 26°C in the morning and stratified at mid-day with surface temperature reaching 34°C and bottom temperature 27°C. Other parameters, such as pH and oxygen, have an identical stratification pattern (Ganf & Horne, 1975).

The lake water is very diluted (conductivity 200 µmhos cm^{-1}) and its composition dominated by calcium carbonate (Viner, 1969a). The nutrient chemistry is closely linked with the daily regime of stratification and mixing since the periodic resuspension of the sediments recirculates the products of mineralization. Viner (1977a) estimates, however, that the sediments play a minor role in nutrient supply. As far as nitrogen is concerned, the main supply channels include the inflows, the rainfall, hippopotamus excretion, bacterial mineralization in the water column, excretion of the consumers and nigrogen fixation by blue-green algae (Horne & Viner, 1971).

The continuous internal recycling is the main process which allows the lake to sustain very high planktonic biomass with constantly low levels of nitrogen and phosphorus in the lake water.

The phytoplankton of Lake George is heavily dominated by blue-greens which makes up 70–80% of the algal biomass (*Anabaenopsis phylippinensis, A. raciborskii, A. tanganikae, Aphanocapsa grevillie, Chroococcus* sp., *Lyngbya limnetica, L. contorta, Merismopedia* sp., *Microcystis aeruginosa, M. flos-aquae, Phormidium mucicola, Spirulina* sp.). Among the Bacillariophyta, *Fragilaria* sp., *Melosira* sp., *Nitzschia microcephala* and *Synedra berolinensis* are common; the genera *Kirchneriella, Pediastrum* and *Scenedesmus* are the most frequently found Chlorophyta.

Measurements over several years show that there is only a two-fold variation in the algal biomass (150–300 mg chlorophyll *a* m^{-3}). In spite of sudden increases or decreases in biomass, probably caused by resuspension or sinking of algae, the algal biomass of Lake George is remarkably stable throughout the year (Ganf, 1972). Similarly, daily gross production, measured by the oxygen method, shows very small variations from 10.4 to 15.2 g O$_2$ m^{-2} d^{-1}. The photosynthetic activity is limited to the upper metre of water because of the rapid attenuation of light at greater depth.

Measurements of the respiration rates of the entire planktonic community yield high values (around $10 g O_2 m^{-2} d^{-1}$) so that the net prodution is low and fluctuates from 1 to $2.6 g O_2 m^{-2} d^{-1}$. It may even have negative values. Photosynthesis is clearly limited by the self-shading effect and a significant inverse correlation exists between the areal concentration of chlorophyll and the average daily photosynthetic rate per mg of chlorophyll. It is moreover dependent on temperature. It has been experimentally determined that the photosynthetic rate of the algal community of Lake George increases with temperature until 35°C. Above 27.5°C however, the respiratory rate increases steeply and the ratio gross photosynthesis:respiration, which determines the net production, is maximum around 27.5°C and decreases sharply above this temperature.

The zooplankton of Lake George is mainly composed of herbivores or detritus-feeders, among them the copepod *Thermocyclops neglectus* (syn. *hyalinus*) which forms 65–95% of the total zooplankton biomass, the rotifers *Brachionus caudatus*, *Keratella tropica* and *Trichocerca* sp., and the Cladocera *Moina micrura*, *Ceriodaphnia cornuta* and *Daphnia barbata*. Carnivorous species are also found, such as the copepod *Mesocyclops leuckarti* and the rotifer *Asplanchna brightwelli* (Burgis 1969, 1971; Dunn et al., 1969). Although there are distinct peaks of copepods in September and February, there are no sudden fluctuations either in biomass or in production which is continuous throughout the year. Consequently, there is no seasonal succession of species and the high values of biomass have a narrow range (270–580 mg dm m^{-3}). More recent evaluation by Burgis (1974) for the year 1969–70 gives a mean standing crop of 828 mg dm m^{-2} for total Crustacea and 559 mg dm m^{-2} for *Thermocyclops hyalinus*, the dominant species.

The fish community is dominated by the herbivores *Sarotherodon niloticus* and *Haplochromis nigripinnis* which feed directly on the phytoplankton and form 60% of the ichthyobiomass. The work of Moriarty (1973) and Moriarty & Moriarty (1973) shows that these fish are able not only to ingest but digest blue-green algae. Other species feed on insects, especially dipteran larvae (*Aplocheilichthys pumilus*, *Barbus kersteni*, *B. neglectus*, *B. perince*, *Haplochromis angustifrons*, *H. elegans*, *H. schubotzi*). *A. pumilus* feeds also on zooplankton. The piscivorous species seem to be restricted to *Bagrus docmac*, *Clarias lazera* and *Haplochromis squamipinnis*.

The planktonic community of Lake George is overwhelmingly dominated by the phytoplankton (95% of the planktic biomass). Most of the energy produced by the primary producers is used directly for their respiration. A smaller fraction is channelled into the herbivorous pathway

The Nile Basin

and the nitrogen and phosphorus released by the consumers represent a non-negligible nutrient source of the phytoplankton. This steady-state system is similar to that which develops in Lake Kinneret in summer (Serruya *et al.*, 1980). In Lake Kinneret, the seasonal events connected with the cold season destroy this equilibrium. In Lake George, the climatological stability of the Equator allows this steady-state equilibrium to perpetuate itself throughout the year. The excretions of the herbivorous consumers allow the continuous algal bloom and the grazing of the herbivores maintains the algal biomass within reasonable limits, thus preventing the crash of the bloom. The nutrient losses of the system through outflow and sedimentation are compensated by the inflows and also by the daily recirculation of nutrients generated by the daily pattern of mixing and stratification.

This 24-hour rhythm of the lake's physical dynamics has imposed an identical rhythm on the biological activities: the diurnal cycle of the phytoplankton activity coincides with an identical cycle of nutrient excretion by zooplankton and fish and of ingestion of phytoplankton by herbivorous organisms (Burgis, 1978).

With its large biomass of primary producers and relatively small secondary production, Lake George can be seen as a rather inefficient ecosystem. However, the lake produces 137 kg fm ha^{-1} fish, a value in the upper range of fish production of natural lakes. This is due to the predominance of herbivorous fish which minimize the metabolic losses. According to Burgis & Dunn (1978), net primary production represents only 28% of gross primary production and only 2.8% of net production is converted into herbivore production.

The very conditions which allow the functioning of the steady state (equatorial location, shallowness, diurnal cycle) may generate severe disturbances causing instability: deoxygenation may be caused by violent storms causing mud-mixing, by calm nights following days of phytoplankton growth, or by persistence of the thermocline for several days followed by its sudden breakdown (Ganf & Viner, 1973). In such cases fish kills have been observed.

The volcanic crater lakes of western Uganda

Melack (1978) describes four clusters of volcanic craters including 89 lakes distributed between 0°42′ N and 0°19′ S. The northern cluster of Fort Portal is located at 1520 m altitude whereas the lakes of the Kasenda cluster are located between 1220 m and 1400 m. The lakes of the Katwe–Kikorongo group are at an altitude of 900 m on the Rift floor and those of the Bunyaruguru lie from 975 to 1220 m.

The largest lakes are Lake Nyamusingire (440 ha), Lake Katwe (245 ha), Lake Nkugute (105 ha) and Lake Kikorongo (103 ha). The deepest ones are Lake Katarida (146 m), Lake Mwengenyi (101 m) and Lake Nkugute (58 m).

These lakes have a very wide salinity range from > 15 000 μmhos cm^{-1} to < 1000 μmhos cm^{-1}.

The mid-day temperatures of the water surface ranged from 22.5 to 26.7°C in dilute lakes and 26 to 33°C in saline lakes at the time of the Melack survey in May 1971. Many of these lakes were stratified with bottom temperatures ranging from 20.9 to 24.8°C. Melack (1978) notes that the density differences (0.0087–0.1307 g l^{-1}) are smaller than those observed in a stratified temperate lake. Melack & Kilham (1972) described a clear case of mesothermy in Lake Mahega which belongs to the Katwe–Kikorongo cluster and is situated at 0°01′ S and 29°58′ E. They visited the lake in May 1971 and on May 30 they observed a thermal profile increasing from 31°C at the surface to 40°C at 1 m and decreasing to 37°C at 3.5 m. The thermal maximum was accompanied by a high concentration of unidentified bacteria and the blue-green alga *Synechoccus bacillaris*. Lake Mahega has very saline water (193 g l^{-1} at 0 m and 415 g l^{-1} at 3 m TDS) which belongs to a sodium sulphate–chloride type. This water type, unusual in East Africa, results from precipitation of salts caused by evaporation in a closed basin. The extremely low concentration of calcium and magnesium are obviously caused by precipitation of calcium carbonate and magnesium carbonate. Moreover, crystals of northupite (Na_2CO_3–$Mg CO_3$–$NaCl$) and thenardite (Na_2SO_4) have been found in surface sediments. Melack & Kilham (1972) note that, in Lake Mahega, the transparent surface water allowed the solar heating of the 1 m turbid layer, since mixing was prevented by the strong ionic gradients and the sheltered position of the lake in the crater.

The volcanic crater lakes of western Uganda can be divided into two groups. The first group includes lakes of low salinity located in the clusters of Fort Portal, Kasenda and Bunyaruguru. To this latter cluster belongs Lake Nkugute (conductivity 86 μmhos cm^{-1}). Other lakes such as Saka Crater, Katanda, Mwengenyi, Mwamba, Lugumbe and Chibwera have conductivities ranging from 276 to 535 μmhos cm^{-1}. Lake Nyamusingire has the highest conductivity of the group with about 900 μmhos cm^{-1}. In contrast, the lakes of the second group, belonging to the Kative Kikorongo cluster, are very saline with conductivities ranging from 16 000 to 160 000 μmhos cm^{-1}. These lakes are very alkaline with pH around 10.

The Nile Basin

Their chemistry has been studied in detail by Arad & Morton (1969). They are all closed lakes where evaporation is the only outlet. Evaporites are deposited in all the lakes with the exception of Lakes Kitagata, Kikorongo and Munyanyange. At Lake Kative, the salts are commercially exploited. The sediments of the lake have a reserve of 20–30 million tons of salts, mainly composed of sodium carbonate and sodium hydrogen carbonate.

The lake water is completely without calcium and magnesium. When expressed in equivalent units, the concentrations of chloride, sulphate and carbonate–hydrogen carbonate of most of these lakes is of the same order of magnitude. Arad & Morton conclude that the water of most of these lakes results from the evaporation of thermo-mineral spring water. These springs emerge at temperatures ranging from 30 to 97°C. They also have a wide range of salinities (364–35 400 mg l^{-1}) but are nearly all characterized by a clear dominance of sodium + potassium over calcium + magnesium and equal amounts of the major anions Cl$^-$, SO$_4^{2-}$ and HCO$_3^-$ + CO$_3^{2-}$. Arad & Morton think that their salts come from a juvenile source of volcanic origin diluted by meteoric waters. The differences observed between the saline waters of western Uganda and those of the Eastern Rift have their origin in the chemical differences of the various volcanic provinces in East Africa.

Lake Albert

Lake Albert is located in the northern part of the Western Rift between 2°15′ and 1°N and between 30°20′ and 31°25′ E (see Fig. 21). The 2000 m escarpment of the graben borders the lake on its western side. The northern part of the lake is an alluvial plain where the Victoria Nile joins the Albert Nile. The southern shore is a swampy area covered with water plants and especially *Papyrus* which developed on the delta of the Semliki River. This river conveys the outflow of Lake Edward, the waters of the Ruwenzori mountains and of the western wall of the graben. On its way from Lake Edward, the Semliki River crosses numerous rapids which are an absolute barrier to faunal exchanges between the lakes. Similarly, the rivers draining the high eastern and western mountains on both sides of the lake join the lake level (616 m above MSL) by series of falls which testify to the juvenile character of this hydrographic network.

Origin of the lake

It is generally admitted that Lake Albert is more ancient than Lake Edward. At the southern end of the present lake and the Lower Semliki, lacustrine formations, 2400 m thick (i.e. 1800 m below MSL), fill the graben. The oldest lacustrine layers are of Miocene age, indicating that Lake Albert already existed in the Miocene. Vestiges of more recent lakes are found at various altitudes above the present lake level: the 'Kaiso series' consist of 600 m lacustrine sediments from early to mid-Pleistocene. In these series were found fossils of fish existing in the present lake, and also of *Clarotes* which belongs to the Nile fauna but is absent from Lake Albert. Molluscs were, however, the most abundant fossils: of 25 species found in the Kaiso beds, 24 are now extinct; *Etheria elliptica*, the Nile mussel, is the only species common to the Kaiso series and the living fauna. The 'Semliki series' (mid- to late Pleistocene) are coarse-grained series without fossils and the later Pleistocene post-Semliki sediments have a mollusc fauna very close to that of the present lake. The destruction of the early Pleistocene fauna of Lake Albert took place in the mid-Pleistocene.

Physical features

The average monthly air temperature varies from 28°C in February to 23.5°C in September. The average rainfall is 950 mm and the western shore of the lake receives more precipitation than the eastern shore. The peak of the wet season is April–May, but >60 mm per month is received from April to October.

The eastern winds are dominant in the basin although, on the lake itself, their direction is unstable. Violent storms with winds exceeding 50 km h^{-1} are frequent.

The morphometric features of the lake are presented in Table 30. From a bathymetrical point of view, the lake is composed of two basins: in the northern one, the western shore has a gentle slope (the 50 m isobath is at 26 km from the western shore) which contrasts with the steepness of the eastern shore. South of Kaiso, both shores are steep but the central part of this southern basin is shallower because of the delta of the Semliki River.

The water temperature (monthly averages) ranges from 29 to 27.5°C in the surface water but, at 40 m, it varies only from 26.7 to 27.2°C. The transparency rarely exceeds 30 cm in the bays but varies from 2 to 6 m in

Table 30 *Morphometric features of Lake Albert*

Altitude	616 m above MSL		
Length	170 km NW–SW	Mean depth	25 m
Width	45 km	Max. depth	58 m
Surface	6800 km²	Volume	$140 \times 10^9 \text{ m}^3$

The Nile Basin

open water. Lake Albert is not stratified and is generally oxygenated at all depths in spite of a 2°C difference in temperature between surface and bottom waters. This obviously frequent circulation was related to the strength and frequency of winds. Fish (1952) found, however, that a significant oxygen depletion could occur in the deep layers. Talling (1963) showed that a layer of cold water may be found on the bottom, for example in November 1960 and August 1961 when its temperature was 25.6°C in comparison with the 26.2–26.4°C of the rest of the profile. Talling explains this deep stratification as the result of a profile-bound density current, produced by night cooling of shallow water.

Chemical features

The main chemical features of Lake Albert are presented in Tables 31 and 32. The high potassium:sodium and magnesium:calcium ratios are distinctive of Lake Albert and other lakes of the western Rift. More specific to Lake Albert are its high concentration of PO_4-P (120–200 μg l^{-1}) and its low level of silica. Talling (1963) assumed that, in addition to diatom uptake, chemical precipitation, favoured by the high pH of the water, was responsible for silica depletion.

Oxygen level was generally low in the deep cold layer but progressive decrease of oxygen with depth was also observed in August 1961 when the deep discontinuity did not exist. Conversely, nitrate concentration was higher in the deep layer than in the upper layers. In August 1961 the profile was nearly homogeneous. Silica behaves as nitrates and Talling does not exclude a possible silica supply from the Semliki River to explain the especially high levels found in the southern part of the lake.

Worthington (1930) reports the presence of warm salty springs at Kibero (north-east shore) with 2500 mg l^{-1} chloride.

Table 31 *Chemical features of Lake Albert*

	(a)	(b)		(a)	(b)
Conductivity (μmhos cm^{-1})	735				
Na$^+$ (mg l^{-1})	91	97	$HCO_3^- + CO_3^{2-}$ (meq l^{-1})	7.33	7.8
K$^+$ (mg l^{-1})	65	66	Cl$^-$ (mg l^{-1})	33	32
Ca^{2+} (mg l^{-1})	9.8	9.3	SO_4^{2-} (mg l^{-1})	32	25
Mg^{2+} (mg l^{-1})	32.1	31.5	pH	–	9.1

(a) Talling, 1965
(b) Verbeke, 1957

Biological features
Very little is known of the algae of Lake Albert; Van Meel (1954) gives a list of species determined from one sample collected in 1907 and including 24 species of Chlorophyta, 14 species of diatoms, one species of Euglenophyta, two species of dinoflagellates and six species of blue-greens. In 1960–61, Talling (1965a) carried out three experiments on primary productivity near the eastern Ugandan shore; *Nitzschia bacata* and *Stephanodiscus astraea* were then the major components of the plankton but flagellates were also common. *Anabaena flos-aquae* is a less regular but sometimes abundant component of the phytoplankton. The vertical distribution of *Nitzschia* was strongly affected by the thermal stratification (Talling, 1963).

The zooplankton is dominated by Copepoda (*Mesocyclops leuckarti*) and the cladoceran *Diaphanosoma sarsi*. Other Cladocera such as *Daphnia*

Table 32 *Modifications of chemical and biological features along the Nile system (Chemical results in meq l^{-1})*

	L. Victoria 2 Mar. 61 (Talling, 1966)	L. Edward 22 June 61 (Talling, 1963)	L. Albert 27 May 54 (Talling, 1965)	Sobat R. 10 Dec. 54 (Talling, 1957)	El Ghazal 4 Dec. 60 (Bishai, 1962)
Conductivity (μmhos cm^{-1})	97	925	675	112	100
pH	8.0	9.1	9.0	6.8	7.4
Na$^+$	0.45	4.78	3.96	–	–
K$^+$	0.10	2.30	1.67	–	–
Ca^{2+}	0.28	0.62	0.25	0.44	–
Mg^{2+}	0.21	3.93	1.33	–	–
HCO$_3^-$+CO$_3^{2-}$	0.92	9.85	7.33	1.52	1.06
SO$_4^{2-}$	0.05	0.65	0.33	<0.03	–
Cl$^-$	0.11	1.01	0.94	<0.06	–
Phytoplankton	*Melosira Nitzschia Anabaena* 1.2–5.5 mg m^{-3} chlorophyll	*Nitzschia Surirella Stephanodiscus*	*Nitzschia Stephanodiscus*		Desmids 300 cells ml^{-1}
Zooplankton	Copepods dominant 300 l^{-1}	Cladocera dominant during mixed period. Copepoda dominant during stratification	*Mesocyclops leuckarti Diaphanosoma sarsi*	low densities: 2 mg m^{-3}	density increases from 1 to 70 l^{-1} from headwater downstream

a Algae in capital letters indicate dominant forms

lumholtzi, *Moina dubia* and *Ceriodaphnia bicuspidata* are also found. The zoobiomass is constantly high from March to October. It is low only during the dry season (Verbeke, 1957).

In the southern area of the lake, a vast area 3 m deep is covered with sand and silt; *Najas marina* and the filamentous alga *Cladophora* are common. The associated fauna includes molluscs (*Bithynia* and *Planorbis*), Ostracoda, Cladocera and Ephemeroptera.

At the mouth of the Semliki, an association of *Panicum* and *Potamogeton* is inhabited essentially by Ostracoda, Copepoda and Diptera. At Kasenyi, on the south-western shore, a vegetation dominated by *Cyperus* and *Phragmites* harbours a rich fauna of Ostracoda, Copepoda, Decapoda, Ephemeroptera and Diptera.

Table 32 (*cont.*)

	White Nile upstream of Gebel Aulia	Gebel Aulia	White Nile Kartoum 1965–67 (Hammerton, 1972)	Roseries + Sennar + Blue Nile Gorge 10 Mar. 64 (Talling & Rzoska, 1967)	Main Nile Cairo 29 Mar. 74 Talling & Heron (in Talling, 1976)
Conductivity (μmhos cm^{-1})			140–390	231	307
pH			8.2–9.1	–	–
Na$^+$			0.2–0.39	0.57	1.20
K$^+$			0.04–0.07	0.57	0.17
Ca^{2+}			0.98–1.41	1.52	1.39
Mg^{2+}			0.41–0.54	0.80	0.82
HCO$_3^-$ + CO$_3^{2-}$			1.63–2.66	2.33	2.70
SO$_4^{2-}$			–	0.48	0.37
Cl$^-$			0.06–0.21	0.11	0.41
Phytoplankton	*MELOSIRA*[a] *Lyngbya* *Anabaena* low densities	*ANABAENA*[a] *LYNGBYA*[a] up to 7000 cells ml^{-1}	*MELOSIRA*[a] *Anabaena* *Lyngbya* *Anabaenopsis* high densities	In Lake Tana: diatoms dominant especially *MELOSIRA*[a] In Blue Nile: *MELOSIRA*[a] *Anabaena*, *Lyngbya* In Roseries: *Microcystis*	Upstream of High Dam: *Melosira* *Volvox*
Zooplankton		Nauplii of Cyclopidae *Ceriodaphnia rigaudi* up to 120 l^{-1}		Cladocera dominant	Cladocera dominant

The sediments of Lake Albert are less organic than those of Lake Edward, probably because of the better conditions of oxidation prevailing on the bottom. Diatom frustules do not exist because of the alkaline pH of the deep water and the pelagic sediments of Lake Albert contain more sand than those of Lake Edward. This does not affect the benthic fauna which is identical to that of Lake Edward, that is dominated by larvae of *Chaoborus* and Chironomidae. Molluscs are represented by a higher number of species than in Lake Edward.

The Albert Nile and Sudd Swamps
The Albert Nile

The Albert Nile is deeply affected by the composition of Lake Albert water. The Albert Nile has a relatively high salt concentration; its high phosphate and sulphate levels are derived from the lake. The impact of Lake Albert on the chemistry of the Albert Nile depends on the fluctuations in Lake Albert discharge (Talling, 1958).

Upstream of the Sudd Swamp area, the Albert Nile conveys an annual discharge of 28×10^9 m^3. In the swamp area, evaporation and evapotranspiration 'consume' 14×10^9 m^3. Just after leaving the swamp, the Nile receives the Bahr el Ghazal River which drains the Marra mountains on the western bank and the Sobat River, fed by the abundant precipitation falling on the 3000 m-high Ethiopian plateau. We note that the water supply from these two tributaries approximately compensates the loss in the swamps. Before describing the Sudd Swamp, however, we should mention the Dariba lakes, two crater lakes of the Marra mountains which have been studied by Green, El Moghraby & Ali (1979). They are located at 2200 m altitude in a 5 km-wide caldera dominated by a peak of 3024 m on its western side. The shallowest lake (11.5 m deep) is saline (conductivity 27 000 μmhos cm^{-1}). The deepest one (108 m deep) is less saline (6000 μmhos cm^{-1}). During the January 1976 survey, the shallow Dariba lake was stratified part of the time and had no oxygen below 4 m. This stratification was easily broken down by strong winds: then, the whole water column had a temperature of 14°C and $<50\%$ oxygen saturation. The same wind mixing caused a nearly complete deoxygenation in the deep lake with oxygen saturation not exceeding 10% at any depth. The shallow lake had an alkalinity of 147 meq l^{-1}, contrasting with a value of only 48 meq l^{-1} m in the deep lake. The saline and alkaline Dariba lake has a dense population of *Spirulina platensis* associated with small numbers of *Anabaenopsis arnoldii*. With the exception of a few small diatoms, the deep lake has no planktonic flora. A group of three rotifers forms the zooplankton of the saline lake: *Brachionus plicatilis, B. dimidiatus* and *Hexarthra jenkinae.* In the deep lake, besides the previously mentioned rotifers, the copepod *Eucyclops gibsoni* is very common, although of unusual proportions (Green *et al.,* 1979).

The poor fauna of these lakes is due to their isolation, their high salinity and their poorly oxygenated water.

The Sudd Swamps
General

The following is based on the extensive paper by Rzoska (1974) on the Upper Nile swamps. Between latitudes 5 and 9° N, a vast area of about 20 000 km² is covered by permanent swamps which are centred along the Gebel River (Main Nile), the Zeraf River, the Ghazal and the Sobat. The Sudd area is a depression surrounded by the highlands of the Nile–Congo waterdivide, the Uganda and the Ethiopian plateaux. The rivers draining these mountain regions have deposited alluvial material and the river channels lie higher than the swamp. This, together with the weak slopes in the river channels (0.008–0.08 m km^{-1}), causes flooding of very extensive areas and, at high water, the swamp area reaches ~90 000 km². The flooding period extends from June to November with peaks in September–November. Evaporation, ranging from 940 to 1643 mm y^{-1}, exceeds water inputs for 6 months a year, causing the fringe areas of the swamp to dry out.

The saturated soils of the permanent swamps are composed of an upper slightly acid and organic layer resting on more mineral and alkaline sediments. The vegetation consists mainly of *Cyperus papyrus, Vossia cuspidata* and *Phragmites mauritanus.*

Standing waters are covered with *Pistia stratiotes, Najas pectinata, Azolla nilotica* and others. Since 1957, *Eichhornia crassipes* has colonized the standing water and spread very rapidly.

In the fringe area, the clayey soils undergo a periodic swelling at high water and drying and cracking at low water. Grasses such as *Echinochloa, Vossia* and *Phragmites* constitute the fodder of the cattle of local tribes.

The emerging ridges have an arborescent vegetation of *Acacia* and *Borassus* and harbour the permanent villages.

Limnological features of the rivers and lakes

The water chemistry of the Bahr el Gebel (Main Nile) is deeply modified by the swamp area. Talling (1957) noted that the oxygen falls from 70–80% to 13–35% saturation, pH drops from 7.9 to 7.1, CO_2

increases from 3 to 18 mg l^{-1} as well as dissolved iron. Sulphate decreases probably as a result of sulphate reduction, although no sulphides were detected. The high conductivity of the Albert Nile ($\sim$ 700 μmhos cm^{-1}) falls to 200 μmhos cm^{-1}. These results were confirmed by a later study by Bishai (1962).

Currents of 0.70 m s^{-1} in midstream allow the drift of *Pistia, Papyrus* and *Eichhornia* floating islands.

The phyto- and zooplankton of the Albert Nile are also modified on their way through the swamps. In general, the high turbidity of the river water maintains phytoplankton mainly composed of *Melosira granulata* var. *angustissima* at very low levels (40–120 cells ml^{-1}). Among the Crustacea, *Diaphanosoma excisum, Daphnia barbata, D. lumholtzi, Ceriodaphnia dubia, C. cornuta, Moina dubia, Thermocyclops neglectus, Mesocyclops leuckarti* and *Thermodiaptomus galebi* are the main species found. Quantitatively, only a few specimens per litre have been observed.

The Ghazal River, originating in Lake Ambadi has a low pH and a low conductivity which increase downstream, a high transparency and a high content of silica and chloride. Lake Ambadi presents characteristics which are unusual in the Nile system: in spite of its shallowness (2.5 m) the water is very transparent and the bottom covered with *Myriophyllum, Utricularia* and *Potamogeton*. The plankton is dominated by desmids (205 species and forms of which 21 species and 48 forms were new). This abundance of desmids is probably related to the slight acidity of the water. These algae decrease in number downstream and nearly disappear in Lake No. Lake Ambadi also harbours species unknown in Sudan waters (*Dinobryon sertularia, Botryococcus braunii, Asterococcus limneticus*). The diversity of species contrasts with the low biomass of phytoplankton (300 cells ml^{-1} in one sampling). Lake Ambadi harbours many herbivorous invertebrates: *Tropocyclops, Microcyclops, Tropodiaptomus processifer, Thermodiaptomus galebi* as well as abundant *Bosminopsis deitersi*. In contrast with the other areas of the swamp, the fringe vegetation is scanty and the low pH is probably responsible for the poor development of *Eichhornia crassipes*. Quantitatively, the zooplankton increases in abundance along the Ghazal River and culminates in Lake No with 71 organisms l^{-1}.

The Sobat River also has low values for conductivity, salinity, alkalinity, chloride and transparency, but a high silica concentration. The zooplankton biomass is very low and fluctuates from 0 to 2 mg m^{-3} (Monakov, 1969).

Mass emergence of insects occurs frequently. Rzoska reports the formation of a cloud of Ephemeroptera in the Zeraf River; 63 species of Culicidae, three species of *Chaoborus* in the plankton, 22 species of

The Nile Basin

Simulium and 68 species of chironomids are known to be subject to periodical mass emergence.

The fish fauna of rivers and lakes of this area is not known in detail but the following groups are represented: Siluroidae, Mormyridae, Cyprinidae, Citharinidae, Cichlidae, Characidae. *Polypterus* spp., *Protopterus aethiopicus, Heterotis niloticus, Gymnarchus niloticus* and *Lates niloticus* are also found.

The diet of 38 species of swamp fish was examined by Sandon & Amin el Tayeb (1953). Most specimens were mixed feeders, six were only herbivorous, one was a zooplankton-eater, one fed on molluscs and six were fish-eaters. Monakov (1969) studied the diet of 24 fish species and found five diet groups:

1. ichthyophagous fishes (*Lates, Mormyrops, Hydrocyon, Bagrus, Schilbe, Eutropius*);
2. detritus-eaters (a major part of the fauna);
3. vegetation-feeders (restricted to *Tilapia galilaea* and *Labes horie*);
4. plankton-feeders which eat zooplankton (*Synodontis* spp.) or insects (*Alestes* spp.); and
5. benthos-feeders which ingest chironomid larvae (*Mormyrus cashive*) or molluscs.

In spite of the scarcity of plankton in rivers, an abundant fish population exploits the numerous other food sources of the swamp ecosystem: vegetal material, detritus and benthos.

Limnological features of the fringe swamp

The slow-moving water of the fringe swamps becomes impoverished in oxygen and enriched in ammonia, iron and CO_2; simultaneously the pH decreases.

The epiphytic algae flourish, such as the red alga *Compsopogon*, and a very abundant fauna, Protozoa, Oligochaeta and Bryozoa, lives on these algae. The shrimps *Caridina nilotica* and *Palaemon niloticus*, Cyclopoda and Ostracoda are abundant. Cladocera have 34 species. Miniature fish are characteristic of the swamp fauna; many of them are endemic (*Cromeria nilotica, Barbus leonensis, Hemichromis fasciatus*).

Antelope, elephant, buffalo, waterbuck, white rhino, giraffe and hippopotamus are the main herbivores; they serve as prey to lion and serval cat. A population of half a million people of Nilotic tribes also lives in this area. The animals and the human population migrate with the floods, towards the rivers at the dry season and in the opposite direction at high water, when the Nilotic tribes come back to their permanent villages.

In conclusion, the Sudd Swamp area strongly affects the composition of the river waters which run through it. Moreover, a significant difference exists between the scanty planktonic populations of the rivers and the abundant communities of standing waters.

The Jonglei project

The considerable water losses in the swamp area led Hurst, Black & Simaika (1946) to propose the construction of a canal bypassing the Sudd Swamps in order to increase the Nile flow downstream by about 8×10^9 m^3. From 1945 to 1954, the Jonglei Investigation Team weighed the ecological implications of the project. Such a bypass may destroy the ecological equilibrium on which the livelihood of the Nilotic tribes is based. For example, the canal would suppress the floods and the dry lands would stop producing the fodder essential to the cattle. The bare and dry land would be eroded at a very high rate instead of being a depositional area. The loss of the flood fisheries would amount to 3000 tons y^{-1} (with present primitive fishing methods) and to 20 000 tons y^{-1} if more modern means were used (Hammerton, 1972). The Sudd Swamps at present play the role of a flood-regulation system which will disappear with the implementation of the project. The elimination of the swamps will also modify the water chemistry of the Nile. In spite of its ecological hazards, the Jonglei Canal is being built jointly by Egypt and Sudan.

The White Nile

We have followed the chemical and biological transformations of the Nile water; a brief recapitulation is presented in Table 32. From a chemical point of view, the diluted water of Lake Victoria is enriched in minerals (especially sodium and alkalinity) by Edward and Albert water. The Sudd Swamp adds CO_2 and organic detritus, decreases the pH and removes nutrients and SO_4^{2-}. The mixing of the low conductivity waters of Bahr el Ghazal and Sobat Rivers operates a considerable dilution especially during the flood period. We note that hydrogen carbonate and carbonate are the major anions all along the river and sodium the major cation.

As far as plankton is concerned, the large phytoplankton and zooplankton inoculum of the head-water lakes is replaced by a sparse river population. Among the algae, diatoms are generally dominant with the exception of the unusual desmid assemblage of Lake Ambadi and Barh el Ghazal. The thin crustacean river populations very often do not include more than a few specimens per litre.

The Nile Basin 167

At approximately midway between Malakal and Khartoum, the White Nile (Fig. 19) widens into the Gebel Aulia Reservoir resulting from the building, in 1937, of the Gebel Aulia Dam.

Influence of the Gebel Aulia Dam on the Nile

The dam was built 44 km south of Khartoum and used for irrigation. In July, the dam is closed and reaches maximum capacity in September (3.5×10^9 m^3). Then, a rise of water level is felt 500 km upstream. The dam is opened in February, in May the reservoir is empty and the river regime is resumed.

Brook & Rzoska (1954) studied the effect of this temporary lake on the production of plankton and on water chemistry. Prowse & Talling (1958) examined the succession of algae in the White Nile and in the reservoir in relation to the seasonal variations of water level and the chemical modifications of the water.

A series of measurements carried out from the upper parts of the lake down to the dam showed that dissolved solids, pH and oxygen content increase towards the dam whereas transparency decreases. This longitudinal study showed also that phytoplankton and zooplankton are very scanty in the upper reservoir and increase suddenly about 140 km upstream of the dam: phytoplankton jumps from 100 to 7000 cells ml^{-1} and zooplankton from 20 to 120 organisms l^{-1}.

The dominant algae were *Anabaena flos-aquae* and *Lyngbya limnetica; Anabaenopsis cunningtonii, A. tanganikae, Raphidiopsis curvata, Oscillatoria geminata* and *O. planktonica* were also present. Among the diatoms, *Melosira granulata* var. *angustissima* dominated but *M. agassizzii, Synedra acus, Nitzschia acicularis* and *N. palea* were also recorded. The Chlorophyceae were represented by *Ankistrodesmus falcatus, Scenedesmus acuminatus, Micractinium pusillum* and *Dictyosphaerium pulchellum.*

The zooplankton was dominated by the nauplii of Cyclopidae and by *Ceriodaphnia rigaudi; Diaphanosoma excisum, Ceriodaphnia dubia, Moina dubia, Daphnia lumholtzi, D. barbata, Thermocyclops neglectus, Mesocyclops leuckarti aequatorialis* and *Diaptomus galebi* were also present.

Abu Gideiri (1969), who also studied the plankton of Gebel Aulia, found a very similar plankton community with two distinct peaks in October–November and in April.

This study indicates clearly that in river conditions the Nile plankton is sparse and dominated by diatoms and copepods. The storage reservoirs, by modifying the hydraulic conditions (decrease in current velocity and in turbulence), allow a significant rise in plankton density and favour the predominance of blue-green algae and Cladocera.

Lake Ambadi

The algae present in Lake Ambadi are very different in comparison with the algal flora of the Nile waters in Sudan. The dominant algae are *Dinobryon sertularia*, *Botryococcus braunii*, *Asterococcus limneticus*, *Nitzschia* spp., *Synedra* spp., *Navicula* spp. and *Stauroneis* sp. *Melosira granulata* is absent from Lake Ambadi. The common Cyanophyta in the White Nile are also absent here (Grönblad, Prowse & Scott, 1958).

Brook (1954) recorded 19 species of desmids in the White Nile, whereas in Lake Ambidi, Grönblad *et al.* (1958) gave a list of 200 taxa. There is evidence that the rich variety of desmids found in Lake Ambadi disappears when the Bahr el Ghazal reaches Lake No.

Lake Tana and the Blue Nile

The Blue Nile originates at Lake Tana, 1829 m altitude, on the Ethiopian plateau. In contrast to the White Nile, the Blue Nile has a very marked flood regime. The Blue Nile contributes a large percentage of the main Nile discharge and sediment load. It is an old river system dating from Eocene times when the Afro-Arabian swell was formed. Later uplifts occurred in the Miocene and in the Pleistocene (Baker, Mohr & Williams, 1972). After leaving Lake Tana, the Blue Nile cuts a volcanic gorge which has provided the sediments of the Nile Valley and delta. The recent work of McDougall, Morton & Williams (1975) indicates that 23–27 million years B.P., i.e. in the Miocene the present gorge was filled with volcanic lava. In its Ethiopian course, the Blue Nile receives numerous tributaries and in its lower course it receives the Dinder and Rahiad Rivers. The total watershed of the Blue Nile is 324 000 km².

Lake Tana

According to Mohr (1962), Lake Tana (Fig. 21) was formed during the Pliocene and the effluent of the lake then started to dig into the volcanic material. This is in good agreement with the findings of Ethiopian

Table 33 *Morphometric features of Lake Tana*

Altitude 1829 m	
Length 78 km	Maximum depth 14 m
Width 67 km	Mean depth 8.9 m
Shoreline 385 km	Surface 3156 km²
Watershed area 13 350 km²	Volume 28×10^9 m³
(lake area not included)	
Watershed area : lake volume ratio 0.48 m² m^{-3}	

The Nile Basin

sediments and pollen of Cenozoic age in the delta (McDougall et al., 1975). Terrasses above the present level testify to a wider extent of the lake in the past. The main morphometric features of the lake are presented in Table 33.

The lake has four island areas: Deck and Daga in the southern part of the lake are the two largest.

The whole watershed of Lake Tana is located above 1500 m and, east of the lake, mountains reach 3000 m. The air temperature ranges from 23 to 30°C during day-time to 6–8°C at night. There is nearly no rain in winter; the wet season extends from June to October and peaks in July–August. The yearly precipitation is about 2000 mm. The watershed is drained by a large number of small rivers. The Little Abbai is the main tributary of the lake; it drains the southern area and is about 100 km long. The Great Abbai or Blue Nile leaves Lake Tana in its south-western corner and 30 km further on flows across the Tississat Falls.

Around the lake, the vegetation is relatively rare, probably because of the rocky nature of a large portion of the shore. In areas with soft sediments, for example in the mouth of tributaries, a littoral vegetation develops. The Italian Mission, which studied the lake in 1936–37, found that *Cyperus papyrus* was particularly abundant in the meridional shores of the lake. In swampy areas of the southern shore, *Echinochloa pyramidalis, E. stagnina, Panicum longi-jubatum, Polygonum barbatum* and *P. senegalense* are found. The free-floating *Pistia stratiotes* is especially abundant at the outlet of the lake. The southern gulf of the torrent Umfras has a vegetation of *Nymphaea caerula, Ceratophyllum demersum* and *Vallisneria spiralis*. In shallow areas with clayey or sandy bottoms, arborescent flora includes *Kanahia laniflora, Tacazzea venosa* and *Sesbania egyptiaca*.

Table 34 *Chemical features of Lake Tana*

	Lake Tana 8 Dec. 65–18 Apr. 66	Tississat Falls 12 Mar. 64	Gorge 10 Mar. 64
Na^+ (meq l^{-1})	0.31–0.41	0.26	0.57
K^+ (meq l^{-1})	0.03–0.05	0.05	0.07
Ca^{2+} (meq l^{-1})	0.70–0.76	0.93	1.52
Mg^{2+} (meq l^{-1})	0.50–0.70	0.52	0.80
$HCO_3^- + CO_3^{2-}$ (meq l^{-1})	1.52–1.92	1.57	2.23
SO_4^{2-} (meq l^{-1})	–	0.08	0.48
Cl^- (meq l^{-1})	0.03–0.07	0.05	0.11
pH	8.1	8.5	–
Conductivity (μmhos cm^{-1})	200–240	156	231

Lake Tana is a shallow water body with a gentle bottom slope on the western, northern and southern shores; the south-eastern shore is steeper. The seasonal rainfall regime results in very large variations in water level (1.90 m y^{-1}); in the dry season, the discharge of the outlet is considerably reduced.

In winter, the temperature recorded by the Italian Mission varied between 21 and 25°C. Temporary thermoclines, with small thermal gradients, form between 1 and 4 m.

The chemical features of Lake Tana waters are shown in Table 34 (Talling, 1976). The prominent feature of the Blue Nile is the predominance of calcium over sodium in comparison with the White Nile.

The phytoplankton is similar to that of other African lakes. The Cyanophyta are poorly represented with *Anabaena flos-aquae*, *A. planctonica*, *Microcystis* spp., *Chroococcus* sp., *Oscillatoria* sp., *Lyngbya* sp., *Phormidium* sp. and *Spirulina* sp. The Chlorophyta include *Botryococcus braunii*, *Oocystis borgei*, *Pediastrum simplex* and *P. clathratum*. The plankton is, however, dominated by diatoms. Brunelli & Cannicci (1940) give a list of 33 species of diatoms among which are *Melosira* sp., *Fragilaria virescens*, *Synedra ulna*, *S. capitata*, *Synedra* sp., *Surirella robusta*, *S. fulleborni*, *S. biseriata*, *S. turgida*, *S. elegans*, *Gyrosygma* sp. and *Navicula* sp.

The various species of *Surirella* seem to be particularly abundant although Talling (Rzoska, 1976) found in 1964 a plankton dominated by *Melosira granulata* var. *jonensis* fa *procera*. Brunelli & Cannicci (1940) stress the absence of dinoflagellates. They reported the following list of crustacean zooplankton in order of abundance: *Bosmina longirostris*, *Thermodiaptomus galebi*, *Diaphanosoma excisuum*, *Ceriodaphnia bicuspidata*, *Daphnia longispina*, *Moina dubia*, *Mesocyclops leuckarti*. *Ceriodaphnia* sp., *C. cornuta*, *Cyclops albidus*. Talling & Rzoska (1967) report a different order of predominance and note the presence of *Daphnia lumholtzi* in samples taken by Talling in 1964. The rotifers found by Brunelli & Cannicci (1940) were cosmopolitan and mostly planktonic; the authors explained the limited number of species (14 species) by the rarity of submerged plants.

The revised list of molluscs of Bacci (1951-52) contains the 15 following species: *Theodoxus africanus*, *Bellamya unicolor unicolor*, *B. unicolor abyssinica*, *Melania tuberculata*, *Radix pereger*, *R. caillaudi*, *Bulinus hemprichi sericinus*, *Biomphalaria ruppelli*, *Unio elongatulus dembeae*, *U. abyssinicus*, *Aspatharia rubens caillaudi*, *Mutela nilotica*, *Etheria elliptica*, *Corbicula fluminalis consobrina* and *Byssanodonta parasitica*.

Neither *Unio* species of Lake Tana is found in the Nile today. These palaearctic elements, probably common in the Nile in the Pleistocene, disappeared from the river when the climate got warmer and became restricted to Lake Tana.

Bulinus and *Biomphalaria* harbour schistosomes which cause bilharzia, a common disease in the area.

The fish fauna includes, according to Greenwood (1976), eight or nine species, only one or two of which are endemic. *Barbus intermedius* is not endemic but not frequently found outside Lake Tana. Other species such as *Varicorhinus beso* do not exist in the Nile. The silurids *Clarias* are abundant and together with *Barbus* sp. form the bulk of the fisheries. The most interesting species is *Nemachilus abyssinicus*, since this is the only species of this Euro-Asiatic genus known in Africa.

Lake Tana, poor in salts and nutrients, does not sustain an abundant planktonic fauna. Its limited, littoral vegetation does not favour the development of epiphytic flora. From a biogeographical point of view, certain species of rotifers and fishes belong to Asiatic genera and this confers on Lake Tana a special place among the tropical lakes of Africa.

The Blue Nile and the impact of reservoirs

The Blue Nile, (Fig. 19) after the falls of Tississat, enters the volcanic gorge which was explored in August–September 1968 by a British team. Morris, Largen & Yalden (1976) gives an interesting description of this unusual environment. The gorge, being warmer than the surrounding highland, has allowed the penetration of Savanna flora and fauna into the Ethiopian plateau in the same way as the Jordan Valley, a finger-shaped arid area penetrating into a region of Mediterranean characteristics, allowed the northward spreading of African species.

As previously mentioned, the Blue Nile has an average base flow of $<200 m^3 s^{-1}$ and an average maximal discharge of $600 m^3 s^{-1}$ in August–September. Then, the green low waters are replaced by the red water loaded with fine sediments; a general dilution of all chemical elements is noted (Talling & Rzoska, 1967). Beam (1906), who studied the river before the construction of the Sennar Dam, did not mention any abundance of plankton. The Sennar Dam was built in 1925 to provide water for the Gezna Irrigation Project. A new survey by Talling & Rzoska in 1954–56 showed a different picture. The dam started to fill in November and the water was released for irrigation from February to May. During the flood, the plankton had very low concentrations. After the flood, the algae developed first with *Melosira granulata* as the dominant species then *Anabaena flos-aquae* fa *spiroides* in January–February. *Anabaena scheremetievi, Anabaenopsis tanganikae, A. cunningtoni, Raphidiopsis curvata* and *Synedra acus* were also present. Observations below the dam revealed that the abundant plankton produced in the reservoir not only did not decrease on the way down to the White Nile but often increased in density. The plankton observed at Khartoum is thus partly produced in river conditions.

The zooplankton, dominated by the Cladocera, behaved very similarly with two peaks in the dry season and low densities during the flood. Hammerton (1976) thinks that the numerous pools remaining after the flood are the sources of a new inoculum. El Moghraby (1977) collected resting stages of crustacean zooplankton in the bottom mud of the Blue Nile and showed experimentally that they are released from diapause at a temperature of 20–30°C. This ability to spend the flood period in diapause would explain the rapid reconstruction of an after-flood zooplankton community.

The construction of the Roseires Dam (1961–66) was accompanied by a survey conducted by Hammerton (1972–1976) aimed at assessing the modifications introduced by the new dam (surface area 290 km² and maximum depth 50 m). The Roseires Dam was aimed at increasing the Blue Nile storage and producing electricity.

The first noticeable modification was an increase in transparency (2.8 m) near the dam in comparison with values 60 km upstream (0.2 m), as a result of silt deposition in the upper expanses of the reservoir. At Sennar and Khartoum transparency also increased considerably.

In the reservoir, the plankton increased by 10- to 100-fold. The average density was lower than at Sennar but the trophogenic zone being deeper, the total areal biomass was greater at Roseires. In contrast to Sennar, only a fraction of the reservoir plankton is carried downstream. This is caused by the temporary stratification which limits the plankton to the upper layers, while the water is flushed from deep layers. The stratification, with a thermocline around 10 m, is accompanied by a complete disappearance of oxygen in the hypolimnion. Primary production experiments at the Ethiopian border (river), the Roseires Reservoir, the Sennar Reservoir and Khartoum have yielded the following respective results: 1.0, 5.5, 6.6 and 6.0 g C m^{-2} d^{-1} (Hammerton, 1976).

Until 1965, the flood was followed by the blooming of *Melosira granulata*. In 1966–67, when the Roseires Dam was filled, *Microcystis flos-aquae* and associated *Phormidium mucicola* became codominant with *Melosira*. Two years later *Microcystis* was the dominant alga from Roseires to Khartoum. In 1969–70, the Chlorophyta *Volvox aureus* and *Pediastrum* spp. became prominent (Hammerton, 1972). In conclusion, the reservoirs on the Blue Nile, in replacing river conditions by lake conditions, have remarkably increased the algal biomass and productivity and created a 'dam eutrophication'. The Nile, which is presently free from cultural eutrophication, has now a seminatural high base level which would amplify any additional ecological stress.

El Moghraby (1972), who studied the zooplankton along the Blue Nile after 1966, found an upstream sparse population mixed with detritus and adventitious species. In the Roseires Reservoir, zooplankton density increased rapidly. Below the dam, the concentration was also higher than in the pre-reservoir period.

Hammerton (1972) reports that the massive silt deposit completely destroyed the population of the mussel *Etheria elliptica* in the reservoir. Conversely, chironomids increased in the Roseires area and *Simulium damnosum*, the vector of 'river blindness' (onchocerciasis), breeds in the turbulent water at the dam fall.

Abu Gideiri (1967), who surveyed the Blue Nile from Roseires to Khartoum, described 45 fish species of 14 families, including *Marcusenius isidori, Gnatocnemys cyprinoides, Mormyrus cashive, Alestes dentex, Distichodus niloticus, Synodontis batensoda, S. membranaceus* and *Gymnarchus niloticus*. The fish population of the Roseires Reservoir includes only 29 species of ten families, among which four species (*Hydrocynus brevis, Distichodus rostratus, D. brevipennis, Mormyrops anguilloides*) were not previously found in the Blue Nile (Hammerton, 1976).

The fishermen claim that the total catch and the fish size are increasing. In the long term however, this trend might not be sustained since the silt deposition in the reservoir damages the spawning areas. It is estimated that, of 130×10^6 tons of silt carried by the Blue Nile, 20×10^6 tons are sedimented in the Roseires Reservoir (Hammerton, 1972).

The Main Nile and Aswan High Dam

The Na^+-dominated White Nile and the Ca^{2+}-dominated Blue Nile mix at Khartoum. The chemical analysis of the Main Nile downstream reflects the dominance of the Blue Nile water since in Aswan and in Cairo the water belongs to the calcium–hydrogen carbonate type. However, Abdin (1948a) emphasizes that Nile water contains more Na^+ and K^+ at low water and more Ca^{2+} and Mg^{2+} during the flood. This indicates the alternate dominance of water from the Western Rift and of water from the Ethiopian plateau.

As far as nutrients are concerned, the White Nile and the Blue Nile at Khartoum seem equally rich in PO_4–P before the main algal bloom (100 μg

l^{-1} PO_4-P; Talling, 1976). The development of *Melosira granulata* in both rivers is followed by a decline in phosphorus. An additional decline in phosphorus has been noted by Entz (1976) from the south to the north of Lake Nasser. In Cairo (Talling, 1976), the concentrations of phosphorus are low (49–90 μg l^{-1}). Nitrate reaching 1000 μg l^{-1} in the flood water of the Blue Nile declines rapidly during the *Melosira* bloom. Ammonium–nitrogen, abundant in the swamp water of the Sudd area, reaches the White Nile. Silica occurs at relatively high levels in waters draining lateritic areas such as the Bahr el Ghazal and consequently the White Nile but this nutrient also declines as a result of *Melosira* growth. Similarly at Cairo, Abdin (1948a) observed a strong decrease of silica in 1941–42 between two flood periods but, here too, this decline was probably related to the diatom dynamics. The vertical distribution of *Melosira* in the river has been studied by Abdel Karim & Saeed (1978).

The abundant phytoplankton which reaches Khartoum generally decreases in density in the upper Cataract area. However, samples taken by Rzoska at Dongala in 1957 had an abundant algal population dominated by *Melosira granulata* indicating the occurrence of 'river blooms'. Observations by Abdin (1948b,c, 1949) revealed that, in 1942, *Melosira granulata* was the dominant alga in the old Aswan Reservoir after the annual filling and was followed by green algae (*Volvox, Pediastrum*). In 1957, *Melosira* and *Volvox* were still preponderant (Elster & Vollenweider, 1961) but several blue-green algae were also well represented. Below Aswan, these authors found in 1957, as in Aswan, an assemblage of *Melosira* and *Volvox*.

Lake Nasser–Nubia

The Aswan High Dam is the last of a long series of measures aimed at taming the mighty river. Large-scale irrigation schemes began in the nineteenth century in Egypt; they led to extensive cotton cultivation and the multi-crop per year system. From 1821 to 1907 the irrigation increased the cultivated area by 76% and the crop production by a factor of 2.5 (Hamdan, 1961). In the twentieth century, Egypt and the British Authority undertook a river management programme which, until 1929, was mainly aimed at the regulation of the annual flow of the river in order to guarantee an adequate summer water supply. To this purpose were built the Delta Barrages of Zifta (1903) and Idfina (1951), the Upper Egypt Barrages of Assiut (1902), Esna (1908) and Nag Hammadi (1908) and the Aswan Dam (1902) which had then an annual storage capacity of 1×10^9 m^3.

After World War I it became obvious that a successful management of the river should be based on cooperation with Sudan, which had built the Sennar Dam in 1925. The 1929 Nile Water Agreement stipulated that, taking into account an average annual discharge at Aswan of 84×10^9 m^3, Egypt would be guaranteed 48×10^9 m^3 y^{-1} water and Sudan, 4×10^9 m^3 y^{-1}. Moreover, the Nile discharge from January to July was assigned to Egypt.

The considerable standard deviation of the annual discharge (20×10^9 m^3) underlined by the drought year of 1912 which brought Egypt only 42×10^9 m^3 water, led to the concept of long-term storage. A first version of this concept was the Century storage scheme promoted by Dr H.E. Hurst from the Egyptian Irrigation Ministry. This was a multi-dam approach with long-term storage in Lakes Albert, Victoria and Tana and the draining of the Sudd Swamps. A second version proposed by the Egyptian–Greek engineer Daninos envisaged the construction of a single dam at Aswan. Because of political and economic considerations, preference was given to the second project which was approved in principle in 1952. Table 35 gives the list and main features of the dams built on the Nile during the twentieth century.

The construction of the Aswan High Dam was initiated in 1960. The reservoir started filling in 1964 and the official inauguration of the dam took place in January 1971. The resulting reservoir is called Lake Nasser in its northern part and Lake Nubia in the south. At full capacity, the reservoir is 500 km long and the lake level reaches 180 m above MSL, whereas the natural river bed altitude ranges from 150 m upstream to 85 m above MSL at Aswan. It extends from 23°58′ to 20°27′ N and from 30°35′ to 33°15′ E (Entz, 1976).

Table 35 *Water storage schemes on the Nile (modified from Mancy, 1981)*

	Date completed	Capacity (10^9 m^3)	Location
Old Aswan Dam	1902	1.0	Aswan, Egypt
Sennar	1925	0.9	Blue Nile, Sudan
Old Aswan Dam[a]	1933	5.3	Aswan, Egypt
Gebel Aulia	1937	3.5	White Nile, Sudan
Owen Falls Dam	1954	200.0	Lake Victoria, Uganda
Khasm el Girba	1966	1.2	Atbara, Sudan
Roseires	1966	3.0	Blue Nile, Sudan
Aswan High Dam	1970	167.0	Aswan, Egypt

[a] The first Aswan dam built in 1902 was raised in 1912 and again in 1933

Lake Nasser–Nubia is a long, narrow and dendritic reservoir bordered on its eastern side by rocky, steep mountains and by the flat and sandy Libyan desert on its western bank (Tables 36 and 37).

A detailed description of the reservoir is given by Entz in *The Nile* monograph (Rzoska, 1976). We shall therefore limit ourselves to a brief outline and develop more extensively the ecological impact of this huge technological intervention on the River Nile.

In the central part of this river-lake, the water flows at a speed of 100–150 cm s^{-1} in the Nubian gorge but the speed drops to 0.3 cm s^{-1} in Lake Nasser. Conversely, in the side, submerged valleys (*khors*) lacustrine conditions prevail. During the flood, the water level increases by 10–20 m. Below the dam, the flood peaks are no longer felt. The lower Nile receives a nearly constant water supply with an additional amount in the irrigation periods.

Table 36 *Morphometric features of Lake Nasser and the total reservoir (Entz, 1976)*

	Lake Nasser		Total reservoir	
Water level	160	180	160	180
Length (km)	292	292	430	406
Surface (km^2)	2585	5248	3057	6216
Volume (km^3)	55.6	132.5	66.9	156.9
Shoreline (km)	5383	7844	6027	9250
Shoreline development	29.9	30.5	30.8	33.1
Mean width (km)	8.9	18.0	7.1	12.5
Mean depth (m)	21.5	25.2	21.6	25.2
Max. depth (m)	110	130	110	130

Additional hydrological details are summarized in Table 37

Table 37 *Main hydrological features of the reservoir (from Waterbury, 1978)*

Maximum recorded flow of the Nile at Aswan	13.500 m^3 s^{-1}
Minimum recorded flow of the Nile at Aswan	275 m^3 s^{-1}
Total capacity	157 × 10^9 m^3
Dead storage capacity	30 × 10^9 m^3
Operational storage capacity	90 × 10^9 m^3
Flood operation storage capacity	37 × 10^9 m^3
Possible constant annual draft from reservoir	84 × 10^9 m^3
Average annual evaporation from seepage	10 × 10^9 m^3

The Nile Basin

Lake Nasser is a warm monomictic lake with a circulation period from November to March. The minimal winter temperature fluctuates between 15 and 17°C. The maximal summer temperature reached 34.5°C in 1972 in Lake Nubia. The thermocline lies between 10 and 15 m. Its maximum of stability is in June–July. In the riverine upper part of the reservoir the water is mixed throughout the year.

Dominant winds blow from north to north-west and the accumulation of superficial warm water on the eastern shore and upwelling on the western side is observed not infrequently. The winds enhance the evaporation which, in this desertous and cloudless area, is normally high (3000 mm y^{-1} whereas the average precipitation is 4 mm y^{-1}).

In April–May, with the onset of the stratification, the oxygen content of deeper layers decreases and, in July, the hypolimnion is devoid of oxygen and H_2S forms but remains at very low concentrations.

The transparency of the water, which reaches 70–140 cm before the flood, drops to 3–30 cm with the flood water which remains separate as far as the II Cataract area. Outside the flood period, the water is green and the depth of 1% incident light does not exceed 2–6 m.

The inflowing water has the Blue Nile chemical print with $Ca^{2+} > Na^+$ and the lake water keeps this basic feature although a slight increase in Na^+ and Cl^- is noted.

Volvox spp. seem to have been dominant in spring previously to 1972. Later, *Microcystis aeruginosa* became very common at this period. Abu Gideiri & Ali (1975) investigated the composition of zooplankton during the period October 1971–September 1972. *Thermocyclops neglectus* and *Mesocyclops leuckarti* represented 18.1%, *Tropodiaptomus* sp. and *Thermodiaptomus* sp. 15.8%, and *Cladocera* 35.7% of the zooplankton. Nauplius larvae, *Brachionus* sp., chironomid larvae and *Keratella* sp. represented ~30% of the total zoobiomass. The peak of zooplankton moves from south to north ahead of the flood water. In areas touched already by the flood, the zooplankton decreases. The flood water does not reach many side-arms which therefore play a major role as a permanent inoculum source.

The molluscan benthic fauna has been destroyed by the lack of oxygen during stagnation; *Tubifex* and chironomids form most of the benthos at present.

In its first years of existence, submerged and floating vegetation was absent and lakeshore plants died during the inundation. In shallow waters, *Potamogeton pectinatus* appeared in 1972 and *Najas* sp. in 1973. Until now, no floating plants have been observed on the lake, not even *Eichhornia crassipes* which is present in other sections of the Nile. During the inundation, 500 000 date palms have been destroyed. Later *Glinus lotoides, Tamarix nilotica* and *Hyoscyamus niloticus* started to grow again along the main channel.

The fishes of Lake Nasser are reviewed by Latif (1976), who found 57 species belonging to 15 families; the most important species are *Tilapia nilotica, T. galilaea, Hydrocynus forskalii, Alestes dentex, A. baremose, Lates niloticus, Bagrus bayad, B. docmac, Synodontis serratus, S. schall, Barbus bynni, Labeo horis, L. niloticus, L. coubie* and *Eutropius niloticus*.

Tilapia spp., *L. niloticus* and Characidae contribute 80–90% of fish landings. In absolute values, the landings increased from 762 tons in 1966 to 12 257 tons in 1974. *T. galilaea* appeared in 1970 and developed very rapidly. In Lake Nubia, Abu Gideiri & Ali (1975) found that *Eutropius niloticus* and *Alestes baremose* represented 51% of the fish population.

The brackish water lakes (Fig. 22)
General

The Nile carries $100-130 \times 10^6$ m^3 y^{-1} of sand, silt and clay mainly originating in the Blue Nile. About 90% of the annual sediment load is delivered during the four months of the flood. The concentration of suspended material rises from 100 mg l^{-1} at low water to 2500–4000 mg l^{-1} at the peak of the flood (Rzoska, 1976). In the pre-Aswan High Dam period, 52% of the sediment reaching Egypt was not deposited in the Nile Valley but transported to the Delta. The alluvial deposits are ~9 m thick in the Nile Valley and ~12 m in the Delta. This 12 m alluvial layer is saturated

Fig. 22. The brackish lakes of the Nile Delta

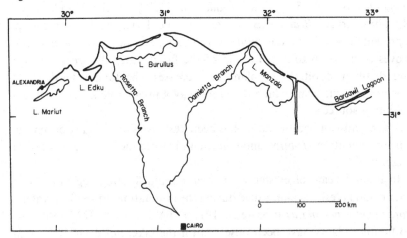

The Nile Basin

with Nilotic freshwater. Underneath is a layer of sand and gravel which is probably of Pleistocene age and local origin and was deposited during a more rainy and erosive period. This layer has a salty aquifer.

This old and recent material has built the Delta area. At present the Nile divides into two branches (Damietta and Rosetta) and two promontories developed in the mouth areas of these two branches. The area between these promontories corresponds to the sediments deposited by an ancient Nile branch. The mouth of this ancient Nile corresponds to the present exit of Lake Burullus. Three other lakes have a similar formation pattern: Lakes Edku and Mariut, west of Rosetta, and Lake Manzala, east of Damietta. These lakes are bordered on the sea side by sand bars of Nilotic origin which were spread on the eastern sides of the mouths of the river by the marine eastward current (Kassas, 1972) which is part of the general counterclockwise circulation of the Mediterranean Sea. These lakes drain the delta waters but most of them are connected to the sea through outlets and surrounded by salt marshes. The fifth brackish lake is not located in the Delta but in the Fayoum province, 100 km south-west of Cairo. Lake Qarun is an endorheic lake, a remnant of a much greater water body.

Lake Mariut

A vestige of a much larger lake in Pharaonic times, Lake Mariut remained dry from the twelfth to the eighteenth centuries except at high water. In the irrigation scheme of the surrounding area, developed at the end of the last century, Lake Mariut became the drainage basin of the irrigated fields. The water level is kept constant (2.8 m below MSL) by a pumping station which discharges the surplus freshwater to the sea ($30-100 \times 10^6$ m^3 per month) since the lake has no connection with the sea.

The lake area is 63 km^2. It is divided by dykes into four basins: the eastern basin (presently receiving untreated sewage and industrial waste), the fish farm and two western basins. The lake depth rarely exceeds 120 cm.

A detailed study was carried out by Aleem & Samaan (1969*a,b*). Wind mixing and high algal densities explain that Secchi-disc transparency varies from 10 to 60 cm. The average monthly temperature of the water fluctuates from 13 to 29°C. No stratification is observed because of the constant mixing caused by the dominant north-west winds. The average annual rainfall is 200 mm but the average evaporation amounts to 1550 mm y^{-1}.

The chloride content of the drainage water ranges from 1.2 to 2.8 g Cl$^-$ l^{-1}. Because of evaporation, however, lake water chloride content varies from 2.5 to 5.6 g Cl$^-$ l^{-1}. A more recent survey in 1972–73 (Wahby, Kinawy, El Tabbagh & Abdel Monein, 1978) reports chloride concentrations ranging from 1 to 5 g l^{-1}. Chloride is the major anion; it is followed by sulphates. Among the cations, sodium dominates followed by magnesium, calcium and potassium (Wahby, 1961). The pH values range between 7.8 and 9.4. Minimum values occur in winter. High concentrations of oxygen are recorded in winter because of permanent wind mixing. In summer, the oxygen consumption increases up to 10 cm^3 O$_2$ l^{-1} d^{-1} in some areas and local complete depletion is observed. In the 1972–73 survey, Wahby *et al.* (1978) showed that, in stations far from pollution sources such as the fish farm, no oxygen depletion was observed whereas polluted areas were depleted of oxygen most of the year. In these latter areas, H$_2$S concentration ranges between 20 and 50 mg l^{-1} but peaks of 116 mg l^{-1} have been measured. The polluted areas are also rich in ammonia (up to 7 mg l^{-1}).

Nitrate content varies from 280 µg l^{-1} in winter to very low values in summer. Silica peaks in winter but its range of variation is from 150 to 300 µg at l^{-1}. Phosphate used to be high in winter and very low the rest of the year with ranges of 0 to 5.6 µg at l^{-1} (Wahby, 1961) and 0 to 3.5 µg at l^{-1} (Aleem & Samaan, 1969a). A recent survey by Wahby & Abd El Moneim (1979), carried out in 1974–75, showed a considerable increase in the phosphorus concentration of Lake Mariut. In the polluted areas, the phosphorus content varied from 80 µg at l^{-1} in winter to 43 µg at l^{-1} in autumn. In unpolluted areas, the respective values were 9.5 and 2.2 µg at l^{-1}. This remarkable increase is due both to direct sewage discharge and phosphorus release from bottom sediments. Wahby & Abd El Moneim estimate the annual loading of the lake to be 13 g P m^{-2}.

Biological features. The phytoplankton of the lake was studied by Salah (1961) and Aleem & Samaan (1969). These two papers indicate a spring maximum of biomass dominated by diatoms (*Cyclotella, Nitzschia, Cocconeis* and *Mastogloia*) and dinoflagellates. The biomass diminishes significantly in summer with a preponderance of diatoms and blue-greens (*Merismopedia, Microcystis, Spirulina* and *Oscillatoria*). A second autumnal peak is dominated by diatoms. Aleem & Samaan (1969) found 35 species of Cyanophyta, 20 species of Chlorophyta, 33 species of diatoms, five species of Euglenophyta and a few species of Cryptophyta and Dinophyta. We note that Salah found that, in Lake Mariut, *Peridinium cinctum* was the only observed dinoflagellate and reported that it was abundant in March; in contrast, Aleem & Samaan did not mention *Peridinium* in their list but did mention *Gymnodinium lohmanni*.

Gross primary production rates amount to 7.5 g C m^{-2} d^{-1} for the planktonic population of the polluted area of the lake, 5.2 g C m^{-2} d^{-1} for

The Nile Basin

lake water and 2.45 g C m^{-2} d^{-1} for the *Potamogeton* area. The net productivity was ~5 g C m^{-2} d^{-1} for the polluted area and 3 g C m^{-2} d^{-1} for other lake areas. These values indicate that Lake Mariut ranks among the most productive lakes of the world.

The zooplankton of Lake Mariut has been investigated recently by El-Hawary (1960), Elster & Vollenweider (1961) and Samaan & Aleem (1972a). Lake Mariut is characterized by a relative poverty in true zooplankton. 'Tychoplankton' constituted 80% of the population. The bulk of the macroplankton is due to *Leander squilla, Mysis* and *Gammarus*. The net-zooplankton includes larvae of *Balanus*, rotifers such as *Brachionus*, nematodes, ostracods, oligochaetes and polychaete larvae. Detailed lists of species can be found in the previously mentioned papers. El-Hawary found that *Moina dubia* and *Daphnia* sp. were lacking in Lake Mariut. *M. dubia* appeared only in 1961 and was especially abundant in summer. The production of zooplankton is high in the lake area and significantly lower in the *Potamogeton* belt, probably because of the abundance in this area of the fish *Gambusia affinis*, a zooplankton-eater. The biomass of zooplankton is very variable from year to year, probably because of the variations of salinity.

It is interesting to note that the production of algae is far in excess of the grazing capacity of the existing zooplankton.

The bottom fauna is poor in the polluted zone because of oxygen depletion. In the *Potamogeton* belt, chironomid larvae and *Gammarus* are the most common animals. The lake area (non-polluted) has the richest fauna with *Nereis diversicolor, Melania tuberculata, Corophium* spp., *Gammarus* and *Chironomus* larvae. In this area, the benthic fauna is estimated to be 76 g wm m^{-2}. The sediments of Lake Mariut have the highest average value of calcium carbonate content (63.3%) due to the accumulation of shell debris of *Balanus improvisus* and *Mercierella enigmatica* (Saad, 1974).

The fish population of the Hydrodrome (the eastern part of the lake transformed in 1939 into a sea-plane base, presently utilized as a naval training base and experimental fish lake), has been investigated by Elster & Jensen (1960). In 1954, this area of 5 km^2 supplied a commercial catch of 31 tons, 84% of which were *Tilapia* spp., *Barbus bynni* and *Anguilla vulgaris*. The analysis of the gut content of the fish gave the following information: *Tilapia nilotica* feeds mainly on diatoms; the *T. gallilaea* alimentary canal contained filamentous green algae, diatoms, pieces of water plants and fragments of *Gammarus; T. zillii* feeds mostly on *Potamogeton* and *Najas* covered with epiphytic diatoms; *Lates niloticus* feeds on small *Tilapia, Leander squilla, Gammarus* etc.; *Barbus bynni* has a mixed diet of plants, amphipods and diatoms; the eels feed essentially on fish; and the digestive tract of *Mugil capito* contained mud and numerous diatoms.

In Lake Mariut, the statistics of fish catch (year 1961) gave the incredible figure of 900 kg fm ha^{-1} y^{-1} (Aleem & Samaan, 1969). Welcomme (1979), in his report on Inland Fisheries of Africa, reported an average annual catch of 3800 tons for the period 1970–76 or 450 kg ha^{-1} y^{-1}.

Lake Edku

Lake Edku has a surface area of 130 km^2 and is connected with the sea; consequently, its water level is in equilibrium with the sea level. The drainage of irrigation water into the lake generates a seaward current and any lake level drop causes a lakeward movement of sea water. Variations in salinity from 0.42% in January to 1.48% in July are therefore important. The water depth varies from 50 to 150 cm.

The seasonal variations in nutrients in Lake Edku depend less on phytoplankton blooms or mineralization in the lake than on the concentrations of nutrients of agricultural origin in the drainage water which enters the lake. Saad (1978) investigated the seasonal fluctuations of the major nutrients in the lake.

The phytoplankton (Salah, 1961) is identical to that of Lake Mariut and showed a similar seasonal pattern although the Dinophyta were less abundant. The algal biomass was approximately one-third that of Lake Mariut.

The gross primary productivity in Lake Edku ranges from 39 to 2034 mg C m^{-2} d^{-1}, the average value amounting to 604 mg C m^{-2} d^{-1} (Samaan, 1974). Compared with the other Egyptian Delta lakes, Lake Edku ranks among the low productive lakes as far as planktonic production is concerned. In contrast, the average production of the hydrophytes *Potamogeton pectinatus* and *Ceratophyllum demersum* is as high as 1320 mg C m^{-2} d^{-1}. The production of organic matter in Lake Edku is carried out mostly by the hydrophytes: the high flushing rate of the lake is unfavourable to the building up of planktonic communities whereas the nutrient renewal is extremely beneficial to the rooted vegetation. El-Hawary (1960) gives a detailed list of the zooplankton of Lake Edku: the most abundant species are *Cyprideis littoralis, Canuella* sp. (marine form), *Acartia latisetosa* (marine form), *Thermodiaptomus galebi, Mesocyclops* sp., *Diaphanosoma excisum, Daphnia* sp. and *Moina dubia*. The relative abundance of Cladocera in Edku is a distinctive feature.

The Amphipoda, with *Gammarus foxi, G. aequicauda* and *Corophium volutator*, and the Polychaeta, with *Nereis diversicolor* and *Merceriella enigmatica*, dominate the benthic fauna (Ezzat, 1972).

Welcomme (1979) reports an average fish catch of 4000 tons y^{-1} for the period 1970–76 (i.e. 308 kg ha^{-1} y^{-1}). For the period 1962–68, the catch varied from 3500 to 7500 tons with 2.4–8.4% contributed by the grey mullet.

Lake Burullus

Lake Burullus has an area of 504 km^2. It is separated from the sea by a sand bar extending from the Damietta–Nile to the natural outlet of the lake. This sand bar was formed from Nile sediments swept eastward by the shore current. The sand bar is 5 km wide near the Nile but only 200 m near the outlet. The depth of the lake does not exceed 180 cm. It is well mixed and salinity varies from 2% in the area receiving the drainage water to 20% at the outlet. The drainage waters are below the lake level and are pumped into the lake. In the southern shore of the lake, small basins have been isolated from the lake for fish farming. These fish farms are bordered southward by a *Suaeda–Salicornia* marsh. In recent years, the high concentration of pesticides and herbicides of the agricultural drainage water caused mass fish killings in this lake. The small drains are colonized by bilharzia-carrying snails; the copper sulphate and other molluscicides added to the drains represent a supplementary hazard for the lake's fish population.

Welcomme (1979) reports an average fish catch of 15 000 tons y^{-1} for the period 1970–76, representing 300 kg ha^{-1} y^{-1}. El Moghraby, Hashem & El Sedfy (1973) reported that, in Lake Burullus, the catch increased from 8200 tons in 1962 to 15 000 tons in 1966. The contribution of *Tilapia* sp. was 60–70%; that of the grey mullet varied from 29% in 1962 to 22% in 1966.

Lake Manzalah

Lake Manzalah is the largest of the Delta lakes (1450 km^2); its depth rarely exceeds 120 cm. It is connected with the sea at El Gamil, west of Port Said, with the Suez Canal at El Kabouti and with the Damietta Nile by three canals. Since 1965 the lake has not received Nile water but only agricultural drainage. The temperature and mixing conditions are similar to those of other Delta lakes.

The concentration of chloride in the south-east region varies from 0.68 to 2.00 g l^{-1} whereas in the north-eastern area, affected by the sea, it fluctuates from 2.9 to 10.1 g l^{-1} (Wahby, Youssef & Bishara, 1972). Shaheen & Yosef (1978) reported a range of 0.88 to 15.7 g Cl$^-$ l^{-1}. pH values are in the range 7.9–8.8; supersaturation of oxygen is frequent and depletion does not occur. Phosphate levels vary from 0.1 to 2.4 μg at l^{-1} PO$_4$-P, nitrates from 1 to 23 μg at l^{-1} NO$_3$–N (Wahby *et al.*, 1972) and silicates from 100 to 460 μg at l^{-1} SiO$_2$–Si. The nutrients originate from the agricultural drainage water; therefore, an inverse correlation is observed between salinity and nutrient correlation (Shaheen & Yosef, 1978).

New series of chemical measurements showed a marked decrease in salinity in 1973 because the sea–lake connection had been blocked since 1970 as a consequence of the War. The drainage water then became a dominant constituent of the lake water and its high content of nutrients compensated the loss of fertilizing substances of the Nile. It follows that Lake Manzalah still has the same level of fertility as before 1964.

Welcomme (1979) reports an annual average fish catch of 21 000 tons (period 1970–75) representing 145 kg ha^{-1} y^{-1}. For the period 1962–68, Hashem, El Moghraby & El Sedfy (1973) report a total catch of 22 600–27 700 tons with 6–16% contributed by the grey mullet.

Lake Qarun

Lake Qarun (Fig. 23) is the only lake of Upper Egypt. It is located in a depression of the Libyan desert at an altitude of 44 m below MSL and separated from the Nile by a low ridge. It is nevertheless connected with the river through the Hawara channel. Lake Qarun is the vestige of a larger lake (Lake Moreis). It is 40 km long on the SW–NE axis and its area is ∼200 km². Lake Qarun is divided into two basins: the eastern and shallower basin is 1–5 m deep with a mean depth of 3 m; the western basin ranges from 2 to 8.5 m. The lake receives mainly drainage water, ∼350 × 10^6 m³ y^{-1}, and an average of 430 000 tons y^{-1} salt (Meshal, 1973).

Fig. 23. Lake Qarun and Qattara depression

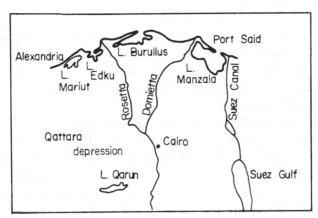

Since the evaporation has been estimated to be 455×10^6 m^3 y^{-1} and the rain is negligible, the lake has probably an unknown underground income (Naguib, 1958).

The temperature of surface water varies from 10°C in January to 32°C in August. The daily variation is 4–7°C. The lake is not stratified. According to Naguib, the absence of stratification is due to evaporation which increases the density of surface layers and induces a vertical mixing which is consequently maximum in summer.

The continuous accumulation of salt in this closed lake causes salinity to increase with time: 13.4‰ in 1901 (Lucas, 1906), 18‰ from 1919 to 1923, 23.4‰ in 1934, 30.6‰ from 1953 to 1955 (Naguib, 1958) and 30.9–34.5‰ in the period 1974–76 (Ishak & Abdel Malek, 1980). These latter authors emphasize the fact that the diminution of the Nile flow into the Mediterranean Sea after the construction of the Aswan Dam (from 47 km^3 y^{-1} up to 1965 to 4.4 km^3 y^{-1} after 1965) caused an increase in the salt concentration of the drainage water feeding the lake and accelerated the process of salinization of the lake water.

The pH of surface water is about 8 and the pH of sediments varies from 7.0 to 7.9. The alkalinity is of an order of magnitude of 200 mg l^{-1} CaCO$_3$. Not only was no oxygen depletion observed near the bottom, but in summer there was an increase in oxygen content in the deep layers related by Naguib to the photosynthesis of submerged plants.

Nitrate concentration ranges from 40 to 50 μg at l^{-1} NO$_3$–N in the whole water column; there is no indication of denitrification. These high concentrations are due to the nitrate-rich drainage water. Phosphate concentrations vary from 0 to 0.7 μg at l^{-1} PO$_4$–P.

The phytoplankton and zooplankton of Lake Qarun were studied by Naguib (1961) during 16 consecutive months (November 1953–February 1955). The phytoplankton was heavily dominated by the diatoms *Nitzschia* and *Synedra* although the blue-greens *Oscillatoria* and *Lyngbya* were occasionally present. The zooplankton was dominated by copepods: *Acartia* sp., *Diaptomus* sp.; *Leander* developmental stages and *Gammarus* were also abundant.

Table 38 shows the monthly distribution of diatoms and copepods. The diatom biomass is maximal from October to April and strongly declines in summer; copepods show less variation and have their peak in spring–summer. The absolute biomass values are relatively low.

The survey conducted by Abdel Malek & Ishak (1980) from 1974 to 1977 shows that the lake was then still dominated by the diatoms, especially *Rhizosolenia setigera, Nitzschia closterium, Melosira granulata, Mastogloia braunii* and *Gyrosigma* sp. The dinoflagellates, less numerous than diatoms, were represented by *Prorocentrum micans, Exuviella apora, Peridinium* sp. and *Gymnodinium* sp. In 1975, the annual average of algal biomass was ~100 mg dm m^{-3}, with peak values of 147 mg dm m^{-3} in spring and minimal values of mg dm m^{-3} in autumn. The zooplankton was still dominated by the copepod *Acartia latisetosa* (5–8 l^{-1}) and copepod nauplii but larvae of Mollusca and Polychaeta represented about 15% of the zooplankton. The rotifers (*Testudinella patina, Asplanchna* sp., *Brachionus* sp. and *Keratella* sp.) had concentrations of 1–3 l^{-1}. The range of zooplankton biomass was from 9 g dm m^{-3} in winter to 40–50 mg dm m^{-3} in spring.

This picture is very different from that described by Wimpenny & Titterington (1936) who observed only three species of zooplankton: *Diaptomus salinus, Leander squilla* and *Moina salinarum*. These animals completely disappeared with the increase in salinity and were replaced by marine species introduced into the lake with the mullet fry from the Mediterranean Sea.

The benthic fauna of Lake Qarun is especially rich in molluscs. The survey by Abdel Malek & Ishak (1980) indicates that the average benthic biomass is 413 g dm m^{-2}, 70% of which is constituted by the bivalve

Table 38 *Monthly averages of diatoms and copepods in Lake Qarun (Naguib, 1961)*

Year	Month	Diatoms (cells l^{-1})	Copepods (individuals l^{-1})
1953	Nov.	68 480	20
	Dec.	76 930	27
1954	Jan.	95 091	31
	Feb.	100 851	34
	Mar.	104 205	40
	Apr.	73 810	44
	May	23 690	47
	June	9 190	51
	July	7 525	56
	Aug.	6 830	42
	Sep.	5 901	30
	Oct.	52 490	17
	Nov.	77 670	8
	Dec.	76 620	23
1955	Jan.	63 530	26
	Feb.	64 320	30

Cerastoderma glaucum (*Cardium lamarcki*).

Four main species of fish inhabit Lake Qarun: *Tilapia zillii, Mugil cephalus, M. capito* and *Solea vulgaris*. The species were artificially introduced in 1928 and the lake has been stocked yearly with these species since then. The feeding habits of these four species were investigated by Naguib (1961). *T. zillii* fed mostly on diatoms, blue-green algae and pieces of higher plants; in summer, during the algal minimum, many stomachs were devoid of algae and the fish compensated for this lack of plant food by eating copepods. *M. cephalus* fed essentially on blue-green algae and diatoms. In summer, during the diatom minimum, this fish fed on copepods and amphipods. *M. capito* fed on diatoms, blue-greens, copepods and *Glochidium* larvae. *S. vulgaris* consumed mainly animal food (amphipods, insect larvae, fish larvae, mollusc larvae) but algae were also present in their digestive tract. Mollusc shells were always found in the diet of *S. vulgaris*.

Naguib (1961) mentions that massive death of molluscs occurs in autumn, probably as a result of water stagnation due to an abundant supply of freshwater drainage water which does not mix with lake water.

Ecological impact of the Aswan High Dam
General

The building of the Aswan High Dam was aimed at increasing the agricultural production of Egypt and Sudan and producing electricity at an order of magnitude of 10 billion kilowatt hours (kW h) per year. The dam was also intended as a flood regulator. Eleven years after its completion, it is clear that the dam has contributed much progress in these three fields. The dam produced 5 billion kW h in 1975, 6 billion kW h in 1976 and 7.15 billion kW h in 1977, which represents about half of the electric-power requirement of Egypt (US Department of Energy, 1979). This production permitted rural electrification and industrial expansion.

Although land reclamation progressed more slowly than expected, the net cultivated area of all reclamation projects in 1982 will be 1.3 million feddans (5500 km^2) (Waterbury, 1978). Moreover, the increase in income due to rice production is impressive: from LE6.7 million in 1963 just before the Nile diversion to LE55 million in 1969. The dry period 1972–73 has emphasized the positive role of the reservoir: traditionally, the low discharge of this year (47×10^9 m^3) would have been accompanied by a dramatic 40% decrease in agricultural income. The dam supplied enough water for summer cotton and rice crops. A series of high floods does not any longer represent a threat to lower Egypt since, as shown in Table 37, the reservoir has a flood capacity of $37 \times 10^9 \, m^3$; recently a relief canal, the Toshka canal, has been built and allows the diversion of water into a depression extending west of the reservoir.

On the occasion of the ninth anniversary of the Nile diversion in May 1973, it was stated that income generated by the dam had already compensated for the cost of the construction of the dam and the power station (*Al-Ahram*, 17 May 1973). This opinion is supported by the cost-benefit analysis of Shibl (1971). The impact of the reservoir is however in direct proportion to its huge size. We shall review the already obvious effects of the dam on its environment.

Seepage and evaporation

Gloomy predictions of high rates of seepage accompanied the construction of the dam and it was even suggested that a net water loss might result from the formation of the reservoir. Since 'vertical seepage' which leads to bed-rock saturation is a slow process, it is difficult to give definitive answers now. 'Horizontal seepage', however, is less than $1 \times 10^9 \, m^3 \, y^{-1}$ and, in 1971, the Ministry of Irrigation estimated that the total losses through evaporation, absorption and seepage amounted to $11.4 \times 10^9 \, m^3$ which compares fairly well with the $10 \times 10^9 \, m^3$ predicted in the initial stages of planning and leaves the water balance very far from a net water loss. Underground seepage and generous irrigation have raised the water table downstream. The rising water level is damaging the foundations of the Karnak temples at Luxor (Mancy, 1981).

The Nile flood and sediment load

Prior to the Aswan High Dam period, the Nile discharged into the Mediterranean Sea an average annual amount of $43 \times 10^9 \, m^3$ freshwater, of which $34 \times 10^9 \, m^3$ was discharged during the August–November period. The Nile also carried ~ 100–$130 \times 10^6 \, m^3 \, y^{-1}$ sediment which was partly deposited in the Nile Valley and Delta; a non-negligible amount (88% according to Waterbury, 1978), reached the sea and was deposited off the Damietta and Rosetta branches of the Nile.

The last Nile flood discharge into the Mediterranean occurred in 1964–65. From 1964–65 to 1967–68, a truncated flood reached the sea and from 1967–68 until now the discharge has been negligible. This drastic change in regime caused scouring of the river bed below the dam and coastal erosion, and deeply affected the brackish-water and marine fisheries.

Problems of a different nature modified the upper reaches of the reservoir. In 1968, when the level of the lake reached 145 m in July and 156 m in December, the current of the river at its entry into the newly formed lake was reduced and the flood load was deposited in the reservoir. Near Aswan, the Nile water, which at the flood period used to carry > 2500 mg l^{-1} matter in suspension, became transparent. The sediment load does not form an homogeneous deposit in the reservoir. The bulk of the silt is sedimented near Wadi Halfa and the Second Cataract where, over a short distance, a turbulent current of 100 cm s^{-1} (flood period) is followed by a laminar flow of 10 cm s^{-1}. A layer of 25 m of sediment now covers the beautiful site of the Second Cataract (Entz, 1976). North of Wadi Halfa the sediments grow thinner and thinner.

Scouring of the river bed. The low silt content of the river water below the dam increases the erosional capacity of the water. The resulting erosion of the river bed endangers the barrages as well as the foundations of bridges and embankments. The construction of weirs in the vicinity of barrages has been envisaged to prevent local scouring (Waterbury, 1978).

Coastal erosion. By comparing the water characteristics and hydrographic conditions off the Egyptian coast before and after 1964, Sharaf El Din (1977) pointed out the major modifications caused by the Aswan High Dam. Before 1964, during the flood period, the estuarine circulation consisted of a two-layer flow at the mouths of the estuaries: an upper, seaward flow of Nilotic freshwater and a deeper, landward flow of sea water; in winter, there was a one-layer landward flow of sea water. After 1964, this latter pattern prevailed throughout the whole year. Before 1964, the velocity of the currents at the mouth of the Nile branches exceeded 4 knots at the surface and 0.5 knots at the bottom; after 1964 this velocity decreased significantly.

As previously described, the Nile sediments reaching the sea used to move eastward with the general anticlockwise circulation generating sand bars and dunes which all contributed to the accretion of the delta. Approximately 100 years ago, an inverse process was initiated; the gradual retreat of the delta has been described in detail by Kassas (1972). For example, the Rosetta mouth of the Nile has lost about 1650 m of its length from 1898 to 1954 at variable rates of 15–81 m y^{-1}. Similarly, villages established on sand bars had to be moved landward. This obvious imbalance between sedimentation and erosion seems to have started with the Delta barrages (1861) which already diminished the amount of Nile water and sediment reaching the sea. The later works (Sennar Dam 1925,

Gebel Aulia Dam 1937, Khasm el Girba Dam 1966, Roseires Dam 1966) enhanced the regression of the Delta. Finally, the Aswan High Dam, which completely suppressed the sediment supply to the Delta, endangers the very existence of the present shoreline.

In a recent study, Manohar (1981) found the 1000 km Egyptian coast 'undernourished' except for 3–4 km in Abu Quir Bay and a 40 km reach east of Lake Burullus which are in equilibrium, and an 8 km reach east of the Damietta outlet which is overnourished. As a result of this overall erosion of the shoreline, the sand bars bordering Lake Manzala and Lake Burullus on the sea side are eroded and likely to collapse. Should such a thing occur, the lakes would be turned into marine bays and saline water would come into contact with cultivated land which is nearly at sea level and with freshwater aquifers.

Similarly, industrial and harbour installations will have to be protected and it is difficult at present to estimate the investments which will be required for such an enterprise.

The spreading of bilharzia

Bilharzia has been called a 'man-made disease' (Weir, 1959). Although the disease has existed in Egypt since Pharaonic times, 'it has taken astronomic proportions since the spread of modern irrigation' (Van der Shalie, 1972). The eggs of the worms *Schistosoma haematobium* (urinary variety of bilharzia) and *S. mansoni* (intestinal variety) hatch in water and are, respectively, hosted by the snails *Bulinus* and *Biomphalaria*, which allow the production of the infective larvae and their release into the water. Humans who come into contact with contaminated waters are infected: the larvae find their way into the blood vessels serving the urinary or intestinal systems and develop into adult worms which produce eggs, which in their turn are released with urine or faeces. Scott (1937) studied the incidence of schistosomiasis in the Nile Valley and discovered a correlation between high incidence of the disease (60% of the population infected) and perennial irrigation, for example in the Delta area; in contrast, the incidence of the disease fell to 5% in areas practising only seasonal irrigation (Aswan to Cairo area). This is of course related to the greater ability of the snails to thrive in permanent water bodies. In December 1968, at the 'Conference of the ecological aspects of international development', Van der Shalie underlined that, since the Aswan High Dam provides water for perennial irrigation allowing four crops a year, 'the prospects are good that this new agricultural development will produce a tremendous increase in schistosomiasis'. According to C.A. Wright (1976) *Bulinus truncatus* and its associated parasite *Schistosoma haematobium* have already settled in the

The Nile Basin

Lake Nasser area. Recent studies, however, (Miller, Hussein, Mancy & Hilbert, 1978) indicate that there has been a general decline in the disease in the rural population of Egypt. This decline coincides with a noticeable improvement in control of water supplies in rural Egypt and the data do not show 'an increase in the overall prevalence of schistosomiasis following the construction of the Aswan High Dam'.

Impact on the fisheries

From 1970 to 1976, the fish catch of inland waters of Egypt increased from 53 700 to 80 660 tons (these figures include aquaculture) (Welcomme, 1979).

In contrast, the marine sardine fishery, which represented 48% of the marine catch, dropped from 18 000 tons y^{-1} before 1965 to 500 tons y^{-1} in 1968 (Worthington, 1972). It is probable that such a rapid decline is due not only to the decline in stock but also to dispersal of the sardine population which used to concentrate off the Nile Delta and feed on the extremely rich plankton which was washed to the sea with the flood. However, Al-Kholy & El-Wakeel (1975), in their report of the Soviet–Egyptian expedition of 1970–71 reported that, in the absence of the freshwater layer of the Nile, the sea water off the Delta area was better mixed. This mixing caused an enrichment of nutrients and a subsequent phytoplankton development which, according to the authors, partly explained the increase in sardine catch observed in 1971 (1083 tons instead of 580 tons in 1970).

In the Delta lakes, the average yearly catch for the period 1970–76 was 43 000 tons (Welcomme, 1979). Many species but especially the *Tilapia* spp. are suffering from the intensive use of pesticides, herbicides and of copper sulphate used against the snail hosts of bilharzia. The increase of sewage discharges, particularly in Lake Mariut, may also affect the fisheries.

Impact on eutrophication

We have seen that the reduced current velocities due to storage reservoirs were favourable to the production of large algal populations. The multiplication of the reservoirs on the Nile River has considerably increased the overall algal biomass of the Nile waters which seems to be much in excess of the requirements of the present herbivorous population. This is a type of eutrophication caused mainly by modification of physical factors. At present, the Nile is fairly free of man-made pollution. This situation will, however, change with the intensification of agriculture and industrial development. The effect of this potential pollution will be superimposed on an already high base level of algal productivity.

In conclusion, it is clear that the Aswan High Dam has not brought doomsday either to Egypt or to the eastern part of the Mediterranean basin; moreover, it has contributed to the economic development of Egypt. Its overall impact on the environment is not negligible and has been the subject of passionate opinions. The long-term studies carried out within the framework of the US-Egypt 'Nile Project' show that most of the new problems can be handled with proper management (Walton, 1981).

The expected shortage of electric power in the next decade has led to the elaboration of the Qattara project, which consists of the filling of the Qattara depression (133 m below MSL) with sea water. Besides the production of electric power, it is expected that the new lake will modify the climate and increase rainfall in north-west Egypt.

LAKES OF THE EASTERN RIFT VALLEY AND ETHIOPIAN RIFT VALLEY

The Eastern Rift begins in North Tanzania and continues northward across Kenya, Ethiopia and the Afar Junction where it joins the Red Sea and the Gulf of Aden. It is a zone of normal faults separating the Horn of Africa from the African Continent. The Rift is a volcanically active graben, 40–65 km wide, crossing the Ethiopian and Kenyan domes. These domes have been uplifted to 2500–3000 m above MSL. Unlike the Western Rift, drained by the Congo River and the Nile River, the Eastern Rift forms an endorheic area with numerous lakes, many of which are highly saline.

Eastern Rift Valley lakes
The lakes of the southern part of the Rift
Lake Eyasi

Located in Tanzania (3°20'–3°90' S; 34°85'–35°20' E) at an altitude of 1117 m, Lake Eyasi has a maximum length of 75 km and a maximal width of 15 km. It is a 'soda lake' of high salinity. The composition of the water, determined by Jaeger (1911), is presented in Table 39. A previous analysis by Lenk (1894) reported 11 400 mg l^{-1} Cl$^-$ and 1210 mg l^{-1} SO$_4^{2-}$.

Lake Kitangiri

Located in Tanzania (4°00'–4°60' S; 34°10'–34°25' E), at an altitude of 800 m above MSL, Lake Kitangiri has an area of 1200 km^2 (maximal length 40 km and maximal width 12 km) and a depth of 5 m. This lake has two small tributaries and an outlet which drains into Lake Eyasi. Its main chemical features are presented in Table 39. This lake has a total fish catch of 4113 tons y^{-1} (Welcomme, 1979).

Lake Manyara

Located in Tanzania (3°25′–3°90′ S; 35°85′ E), at an altitude of 1045 m, Lake Manyara has an area of 470 km² (maximal length and width are, respectively, 40 and 15 km). Its chemical features are presented in Table 39.

Lake Magadi

Located in Kenya (1°43′–2°00′ S; 36°13′–36°18′ E) (Fig. 21), Lake Magadi is divided by a ridge into two lakes at respective altitudes of 660 and 683 m above MSL with a total surface of 108 km². The composition of its water is shown in Table 40.

Table 39 *Chemical features of a few East-African lakes*

	L. Eyasi (Jaeger, 1911)	L. Kitangiri (Talling & Talling, 1965)	L. Manyara (Talling & Talling, 1965)	L. Natron (Guest & Stevens, 1951)
Conductivity (μmhos cm^{-1})	–	785	94 000	340 000
TDS (mg l^{-1})	–	404–432	–	–
Ca^{2+} (mg l^{-1})	traces	24.1	10	–
Mg^{2+} (mg l^{-1})	4	6.7	30	–
Na$^+$ (mg l^{-1})	4480	155.0	21 500	–
K$^+$ (mg l^{-1})	55	4.8	94	3000
HCO$_3^-$ + CO$_3^{2-}$ (meq l^{-1})	–	6.65	806	2600
Cl$^-$ (mg l^{-1})	4366	64.0	8670	65 000
SO$_4^{2-}$ (mg l^{-1})	340	5.0	1056	3100
pH	–	8.0–8.9	–	–
Total P (μg l^{-1})	–	1020	65 000	29 000
SiO$_2$ (mg l^{-1})	–	–	19	–

Table 40 *Chemical features of Lake Magadi (Talling & Talling, 1965)*

Conductivity 160 000 μmhos cm^{-1}
Surface temperature 28–43°C

	Lake Magadi	Hot springs		Lake Magadi	Hot springs
pH	10.5				
Na$^+$ (mg l^{-1})	38 000	12 750	HCO$_3^-$ + CO$_3^{2-}$ (meq l^{-1})	1180	411
K$^+$ (mg l^{-1})	537	188	Cl$^-$ (mg l^{-1})	22 600	6200
Ca^{2+} (mg l^{-1})	<10	<10	SO$_4^{2-}$ (mg l^{-1})	900	790
Mg^{2+} (mg l^{-1})	<30	<30	SiO$_2$ (mg l^{-1})	250	72
Total P (mg l^{-1})	11				

In this lake, the overwhelming predominance of Na^+ and Cl^- over other cations and the absence of divalent cations is probably due to the inflowing hot springs and not to secondary modifications within the lake. The composition of the hot springs, measured by Talling & Talling (1965) in February 1961, is also shown in Table 40.

The hydrochemical model, elaborated by Eugster (1970) and refined by Jones, Eugster & Rettig (1977) to explain the composition of the waters of the Lake Magadi basin takes into consideration five stages in the evolution of the water composition: dilute streamflow, dilute groundwater, saline groundwater or hot spring reservoir, saturated brines, and residual brines. The transformation from one stage to another is through an evaporative process.

Since no surface drainage is observed in the Rift between Lake Naivasha, at 1885 m altitude, and Lake Magadi at 660 m, Jones *et al.* assume that a large amount of freshwater infiltrates into the alkaline lavas forming the Rift floor. The weathering of these rocks results in a dilute groundwater (900–2000 mg l^{-1}) rich in sodium and fluoride which directly feeds the hot springs. These, in their turn, flow into the lake and concentrate by evaporation.

The Lake Magadi basin is composed of the recent Evaporite Series, the High Magadi beds (10 000 years B.P.) and the Oloronga beds (>800 000 years B.P.). The Evaporite Series contains very concentrated brines ($\sim$270 000 mg l^{-1}) in equilibrium with thick trona beds. Below, in the older formations, are less saline brines (130 000 mg l^{-1}) which partly find their way up to the hot springs.

This model has been confirmed by various boreholes drilled in the Magadi formations. It appears then that the saline waters of the lake are part of a vast hydrological complex and deep circulation within the lava formations.

In the second half of 1962, a population of 2.2×10^6 lesser flamingoes was observed (Brown & Root, 1971). In March 1969, only 6580 animals were counted (Bartholomew & Pennycuick, 1973). *Tilapia grahami* is endemic to Lake Magadi; it lives in hot springs around the lake (Coe, 1966).

Lake Natron

Located in Tanzania and Kenya (2°10′–2°35′ S; 36° E), at 675 m above MSL, Lake Natron has a surface of 1000 km² and its chemical features are presented in Table 39.

Countings of the flamingo population of Lake Natron from 1954 indicate that the number of animals generally exceeds one million (Brown

& Root, 1971). In 1969, however, their number diminished to 25 125 specimens (Bartholomew & Pennycuick, 1973).

Tilapia alcalica is endemic to Lake Natron and lives in hot springs around the lake (41°C and 30% salinity) (Coe, 1969).

Lake Elmenteita

Located in Kenya (0°25′ S and 36°15′ E), at an altitude of 1775 m above MSL, Lake Elmenteita has a surface area of 16 km² and a maximal depth of 1.9 m. The lake has a pH of 10.9 and chloride concentration exceeds 4000 mg l⁻¹. Lake Elmenteita is fed by alkaline springs (Richardson & Richardson, 1972).

The number of flamingoes of Lake Elmenteita ranges from 3000 to 100 000 (Vareschi, 1978), however, Brown (1971) found only 500 in April 1957. The algal population is dominated by *Spirulina platensis* and *Anabaenopsis circularis*.

Lake Naivasha

General geography

Located in Kenya (0°46′ S and 36°22′ E), at 1890 m above MSL, Lake Naivasha has an area of 145 km², a maximum depth of 7.3 m, a mean depth of 4.7 m and a volume of 680×10^6 m³. It is situated in a depressed area together with Lake Nakuru and Lake Elmenteita.

Around the lake, the vegetation is dominated by *Cyperus papyrus*, which may form small clumps, fringe swamps parallel to the shore, papyrus reefs forming lagoons behind them and large floating swamps (Gaudet, 1977). In shallow waters, *Ceratophyllum demersum* covers the bottom; *Najas pectinata*, *Utricularia* spp., and *Potamogeton* spp. are also common. Three Charophyta are found (*Nitella knightiae*, *N. oligospira* and *Chara braunii*). The floating vegetation is dominated by *Nymphaea caerulae*. The quiet waters near the papyrus fringe swamp are covered with *Limna* and *Wolfiopsis*. In 1978 and 1979, the lagoons and shallow water became colonized with the floating *Salvinia molesta*. When the level decreases, wet mud detritus is actively eaten by *Myocastor coypus*, and crayfish use the wet mud area as breeding grounds. Young *Cyperus* sp. colonize the wet mud and further inland *Conyza*, *Gnaphalium* and *Sphaerantus* occur but die during the next inundation. The drawdown flora serves as food for grazers such as hippos. At reflooding, the dead vegetation and the grazers' dung cause an increase in productivity and a considerable development of Ostracoda.

Origin of the lake

Richardson & Richardson (1972) have studied the past history of Lake Naivasha on a 28 m-long core and raised beaches. A dry period started 25 000 years B.P. but ~12 000 years B.P. the lake level rose and the resulting large lake overflowed through the Njorova gorge. Around 5700 B.P., a new dry phase began, the level dropped and the lake became a closed basin. For the past 3000 years a small lake, often smaller and more concentrated than the present lake, occupied the closed depression.

Physical and chemical features of Lake Naivasha

Because of altitude, the maximal temperatures of the lake water are relatively low. Seasonal variations range from 19.5 to 23°C at the surface and 19.2 to 21.5°C near the bottom (Litterick, Gaudet, Kalff & Melack, 1979). Temperature profiles (Melack, 1979) indicate that strong winds and night cooling cause a daily mixing; even superficial thermoclines, developing as a result of mid-day heating, generally disappear at night. Consequently, the lake is usually well oxygenated down to the bottom.

The Malewa River, draining a watershed of 1730 km^2, is the main inlet of the lake. It reaches the lake after passing through the swamps which extend in the northern part of the lake and consist mostly of floating mats of *Cyperus papyrus*.

The hydrological balance of the lake was studied by Gaudet & Melack (1981) and Litterick *et al.* (1979) and is presented in Table 41. The northern shores are the main areas of seepage of the lake. The main chemical features of the lake and the Malewa River are presented in Table 42. Lake Naivasha has exceptionally fresh water for a lake which had no outlet during the last 5000 years. The reasons suggested by Gaudet & Melack concern the low concentrations of the main inlet and losses through seepage and sedimentation. Phosphate is generally high and does not limit growth: 70% of total phosphorus was found to be PO_4-P. Total nitrogen is also abundant but mostly organic (algae, bacteria and excreta of birds). A nitrogen concentration of 400 μg l^{-1} was measured in Lake Naivasha, 93% of which was in organic form (Millbrink, 1977).

Biological features

Blue-green algae dominate the phytoplankton: *Lyngbya contorta, Anabaenopsis tanganikae, Spirulina laxissima, Oscillatoria* spp., *Microcystis* spp., *Aphanocapsa delicatissima, Merismopedia* spp., *Chroococcus* spp., but diatoms (*Synedra acus, Surirella linearis, Nitzschia* sp., *Melosira ambigua*) and chlorophytes (*Pediastrum duplex, Scenedesmus* sp., *Staurastrum* sp.) are also frequently found (Melack, 1979). As a result of frequent mixing, the algal biomass is evenly distributed with depth. The concentration of chlorophyll *a* ranges from 1.0 to 17 mg m^{-3}; it is minimal in November–December and maximal from April to July. The primary production rates measured by Melack (1979) in 1973–74 ranged from 3.7 to 6.2 g O_2 m^{-2} d^{-1}. Similar values (3.5–6.8 g O_2 m^{-2} d^{-1}) were measured by Källqvist (1980) in 1976–77.

The zooplankton is dominated by six species: the Cladocera *Diaphanosoma excisum* and *Simocephalus vetulus*, the Copepoda *Thermocyclops schurmanni* and *Mesocyclops leuckarti*, and the Rotifera *Brachionus calyciflorus* and *B. caudatus*. Copepods are numerically dominant throughout the year. Cladocera and Rotifera show a minimum in November–December. The number of individuals of total zooplankton varies from 70 to 250 per litre (Litterick *et al.*, 1980).

Table 41 *Hydrological balance of Lake Naivasha for the year 1974 (results in 10^6 m^3)*

	IN		OUT
Surface runoff	0.7	Evaporation (swamps)	13.2
River discharge	204.0	Evaporation	276.0
Rainfall	114.2	Seepage out and irrigation	50.6
INPUT	361.2	OUTPUT	339.8
Change of storage	+21.4		

Table 42 *Chemical features of Lake Naivasha and the Malewa River (Gaudet & Melack, 1981)*

	Lake Naivasha	Malewa River		Lake Naivasha	Malewa River
Na^+ (mg l^{-1})	40	9.0	HCO_3^- (mg l^{-1})	192	70
K^+ (mg l^{-1})	20	4.3	CO_3^{2-} (mg l^{-1})	10.6	0
Ca^{2+} (mg l^{-1})	21	8.0	SO_4^{2-} (mg l^{-1})	6.2	6.2
Mg^{2+} (mg l^{-1})	6.4	3.0	Cl^- (mg l^{-1})	14.0	4.3
SiO_2 (mg l^{-1})	34	17.2	F^- (mg l^{-1})	1.5	0.4
Conductivity (μmhos cm^{-1})	311–353	88–179			

Chironomid larvae are abundant ($4000\,m^{-2}$) (Millbrink, 1977).

Prior to 1925, the fish fauna of Lake Naivasha was limited to *one* endemic species, the carp *Aphocheilichthyes antinorii*. This unusual feature is probably related to the destruction of the former fauna of the lake during dry periods. Between 1925 and 1970, eight species of fish were introduced (Litterick *et al.*, 1980): *Sarotherodon spirulus nigra, Micropterus salmoides, Tilapia zillii, T. nilotica, Gambusia* sp., *Poecilia* sp., *Sarotherodon leucosticta* and *Salmo gairdneri*. Other organisms were also introduced: *Procambarus clarkii* (invertebrate), *Myocastor coypus* (mammal) and *Salvinia molesta* (macrophyte). This increase in potential food attracted a large number of fish-eating birds (pelicans, cormorants, fish eagles and herons).

The fisheries

The crayfish *Procambarus clarkii* and three fish species (*Sarotherodon leucosticta, Tilapia zillii* and *Micropterus salmoides*) constitute the fishery. The total *Tilapia* catch reached 1000 tons in the 1960s and early 1970s then dropped to 50 tons from 1973 to 1977. Litterick *et al.* (1980) explain this by a combined effect of decrease in water level, expansion of *Salvinia molesta* and the build up of a large population of crayfish in the north-east area of the lake as well as overfishing and illegal fishing.

Lake Nakuru

Lake Nakuru is located in Kenya (0°22′ S and 36°05′ E), at 1759 m above sea level. The main characteristics of the lake are presented in Table 43. From 1961 to 1974, a portion of the lake, then the whole lake and finally an area of 160 km² including the lake was transformed into a protected National Park.

Water chemistry

The chemical features of Lake Nakuru are shown in Table 44. Lake Nakuru is characterized by the nearly total absence of divalent

Table 43 *Morphometric features of Lake Nakuru*

Surface area 40 km²	Max. depth 2.8 m
Shore length 27 km	Mean depth 2.3 m
Development of shore line 1.2	Volume $92.0 \times 10^6\,m^3$
Watershed area $1800 \times 10^6\,m^2$	
Watershed area : lake volume ratio $19.6\,m^2\,m^{-3}$	

cations. In 1960-61, the lake nearly dried and this was at least the third time during the last 30 years (Millbrink, 1977). It is clear that the absence of outlets from Lake Nakuru is one of the reasons for its high salinity; the extremely high watershed area: volume ratio is an additional and non-negligible factor of salt concentration. Moreover, the lake is fed by alkaline springs (Richardson & Richardson, 1972). Lake Nakuru has nearly no rooted vegetation except in the north-east area where freshwater riverlets allow the growth of *Papyrus*.

Biological features

The phytoplankton of Lake Nakuru is heavily dominated by *Spirulina* spp. and especially *S. platensis*. Their biomass is of an order of magnitude of 180 mg dm l^{-1}. In early 1974, *Spirulina* decreased and other algae became dominant: *Anabaenopsis arnoldii, A. elenkinii* and diatoms.

The zooplankton and zoobenthos include copepods (*Lovenula africana*), chironomid larvae (*Leptochironomus* sp. and *Tanytarsus* sp.) and the rotifers *Brachionus dimidiatus, B. plicatilis* and *Hexartha jenkinae*. The zoobenthos is dominated by the larvae of *Leptochironomus* which reach 60 g wm m^{-2} (Millbrink, 1977).

The biology of Lake Nakuru is best characterized by the impressive number of lesser flamingoes (*Phoeniconaias minor*) varying from 300 000 to 1 500 000 although, in 1974, numbers as low as 32 000 were recorded. These birds feed essentially on algae. Vareschi (1978) estimated the average flamingo biomass to be 41.2 g m^{-2}. The lesser flamingo is a filter feeder whose specialized bill contains more than 10 000 small plates filtering plankton from the stream of water. Experiments showed an average filtration rate of 32 l $bird^{-1}$ h^{-1} (adult bird). The feeding rates were 5.6 g dm

Table 44 *Main features of Lake Nakuru*

	(a) Feb. 1961	(b) Dec. 1930		(a) Feb. 1961	(b) Dec. 1930
Na^+ (mg l^{-1})	38.000	5.550	$HCO_3^- + CO_3^{2-}$ (mg l^{-1})	1440	205
K^+ (mg l^{-1})	1.312	256	Cl^- (mg l^{-1})	13.000	1375
Ca^{2+} (mg l^{-1})	10	10	SO_4^{2-} (mg l^{-1})	1.800	253
Mg^{2+} (mg l^{-1})	30	0	SiO_2 (mg l^{-1})	730	–
Conductivity (μmhos cm^{-1})	162.500		Total P (μg l^{-1})	12.200	–

(a) Talling & Talling, 1965
(b) Beadle, 1932

bird^{-1} h^{-1}. Since the flamingo spends $\sim 50\%$ of its time feeding, an adult bird ingests daily ~ 72 g dm *Spirulina*, that is, for Lake Nakuru conditions, 60 tons dm *Spirulina* material, i.e. 0.7 g dm m^{-3} d^{-1}. The flamingoes can also feed on rotifers.

Observation of flamingo migrations seems to indicate that, in some cases, these migrations are motivated by algal concentrations; however, certain lakes of the rift have high concentrations of *Spirulina* and have no flamingoes; in other cases, they migrate from one lake to another in search of suitable breeding grounds.

For the period 1972–74, the mean *Spirulina* biomass was 180 g dm m^{-3} and its production was 1.2 g dm m^{-3} d^{-1}. The lesser flamingo ingests 0.7 g dm m^{-3} d^{-1} that is 60% of the daily production but no more than 0.4% of the biomass. The flamingo is the main grazer of Lake Nakuru; all the other grazers are left with only 40% of the primary production.

In fact, the second important grazer is the cichlid *Tilapia grahami*, introduced from Lake Magadi in 1953, 1959 and 1962 (Vareschi, 1979) and it is the only fish existing in Lake Nakuru.

In Lake Magadi hot springs, *T. grahami* fed on filamentous blue-green algae, insect larvae and copepods but, in Lake Nakuru, they feed on *Spirulina* which seems to be a very suitable food since the fish grow larger in Nakuru than in Magadi. The fish show interesting daily migrations: at noon, they concentrate near the shore and at night they move to offshore regions, probably as a result of their affinity for warm waters: 70% of the population was found in the upper 50 cm and 80% in the upper metre. The total mass of the fish was 90 tons dm (2.1 g m^{-2}) in 1972 and 400 tons dm in 1973 (10.2 g m^{-2}). The low values of 1972 were imputed to a mud-mixing which created anoxic conditions and fish killing.

The first consequence of the introduction and successful growth of the cichlid was the appearance of a fish-eater, the pelican. Moreover, *Tilapia* became a competitor of the flamingo: in 1972–73, the *Tilapia* biomass was 8–40 g fm m^{-2} and the biomass of the lesser flamingo was 20–61 g fm m^{-2}. Since all the consumers graze about 1% of the algal biomass per day, however, the introduction of the fish did not put any unbearable pressure at the primary consumer level. The presence of the fish increased the diversity of the bird population. Pelicans, which numbered 2500 in 1971–72, increased to 10 000 in 1974; they breed at Lake Elmenteita, 14 km south-east of Nakuru. Such a predator eats 16–20 tons fm fish per day meaning a loss of 72 kg of PO$_4$-P and 486 kg of nitrogen per day. Since they breed at Elmenteita but come daily to feed at Nakuru, there is an 'export' of nitrogen and phosphorus from Nakuru to Elmenteita through the droppings of the birds.

Lake Baringo

Located in Kenya (0°46′ N; 36°15′ E), at an altitude of 915 m, Lake Baringo has a surface area of 130 km², a mean depth of 5.6 m and a volume of 700×10^6 m³.

The lake receives two small streams at the southern end of the lake and a few hot and very alkaline springs. The shores are generally rocky with the exception of the southern part of the lake, covered with swamps and their associated vegetation of *Papyrus, Pista stratiotes, Azolla* and *Nymphaea*.

The chemical features of Lake Baringo water are presented in Table 45. The water of Lake Baringo has a high content of silt which reduces light penetration. The dominant algae are *Microcystis aeruginosa* and *Anabaena circinalis*. These algae accumulate at the surface and form patches in calm weather. The Secchi-disc readings did not exceed 7 cm in the 1976–77 survey of Källqvist (1980). These conditions explain the relatively low primary production rates (0.3–1.0 g O_2 m⁻² d⁻¹) (Källqvist, 1980). This limited food supply from primary producers does not seem to affect fish yield, which in 1977 was 246 tons, about three times more than in Lake Naivasha which has a much higher primary productivity.

Worthington & Ricardo (1936) believed that a vast area of the Eastern Rift from Lake Magadi to Lake Baringo was in the past occupied by one unique large lake, Lake Kamasia, and they expected Lake Baringo to have a Nilotic fish fauna similar to that of Lake Rudolf. In fact, Lake Baringo has only *Tilapia nilotica* in common with Lake Rudolf. The other five known species of fish in Lake Baringo are *Barbus gregorii, Clarias mossambicus, Labeo cylindricus, Aplocheilichthys* sp. and *Barbus lineomaculatus*. *T. nilotica* is the main commercial species.

Table 45 *Chemical characteristics of Lake Baringo water*

	(a) Dec. 1930	(b) May 1961	(b) May 1962		(a) Dec. 1930	(b) May 1961	(b) May 1962
Na^+ (mg l⁻¹)	126	64	95	$CO_3^{2-} + HCO_3^-$ (meq l⁻¹)	5.6	3.92	4.44
K^+ (mg l⁻¹)	15	14	13	Cl^- (mg l⁻¹)	36	18.0	25.0
Ca^{2+} (mg l⁻¹)	2	10.2	11.6	SO_4^{2-} (mg l⁻¹)	40	–	19.0
Mg^{2+} (mg l⁻¹)	36	9.8	3.15	SiO_2 (mg l⁻¹)	–	47.0	23.5

(a) Beadle, 1932
(b) Talling & Talling, 1965

Lake Rudolf

General geography

Located in Kenya and Ethiopia (2°25′–4°35′ N and 35°50′–36°45′ E), at an altitude of 406 m above MSL, Lake Rudolf has a surface area of 7200 km². It is long (296 km in the north–south axis) and narrow (59 km maximal width). The maximal depth is 120 m according to Beadle (1974) but only 73 m according to Welcomme (1979). The lake has no outlet.

The area surrounding Lake Rudolf is composed of metamorphic rocks, with the exception of the mountains of the southern part of the watershed which are of volcanic origin and exceed 2000 m in altitude. The western shore is bordered by sandy plains whereas, in the north-west part of the lake, Mount Labur (1560 m) is very close to the lake. In the northern part of the lake, Sanderson's Gulf and the mouth of the Omo River which were, in the past, the northern prolongations of the lake dried up between 1903 and 1933 (Arambourg, 1933).

The lake has three main islands: North Island, Central Island and Elmolo Island, of volcanic nature, in the southern part of the lake. The macrophyte vegetation in shallow water is scarce, probably because of strong wind-induced currents.

The climate of this area is very hot and arid and evaporation has been estimated at 3.5 m³ y⁻¹ (Beadle, 1974).

General history of the lake

The past history of Lake Rudolf was investigated by Butzer *et al.* (1972). These workers found that the period extending from the mid-Pleistocene to mid-Holocene was represented by lino-fluvial and tectonically undisturbed sediments known as the Kibish Formation. A terrasse, 60 m higher than the present level was dated 130 000 years B.P.; another high level of the lake was found to date from around 35 000 years B.P. No deltaic formation having been found for the time-span 35 000–9500 B.P., it is probable that the lake level was very low. At 9500 B.P., the lake level was between 60 and 80 m above the present level but, about 7500 B.P., the lake was reduced to its present dimensions. Between 6600 and 4400 years B.P., the level was again at 70 m above the present level. This was followed by a regression and a new level rise at 3000 B.P. Then the level fluctuated with a general trend toward low levels. From 1877 to 1955, the level dropped from +15 m to −5 m in comparison with the present level. A rise of 4 m from 1962 to 1965 was caused by the intensification of autumnal rains over East

Africa. During the high levels of 9500–7500 B.P., the lake probably overflowed to the Sobat Basin and Nile River.

Water chemistry

The chemical features of Lake Rudolf are presented in Table 46. Simple salt balances show that the sole input from the freshwater of the Omo River (conductivity 80 μmhos cm^{-1}) to the closed Lake Rudolf would have increased the conductivity by 2000 μmhos cm^{-1} in 160 years. If we add the increase in salinity due to the excess of evaporation over inflow (since the level has decreased) and the fact that the lake has not been closed for 160 years but for 7000 years, we find that the lake water should be much more saline than it is (Beadle, 1974). It is clear that there is some 'freshening process', perhaps similar to the mechanism which is keeping Lake Chad its low salinity.

Biological features

It seems that the limit of salinity for most freshwater fishes is around 5‰ (Beadle, 1974). Lake Rudolf, with a salinity of 2.5‰, is far below this limit. However, Beadle reports that in one of the small crater lakes located in the Central Island of Lake Rudolf, the salinity and alkalinity were, respectively, twice and three times the values of the main lake. Among the 37 species of fish existing in the main lake, only two species survived in the crater lake (*Clarias lazera* and a *Tilapia* derived from *T. nilotica* of the main lake). We note that the Lake Kinneret (Israel) *Clarias lazera* populations concentrate near a group of hot salty springs; this confirms their low sensitivity to high salinity.

In the main lake, one of the most common fish is the cyprinid

Table 46 *Chemical characteristics of Lake Rudolf*

	(a) Jan. 1931	(b) Jan. 1961		(a) Jan. 1931	(b) Jan. 1961
Na$^+$ (mg l^{-1})	770	810	HCO$_3^-$ + CO$_3^{2-}$ (meq l^{-1})	21.7	24.5
K$^+$ (mg l^{-1})	23	21	Cl$^-$ (mg l^{-1})	429.0	475
Ca^{2+} (mg l^{-1})	5	5.7	SO$_4^{2-}$ (mg l^{-1})	56	64
Mg^{2+} (mg l^{-1})	4	3.0	Conductivity (μmhos cm^{-1})	2860	3300
pH	9.5–10				

(a) Beadle, 1932
(b) Talling & Talling, 1965

Engraulicypris stellae, feeding on zooplankton. Predator fish include *Hydrocynus* sp. and *Lates* sp. There are 31 non-cichlid species, four of which are endemic, and five cichlids including *Tilapia galilaea* and *T. zillii*. An endemic cichlid species *Pelmatochromis exsul* poses an interesting problem of zoogeography since the genus is unknown in the Nile, Lake Chad and Niger; it is known in the Zaire and South Cameroons. The nearly complete absence of mormyrids is another unsolved question.

The algal community is dominated by green algae (*Botryococcus braunii*) and diatoms (*Navicula* sp., *Cymbella* sp., *Cyclotella meneghiniana*, *Stephanodiscus astraea* and *Rhopalodia* sp.). Blue-green algae are also found: *Anabaenopsis arnoldii*, *Spirulina platensis* and *Lyngbya hetea*. This is according to Rich (1932) reported by Van Meel (1954).

Ethiopian lakes

The main Ethiopian Rift is the north-eastern extension of the Eastern Rift which crosses Tanzania and Kenya in a north–south direction. It is a downfaulted structure with a silicic vulcanism. North-east of the Ethiopian Rift, the Afar area is the meeting place of the Main Rift with the Red Sea and Aden Rift systems. It is a very complex geological unit with basaltic vulcanism. The altitude exceeds 4000 m in the Bale mountains and ranges from 1800 to 1500 m in the Galla basin, while the Lake Assal area is below MSL.

The air temperature varies with altitude; annual rainfall decreases from an average value of 600 mm at Lake Ziwag to a few millimetres in areas below sea level (Gasse & Street, 1978). In areas below 1700 m, evaporation largely exceeds rainfall and the very existence of lakes depends on runoff and groundwater recharge.

The Ethiopian Rift has two groups of lakes, one known as the Galla lakes and another group, south-east of Addis Ababa, known as the Bishoftu lakes. Three saline lakes (Abhe, Afrera, Asal) drain the Afar area and a fourth group of lakes is located above 4000 m altitude, in the Bale mountains.

The Galla lakes

The morphometric and chemical data concerning seven Galla lakes are presented in Tables 47 and 48. The morphometric data are from Welcomme (1979).

A detailed study of the thermal structure of Lake Awasa was carried out by Baxter, Prosser, Talling & Wood (1965). Only a superficial thermocline was observed in February and June 1964. Conversely, in June 1964, a deep

Table 47 Morphometric features of the Galla lakes, Ethiopia

	Shamo (Ruspoli)	Abaya (Margharita)	Awasa	Shala	Langano	Abiata	Ziway
Location	5°40′ N; 37°37′ E	6°2′–6°35′ N; 37°40′–38°5′ E	7°03′ N; 38°24′ E	7°30′ N; 38°30′ E	7°36′ N; 38°45′ E	7°36′ N; 38°48′ E	
Altitude (m)	1280	1285	1708	1567	1585	1573	1848
Surface (km^2)	551	1161	130	409	230	205	440
Max. length (km)	36	70	17	27	23	21	32
Max. width (km)	23	28	11	17	16	12	20
Max. depth (m)	12.7	13	22	226	46	14	7
Mean depth (m)	—	7	11	86	17	7.6	2.5
Volume (m^3)	—	8.2×10^9	1.3×10^9	37×10^9	3.8×10^9	1.6×10^9	1.1×10^9
Shoreline (km)	118	225	52	110	78	62	102
Shore development	—	1.86	1.29	1.53	1.45	1.22	1.37
Watershed area (km^2)	—	17 300	1250	3920	1600	1630	7025
Watershed area : lake volume ratio (m^2 m^{-3})	—	2.11	0.96	0.11	0.42	1.02	6.39

Table 48 *Chemical features of the Galla lakes*

	Shamo (Ruspoli) May 1961 (a)	Abaya (Margharita) Dec. 1937–Feb. 1938 (a)	Abaya May 1961 (b)	Awasa May 1961 (b)	Shala Apr. 1938 (a)	Shala May 1961 (b)	Langano May 1961 (b)	Abiata Apr. 1938 (a)	Abiata May 1961 (b)	Ziway May 1961 (b)
Na^+ (mg l^{-1})	–	–	206	235	5894	6250	500	3000	6375	64
K^+ (mg l^{-1})	–	14.1	16	45	440	252	27.5	75.3	331	14
Ca^{2+} (mg l^{-1})	–	15.9	12.1	4.4	10	3	2.5	8.5	3	10.2
Mg^{2+} (mg l^{-1})	–	6.3	5.7	4.7	8.8	7.5	2.7	6.0	7.5	9.8
$HCO_3^- + CO_3^{2-}$ (meq l^{-1})	–	–	8.5	10.5	–	200	15.0	–	210	3.92
Cl^- (mg l^{-1})	–	52	53	34	3136	3300	216	1500	3240	18
SO_4^{2-} (mg l^{-1})	–	24.6	28	27	128.7	275	28	67.4	210	–
Conductivity (μmhos cm^{-1})	927	670–766	900	1050	20 400	29 500	19 000	10 700	30 000	370
pH	–	–	–	–	–	–	–	–	–	–
Total P (μg l^{-1})	–	–	290	98	–	900	165	–	890	170

(a) Loffredo & Maldura, 1941
(b) Talling & Talling, 1965

thermal discontinuity existed between 5 and 12 m. Occasional accumulation of cold and well-oxygenated water at the edges of the lake has been interpreted by the authors as local cooling of superficial water which subsequently sinks down, flows on the lake bottom and perpetuates stratification. The associated transfer of oxygen to deep layers is underlined by the fact that the oxygen profiles frequently show an oxygen minimum at intermediate depths.

Four of these lakes (Ziway, Langano, Abiata and Shala) have probably formed one large lake during wet periods which may have overflowed northwards into the Awash River. At present, Lakes Ziway and Langano drain into Lake Abiata but there is no connection between Lake Abiata and the deep Lake Shala.

Ziway, fed by direct runoff from the high plateaux, has conductivity of 370 μmhos cm^{-1}. Langano, receiving water only from the south-eastern plateau, is already more saline (1900 μmhos cm^{-1}) and the closed Lake Abiata has a conductivity of 30 000 μmhos cm^{-1}, corresponding to a salt content of 16 g l^{-1}. Lake Shala lies in a volcanic caldera; with its maximal depth of 266 m, it is the deepest lake of the part of Africa located north of the Equator. Shala has the same salt content as Abiata and deposits trona (Na$_2$CO$_3$. NaHCO$_3$. 2H$_2$O) on its shores.

The Bishoftu lakes

Five lakes in this group have been studied by Baxter *et al.* (1965), Prosser, Wood & Baxter (1968), Talling, Wood, Prosser & Baxter (1973) and Wood, Prosser & Baxter (1976). Lake Biete Mengest is described by Prosser *et al.* (1968). The main characteristics of these lakes are shown in Tables 49 and 50.

According to Mohr (1961), the craters which harbour the Bishoftu lakes were formed around 7000 years ago.

The degree of stratification of the Bishoftu lakes depends on depth and wind exposure (Wood *et al.*, 1976). Lake Pawlo has a surface temperature of 19.2–24.5°C and an almost constant bottom temperature (19.2–19.4°C). In 1965, a stable stratification was observed from March to May. In June, a 3°C difference in temperature was noted between surface and bottom but without any clear thermal gradient. In July, a new, well-marked thermocline appeared at 5 m and deepened until reaching 25 m in December. We note that well-defined thermoclines may develop during the cooling period.

Lake Bishoftu displays very similar behaviour to Lake Pawlo. Lake Aranguadi, although shallower, shows a pronounced stratification. The surface temperature is higher than in the former lakes and the profile is characterized by a very sharp increase in temperature near the surface, probably related to the lesser wind input in this lake. A net thermocline exists at 2–4 m. Lake Aranguadi underwent complete mixing from November 1964 to January 1965.

The shallow and exposed Lake Kilotes is generally well mixed with only short periods of stratification.

The heat budgets of Lakes Pawlo, Bishoftu and Aranguadi are, respectively, 5700 cal cm^{-2}, 4950 cal cm^{-2} and 3100 cal cm^{-2}, which are very low values when compared with temperate lakes. Net daily changes in heat content frequently represent only 3% of the diurnal heat flux. Heat content variations depend upon solar radiation and relative humidity. Very high solar inputs, by coinciding with very low relative humidity and thus, enhancing evaporation, frequently result in heat losses. In contrast, the high relative humidity which accompanies the rainy period does not favour

Table 49 *Morphometric features of the Bishoftu lakes*

	Altitude (m)	Area (km^2)	Max. depth (m)	Mean depth (m)	Volume (10^6 m^3)
Pawlo	1890	0.60	62.5	38	22
Bishoftu	1890	0.90	87.0	55	52
Aranguadi	1910	0.50	28.3	18.5	10
Kilotes	2000	0.77	6.2	2.6	2
Biete Mengest	–	1.03	–	17.5	18
N. Crater	–	–	38	17.5	–
S. Crater	–	–	31		

Table 50 *Main chemical features of the Bishoftu lakes (Prosser et al., 1968)*

	Bishoftu	Pawlo	Aranguadi	Kilotes	Biete Mengest
pH	9.2	9.2	10.3	9.6	9.2
Alkalinity (meq l^{-1})	20.0	10.2	51.4	63.4	26.8
Cl$^-$ (meq l^{-1})	4.0	0.9	22.0	13.6	5.67
SO$_4^{2-}$ (meq l^{-1})	0.35	<0.1	0.7	0.4	0.4
Na$^+$ (meq l^{-1})	16.0	5.5	67.0	70.5	23.9
K$^+$ (meq l^{-1})	1.5	1.05	8.1	4.5	1.31
Ca^{2+} (meq l^{-1})	0.37	0.55	0.67	0.7	0.34
Mg^{2+} (meq l^{-1})	5.8	4.65	<0.6	<0.6	3.94

evaporation and, in spite of the low solar radiation, the overall result is a significant heat gain.

The small thermal stability of these lakes and its strong dependence upon local and temporary meteorological changes turn the lake mixing into a limnological event difficult to predict or at least far less periodical than in temperate lakes.

The Afar lakes

The Awash River, which originates in the western plateau, south of Addis Ababa, flows into Lake Abhe. This lake has no outlet; it has a total salt content of 160 g l^{-1} and a pH of 10–11. It has a sodium-carbonate type of water.

Lake Afrera (13°20′ N, 40°55′ E) is fed by groundwater springs from the Ethiopian plateau. It has a sodium-chloride type of water with a total salt content of 159 g l^{-1}.

Lake Asal (11°40′ N, 42°25′ E) is located at 155 m below MSL. The lake bottom is made of recent basalt emitted by the Rift fissures. The lake has a surface area of 54 km^2, a mean depth of 7.4 m and a volume of $400 \times 10^6 \text{ m}^3$. Its watershed has a surface area of 600 km^2 with a mean annual rainfall of 200 mm; the evaporation of the lake surface is estimated to be 1900 mm y^{-1} (Fontes, Florkowski, Pouchan & Zuppi, 1979). The lake is fed both by groundwater and sea water infiltration. Its salt content reaches 300 g l^{-1} and gypsum and halite are deposited.

Fontes *et al.* (1979) carried out an isotopic survey utilizing stable isotopes (^{18}O and ^{2}H) as well as tritium in order to clarify the hydraulic behaviour of the system. The results of the survey together with information derived from chemical balances indicate that Lake Asal is leaking $0.8 \text{ m}^3 \text{ s}^{-1}$, whereas the total water input is $6.7 \text{ m}^3 \text{ s}^{-1}$. This means that $\sim 88\%$ of the total supply is lost by evaporation.

The Bale Mountain lakes (Löffler, 1978)

These lakes are located in southern Ethiopia at 6°30′–7°20′ N and 39°–40°30′ E between 4000 and 4200 m altitude. The environment is volcanic, dominated by olivine-basalt. These lakes are above the timber line but no permanent snow is observed. The largest lakes are Hora Orgona ($400\,000 \text{ m}^2$ and 45 cm depth) and Garba Guratsch ($200\,000 \text{ m}^2$). The annual precipitation is around 1500 mm and is distributed during two rainy seasons, March–May and October–November, so that many lakes are periodically dry. The large lakes have temperature ranging from 10 to 12°C. The chemical features of these two lakes and of two ponds are summarized in Table 51.

Table 51 Chemical characteristics of the Bale Mountain lakes (Löffler, 1978)

	pH	Conductivity (μmhos cm^{-1})	Ca^{2+} (mg l^{-1})	Mg^{2+} (mg l^{-1})	Na$^+$ (mg l^{-1})	K$^+$ (mg l^{-1})	Cl$^-$ (mg l^{-1})	SO$_4^{2-}$ (mg l^{-1})	Alkalinity (meq l^{-1})
Tarn 1	6.4	57	2.8	1.92	2.99	0.55	3.82	3.36	0.4
Tarn 2	6.20	55	6.0	1.32	2.30	1.05	2.94	7.68	0.3
Hora Orgona	3.20	718	75.0	42.36	65.78	14.47	143.12	32.64	5.5
Garba Guratsch	7.40	131	4.6	3.00	6.67	0.74	1.75	0.56	0.6

These lakes and ponds had neither net phytoplankton nor rotifers but only nanoplankton and palaearctic crustacean zooplankton (*Daphnia obtusa, Arctodiaptomus* n.sp. and *Megacyclops viridis*). *D. obtusa* was pigmented as a result of the low oxygen content at this altitude. In contrast, the benthic Cladocera are of African origin.

The investigations carried out on a 733 cm core taken in one of the shallow lakes revealed the presence of seven upper metres composed of organic sediments lying on inorganic gyttja. Most of the remains of organisms found belonged to living species such as *Daphnia obtusa, Chydorus* sp., and the diatom *Campylodiscus noricus*.

The fauna of the Bale Mountain lakes has close affinities with the fauna of East-African highlands and even southern Africa. The palaearctic character of this fauna and its occurrence in many lakes of high altitude indicate that in the Pleistocene when the snowline was much lower (~ 3600 m), the cold southern African fauna had a continuous distribution along the mountainous backbone of East Africa, and that the recent warming of the climate has caused the isolation of these animals in the highest lakes of the area.

THE ZAIRE BASIN

The Zaire basin is the African equivalent of Amazonia in many respects: a huge catchment area, located across the Equator on Precambrian stable baserock and covered with tropical forest, is drained by a central river of equatorial regime, having a dense hydrographic network and connections, through swampy watershed divides, with neighbouring basins.

Origin and hydrography

Together with the catchment areas of Lakes Tanganyika and Kivu, weakly connected to the Zaire basin via the Lukuga River, the Zaire basin has a surface of $> 4 \times 10^6$ km² and extends from 7°N to 12°S. Some 900 million years ago, the central part of the basin became a stable continental area (Congo craton), later transgressed by the Palaeozoic and Triassic Seas (Robert, 1946). In Cenozoic times, the uplifting of the coastal region prevented the drainage to the sea and a large lake covered the basin during the Pliocene, in the same way as Amazonia was transformed into a lake when the incipient Andes blocked the drainage to the west. In the late Pliocene, the lake cut its way to the sea through a series of falls and rapids. It follows that as a lake or as a river, the Zaire basin has existed without dramatic changes since at least the Pliocene and probably much before. The turbulent Plio–Pleistocene history of East Africa did not have much effect upon the Zaire basin. At present, the basin is composed of central lacustrine Plio–Pleistocene sediments, bordered by Palaeozoic and Triassic formations, surrounded in their turn by Precambrian rocks.

The Zaire River has a total length of 4700 km and the highest headwaters are the springs of the Lualaba River at 1540 m. The longest continuous stream runs from the Chambezi River near Lake Malawi, to Lake Bangweulu, the Luapula River, Lake Mweru, the Luvua River, the Lualaba River and the Zaire River. In its central part, the river flows slowly: there is only 100 m fall in 2000 km between Kisangani–Stanleyville and Kinshasa–Leopoldville. This strongly contrasts with the 275 m drop in 340 km in the rapid and fall areas between Kinshasa and the sea. Similar rapids exist on the Lualaba River (Portes d'Enfer and Stanley Falls). The width of the river varies from 1 to 15 km.

A tropical humid climate characterizes nearly the whole Zaire basin. As in Amazonia, relatively poor soils are covered with forest. In the Zaire basin more than in Amazonia however, the cultivations have replaced the rain forest. Papyrus is frequently found in the river valley, with an abundant floating and submerged flora in quiet waters. *Eichhornia crassipes* invaded Zaire in the 1950s (Rzoska, 1978).

The Zaire basin seems to have grown by stream captures. It is likely that the Lualaba River was previously connected to the Nile and was captured by the Zaire at 'Les Portes d'Enfer'. There are occasional connections with adjacent basins: with the Nile basin by affluents of Ubangi and Bahr el Ghazal, with the Chad basin by affluents of Ubangi, Gribinqui and Ouham and with the Zambezi basin by affluents of the Lualaba.

The Zaire River
Hydrology

The upper course of the Zaire consists of the Lualaba and Lufira Rivers which have their maximum in February. These rivers join in the Upemba Swamps which delay their maximum to April–May. The Luapula–Luvua branch has a similar regime after crossing the lakes Bangweulu and Mweru. Lakes Tanganyika and Kivu, which flow into the Lualaba River via the Lukuga River also have their maxima in May, so that, after the Lukuga junction, the Lualaba has a clear maximum in May and a clear minimum in September–October. From this point to Kisangani, the tributaries have an equatorial regime which slowly modifies the Zaire regime. At Kisangani, the Zaire has two maxima (November–December and April–May), a small minimum in

February–March and a main minimum in August–September. From Kisangani to Coquilhatville, the Zaire receives three main tributaries from the Northern Hemisphere and four from the Southern Hemisphere. This balanced supply explains why the Zaire still has an equatorial regime at Coquilhatville. Then the Oubangui and the Sanga Rivers, with their October maxima and March minima, modify the Zaire regime by enhancing the October maximum and the March minimum.

The fluctuations in water level are ~ 3 m (compared with 15 m for the Amazon). This relatively modest variation is due partly to the buffering effect of the upstream lakes and swamps and to the inundation of a wide flood plain but it is mainly due to the lower amount of precipitation received by the Zaire basin than by Amazonia (1.5 and 2.5 m, respectively).

Then, the Zaire receives the Kasai River which inverses the minima; in its lower course the Zaire then has a secondary minimum in March, a secondary maximum in May, a main minimum in August and a main maximum in December (Guilcher, 1965).

The Zaire has an average discharge of $40\,000$ m^3 s$^-$ or 1250×10^9 m^3 y^{-1}. In spite of its equatorial regime, its specific yield is only 310 m^3 km^{-2} y^{-1}, nearly four times less than the Amazon.

Water chemistry

Black and acid water (pH 3.5–5.2), with a high content of humic acids, characterize the southern affluents, and for part of the year the water of the Zaire River itself is brown-black. The Ubangui River is the only main white-water tributary (Roberts, 1973). In 1976 and 1978, the Committee on Oceanic Research of the Netherlands Academy of Sciences carried out investigations aimed at determining the contribution of the Zaire to the ocean. This considerably improved our chemical knowledge of the river which had been based until then on one water analysis by Symoens (1968). The analysis of Zaire water (Meybeck, 1978) was carried out in November 1976 (Table 52). The study of nutrients in the estuary of the Zaire (Van

Table 52 *Water characteristics of the Zaire River (Meybeck, 1978)*

SiO_2	9.60 mg l^{-1}	SO_4^{2-}	2.95 mg l^{-1}	Al	36 µg l^{-1}
Ca^{2+}	1.75 mg l^{-1}	HCO_3	11.2 mg l^{-1}	B	10 µg l^{-1}
Mg^{2+}	1.04 mg l^{-1}	NO_3-N	98 µg l^{-1}	Ba	18.3 µg^{-1}
Na^+	0.96 mg l^{-1}	NO_2-N	3.8 µg l^{-1}	Cu	0.3 µg^{-1}
K^+	1.36 mg l^{-1}	NH_4-N	5.6 µg l^{-1}	Fe	250 µg^{-1}
Cl^-	1.30 mg l^{-1}	PO_4-P	1.8 µg l^{-1}	Mn	8.3 µg^{-1}
				U	0.075 µg^{-1}

Bennekom, Berger, Helder & de Vries, 1978) shows that Zaire waters are more fertile than South-American waters, probably because of the higher demographic density in the Zaire watershed and the higher PO_4–P content of African waters.

Biological features

The phytoplankton of the Zaire River has been studied by Kufferath (1956b), chlorophyll measurements were made by Berritt (1964) and primary production was measured by Dufour and Merle (1972). A recent investigation covering all these parameters was carried out by Cadée (1978) in November 1976 in the Zaire estuary. The phytoplankton was rather poor and dominated by diatoms (*Melosira granulata* var. *angustissima*) and small unidentified cells; green algae were represented by *Scenedesmus, Ankistrodesmus, Crucigenia*. The corresponding productivity was 100 mg C m^{-2} d^{-1}. The primary productivity of the river plume in the sea was lower than that of the river and of the adjacent ocean. This strongly contrasts with the observations on the Amazon River where primary productivity is one order of magnitude higher in the plume than in the ocean. This difference is caused by the extensive mixing occurring in the Amazon estuary which allows a progressive development of phytoplankton, whereas the well-defined freshwater tongue of Zaire water in the sea causes the death of phytoplankton.

Much has been published on the fish fauna of the Zaire (Poll, 1959; Gosse, 1963; Matthes, 1964). These results have been reviewed by Beadle (1974) and Lowe-McConnell (1975).

The long period of stability which characterizes the geological history of the basin and its relative isolation account for the large number of species (669 known species belonging to 25 families and 168 genera) and the remarkable number of endemic species (558).

Comparing the fish fauna of Amazon and Zaire basins, Roberts (1973) underlines that, in both basins, most of the fishes belong to the order Ostariophysii: 85% in the Amazon and 54% in the Zaire basin. In the Amazon, characoids represent 43% of the fishes, siluroids 39% and gymnotoids 3%; in the Zaire basin characoids form only 15% of the fish population, siluroids 23% and cyprinoids 16%. The non-Ostariophysan primary freshwater fishes of Amazonia include a few Lepidosirenidae, Osteoglossidae and Nandidae and the secondary freshwater fishes comprise Cichlidae and Cyprinodontidae. The non-Ostariophysan primary and secondary freshwater fish fauna of the Zaire basin includes three main groups:

1. families shared with Amazonia: Lepidosirenidae, Osteoglossidae, Cichlidae and Cyprinodontidae;
2. families shared with Asia: Notopteridae, Cyprinidae, Mastacembelidae, Anabantidae and Ophiocephalidae; and
3. archaic families known only in Africa: Polypteridae, Pantodontidae, Phractolemidae, Kneriidae and Mormyridae.

The two branches of the Upper Lualaba River and the Upper Kasai River are faunistically separated from the main Zaire River by falls and rapids; consequently, these two rivers have 52 endemic species and various Zambesi species which are not found in the middle Zaire River. The fish fauna of the Upper Lualaba contains several species common in the Nile system and unknown in the Zaire basin (*Protopterus aethiopicus congicus, Polypterus bichir katangae, P. senegalus meridionalis, Mormyrus kannume, Tilapia nilotica* and *Ctenopoma murei*) (Beadle, 1974). This strengthens the idea of a previous connection with the Nile system.

In the central part of the Zaire River, the fish population is dominated by the Mormyridae whereas *Protopterus dolloi* is common in the adjacent swamps. The rapids below Kinshasa have a very specialized fauna with anguilliform bodies and, in certain species, reduction of the accessory breathing organs and the eyes.

In Malebo Pool (Stanley Pool), a 500 km^2 enlargement of the river, the Mormyridae dominate with 38 species out of 193 (Lowe-McConnell, 1975). *Petrocephalus* and *Stomatorhinus* are the most common species.

In the pelagic zone of the rivers are found the planktivorous fishes: the clupeid *Microthrissa*, the characid *Clupeopetersius*, the cyprinids such as *Barbus* and predators such as *Hydrocynus vittatus*. The benthic area of the rivers is mainly populated by bottom insect feeders including numerous mormyrids (*Chrysichthys, Synodontis, Tylochromis* and *Barbus*), mud-eaters, detritus-feeders and piscivorous fishes such as *Polypterus, Mormyrops, Chrysichthys* and *Parophiocephalus*.

Except the cichlids which breed nearly all the year round in equatorial water bodies and certain pelagic and benthic fishes bound to their habitats, the fishes of the Zaire basin migrate upstream at the onset of the flood period and then to the flood plain where they spawn.

Lake Bunyoni

Lake Bunyoni is located in the north-eastern part of the Virunga volcanic field at 1°18′ S and 29°54′ E and 1950 m altitude. The lake was formed 18 000 years ago by a volcanic eruption which dammed a narrow valley.

Africa

The lake has been investigated in the 1930s by Worthington (1932), Worthington & Ricardo (1936) and more recently by Green (1965), Beadle (1966) and Denny (1972, 1973).

Lake Bunyoni has the dendritic shape of a man-made lake. It is 22 km long and 6 km wide at the widest section. Its area is ~60 km^2 and its maximum depth 40 m. It has no major inflow but numerous streams draining a catchment area exceeding 2500 m altitude. A north-westerly outflow drains the lake into the Ruhuma Swamp to Lakes Mutanda and Muhele. The lake level rose 2 m during the 1962–63 wet period. Lake Bunyoni was stocked with *Tilapia* in the 1930s.

Thermal profiles by Denny (1972) in April–May 1969 showed a surface temperature up to 25.1°C, but generally less (23–24°C) because of wind mixing, and a bottom temperature of 19.3°C. The thermocline, between 5 and 7 m, generates a strong chemocline below which there is no oxygen. The hypolimnion is also characterized by a decrease in pH (7.1 instead of 8.0 in the epilimnion) and by an increase in conductivity and alkalinity.

The chemical features of the lake are summarized in Table 53. Unlike many East-African lakes, Lake Bunyoni water belongs to the calcium–magnesium–hydrogen carbonate type. Release of phosphate from the mud seems very active and may indicate that hypolimnic anoxia is semipermanent.

With a transmission coefficient for white light of 0.54, Lake Bunyoni is classified among the clear lakes. Algal blooms are unknown and the phytoplankton community is poor in species and number, with *Anabaena* spp. and the desmid *Closterium aciculare* var. *subpronum* as dominant forms. The lake has a reed swamp dominated by *Phragmites* and *Nymphaea coerulea*. *Potamogeton pectinatus* is common. Grönblad, Scott & Croas-

Table 53 *Chemical features of Lake Bunyoni*

Depth (m)	0	30		0	30
pH	8.0	7.1	Ca^{2+} (meq l^{-1})	0.87	1.02
Conductivity (μmhos cm^{-1})	258	284	Mg^{2+} (meq l^{-1})	0.85	0.87
PO_4–P (μg l^{-1})	<26	480	Na^+ (meq l^{-1})	0.66	–
NH_3–N (mg l^{-1})	0.12	3.4	K^+ (meq l^{-1})	0.15	–
Total N (mg l^{-1})	0.79	–	Cl^- (meq l^{-1})	0.82	0.78
SiO_2 (mg l^{-1})	2.15	–	SO_4^{2-} (meq l^{-1})	0.14	0.13
			Alkalinity (meq l^{-1})	1.94	2.35

The Zaire basin

dale (1964) recorded 10 taxa of desmids epiphytic on these plants; the algae belonged to the genera *Euastrum, Cosmarium* and *Staurastrum*.

Denny (1972) estimates that wind-induced seiches cause irregular partial mixing between epi- and hypolimnion. Such events are probably responsible for mass fish killings, the last of which occurred in 1964 and affected mostly *Haplochromis* and *Tilapia*. Further stocking of the lake was unsuccessful and commercial fishing has been abandoned.

It has been suggested that the decaying vegetation of the swamp areas through which the inflowing streams reach the lake might cause a reduction in the oxygen level of the inflowing water. Oxygen profiles taken at the edge of the *Cyperus papyrus* and *Cladium jamaiscense* swamps, however, indicated that oxygen levels are near saturation in these waters. Therefore, upwelling of oxygen-free hypolimnic water is the most probable reason for sudden drops in oxygen in upper waters and for fish killing.

Lake Kivu

Lake Kivu is located between 1°34′ and 2°30′ S and 28°10′ and 29°25′ E in Rwanda and Zaire (Fig. 21) at an altitude of 1463 m above MSL.

Origin

The area presently occupied by Lake Kivu was, in pre-Pleistocene time, drained to the Nile system. In late Pleistocene, the eruption of the Virunga volcanoes set a topographical barrier exceeding 4000 m and separated Lake Kivu from the Nile drainage.

Lake Kivu was not, however, formed as a result of the Virunga volcanic activity; a lake, smaller than the present one, existed prior to these events: Degens, Deuser *et al.* (1971) and Degens *et al.* (1973) report a maximum sediment thickness of 500 m in the northern basin whilst in the more southern areas the sediment cover is thin or non-existent. Considerations on average rates of sedimentation led these authors to attribute a Pliocene age to the northern basin.

The last 15 000 years of the lake's history have been studied using core material (Degens *et al.*, 1973). A top unit, 130 cm thick, is composed of fine alternate brown and white laminae; a second unit, also finely laminated, lacks the dark brown bands; a third unit, below 330 cm, is similar to the uppermost one. A beach deposit was found in one of the cores, at a depth corresponding to a radiocarbon dating of 6000 years B.P. and located 310 m below the present level of the lake. This indicates that until 6000 years ago the lake had a much lower level and was probably limited to the northern basin. From 10000 to 6000 years B.P., the diatom assemblage was dominated by *Stephanodiscus astrea* and *Nitzschia fonticola*. From 6000 years B.P. to the present, the diatom community was composed of many uncommon species of *Nitzschia*. In the organic, dark-brown layers, the diatoms are not abundant, in contrast with siliceous flagellates belonging to the Chrysophyta.

At the base of several cores, the presence of thick tuff beds and volcanic ash layers indicates that intense volcanic activity occurred some 12000 years B.P. and may correspond to the main phase of Virunga eruption.

The magnetic survey of the lake (Degens *et al.*, 1973) shows a smooth magnetic field. North-east to south-west anomalies are noted along the northern coast and are probably connected to the magnetic effect of basaltic layers; it is known that, in the period 1938–40, lava flows of large dimensions poured into the north-west part of Lake Kivu. Similar anomalies, caused by similar events, have been found in Lake Kinneret, Israel (Ben Avraham *et al.*, 1980).

The seismic survey of the lake revealed a sharp echo 400 m wide and 80 m high in the southern part of the northern basin. This has been interpreted by Degens *et al.* (1973) as a 'hydrothermal jet' on the lake floor.

General features

The main morphometric features are shown in Table 54. On its western side, the lake is bordered by the Rift escarpment; the northern part of the lake is occupied by the immense lava field of the Virunga. The highlands of Rwanda limit the lake in the east whereas the volcanic massif of Kahusi closes the lake's southern area.

The lake has about 150 islands of total area 315 km², the main one being Idjwi Island. The lake itself is formed of five distinct basins. The northern basin, north of Idjwi Island reaches a depth of 485 m; east of Idjwi Island the eastern basin reaches 400 m; west of the island, the Kalehe basin is 300 m deep; and south-west of the island extend the shallow Ishungu and Bukavu basins, draining into the Ruzizi River at an average discharge of 70 m³ s⁻¹ (Verbeke, 1957).

Table 54 *Morphometric features of Lake Kivu*

Max. length	96 km	Mean depth	240 m
Max. width	48 km	Max. depth	489 m
Surface area	2700 km²	Volume	583×10^9 m³
Shoreline	1196 km		

The Zaire basin

Lake Kivu is an equatorial lake of relatively high altitude; the average yearly temperature is about 20°C with 27.5°C and 12°C as respective maximal and minimal temperatures. The average amount of precipitation in the lake is 1300 mm y^{-1}; the dry season is centred around July and the wet season around September. The predominant winds blow from southeast to east. In the northern basin, cold winds blow from the Virunga area to the lake during the night and from the lake during the day-time.

Thermal structure

In February 1936, Damas (1937), studying the thermal structure of Lake Kivu, discovered that below the regular epi-meta-hypolimnion structure of the upper 70 m, the temperature rises again from 22.3°C at 70 m to 25.3° at 375 m. Moreover, he found this layer loaded with H_2S and gases. He concluded that this deep layer never mixes with the upper water and called it the 'dead layer'. Damas found also that at a given depth, in the dead layer, the temperature is uniform all over the lake. Similar observations were made by Schmitz & Kufferath (1955) and Degens *et al.* (1973). Damas explained this unusual thermal structure by hypothesizing that, during the volcanic eruptions, the water, by contact with the lava, became hot and saline, accumulated in deep layers and kept part of the stored heat. More precise and continuous water temperature profiles by Degens *et al.* (1973) showed the existence of sharp stepwise changes in temperature below 200 m, characteristic of convection caused by an increase of temperature and salinity with depth. This process leads to the formation of distinct layers well mixed within themselves but clearly separated one from another. The 'hydrothermal jets', suggested for the interpretation of special echoes observed in seismic profiles of the lake bottom, seem to be the driving force of the vertical thermal and chemical gradients and the resulting convection. Hot springs are found around the lake. According to Degens *et al.* (1973), the hydrothermal waters originate from meteoric water percolating downward into volcanic material and emerging as hot saline springs. This hypothesis is strengthened by the fact that the ^{18}O content of the lake's surface water is much higher (because of evaporation enrichment) than that of the 'dead layer', which has the same ^{18}O content as meteoric water. If this hypothesis is correct, the hydrothermal discharge should vary as a function of long-term amounts of precipitation. Degens *et al.* (1973) related the dark-brown laminae in the sediments with hydrothermal discharge and correlated the absence of these laminae in the sediments between 9000 and 5000 years B.P. with a dry episode identified in many lakes of East Africa.

Vertical profiles of temperature microstructure by Newman (1976) reveal that up to 150 isothermal layers, 0.25–2 m thick, may appear in a single profile with a 0.01–0.03°C increment from layer to layer. Such results suggest an upward heat flux of 0.71 to 1.6 W m^{-2} and a corresponding upward salt flux corresponding to 20% of the salt output of the Ruzizi river. Confirming Degens' views, Newman suggests that thermo-mineral springs, located on the lake bottom, are the source of the heat and salt flux. The layering is generated by the process of double diffusion described by Turner (1973).

Chemical features

The main chemical features of Kivu water are presented in Table 55. The pH of Kivu waters varies from 9 at the surface to 6.5 in the anoxic layer. The oxygen disappears around 60 m and sulphide reaches 9 mg l^{-1} in the deep layer which also contains up to 20 mmoles l^{-1} methane. This unique accumulation of methane is not from volcanic origin, since hot springs do not contain the gas; it comes from anaerobic decomposition of the organic matter by methane-producing bacteria which have been found in abundance in surface sediments. A large variety of methane-oxidizing bacteria has been observed. Jannasch (1975) estimated the rate of methane oxidation to be 116×10^6 m^3 y^{-1} for the whole lake in a 15 m thick layer above the oxic–anoxic interface. Assuming steady-state conditions between methane production and methane oxidation, 461 m^{-2} methane are annually produced in the lake. Methane and CO_2 concentrations in deep layers remain below the solubility of these gases at the pressures corresponding to the deep layers so that no spontaneous gas bubbling is likely to occur.

The concentration of major ions in the anoxic layer shows the same stepwise increase with depth as temperature.

The concentration of dissolved organic carbon is 10–50 times that of the ocean.

The Kivu water belongs to the rare sodium- and magnesium-hydrogen carbonate type.

Biological features

Van Meel (1954) published a list of algae determined from the material collected by Damas in 1935–36. The Cyanophyta comprise *Microcystis flos-aquae, Microcystis* sp., *Chroococcus minutus, Lyngbya* sp., *Oscillatoria* sp., *Spirulina* sp. and *Anabaenopsis tanganikae*. The green algae are mainly represented by *Cosmarium* sp., *Chlorella vulgaris*,

Table 55 *Chemical features of Lake Kivu waters*

	Na$^+$ (mgl^{-1})	K$^+$ (mgl^{-1})	Ca^{2+} (mgl^{-1})	Mg^{2+} (mgl^{-1})	HCO$_3^-$ (mgl^{-1})	Cl$^-$ (mgl^{-1})	SO$_4^{2-}$ (mgl^{-1})	NH$_4$-N (μg at l^{-1})	P (μg at l^{-1})
Surface water, Kisenyi Bay (Hundeshagen, 1909)	202.8	30.7	8.1	122	1108	42.4	32.4	–	–
Surface water (Degens et al., 1973)	121.6	97.4	4.8	87	750	55	23.8	18	0.8
Surface water Kabuno Bay (Delhaye, 1941)	210	18	37	82	750	35	6	–	–
Deep water, 200 m (Degens et al., 1973)	244.6	178.8	83.1	182	2520	–	166.4	1314	32.7
Deep water, 440 m (Degens et al., 1973)	487.4	338.0	112.6	417	7260	–	220.0	7105	54.8
Hot spring, south of Kisenyi Bay (Hundeshagen, 1909)	558.5	54.2	24.2	16.6	1281	230	31	–	–
Hot spring, Kakondo (Degens et al., 1973)	179.6	56.6	82.2	53	1782	80	16.8	109	3.6

Chodatella longiseta, Crucigenia tetrapedia, Oocystis sp., *Scenedesmus cristatus, Tetraedron* sp. and *Botryococcus braunii*. The diatoms are represented by 132 species, strongly dominated by *Nitzschia* (*N. confinis, N. lancettula, N. tropica*); *Melosira ambigua* is common whereas *M. granulata* is rare. This list is not exhaustive since the material was collected by large mesh nets. The primary productivity has been found to be 375 g C m^{-2} y^{-1} (Degens *et al.*, 1973).

The calcareous formations common on the lake shores are covered with *Cladophora* down to 8 m depth. The photosynthetic activity of *Cladophora* causes the precipitation of calcium carbonate in the vicinity of the algae.

Copepods clearly dominate the zooplankton with *Mesocyclops leuckarti* and *Tropocyclops confinis*. In August the Cladocera, which generally represent 1–20% of the zooplankton, grow rapidly and form 49% of the zooplankton, with mainly *Moina dubia, Daphnia pulex* and *Ceriodaphnia rigaudi*.

The benthic organisms are limited to the sediments in contact with oxygenated water, ~12% of the surface of the lake. Benthic diatoms belonging to the genera *Surirella, Synedra, Cyclotella* and *Pinnularia* are abundant. The benthic fauna is poor in species and has a low biomass. Chironomid larvae have a maximal density at 20 m (1985 larvae m^{-2} as an average value). Tubificidae are, with the Chironomids, the only group living in the oxygenated sediments of the Kivu. The sediments periodically carried by the rivers form the only substrate where the benthos can live, since the shores are continuously covered with new layers of calcium carbonate. *Chaoborus* larvae are completely absent from Lakes Kivu and Tanganyika. This, according to Verbeke (1957), may be due to the composition of the water or to the absence of appropriate food for the first larval stages.

The fish fauna, characterized by Beadle (1974) as 'an attenuated version of that of Lake Edward', is composed of 16 species belonging to the Cyprinidae, Clariidae and Cichlidae. The fish fauna of Kivu has little in common with that of Lake Tanganyika and only three Kivu species are also found in Tanganyika; the cyprinid *Barilius moori* which is otherwise endemic to Tanganyika is the only species of Lake Tanganyika which overcame the Pauzi Falls separating the two lakes. However, other lakes which formed as a result of blockage of rivers by volcanic lava during the Virunga eruption have a much poorer fish fauna than Lake Kivu. This and the existence of six endemic species of *Haplochromis* and one endemic subspecies of *Tilapia nilotica* support the idea of a proto-Kivu which was deepened by the Virunga volcanic event. These signs of biological maturity

The Zaire basin

are not in good agreement with the absence of large predator fish such as *Bagrus*, *Lates* or *Hydrocynus* and of pelagic zooplanktivorous fish. The ungrazed zooplankton is therefore sedimented and possibly contributes to methane formation.

Lake Tanganyika

Lake Tanganyika (Fig. 21), elongated in the N–S axis, extends from 3°20′ to 8°45′ S and from 29° to 31° E at an altitude of 773 m above MSL at the border of Zaire, Burundi, Tanzania and Zambia.

General features

Lake Tanganyika is the second deepest lake of the world after Baikal. It is formed of four basins (Capart, 1949). The northern basin of Usumbura has a maximum depth of 450 m and is filled with the material of the Ruzizi sublacustrine delta. The Kigoma basin, limited in the north by the sill of Ubwari and in the south by the sill of Kungwe, has a maximum depth of 1310 m. Its central part is a flat area, at 1250 m depth, with steep shores. In the southern part, the delta of the Malagarasi River reaches the central part of the lake. The Kungwe sill is at 700 m depth and any drop of level greater than 700 m would cause the formation of two distinct lakes without any connection between them. The Alberville basin is limited to the south by the Marungu sill at 800 m depth. This basin does not exceed 885 m depth. South of the Marungu sill, the basin of Zongwe is the deepest part of the lake with a maximum of 1470 m.

In many places the lake is limited directly by the Rift walls, 1000–2000 m high, which plunge steeply at great depths. This results in a difference in altitude of > 3000 m between the bottom of the lake and the rim of the plateau. This topographic feature, together with the thick sediment layer on the lake floor gives the lake bottom the characteristic 'U' shape frequently recorded in Capart's echosounding profiles.

These profiles revealed another interesting feature: numerous sublacustrine canyons, continuing the present aerial valleys, can be traced down to 550 m. Capart (1949, 1952) concluded that the Lake Tanganyika level was much lower in the past. The main morphometric features of the lake are shown in Table 56.

Hydrology

The lake is fed by three main rivers: the Malagarasi River, the Ruzizi River and the Mutembala River, which have built large deltas in the lake. The only outlet is the Lukuga River which drains the Tanganyika into the Zaire basin. Like the Shire River at the outlet of Lake Malawi, the Lukuga River is blocked at low water by sand and vegetation. Then, the outflow is interrupted until a substantial increase in the lake level flushes away the obstacles. Such sequences were observed at the end of the last century.

The wet season extends from November to April and the average yearly rainfall is about 1 m. The wind blows from land to lake during the night and from lake to land during day-time. This cycle, periodically observed during the wet season, may be interrupted by strong northern winds reaching 90 km h^{-1} and causing violent storms. Capart also observed waterspouts several times. In the dry season dominant winds are from the south-east.

The hydrological budget of the lake can be expressed in a very simple manner: 10% of the yearly water income is carried to the lake by rivers and 10% is evacuated by rivers; approximately 90% of the income is directly supplied by the rainfall on the lake and an identical amount is lost by evaporation from the lake surface (Livingstone & Melack, 1979).

Thermal structure

The thermal structure of Lake Tanganyika was studied by Capart (1952) who distinguished three layers: the upper 100 m, affected by daily thermal variations, the intermediate 100–250 m layer, affected by seasonal variations of temperature, and the deep layer which is thermally stable. In the upper two layers, one can distinguish the well-mixed epilimnion from surface to 60 m, where temperature varies from 27°C in April during the stratified period to 25°C in August when the stratification is weaker, the metalimnion around 60 m where temperature drops to 24°C, and the hypolimnion from 80 to 200 m where temperature varies little (23.43–23.48°C). The oxygen layer extends far below the seasonal thermocline.

In 1961, the thermocline remained stable around 50 m until July. In July–August, the thermocline disappeared and, in August, the zero-oxygen

Table 56 *Morphometric features of Lake Tanganyika*

Max. length	673 km	Mean depth	~700 m
Max. width	80 km	Max. depth	1470 m
Surface area	33 000 km²	Volume	19 000 × 10⁹ m³
Shoreline	1500 km	Watershed area	250 000 km²
Watershed area:lake volume ratio 0.013 m² m^{-3}			

curve was as high as 80 m (JFRO, 1964). This is interpreted as the effect of an internal wave with a resulting local upwelling.

Below 200 m, the temperature variation is very small: it decreases by 0.2°C down to the thermal minimum located between 500 and 800 m. Near the bottom the temperature rises slightly (23.32–23.35°C).

The annual limnological cycle is described by Coulter (1968a) who distinguishes the two following phases: 1) strong southerly winds between April and September cause a downward tilt of the thermocline in the northern area; the resulting weakening of the thermocline in the southern region allows upwelling and mixing; and 2) stratification is re-established in September–October with oscillations of the thermocline suggesting the occurrence of internal waves.

The stability of 70% of the lake volume and the great age of the lake should logically lead to an accumulation of salts in the deep layers. Surprisingly, the measurements of dissolved salts by Kufferath (1952) showed remarkable homogeneity confirmed by the more recent profiles of Degens et al. (1973) and suggesting some kind of circulation.

Capart had already observed that the water of the Ruzizi River, cooler than the lake water, sinks down and can be traced at 53 m depth by its turbidity and salinity. On various occasions rain water could be traced chemically at great depths (400 m). Coulter (1968b) showed that downward flow of dense water is probably a daily process occurring not only at the river mouths but every night along the lake margins. Moreover, Coulter observed that, during the wet season, the torrential, cool flow of the small rivers sinks rapidly via the sublacustrine valleys which may explain the absence of sediments. Upwelling and internal waves represent an additional factor of circulation (Coulter, 1963, 1968a). A first model of circulation, envisaged by Coulter, consists of two cells: an upper one, down to 500–800 m, the thermal minimum, where mixing results from 'turbulence and mass displacement in the epilimnion and metalimnion caused by wind stress and at deeper levels by internal waves', and a lower cell where heating from compressions and terrestrial heat flux will be greater near the bottom and generate convection currents which will produce mixing in the boundary layer with the upper cell.

Chemical features

The Tanganyika waters are strongly marked by the composition of the Kivu–Ruzizi water and belong also to the sodium–magnesium-carbonate type (Table 57). The absolute concentrations are surprisingly low (420 mg l^{-1}) for a lake which has been isolated for a long period, which

Table 57 *Chemical composition of Lake Tanganyika water (results in mg l^{-1})*

	Ca^{2+}	Mg^{2+}	Na$^+$	K$^+$	CO$_3^{2-}$	Cl$^-$	SO$_4^{2-}$	SiO$_2$	TDS
Tanganyika 700 m (R. S. A. Beauchamp, 1939)	15.2	43.7	64.2	33.5	207.6	28.0	4.0	13.5	413.3
Inflows (R. S. A. Beauchamp, 1939)	19.0	17.2	25.0	9.0	100.0	14.0	8.2	26.0	220.6
Tanganyika (Kufferath, 1952)									
surface	13.0	42.6	—	—	204.3	27	3.0	0.3	—
1300 m	17.6	43.2	—	—	208.8	27.9	3.0	12.0	—
South lake average (Hecky et al., 1978)	11.0	39.1	62.9	32.3	—	27.2	7.6	—	—
Central lake average (Hecky et al., 1978)	11.0	38.8	63.1	32.8	—	26.4	6.3	—	—
North lake average (Hecky et al., 1978)	10.7	37.9	61.7	32.0	—	26.4	6.3	—	—

receives relatively salty water from Lake Kivu, and where evaporation accounts for 90% of its losses. Beauchamp (1946) relates this low salinity to the predominance of rain water in the lake water income. The homogeneity of concentrations in such a huge water body where 75% of the volume is permanently devoid of oxygen and loaded with H_2S and CH_4 is rather paradoxical but we have seen that complicated patterns of circulation allow a slow but efficient mixing.

The fluctuations observed in the vertical variations of nutrient concentrations can be explained by the activity of organisms. In Degens, Von Herzen & Wong's (1971) profiles, a superficial peak of ammonia is followed at around 100 m by a sharp peak of nitrates which disappear at 170 m. Viner (1979) interprets this decrease as resulting from denitrification. In contrast, SiO_2 and PO_4-P concentrations increase significantly from 0 to 400 m and then remain stable down to the bottom. This type of curve is explained by Degens *et al.* by the active uptake of SiO_2 and phosphorus in the upper layers and by the dissolution of biological remains while sinking to the bottom. It is however surprising that the increase in SiO_2 and PO_4-P in the deep layer is not accompanied by an increase of ammonia; the eventual nitrification of ammonia is excluded since this is an anaerobic environment. Total CO_2 increases with depth (Craig & Craig, 1979) but the $^{13}C:^{12}C$ ratio is minimum at 200 m depth. Craig & Craig conclude that low $^{13}C-CO_2$ is produced around 200 m by dissolution of carbonate and oxidation of organic carbon, whereas ^{13}C-rich CO_2 comes from the deep layer where it has been enriched in ^{13}C by methane production.

Biological features

The great transparency of the lake (up to 18 m in April–May, Hecky, Fee, Kling & Rudd, 1978) suggests that the lake is unproductive, whereas the abundance of the sardine population points to a fairly high productivity.

Phytoplankton, Protozoa and bacteria. Van Meel (1954) found that the Lake Tanganyika phytoplankton was dominated by Chlorophyta and diatoms (93 and 88 species, respectively) followed by Cyanophyta (41 species) and Dinophyta (four species). During the mixed period of 1962 (June–September), the plankton was dominated by *Nitzschia nyassensis* and *N. asterionelloides*. *Anabaena flos-aquae* became dominant from October to January (JFRO, 1964).

Hecky *et al.* (1978), during two cruises in April–May 1975 and October–November 1975, measured the biomass of plankton at 55 and 59 stations, respectively; a summary of their results is presented in Table 58.

The plankton of Lake Tanganyika is characterized by the abundance of nanoplankton and Protozoa. In the northern station, three seasonal phases were observed in algal and protozoan biomass. The first period (mid-February–late April) was characterized by low algal biomass (50–100 mg m^{-3}) and high protozoan biomass which was in many instances higher than the algal biomass. For example, in February 1975 the algal and protozoan biomasses were, respectively, 160 and 460 mg m^{-3} (Hecky & Kling, 1981); the Protozoa were mainly represented by *Strombidium* cf. *viride* which always contained green algal cells. The Chlorophyta dominated the phytoplankton with species of *Kirchneriella*, *Treubaria*, *Lagerheimia*, *Lobocystis*, *Quadricoccus*, *Dictyosphaerium* and *Chlorella*. The Cyanophyta, secondary in abundance, were represented by *Chroococcus limneticus*.

During the second seasonal phase (May–mid-September) phytoplankton biomass exceeded 100 mg m^{-3} and was always greater than protozoan biomass. In the early stage the increase in biomass was caused by the development of chrysophyceans such as the small *Chrysochromulina parva* and *Chromulina* sp. and the larger *Ochromonas* spp. and *Spumella* sp. In late May this population was replaced by the diatom *Nitzschia* and the cryptophytes *Katablepharis ovalis* and *Cryptaulax* cf. *conoidea*. In July–August small *Nitzschia* spp. and *Stephanodiscus* sp., together with the green alga *Coelastrum reticulatum*, replaced the previous assemblage. During this period *Strombidium* declined whereas *Tintinnidium* spp. developed.

The third period (mid-September–November) was the period of algal blooms: an early October peak of *Nitzschia* spp. was followed by a peak of chlorophytes with biomass values of 300 mg m^{-3} and a bloom of *Anabaena* reaching biomass values of 930 mg m^{-3}. The Protozoa *Strombidium* cf. *viride*, *Halteria* sp. and *Vorticella* sp. were also abundant in November.

The primary productivity was unusually low during the April–May survey (600 mg m^{-2} d^{-1}) and very close to the value obtained by Melack (1980) in April 1971 (500 mg m^{-2} d^{-1}). In October–November, production was much higher (1400 mg m^{-2} d^{-1}). The mean annual daily rate of primary production for the whole lake was estimated to be 1 g C m^{-2} d^{-1} (Hecky & Fee, 1981). Since algal biomass is never very high, the rates of primary production observed imply rapid turnover of the algal population. High turnover which is not accompanied by drastic decrease in biomass means high loss rates, probably through zooplankton grazing.

Calculation of the fish yield obtainable from the measured levels of

Table 58 *Phytoplankton and protozoan biomass and relative composition*

Station		Biomass (mg m^{-3}) Phytoplankton	Protozoa	Per cent composition of total biomass					
				Cyano-phyta	Chloro-phyta	Chryso-phyta	Bacillario-phyta	Crypto-phyta	Pyrro-phyta
Kipanga (southern station)	AM	169	89	37	27	18	6	10	3
Kipili	AM	138	144	35	16	25	15	7	2
(south-eastern shore)	(ON)	(105)	(24)	(8)	(66)	(15)	(–)	(10)	(–)
Malagarasi	AM	362	143	9	15	32	3	5	36
(north-eastern river)	(ON)	(105)	(38)	(53)	(30)	(1)	(10)	(5)	(0)
Rumonge	AM	108	72	21.6	38.2	25.2	–	13.2	2
(northern station)	(ON)	(110)	(27)	(18)	(34)	(11)	(28)	(1)	(–)

AM = April–May survey
(ON) = October–November survey

productivity gives values which are one order of magnitude lower than the actual fish catch. It seems therefore that in Lake Tanganyika other food sources are available to zooplankton and fish; methane-oxidizing bacteria are likely to be one of them.

Methane-oxidizing activity was found in a 10 m-thick layer at the oxic–anoxic interface. The highest rates of oxidation occurred at both ends of the lake. The low rates, in the central part of the lake, are usually found during the dry windy season when stratification in the oxygenated layer is weak. Each mixing episode destroys the methane oxidation layer. Therefore, methane oxidation is maximum during the wet and well-stratified period. The rates for dry and wet period are, respectively, 7.2 and 15.0 mg C m^{-2} d^{-1}. These rates are similar to those found in Kivu (Jannasch, 1975). The CH_4 concentrations being much higher in Kivu, it follows that the rates of vertical transport are much higher in Tanganyika than in Kivu, which is consistent with the absence of accumulated salts in deep layers of Tanganyika. The respiration rate of the methane oxidizers is unusually high: 74% of the methane was oxidized whereas only 26% was converted into bacterial cells. It follows that, as stated by Rudd (1980), the net production of the methane-oxidizing bacteria represents only a few per cent of algal primary production.

Zooplankton. According to Leloup (1952), the zooplankton is abundant and consists mostly of *Cyclops* and *Diaptomus simplex* but also includes the jelly-fish *Limnochida tanganika*, called by local people 'the eyes of the lake'. During day-time, the zooplankton is found between 50 m and 125 m. Leloup mentions that the respiration of the zooplankton in this layer consumes more oxygen than the phytoplankton can supply and that consequently the oxygen concentration is often lower between 75 and 125 m than at 150 m. This early finding that the primary production does not compensate the respiratory needs of the aquatic community was fully confirmed by Hecky *et al.* (1978). At night, the zooplankton concentrates around 10 m depth.

A peculiar feature of Lake Tanganyika is the complete absence of Cladocera from the lake water whilst they are found in the lake tributaries. The high content of magnesium of the lake water was suggested to be the cause but this was rapidly rejected when Cladocera were successfully cultured in Tanganyika waters. Leloup (1952) claimed that these filter feeders did not find adequate food because of the paucity of nanoplankton and bacteria in the lake. We now know that bacteria and nanoplankton are abundant in the lake and that a more plausible reason for the absence of

The Zaire basin

Cladocera is the grazing pressure exerted by the planktivorous fish (Harding, 1957).

Fish. The Lake Tanganyika fish population is famous not only for its great number of species but also for their considerable degree of speciation. The lake has 193 species representing 13 families (Lowe-McConnell, 1975). The cichlids, which represent 62% of the fauna, are all endemic at the species level and 33 of 37 genera are endemic (Fryer & Iles, 1972).

Among the non-cichlids, the Cyprinidae and Bagridae dominate with 11 species each, then the Mastacembalidae with seven species, Clariidae, Mochokidae and Centropomidae with five species each, Characidae (four species), Clupeidae and Distichodontidae (two species each) and Protopteridae, Polypteridae, Citharinidae, and Malapteruridae (one species each). The non-cichlid fish have eight endemic genera in comparison with one in Lake Victoria and none in other lakes. This feature testifies to the longer isolation of Lake Tanganyika.

A few species living in the lake are identical with those of the Nile basin. Poll (1953) and Marlier (1953) consider that *Melapterurus electricus, Tilapia nilotica, Bagrus docmac, Ctenopoma muriei* are relicts of Pliocene fauna which remained unchanged. *Hydrocynus lineatus*, another Sudanese species found also in the Zaire may have penetrated into Tanganyika in the late Pleistocene via the Lukuga River. There have apparently been no faunal connections between Lake Tanganyika and the Kivu basin. Only one species of Lake Tanganyika has recently reached Lake Kivu; moreover, the Ruzizi River has a Tanganyikan fish fauna in contrast with the inflows to Lake Kivu (Marlier, 1953).

Special mention should be made of the two most abundant clupeids of the lake, *Limnothrissa* and *Stolothrissa*. This latter species is truly pelagic and feeds exclusively on zooplankton. The sardines follow the diel migration of the zooplankton. Coulter (1979) has followed their schooling behaviour by echosounding. He concluded that the fish prefer to remain in low light intensity and form schools when a light threshold is exceeded, which seems to be an adaptation to severe predation pressure in very clear water. According to a population dynamics study by Roest (1978), carried out in the Burundi waters, 65% of the sardine biomass could be annually harvested instead of the present 50%. Spawning periods of *Stolothrissa* are flexible with a frequent lapse of 14 months between two peaks, corresponding to the time for two generations to reach maturity.

The sardines represent a substantial part of the fisheries which fluctuated between 70 000 and 90 000 tons y^{-1} in the last 20 years. Acoustic estimations of pelagic fish biomass range from 2.8 million tons to 470 000 tons (Rufli, 1979). Coulter (1970) has followed the population changes of the sardines and their predators. From 1963 onwards, the biomass of the predators decreased and this was followed by a substantial increase in the sardine catch.

Deelstra (1979) underlines that this abundant source of protein is already seriously contaminated by insecticides utilized for the control of tropical diseases.

It seems that Lake Tanganyika, for purely tectonic and topographical reasons, has had very few contacts with other water bodies. Moreover, its enormous volume preserved it from drying during the dry period of the mid-Pleistocene which has allowed a degree of speciation unknown in any other lake. This great age and lengthy isolation also explains the original food chain it has developed which is based on delicate equilibria concerning vertical transport of organic and inorganic nutrients, allowing the development of bacteria and primary producers which are probably both used as food sources by protozoa and zooplankton.

Origin of the lake

Gunther (1898), thinking that the thick-shelled molluscs of the lake were of marine origin, put forward the hypothesis that Lake Tanganyika was the remnant of an ancient sea. This idea was not confirmed by later geological investigation.

Seismic profiles (Degens, Von Herzen & Wong, 1971) revealed that the sediment layer is more than 1500 m thick, confirming the Pliocene origin of the lake. During the Miocene, the area had a general westward drainage and the Malagarasi River possibly joined the present Lukuga River which drained into the Lualaba River which, in its turn, was connected to the Nile basin. Some time during the Pliocene, the Lualaba was captured by the Zaire and the formation of the Rift cut the Malagarasi–Lukuga River into their present two portions. This would explain the insolite presence, in the Malagarasi River, of riverine Zairean species which are common in Zaire River but do not exist in the lake (Poll, 1950).

We know that the lake water level has fluctuated by a few metres during the last century but there is much uncertainty concerning water level changes in a more remote past. Capart (1949) found by echosounding that the bottom of the lake was not covered by sediment down to 150 m. Below this depth, he found several reflective layers the number of which decreased with depth. He interpreted these layers as beach formations which developed at lower than present levels of the lake. Since the deepest hard

The Zaire basin

layer was registered at 850 m depth, he concluded that the lake had reached its present level only in modern times which would explain the absence of sediment in shallow water. Livingstone (1965) however, working on a core taken at 440 m depth in the southern basin showed that a volcanic ash layer was responsible for the reflection signal at this particular locality. The core material covered a time-span of 22 000 years and did not show any unconformity or presence of erosion or terrestrial material. It is therefore very likely that the water level has not dropped below 440 m during the last 22 000 years.

It is however the general opinion that, prior to the Virunga eruption in the early Pleistocene, the level was lower and the Tanganyika had no outlet. It is assumed that the inversion of flow pattern of the Kivu via the Ruzizi River which followed the Virunga eruption caused a significant rise in the lake level and the activation of the Lukuga River as an outlet. The discharge of the Ruzizi River seems to have been intermittent since, in the late Pleistocene, the Kivu levels were lower than the Ruzizi outlet (Hecky *et al.*, 1978). It is likely then that the Tanganyika level has fluctuated considerably during the Pleistocene.

Water bodies of Upper Zaire
Lake Bangweulu

Lake Bangweulu is located in Zambia, between 10°15′–12°30′S and 29°30′–30°05′ E at 1160 m altitude. Its depth does not exceed 5 m. The lake itself, with a surface area of 2800 km², is bordered on the eastern and southeastern sides by a large swampy area of 7000 km² (Table 59). The Chambezi River is the main inlet of the complex lake–swamp which is drained by the Luapula River.

Table 59 *Morphometric features of the Upper Zaire lakes*

	Bangweulu	Mweru	Upemba
Max. length (km)	72	124	40
Max. width (km)	39	51	20
Surface area (km²)	2800	4580	530
Mean depth (m)	4	3–10	0.3
Max. depth (m)	10	37	35
Volume (10^9 m³)	11.2	36.3	0.9
Watershed area (km²)	100 800	–	–
Watershed area : lake volume ratio (m² m^{-3})	9	–	–

The water is weakly acid and has a very low content of salts and nutrients (Table 60).

Thomasson (1965) gave a detailed list of algal species. The phytoplankton is dominated by Desmidiales: among the 52 species reported by Thomasson (1965), *Closterium, Micrasterias, Cosmarium* and *Staurodesmus* are the genera most frequently encountered. Chlorophyceae are represented by 17 species with *Botryococcus braunii, Pediastrum* spp., *Coelastrum* spp., *Dictyosphaerium pulchellum* and *Scenedesmus* spp. as the dominant species. Cyanophyta are mostly represented by *Microcystis* spp. and *Gomphosphaeria* spp. The chrysophycean *Dinobryon cylindricum* was also recorded. In another study, Thomasson (1960) reported the presence of species belonging to the genera *Euglena, Phacus* and *Trachelomonas* and the occurrence of the dinoflagellates *Ceratium brachyceros, Peridinium volzii* var. *cinctiforme* and *Stylodinium truncatum*. Part of the swamp area is permanently flooded; the swamp water has a low oxygen content but high concentrations of phosphate, iron and humic acids. At low water, the fish avoid the swamp water and congregate in the oxygenated open channel crossing the swamp. At high water, the swamp water is flushed by well-oxygenated river water and the fish move into the flooded area for spawning.

Phytoplankton represents only a small fraction of the zooplankton food. It is mainly consumed by pelagic zooplankton which is preyed upon by the only pelagic fish (*Engraulicypris* sp.) which is not abundant enough to fully exploit the pelagic zooplankton. Shallow-water crustacea and chironomids are mostly detritivores and preyed upon by *Tylochromis* sp., *Gnathonemus*

Table 60 *Chemical features of the Upper Zaire lakes (results in mg l^{-1})*

	Bangweulu		Mweru	Upemba
	(a)1960	(b)1936		
Ca^{2+}	1.5	1.1	7.5	20.6–42.3
Mg^{2+}	0.8	0.1	5.1	2.9–17.3
Na^+	2.6	5.1	4.6	–
K^+	1.3	2.2	1.25	–
$CO_3^{2-} + HCO_3^-$ (meq l^{-1})	7.8	0	0.83	54–99.9
Cl^-	0.3	0.8	5.0	–
SO_4^{2-}	1.0	2.3	3.7	–
pH	–	7.0	–	6.4–8.0
SiO_2	–	16.9	10.5	68–144
PO_4–P ($\mu g\ l^{-1}$)	–	–	–	31–65

(a) Harding & Heron, in Talling & Talling (1965)
(b) Ricardo (1939)

The Zaire basin

sp. and *Synodontis* spp., whereas aquatic insects are consumed by the juveniles of *Haplochromis* and *Barbus*. These five fish are eaten by a group of primary predators including *Chrysichthys*, *Schilbe mystus*, *Mormyrops*, *Serranochromis* and *Sargochromis*. A few secondary predators complete this food chain (*Hydrocyon vittatus*, *Clarias mossambicus* and *Heterobranchus longifillis*). *Tilapia macrochir* is one of the few phytoplankton-eaters of the lake (JFRO, 1958).

In absolute terms, the lake is not very productive in spite of its shallowness and strong wind-induced mixing because of slow decomposition rates or because of lack of benthic organic material in the lake area.

Lake Mweru

Lake Mweru is located in Zaire and Zambia between 8°27′–9°31′ S and 28°25′–29°10′ E, at 927 m altitude. It has a surface area of 4580 km². Its mean depth is 3 m in the southern basin and 10 m in the northern one; the maximum depth is 37 m (Table 59). The Luapula River drains into the lake, the swamp area located south of the lake. This influx of nutrient-rich water is apparently the direct cause of the high productivity of this lake. In Bangweulu Lake, the swamp is drained directly into the river, not via the lake which does not benefit from this important nutrient source. This topographic favourable situation allows the development of an abundant phytoplankton and bottom fauna although the waters are poor in salts. The molluscan fauna comprises 20 species, ten of which are endemic. The fish population is composed of 70 species among which ten are endemic. Lake Bangweulu has only 54 fish species of which five are endemic (Marlier, 1963). The total annual fish catch of both lakes reflects their difference in productivity: 9000 tons in Lake Bangweulu (1969–70) in comparison with 31 000 tons in Lake Mweru (1966–70 average) (Welcomme, 1972).

Lake Upemba

On the western branch of the Lualaba River extends a vast flood plain, the Kamelondo depression, in which more than 50 lakes can be found. Lake Upemba, which is the largest, is located in Zaire at 1000 m altitude. Its surface area is 530 km² and its mean and maximal depths are 0.3 and 3.5 m, respectively. The lake was studied by Van Meel (1953). Its salt concentration is higher than that of Bangweulu and Mweru (Table 60).

Mwadingusha Reservoir

The Mwadingusha Reservoir results from the damming of the Lufira River in 1930. It is located in Zaire at 1105 m altitude, at 10°54′ and 27°05′ E; it has a surface area of 410 km², a mean depth of 2.6 m, a maximal depth of 14 m and a volume of 1063×10^6 m^3. The reservoir is mixed from May to August with homogeneous temperature decreasing from 24°C to 18°C. From September to April the reservoir is weakly stratified (Magis, 1962). The waters of the Lufira are nearly always colder than the reservoir water and sink in deep layers: this may slightly increase the stability of the thermocline during the wet season.

Nzilo Reservoir

Located on the Lualaba River at 10°30′ S and 25°30′ E in Zaire, at 1200 m altitude, the Nzilo Reservoir has a surface area of 280 km^2 and a maximal depth of 45 m. The thermal cycle has been studied by Magis (1962) who has shown that, in contrast with Mwadingusha Reservoir, the Nzilo Reservoir has a very thick hypolimnion (30–35 m). The precipitations which occur when the lake is stratified are colder than the lake water, causing mixing in the epilimnion and deepening of the thermocline.

Lubumbashi Reservoir

The reservoir is located at 11°39′ S and 27°27′ E in the province of Katanga, on the Lubumbashi River. It was completed in 1962; its surface area is 40 ha and depth 4.6 m.

The reservoir has been studied by Freson (1972). Its water belongs to the calcium–magnesium hydrogen carbonate type and undergoes considerable dilution during the rainy season (Table 61).

In these waters, Freson found 72 species of algae dominated by diatoms and chlorophytes. The most common species are *Melosira ambigua, M. varians, Synedra ulna, Sphaerocystis schroeteri, Anabaena flos-aquae, Oscillatoria agardhii, O. limosa* and *Microcystis aeruginosa*. The chlorophyll concentration varies between 8 and 34 mg m^{-2}. The highest values occur in June–July at the beginning of the dry season when the temperature of the water is at its lowest. The primary productivity varies between 86 and 1204 mg O$_2$ m^{-2} d^{-1} with an annual average of 686 mg O$_2$ m^{-2} d^{-1}. An average specific productivity of 20.2 ± 6 mg O$_2$ per mg chlorophyll *a* per

Table 61 *Chemical composition of Lubumbashi Reservoir*

	pH	Conductivity (μmhos cm^{-1})	Ca^{2+} (mg l^{-1})	Mg^{2+} (mg l^{-1})	Na$^+$ (mg l^{-1})	K$^+$ (mg l^{-1})	Cl$^-$ (mg l^{-1})	SO$_4^{2-}$ (mg l^{-1})
Max.	8.2	309	49.8	25.8	5.9	1.5	2.5	5.0
Min.	7.1	128	11.5	6.7	3.8	0.6	0.8	0.2

The Zaire basin

hour has been measured slightly below the average value of 25 ± 6 found by Talling in Lake Victoria.

Water bodies of Lower Zaire
Lake Tumba

Located in Zaire at 1°S and 18°E at an altitude of 350 m, Lake Tumba has a surface area of 720 km² and a depth of 3–5 m. It has been studied by Marlier (1958) and Dubois (1959).

The lake has numerous tributaries from east and south and drains into the Zaire by the 35 km-long Irebu channel. The shores are rocky where the alios (pseudolateritic layer) reaches the lake; elsewhere they are flat and swampy. The lake is surrounded by dense forest; in flat areas, this forest is flooded at high water and resembles the Amazonian *igapo*.

The lake bottom consists of a thick layer of kaolinite covered by organic debris from the forest. The steep shores and flat bottom give a characteristic 'U' shape to the lake bottom.

The lake level follows the equatorial regime of the Zaire with two high-water periods (November–December and April–June) and two minima (January–March and July); the amplitude of level variation is about 3 m. The shallow water body is thoroughly mixed by strong and frequent storms and the thermal profile of the lake is uniform from surface to bottom.

The waters of Lake Tumba are blackwaters, rich in humic acids. They resemble Rio Negro waters and have a similar low pH (4.5–5.0). They have a very low salt content (72–90 mg l^{-1}); Ca^{2+} and Mg^{2+} concentrations are 0.7 and 0.3 mg l^{-1}, respectively. The concentration of nutrients in the waters is also very low. Oxygen is present at all depths but the concentration always remains far from saturation.

As in blackwaters of Amazonia, algae are not abundant and their primary productivity is limited. Zooplankton is not abundant; a colonial ciliate has frequently been observed in surface waters however. The benthic fauna is limited to the littoral zone where the products of decomposition of the forest offer a reasonable food source. Molluscs, however, have never been found: their absence is probably related to the absence of calcium and the low pH. The sponge *Metania lissostrongyla* is common on stones and harbours hydracarids. Cladocera (*Euryalona orientalis*) and copepods (*Paracyclops affinis*, *Microcyclops varicans*, *Ectocyclops rubescens*, *Macrocyclops albidus*, *Eucyclops fragilis* and *Cryptocyclops hinjanticus*) are very abundant in the littoral vegetation. Larvae of *Chaoborus* seem to be completely absent from the fauna. The ichthyological population has been studied by Poll (1942) and Matthes (1964). The fish fauna comprises more than 60 species, three of which are endemic: *Clupeopetersius schoutedeni*, a small pelagic characin, *Tylochromis microdon* and *Eutropius tumbanus*. Few fishes feed on planktonic material (*Microthrissa* and *Clupeopetersius*); insectivores such as *Lamprologus, Xenomystus, Gnathonemus, Anchenoglanis* and *Tylochromis* and insectivorous–piscivorous fish (*Hemichromis, Clarias, Alestes*) are more frequent. The piscivores feed upon *Chrysichtys, Mormyrops, Hydrocyon, Clarias* and *Alestes*.

Lake Tumba and Maji Ndomba (Lac Leopold II) have been interpreted as the remnants of the Pliocene lake of Central Zaire. However, the fauna is of riverine character and its low endemism indicates a recent origin. Marlier attributes the formation of the lake to a southern tributary of the Zaire separated from the main river by sand bars.

Lake Maji Ndombe (Lac Leopold II)

Located in Zaire at 0°38′ S and 16°33′ E 60 km south of Lake Tumba, Maji Ndombe has a surface area of 2300 km^2 and a mean depth of 5 m. It is drained by the Fimi River into the Zaire River. As in Lake Tumba, the water is acid (pH 4).

Malebo Pool (Stanley Pool)

Located in Zaire and Congo at 4° S and 15°28′ E, Malebo Pool has a surface area of 700 km^2, a mean depth of 3 m and a maximal depth of 20 m.

THE ZAMBEZI BASIN (Fig. 24)
The Zambezi River

The Zambezi River originates at 1400 m altitude at approximately 9° S and, after a 2500 km course, flows into the Indian Ocean at 18°47′ E. The river basin is composed of three main regions.

1. *The Upper Zambezi* River flows mainly on Pleistocene Kalahari sands and sandstones. The Precambrian basement forms outcrops in the eastern corner of the watershed and north of the Victoria Falls which are themselves made of basaltic rocks.

There is affinity between the fishes of Upper Zambezi, the southern tributaries of the Zaire and the Okavango River. It has therefore been assumed that, in pre-Pleistocene times, the Upper Zambezi belonged to the Zaire drainage system (Bell-Cross, 1972) or to the Limpopo watershed via the Chobe and Okavango Rivers, Lake Ngami and Lake Dow (Balon, 1974). The change of flow direction of the Upper Zambezi was caused by its

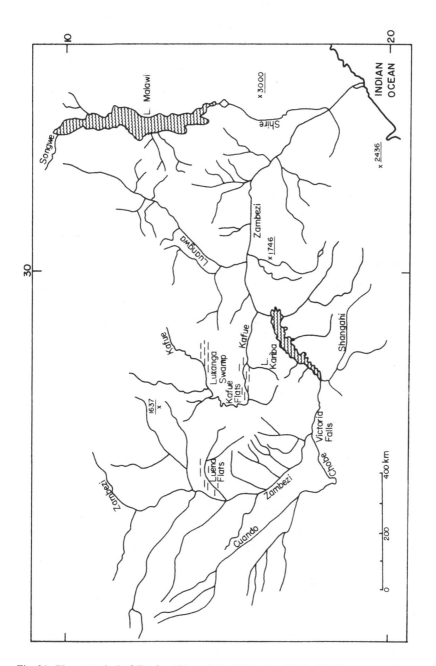

Fig. 24. The watershed of Zambezi River, Lake Malawi and Lake Kariba

capture by the Victoria Falls or by an upward movement in the area of the Chobe River or by both processes (Balon, 1974a).

The Upper Zambezi basin is separated from the Middle Zambezi region by a basaltic area 100 km long and of Jurassic age (Karoo basalt). Approximately 500 000 years ago, tectonic movements favoured the erosive recession of the Victoria Falls. The Middle Zambezi receded westward, etching a deep gorge along pre-existing faults, in the basaltic mountains and finally captured the Upper Zambezi. The water, which may reach 30 m depth in the gorge, flows between two vertical walls which are 200 m high in certain locations.

The connections of the Upper Zambezi with the Okavango system have not completely disappeared however, since, in rainy years, the Selinda spillway which separates both watersheds allows overflowing of the Okavango into the Zambezi (Beadle, 1974). At high water, the Upper Zambezi overflows into the Barotse flood plain, the surface area of which increases from 700 km² to 7800 km².

2. *The Middle Zambezi* runs from the Victoria Falls to the Cabora Bassa Rapids. This portion of the river, part of which is called Gwembe Valley, is located in a synclinal and faulted structure of the Precambrian basement. Karoo formations filled this depressed region and the Middle Zambezi river flows on the soft Karoo sediments. East of the river, a series of thermal and mineral springs illustrates a major NE–SW fault which moved the eastern part of the Middle Zambezi basin upwards.

Some 350 km downstream from the Victoria Falls, the damming of the Zambezi caused the formation of one of the largest man-made lakes, Lake Kariba. Below the dam, two main tributaries, the Kafue and Luangwa Rivers, join the Zambezi.

3. *The Lower Zambezi* runs from the Cabora Bassa Rapids to the delta. On this portion of its course, the Zambezi receives the Shire River which drains Lake Malawi (Nyassa) and Lake Malombe. The Murchison Falls, located on the Shire River between Lake Malombe and the Zambezi, have a length of 80 km and a topographic drop of about 400 m. They constitute a fish barrier which has efficiently isolated the ichthyological fauna of Lake Malawi.

The climate is characterized by a rainy season from November to March and a dry season during the rest of the annual cycle. The yearly rainfall ranges from 1420 mm in the Zambezi headwaters to 610–813 mm in the area of Lake Kariba and only 406–610 mm in the Cabora Bassa area.

The Zambezi River has an average discharge of $221 \times 10^9 \, m^3 \, y^{-1}$ or $7000 \, m^3 \, s^{-1}$ (*The water encyclopedia*: Todd, 1970). Its drainage area extends

The Zambezi basin

over 1.3×10^6 km^2 and its average specific yield is 170 656 m^3 km^{-2} y^{-1} or 5.41 km^{-2} s^{-1}. According to the long-term (1925–66) hydrological record of Livingstone Station, near the Victoria Falls, the minimal discharge occurs in October (331 m^3 s^{-1}) and the maximal yield is observed in April with about 3500 m^3 s^{-1} (Balon & Coche, 1974). The Shire River, 520 km long, has a discharge fluctuating from 283 to 566 m^3 s^{-1}.

The high-water period extends from December to July with a maximum in March–April. The Kafue River has its maximum in May because of the slow draining of the swamps of its upper valley. The water quality of the river depends on the flow regime and the available analyses have been reported by Balon & Coche (1974). According to these authors, the general range of concentration is as shown in Table 62. The Zambezi water belongs to the calcium–hydrogen-carbonate type and its low salinity reflects the low solubility of Precambrian rocks and Karoo formations. The salinity increases at low waters. The Zambezi water has a composition similar to the world inland water average and completely different from the composition of East-African waters.

The Barotse flood plain and the Chobe Swamps play the role of sediment traps; consequently, the concentration of silt in the flood period does not exceed 103 mg l^{-1}; it drops to 36 mg l^{-1} at low waters.

Lake Kariba
General

Lake Kariba (Fig. 24) was formed by damming the Zambezi River at Kariba Gorge in December 1958. The advantage of damming the river at this point for irrigation had been pointed out as early as 1912. When the

Table 62 *Chemical features of the Zambezi River water (data from Balon & Coche, 1974)*

			mg l^{-1}	meq l^{-1}
Temperature (°C)	17–30	Salinity	20–69	–
pH	6.4–8.0	HCO$_3^-$ + CO$_3^{2-}$	9–35.4	0.3 –1.18
Conductivity	36.121	Hardness	17–50	–
(μmhos cm^{-1})		Ca^{2+}	4.7–12.8	0.16–0.43
		Mg^{2+}	0.9–7.3	0.08–0.61
		Na$^+$	1.0–8.0	0.04–0.35
		K$^+$	0 –2.0	0 –0.05
		SO$_4^{2-}$	3.0–3.8	0.06–0.08
		Cl$^-$	0 –3.0	0 –0.09

project was approved in 1955, its main purpose was to provide electricity to the Federation of Rhodesia and Nyasaland but also to fulfil other aims such as flood regulation, irrigation, recreation and fish production.

In 1956–57, studies were carried out on fish populations and a committee was set up to examine the potential uses of the future lake. The closing of the dam was accompanied by a large-scale animal rescue operation which was completed in 1963. Then the level had risen from 390 to 487 m above MSL. Considerable modifications took place, the most conspicuous of which was the colonization of the lake by *Salvinia auriculata*, and the development of fish populations.

Prior to the filling of the reservoir, the area was seismically quiet; in 1959, when the lake level had already risen by 60 m, seismic events were registered. In later years, a weak but continuous seismic activity was observed; it culminated in 1963–64 when stronger shocks were superimposed on the basal activity. The weight of accumulating water is assumed to have reactivated existing faults (Van der Lingen, 1973).

In April 1963, the lake reached its average operating level (485 m). In following years, the lake level fluctuated between 490 and 480 m.

Morphometric and physical features

The main morphometric features of Lake Kariba are presented in Table 63. Lake Kariba is divided into four basins by topographical entities. The upper basin (No.1) is limited by the Sebungwe narrows and the following one (No.2) by the Chete gorge; basin No.3 ends at the Sibilobilo narrows and basin No.4 extends downstream to the dam.

Lake Kariba has 293 islands with a total shoreline of 604 km and a surface area of 147 km².

Details of the hypsometric curve of the lake can be found in Balon & Coche (1974).

Table 63 *Main morphometric features at water level 485 m above MSL (from Balon & Coche, 1974)*

Length (NE–SW axis)	277 km	Mean depth	29.2 m
Mean width	19.4 km	Max. depth	93.0 m
Surface area	5364 km²	Watershed surface	663 820 km²
Shoreline	2164 km	Volume	156.5×10^9 m³
Shoreline development	8.3	Watershed surface:volume ratio	0.24 m² m⁻³

The Zambezi basin

Hydrology

The average annual budget for the period 1963–66 (Balon & Coche, 1974) indicates that the rainfall on the lake area (3.5×10^9 m^3) contributes only 7% of the water income whereas the discharge of the rivers (Zambezi and tributaries) contributes 47×10^9 m^3, i.e. 93% of the income. The evaporation from the lake surface accounts for 14.3% of the losses (7.2×10^9 m^3 y^{-1}) whereas the outflow through the turbines and spillway amounts to 43.3×10^9 m^3, i.e. 85.7% of the annual losses. The replacement time is approximately three years.

Sedimentation

The Zambezi River has a relatively low sediment load mainly consisting of sand. The sudden decrease of current velocity at Devil's Gorge causes the rapid sinking of suspended matter at this point. Identical phenomena take place at the mouths of the tributaries.

Surface seiches, tides and wind set up

Records of water levels at various stations along the lake provide information on external seiches and tides. The longitudinal external seiches have periods ranging from 2.73 h to 9.74 h for the first four modes. In the absence of seiches, semi-diurnal tides are observed especially at times of new and full moon during the spring and autumn equinoxes (Ward, 1979).

Light transmission

The Secchi-disc readings increase from basin No.1 to basin No.4. In basin 1, the depths of visibility range from 60 cm to 5 m; in basin 4, the respective values are 2.2 and 10.6 m. The 1% of incident light was reached between 2 and 5 m in basin 1, between 5 and 10 m in basin 2, between 7 and 20 m in basin 3 and between 11 and 25 m in basin 4 (Coche, 1974). Green light was generally the most penetrating wavelength as in Lake Victoria, Lake Albert and Lake Edward (Talling, 1965a) and in Lake Kinneret (Rodhe, 1972; Berman, 1976). Deeper penetration of red light in basin 1 is related to absorption of lower wavelengths by colloids. Studies of light penetration have shown the euphotic zone to be about 10 m in basins 1 and 2 and 16–24 m in basins 3 and 4.

Water temperature

Lake Kariba is a monomictic lake with full circulation in mid-July, at least in the lower basins. In the upper basins, the Zambezi water, entering at a temperature of 19.2°C, flows as a density current in the lower layers of the lake. In mid-October, stratification appears and the warm Zambezi water flows in the upper layers. From February to late March, a strong stratification persists in basins 2, 3 and 4, whereas basin 1 is gradually flushed with Zambezi water. The thermal structure of the different basins is described in detail by Coche (1968, 1974).

The minimal temperatures of the hypolimnion during the stratified period range from 21°C to 24°C; homothermy occurs at 22°C. In January–February, the surface water reaches a temperature of 29–31°C. As previously mentioned, the lake level is affected by motions due to external seiches.

The thermocline is located between 15 and 25 m depth, which means that 40–65% of the volume of the lower three basins is mixed and available to biological aerobic activity. It is interesting to note that the ratio of mixed depth to illuminated depth is approximately one.

In January, the metalimnion is ~ 10 m thick and the thermal gradient is $0.5°C\,m^{-1}$.

In most parts of Lake Kariba, the annual heat budget is about 15 000 cal $cm^{-2}\,y^{-1}$, a low value characteristic of tropical areas. Similarly characteristic of such regions, the Residual Heat (amount of heat above 4°C present in the lake) is high (40 000 cal $cm^{-2}\,y^{-1}$).

Chemical features

Oxygen is uniformly distributed in the water mass during the mixed period. In February when the stratification is well established, the oxygen disappears from the hypolimnion. Deoxygenation, as thermal stratification, proceeds from basin 2 towards basin 4. Similarly, the rate of oxygen consumption is higher in basin 4 than in basin 3: for the period 1965–69, the average rate was 0.107 mg O_2 cm^{-2} d^{-1} for basin 3 and 0.140–0.227 mg O_2 cm^{-2} d^{-1} for basin 4.

The thermal and oxygen distribution patterns became established the first year after filling the lake. The spatial extension and temporal duration of the lack of oxygen decreased with time however. In November 1959, oxygen was absent below 10 m depth. Already in 1961–62, oxygen was present in the hypolimnion in December and the situation continued to improve in following years. The first year after filling, the oxygenless period lasted eight months, then in following years it shortened and stabilized at around four months. This evolution is related to the amount of available and readily decomposable organic matter flooded during the lake filling.

Taking into account a limit value of 2 mg l^{-1} oxygen for fish life, Coche

(1974) defined the depth at which this concentration is reached as well as its variation in time. In basin 1, the anoxic zone is negligible; in basin 2, only depths below 30 m are unavailable to fish in February. In basin 3, the critical period extends from January to March below 35 m. In basin 4, especially in its eastern part, the critical value of oxygen is already reached in January at 15 m or even 10 m. Unfavourable conditions to fish life are found from January to May.

Hydrogen sulphide was present in the whole deoxygenated hypolimnion after the filling of the dam. From 1963 onwards, the extension and duration of the presence of H_2S were reduced; in 1965–66, it was detected only in deep gorges (Coche, 1968). This evolution was parallel to the progressive disappearance of flooded vegetation. The accumulation of dead *Salvinia* weed generated a new cycle of H_2S, especially in the infested tributaries (Begg, 1970).

The composition of Lake Kariba water (average for the whole lake) is shown in Table 64. The composition of Lake Kariba water, very similar to that of Zambezi River, is dominated by calcium hydrogen carbonate and is very close to the world average value for inland waters. The absolute values of carbonates rarely exceed 1 meq l^{-1} and Kariba waters are very soft. Sulphates and chlorides are low since they are not abundant in igneous and metamorphic rocks. Silica is abundant (10–15 mg l^{-1}). Nitrate–nitrogen and PO_4–P are variable but rarely exceed 25 μg l^{-1}. In general, salinity increases from the upper basins towards the dam. Hypolimnic water has lower conductivity and alkalinity than epilimnic water as well as lower concentrations of sodium and potassium. In contrast, nitrate is more abundant in the hypolimnion. Just after the closure of the dam, there was a clear increase of dissolved salts. From 1959 to 1963, the salinity decreased to two-thirds of its peak value. In 1963, when the lake was full, the salinity increased until 1966 and stabilized to about 150% of the pre-dam period.

The passage from lotic to lentic conditions was accompanied by a

Table 64 *Chemical features of Lake Kariba water. Average for the whole lake. Measurements of 1965. (Data from Coche, 1974)*

TDS (mg l^{-1}) 58	Ca^{2+} (mg l^{-1}) 9.32	$CO_3^{2-} + HCO_3^-$ (mg l^{-1})	22.68
Conductivity (μmhos cm^{-1}) 73	Mg^{2+} (mg l^{-1}) 1.95	Cl^- (mg l^{-1})	1
Salinity (mg l^{-1}) 42	Na^+ (mg l^{-1}) 3.52	SO_4^{2-} (mg l^{-1})	3
pH 7.5–8.5	K^+ (mg l^{-1}) 0.97		

relative increase in calcium and potassium, a decrease in magnesium and sodium and an increase in alkalinity. Nitrate and phosphate dropped significantly.

Biological features

Thomasson (1965) who sampled the lake near Songwe in May 1959 gives a detailed list of the 170 algal species he found at this location. The Cyanophyta include only *Microcystis flos-aquae* and the Pyrrhophyta, *Peridinium gatunense*; the Bacillariophyta are composed of *Melosira* spp., *Eunotia* spp. and *Synedra ulna*; the Chrysophyta are represented by *Dinobryon cylindricum*; and among the Chlorophyta are *Eudorina elegans, Botryococcus braunii, Pediastrum* spp., *Closterium* spp., *Microasterias* sp., *Cosmarium* and *Staurodesmus*. In a later examination of the phytoplankton of Lake Kariba, Thomasson (1980) found *Ceratium hirudinella brachyceroides* in July–August, *Peridinium cunningtonii* from January to April, *P. willei* and *P. volzii* var. *cinctiforme* in April and *Peridinium* sp. in January.

Hancock (1979), studying the diatom succession, found a dominant association formed by various species of *Melosira; M. granulata* dominated during homothermy and early stratification in the lower lacustrine type basin whereas *M. granulata* var. *angustissima* dominated during overturn and was the main species in the upper basins. A second association composed of *Cyclotella, Fragillaria* and *Synedra* was subdominant to the *Melosira* association near the dam wall but decreased up the lake.

During the filling phase, the increase of nutrients and general salinity was accompanied by thick blooms of *Microcystis* (Begg, quoted in Van der Linden, 1973; Mitchell, 1973). After the drop in nutrient and salinity, the algal blooms became rare and of short duration.

The most conspicuous event in the flora of the lake was the explosive development of the floating fern *Salvinia* (*molesta* or *auriculata*). *Salvinia* had been collected in 1949 at Victoria Falls. Its sudden growth in 1959 resulted from the combination of various factors: the increase in the nutrient concentration in the lake due to the decay of flooded vegetation and nutrient release from ash originating from bush-clearing operations and the changing lake level which conferred a clear ecological advantage upon free, floating plants. In 1962, *Salvinia* covered 1000 km^2 and had a mass of 8500×10^3 tons wm. This amount contained 7900 tons nitrogen, 417 tons phosphorus, 16 445 tons potassium, 6435 tons calcium and 2800 tons magnesium (Mitchell, 1973). *Salvinia* developed mainly in quiet areas in thick mats on which other plants grew. The *Salvinia* mats prevented plankton photosynthesis, strongly diminished turbulence and subsequent mixing and reoxygenation of water and enhanced the production of H_2S; they served as nursery grounds for young cichlids but hampered the migration of adult fish to spawning areas.

The zooplankton, investigated by Thomasson (1965) in 1959 was dominated by *Brachionus falcatus* and *Bosmina longirostris*. Begg (quoted in Van der Linden, 1973) in his observations of 1967, found a small number of species dominated by *B. longirostris* and *Tropodiaptomus kraopelini*. Mills (1977) gave a detailed list of zooplankton and Magadza (1980) carried out a multivariate analysis on a population including one protozoan, five rotifers and 25 crustaceans sampled in Sanyati Bay. This analysis revealed the presence of five main groups, each composed of a defined community of animals. The underlying cause of this structure is not well understood but there seems to be a relation between the density of zooplankton and the river inflows. The average standing crop was found to be 2.76 mg dm m^{-3} with values ranging from 0.26 to 15.9 mg dm m^{-3}. Begg (1976) studied the relation between the diurnal movements of certain species of zooplankton and the sardine *Limnothrissa miodon*. Adults of *Mesocyclops leuckarti* were the most active migrators; *B. longirostris* performed similar but less extensive migrations. The rotifers were much more static. The overall movements of the zooplankton consist of an upward migration at night. Coulter (1960) noted that the sardine performed identical diurnal migrations and McLaren (1963) believes that members of the zooplankton migrate upwards when their predators cannot see them. The study of Begg indicates that the migrations of fish and zooplankton are 'independent responses to light-intensity stimuli'.

McLachlan (1974) investigated the evolution of the population of invertebrates during the filling and postfilling phases of the lake. The disappearance of oxygen from the hypolimnion limited the dwelling area of invertebrates. Certain groups, such as the chironomids, started colonizing the deeper parts of the lake as soon as the oxygen deficit decreased. The submerged trees were occupied by chironomids and oligochaetes (22 mg dm m^{-2}). During periods of low levels, the wood-boring beetle *Xyloborus torquatus* bored cavities under the bark which were occupied, after the next flooding, by additional fauna which then reached 97 mg m^{-2}.

The floating fern *Salvinia* served as a habitat for *Bulinus* and *Planorbis* and played a major role in the spread of bilharzia.

The fluctuations in water level created a pattern of animal succession. At static level, shallow mud is occupied by dense populations of chironomids, up to 200 mg m^{-2}. When the water level drops, it uncovers a large number of dead chironomids rapidly eaten by the fly *Lispe nuba*, preyed upon in its turn by birds. The invertebrate fauna of Lake Kariba has not yet reached a stage of equilibrium.

Fish. Prior to the creation of the dam, Jackson (1961) listed 28 species from the Middle Zambezi and Harding (1964) found 31 species. In two experimental fish population assessments (1968 and 1970), carried out at 15 points, Balon (1973) established a list of 39 species. The most abundant fish was *Alestes lateralis* (58.8% of total number); then were found in order of abundance (expressed as % of number of specimens) *Tilapia mossambica* (*ortimer*) (10%), *Cyphomyrus discorhynchus* (7.9%), *Haplochromis darlingi* (5.2%), *Barbus fasciolatus* (4.6%), *T. melanopleura* (*rendalli*) (3.6%), *Hydrocinus vittatus* (2.5%), *Clarias gariepinus* (2.0%), *B. unitaeniatus* (1.8%) and *Hemihaplochromis philander* (1.3%).

The fish population, being of riverine origin, was concentrated in shallow inshore waters, corresponding to the 0–20 m depth area. In other words, in 1968–1970 the fish lived in 2020 km^2, i.e. only 37.7%, of the lake area (Balon, 1974b). There were practically no fish in open water except *Alestes lateralis* which started to explore this new environment, the eel *Anguilla nebulosa labiata*, and the newly introduced clupeid *Limnothrissa miodon*. For the inshore area, the standing crop was estimated by Balon as 110 000 tons. The value of 18 000 metric tons given by Balon (in Balon 1973, 1974b) is an error of calculation. This author utilized a value of 33 000 ha for the area inhabited by the fish (printed in Balon, 1973), instead of the exact value of 202 000 ha, and the fish standing crop of Lake Kariba was then 540 kg ha^{-1}.

The eels showed an incredible ability to surmount obstacles. Jubb (1964) predicted that the eels would slowly disappear from the lake since he estimated that they were quite unable to negotiate the 143 m high wall of the dam on their way back from the ocean. Investigations by Balon (1974a) show that juvenile eels (6–10 cm long) move over the dam and migrate into the tributaries of Lake Kariba and spend seven years in the streams. Then, they drift back into the lake and spend another one to nine years at a depth of 25–40 m. The migration from streams into the lake is probably the first step of the catadromous migration.

Interesting modifications took place in the fish population after the dam was closed, for example the replacement of *Brachyalestes imberi* by *Alestes lateralis*. Before the construction of the dam, *A. lateralis* was a fish of the Upper Zambezi whereas *B. imberi* was the most abundant fish of the Middle Zambezi and was unknown in the Upper Zambezi. Lake Kariba was inhabited exclusively by *B. imberi* until 1962. In 1963, the first specimen of *A. lateralis* was caught in the lake and in 1964–65 Matthes (1968) noted that *A. lateralis* was largely dominant; finally, the latter species completely eliminated the former one. It appears that *A. lateralis* reached the Middle Zambezi area via Victoria Falls. The reason why *A. lateralis* did not colonize the Middle Zambezi earlier seems to be that in lotic conditions, *B. imberi*, spawning in the flood plains, had a high survival rate. In contrast, *A. lateralis* was significantly favoured by the lentic conditions which provided it with adequate spawning grounds among submerged plants.

A. lateralis is far from being the only species surviving the shock of the Victoria Falls. A comparative study of the species of fish at the edge of Victoria Falls with the lake species (Balon, 1974*a,b*) has shown that *Marcusenius macrolepidotus*, *Labeo forskalii lunatus*, *Schilbe mystus*, *Serranochromis macrocephalus*, *S. robustus jallae*, *Sargochromis giardi* and *Haplochromis carlottae* present at Victoria Falls are recent invaders of Lake Kariba. Another 10 species may have invaded the Middle Zambezi prior to the filling of Lake Kariba but 11 species present at Victoria Falls were never found in the lake. It follows that the assumption that Victoria Falls forms an impassable barrier to fish in both directions should be revised.

The lack of fish, together with the abundance of zooplankton in open water, justified the stocking of the lake with two clupeids feeding on zooplankton and endemic to Lake Tanganyika, *Limnothrissa miodon* and *Stolothrissa tanganicae*, in 1967 and 1968. Experimental fishing surveys, carried out in 1969 and 1970, revealed that the anchovy *L. miodon* was present in all parts of the lake. A 1971 survey indicated that the tiger fish *Hydrocynus vittatus* was moving offshore to exploit the anchovy which was found in abundance in its stomach contents. However, the density of anchovies was estimated at 8 m^{-3}, at least one order of magnitude lower than in Lake Tanganyika.

Tilapia macrochir was introduced from 1959 to 1962 but did not develop and disappeared completely. Conversely, *T. mossambica* (*mortimeri*) and *T. melanopleura* (*rendalli*) experienced a rapid growth.

The potential fish production of Lake Kariba had been estimated at 28 000 tons y^{-1}. On the southern side, total production reached 2700 tons in 1964, then declined in following years to 1500–1800 tons y^{-1}. In 1969, the landings included 62% cichlids, 8% *Labeo* and *Distiochodus*, 14% *Hydrocynus* and 16% *Clarius* (Van der Lingen, 1973). The north shore production was 820 tons in 1967 and 1550 tons in 1970 (Joeris, 1973) although they had exceeded 3000 tons in 1963 (Harding, 1966). More recent data by Welcomme (1972) give a total annual catch of 4080 tons for the whole lake (Rhodesia (Zimbabwe) 3000 tons, 1970 and Zambia 1080 tons, 1966–70). Data for Zambia only (Welcomme, 1979) give an average of 2418 tons for the period 1970–74. In the best years, the fish production of Lake Kariba is therefore 5000 tons, i.e. ~20% of the estimated potential.

Some aspects of the impact of Lake Kariba

1. Although Harding (1966) pointed out that the lentic conditions were favourable to all species by preventing overcrowding in the river, and consequent predation at low water, the modifications which occurred in the fish population did not produce a profitable fishery, in spite of expensive fish stocking performed by air lifting of juveniles. The closing, in 1966, of an ice plant with a daily capacity of 7.4 tons, built according to the estimation of potential catch, illustrates well the divergence between hopeful planning and reality.

2. The explosion of the weed *Salvinia*, which by itself was a nuisance, enhanced considerably the extension of molluscs and particularly of the bilharzia host snails and, in 1967, heavy infestation of human population was reported.

3. The effect of the lake on downstream flood plains has been studied by Atwell (1970). The flood plain of Mana, below the dam, harbours many big mammals. At present it does not receive the silt which allowed the regeneration of vegetation in the past and overgrazing is already obvious. The amounts and timing of water release from the dam are monitored by the technical requirements of the hydroelectric plant. The out-of-season release of water affects the reproduction of many species (crocodile, Nile monitor (*Varanus niloticus*) and birds have their eggs destroyed). The lack of flooding makes the flood plain available to grazers during a longer period and enhances overgrazing and flood-plain degradation. The effect of damming on the downstream flood plains has been aggravated by the construction of smaller reservoirs on the tributaries of the Zambezi. For example (White, 1973), the Kafue Gorge Dam, completed in 1971, removes from the downstream flood plains a substantial supply of silt and water. Conversely, the dam causes the permanent flooding of the Kafue flood plain (Dudley, 1979) which limits reproduction and year class size of the *Sarotherodon* species.

4. The response of the tsetse fly to inundation and migration of the 57 000 inhabitants of the valley has been described by Scudder (1972). The farmers settled on new lands in the vicinity of the lake and started raising cattle on a

large scale. The tsetse fly, which was present in certain parts of the valley, moved up with the rising of the water level and reached the new settlements causing much damage among the farmers and the cattle, by infecting them with trypanosomiasis.

Lake Malawi (Nyasa)
General

Lake Malawi (Fig. 24) is located in Malawi, Mozambique and Tanzania at 9°30′–14°40′ S and 34°–35° E, at an altitude of 469–475 m, depending on fluctuations in the water level. The lake is located in a deep graben which results from the confluence of the Western Rift Valley and the smaller Rift Valley drained by the upper part of the Ruaha River. The Malawi Rift Valley continues southwards along the course of the Shire River. The shores of Lake Malawi are bordered, especially in the north, by steep, fault-like escarpments leading to high plateaux exceeding 2000 m in altitude.

In Zambia, a certain number of east–north-east transcurrent dislocation zones in the Precambrian rocks has been observed. The best defined is the Mwembeshi zone which enters Malawi at the 12°45′ S parallel, runs eastward and crosses Lake Malawi in its central area. It is interesting to note that the Kazali Hot Springs on the western shore of the lake and the Likoma and Chizumulu Islands lie on this line. These granite islands have been considered by Bloomfield (1966) as the continuation of the Mwembeshi dislocation zone. Bloomfield also suggested that the dislocation is a transform fault offsetting the Nyasa Rift Valley. This would support the hypothesis that the African Rift Valleys are mid-oceanic ridges in formation. A heat-flow survey of the floor of Lake Malawi by von Herzen & Vacquier (1967) showed regions of low heat flow at the northern and southern ends of the lake (0.54–0.070 μcal cm^{-2} s^{-1}) separated by a central area of high heat flow (2.30 μcal cm^{-2} s^{-1}). Since Lake Malawi is located on Precambrian rocks, generally associated with lower than normal heat-flow values, it has been assumed that the high values measured in the central part of the lake resulted from magma intrusions of relatively shallow depths of the type occurring in other parts of the mid-oceanic ridge.

A dry and relatively cold season (April–September) with regular winds from the south-east is followed by a warm, wet season from October to March. Rainfall is maximum in the second half of the warm season, characterized by irregular and violent storms. Yearly precipitations are of an order of magnitude of 900 mm in the lake area (from 650 mm at Mangochi on the southern part of the lake to 2500 mm at the northern extremity) (Eccles, 1974), and exceed 1500 mm in the nearby mountains which supply most of the inflowing water of the lake. The total annual water income of the lake is equivalent to a lake level rise of 1.73 to 2.5 m; annual losses through evaporation represent about 1.9 m whereas outflow down the Shire River corresponds to only 45 cm at present high-water levels (Pike, 1964). The Shire River which drains the lake at its southern end is an intermittent outlet and during periods of low rainfall Lake Malawi becomes a closed basin. Pike & Rimmington (1965) showed that the lake level dropped continuously from 1882 to 1915 when it reached 469 m above MSL, and the outflow ceased. The outflow was resumed only around 1935 at a level of 474 above MSL because of the presence of sand bars which built up after the flow had stopped. A barrage was constructed in 1965 at Liwonde on the Upper Shire and is operated to maintain the water level required by hydroelectric production. The analysis of the temporal behaviour of the level of Lake Malawi allows us to predict future levels (Dyer, 1976). External seiches of 12 cm amplitude and periods of 6 and 24 h have been observed, especially at the southern extremity of the lake.

Morphometric and physical features

The main morphometric characteristics of the lake are presented in Table 65. Beauchamp (1953) made the first detailed limnological investigation in 1939–40. In 1954, the Joint Fisheries Research Organization of Northern Rhodesia and Nyasaland started a programme of measurements on the lake (Iles, 1960; Eccles, 1962a,b, 1965, 1974; Jackson, Iles, Harding & Fryer, 1963).

Below 250 m, the water of Lake Malawi is anoxic, rich in H_2S and homothermal at 22.5°C. In the upper 250 m, an annual pattern of stratification and partial mixing takes place. Stratification develops from September to December with an upper-layer temperature of 28°C. The thermocline moves progressively deeper and, in May, a 60 m thick epilimnion is homothermal at 27°C (Eccles, 1974). In March–April, with the onset of the dry and relatively colder season, the upper layers lose heat

Table 65 *Morphometric features of Lake Malawi*

Max. length (N–S)	603 km	Mean depth	426 m
Max. width	87 km	Max. depth	758 (685) m
Surface area	30 800 km²	Volume	8400 × 10⁹ m³
Shoreline length	1500 km	Watershed area	65 000 km²
Shoreline development	2.4	Watershed area : volume ratio	0.0077 m² m⁻³

and the destratification process begins. Complete mixing has never been observed although Eccles described a nearly homothermal condition in August 1958 after an 8-day period of continuous strong southerly winds.

It seems that, besides the seasonal warming of upper layers, the stratification is maintained by topography-bound density currents occurring in the south-eastern part of the lake. In this area, the strong and periodic south-easterly winds are dry and cool and their impact on the shallow water of the south-eastern area of the lake explains that, off Monkey Bay, the temperature of bottom water in the dry season is often lower than that of the hypolimnion of the open lake. This topographical and meteorological feature of the lake generates a northward bottom current involving large volumes of water which enter the deep hypolimnion. This mechanism prevents the deepening of the thermocline and causes a very stable stratification.

The numerous undulations of the isotherms led Beauchamp (1953) to assume the existence of internal waves having considerable amplitude. These were demonstrated by Iles (1960) and Eccles (1962a,b) showed that their amplitude reaches up to 20 m and their period ranges from 20 to 30 days. The southerly wind causes the accumulation of warm water in the northern part of the lake where the isotherms are 50 m deeper than in the south and a consequent upwelling of cold and nutrient-rich water at the southern extremity of the lake. This accounts for the fact that the southern area represents the most productive fishery of the lake and is the site of thick *Anabaena* blooms.

The increase in volume of the deep hypolimnion because of cold density currents prevents the deepening of the thermocline whereas the internal waves exert a downward pressure on the thermocline in the northern part of the lake. These antagonistic processes create turbulence in the thermocline area and probably a certain mixing which may explain the rise of temperature which occurred at 300 m between August 1939 (22.1°C measured by Beauchamp) and 1964 (22.56°C measured by Eccles).

Upwelling due to internal waves and turbulence and mixing in the metalimnion are, in the absence of complete and regular mixing, the mechanisms which recirculate nutrients from the deep hypolimnion to the epilimnion (Eccles, 1974). The isotope investigation of Lake Malawi in 1976 (Gonfiantini, Zuppi, Eccles & Ferro, 1979) indicated that the water mixed slowly but much more rapidly than in Lake Tanganyika. The tritium profile shows that about 25% of the water is yearly exchanged between the epilimnion and the metalimnion and about 20% between the metalimnion and the hypolimnion.

Chemical features

The mechanism which transfers heat to the hypolimnion without destroying the thermocline allows a slow but continuous supply of oxygen from the epilimnion to the layers located between 60 and 250 m. This is inferred from the fact that these layers have a more or less constant concentration of oxygen during the stratified period: for example Beauchamp (1953) measured a concentration of 4–5 mg l^{-1} oxygen from October to late February at a depth of 110 m (isotherm 23°C). Then, a general decrease of oxygen is observed resulting probably from stronger mixing caused by storms. At the period of maximal mixing (August), the upper 100 m contain 6–7 mg l^{-1} oxygen.

The composition of Lake Malawi water is presented in Table 66.

Biological features

Phytoplankton. Van Meel (1954) published a list of 331 species based on the works of Dickie, Muller & West and including 28 species of Cyanophyta, 167 species of Bacillaryophyta, seven species of Euglenophyta, 124 species of Chlorophyta and five species of Dinophyta.

Fryer (1957) found the zooplankton of Lake Malawi 'neither qualitatively nor quantitatively rich'. In the north, the common species were: *Diaptomus (Tropodiaptomus) kraepelini, Mesocyclops leuckarti, M. neglectus, Diaphanosoma excisum, Bosmina longirostris* and *B. deitersi. D. kraepelini* and *M. leuckarti* are the main components of the zooplankton and constitute the bulk of the food of plankton-eating fishes. In the south, the zooplankton is more abundant and includes two additional species: *Daphnia lumholtzi* and *Diaptomus mixtus*, this latter species being endemic to the lake.

Table 66 *Composition of Lake Malawi water, September 1961 (from Talling & Talling, 1965)*

	mg l^{-1}	meq l^{-1}		mg l^{-1}	meq l^{-1}
Na^+	21.0	0.91	$HCO_3^- + CO_3^{2-}$	70.8	2.36
K^+	6.4	0.16	Cl^-	4.30	0.12
Ca^{2+}	19.8	0.99	SO_4^{2-}	5.50	0.11
Mg^{2+}	4.7	0.39	SiO_2	1.10	–
Conductivity (μmhos cm^{-1})	210				

Benthic fauna. The presence of oxygen below the thermocline allows a rich benthic life; chironomids are especially abundant. Lake Malawi has a rich fauna of molluscs which has been investigated by Mandahl-Barth (1972). It includes 40 species of snails and bivalves of which 50% are endemic and live in the lake itself while non-endemic species inhabit the marginal swamp. The endemic species living in the lake belong mostly to the following gastropod genera: *Bellamya* (3 species), *Lanistes* (3), *Gabbiella* (1), *Melanoides* (8), *Bulinus* (2). The aquatic gastropods of Lake Malawi (revision of Mandahl-Barth 1972) are the following (E = endemic): *Bellamya capillata, B. jeffreysi* (E), *B. robertsoni* (E), *B. ecclesi* (E), *Neothauma ecclesi, Lanistes ovum, L. ellipticus, L. solidus* (E), *L. nyassanus* (E), *L. nasutus* (E), *Gabbiella stanleyi* (E), *Melanoides tuberculata, M. hodicincta* (E), *M. pergracilis* (E), *M. pupiformis* (E), *M. turritispira* (E), *M. polymorpha* (E), *M. truncatelliformis* (E), *M. magnifica* (E), *Lymnaea natalensis, Biomphalaria pfeifferi, Ceratophalus natalensis, Gyraulus costulatus, Bulinus globosus, B. nyassanus* (E), *B. succinoides* (E) and *B. forskali.*

A few endemic species of large bivalves belong to the genera *Caelatena, Aspatharia* and *Mutela.* The endemic molluscs are found down to 27 m depth, with the exception of *Bulinus nyassanus* which lives down to 95 m depth. The molluscan fauna of Lake Malawi is very different from the fauna of Lake Tanganyika and the complete absence of the prosobranch *Pila* is an unexplained feature. Similarly, the absence of molluscs from rocky shores, which in Lake Tanganyika support a rich assemblage of browsing prosobranchs, is typical of Lake Malawi. The taxonomic characteristics of the snail fauna of Lake Malawi consist of the presence of the endemic species of *Bellamya* and *Lanistes* which do not exist in Lake Tanganyika and of seven endemic species of *Melanoides* which is represented by only one endemic species in Lake Tanganyika. The gastropod fauna of Lake Malawi, with 17 endemic species, shows less radiation in ecology and morphology than that of Lake Tanganyika with its 30 endemic species.

Fish: zoogeographical relationship. Lake Malawi was formed one or two million years ago in the Nyasa Rift which developed along the axis of a north–south valley drainage into the Zambezi River. It is also possible that the downfaulted area captured rivers which were previously connected with the Bangweulu Lake and the Upper Zambezi via the Chambezi River. This would explain why the Malawi fish fauna is closely related to that of the Upper Zambezi and has little in common with the northern fish population of the Tanganyika–Zaire basins. For example, the Sudanese predator fish *Lates* and the clupeids which originated in the Zaire basin are absent from Malawi; in contrast, Malawi has 15 species in common with the Upper Zambezi and Lake Bangweulu. Little affinity exists between Malawi fish and the Lower Zambezi ichthyological fauna. At present, the Murchison Rapids, located on the Shire River south of Lake Malombe, is an absolute physical barrier to upstream movement of fish species of the Lower Zambezi. The barrier to downstream movement by Malawi species is mainly ecological: the lacustrine species do not adapt easily to the riverine conditions of the Shire, the turbulence of the cataracts and the competition with the Lower Zambezi fauna (Tweddle, Lewis & Willoughby, 1979). It seems that the Murchison Rapids have also been a major obstacle to upstream migration of fish in the past.

There are more than 245 species of fish in Lake Malawi (Beadle, 1974) but they belong to only eight families: Anguillidae (1 species), Mormyridae (5), Characidae (1), Cyprinidae (10), Bagridae (1), Clariidae (13), Mochokidae (1), Cichlidae (>200) (Lowe-McConnell, 1975). Additional new species are frequently found: at the occasion of exploratory trawl surveys, conducted in the southern fifth of the lake (Turner, 1977) 160 species were found in the catch, including 80 species of *Haplochromis*, 30 species of *Lethrinops* and other species of cichlids so that cichlids represented the bulk of the catch by number and weight. About 35 species of cichlids were unknown before the survey.

During the long isolation of the lake, the cichlids have undergone a very active speciation and presently all but four species of cichlids are endemic whereas only 65% of the non-cichlid species are endemic. Of 23 genera of cichlids, 20 are endemic (Fryer & Iles, 1972). In contrast, there is no endemic genus among the non-cichlids. The genus *Haplochromis* deserves special mention since it has given more than 120 species, all endemic and covering a wide range of habitats and feeding habits.

Distribution of the fish fauna in the lake. The ecology of the fishes in the lake was studied by Bertram, Boley & Trewavas (1942), Lowe (1952) and Jackson *et al.* (1963).

In the pelagic zone the abundant zooplankton is consumed by several species, particularly the cyprinid *Engraulicypris sardella* which feeds in the open water and has planktonic eggs and larvae. Various endemic *Haplochromis* species (the Utaka of local fishermen) also feed on pelagic zooplankton. These relatively small pelagic fishes are preyed upon by *Ramphochromis* sp., the catfish *Bagrus meridionalis* and the cichlid *Diplotaxodon*.

In deep bottom water from 100 to 300 m, Utaka species, *Ramphochromis* sp., *Bagrus meridionalis*, *Mormyrus longirostris*, the mockokid *Synodontis nyassae* and *Haplochromis heterotaenia* were found, sometimes very near to the oxygen-free layer. Special mention should be made of the catfish belonging to the genus *Dinotopterus* of which four species have been found at about 300 m deep and which are characterized by a reduction in the air-breathing organs and an increase of the gill area.

The rocky shores have a rich flora of blue-green algae covering the stones (*Calothrix* sp.) on which feed many invertebrates (copepods, ostracods, chironomid larvae and other insect larvae). As previously mentioned, the absence of molluscs from rocky shores has not received a satisfactory explanation. The density of fish in rocky areas is very high and this population locally called 'Mbuna' includes 27 endemic species of cichlids and five species of other families. They feed on the algal mat covering the rocks by scraping or sucking, on the small invertebrates living in the algal mat, on the littoral zooplankton, on fish and insect larvae, and even on scales and fragments of fins.

The sandy shores are commonly occupied by the water weed *Vallisneria*, its associated epiphytic algae and its abundant invertebrate fauna, including numerous molluscs. *Haplochromis mola* feeds on molluscs and *H. similis* and *Tilapia melanopleura* on water weeds. Other species exploit phytoplankton, detritus and chironomid larvae. Fryer (1959) showed that, in spite of spatial proximity, the rocky areas and sandy shores constitute two different habitats which have led to different specializations and allopatric speciation within the same ecosystem. Substantial differences may exist inside a given species in different parts of the lake. Fryer & Iles (1972) distinguish two subspecies of *Haplochromis marginatus* at the north and south of the lake.

Lake Cabora Bassa

The second largest artificial lake on the Zambezi River resulted from the damming of the Cabora Bassa Gorge with a 171 m-high wall. The lake began to fill in late 1974 and is now functional. With the experience of the problems which arose at Lake Kariba, Davies, Hall & Jackson (1975) reviewed in advance the problems to be expected at the new lake. They pointed out that the Zambezi below Kariba is richer in nutrients than the Upper Zambezi because of the reservoir itself, the nutrient-rich Kafue River and Kafue Dam and the polluted Hunyani River which drains the eutrophic Lake McIlwaine. The composition of the Zambezi water near the Cabora Bassa Gorge before the impoundment is presented in Table 67.

The Lake after the impoundment was studied by Bond & Roberts (1978).

The main characteristics of the lake are presented in Table 68. The lake is formed of five basins separated by promontories. The northern shore is generally steeper than the southern one and, since there was no tree-clearing prior to the filling, the southern part of the lake is covered with a thick semi-submerged vegetation.

In May 1975, the lake level was already at 314 m above MSL, 12 m below the level at full capacity; it later dropped and reached 302 m in December 1975. Stratification developed in the eastern basins, where deep layers contained H_2S, and spread westward. The composition of surface water was comparable with pre-filling values except for an increase in sulphates.

A list of zooplankton established on samplings of December 1973 and March 1974 contained Rotifera with *Brachionus quadridentata*, *Brachionus* sp., *Keratella* sp., and *Tetramastix* sp.; the Cladocera *Ceriodaphnia* sp., *Diaphanosoma* sp., *Bosmina longirostris*, *Alona* sp., *Daphnia lumholtzi*, *Daphnia* sp. and *Chydorus* sp.; and the Copepoda *Tropodiaptomus* sp.; and *Mesocyclops* sp. Many non-planktonic animals appeared in the samples (the coelenterate *Hydra* sp., Oligochaeta, Nematoda, the decapod *Caridina* sp., Ephemeroptera, Trichoptera, Diptera and Hemiptera).

Lake Cabora Bassa receives water from the outflows of three man-made

Table 67 *Chemical features of the Zambezi River near the Cabora Bassa Gorge before the impoundment. Average 1973–1974 values (from Hall et al., 1976)*

pH	8.00	Ca^{2+} (mg l^{-1})	10.6
Conductivity (μmhos cm^{-1})	118	Mg^{2+} (mg l^{-1})	4.1
NH_4–N (mg l^{-1})	0.09	Na^+ (mg l^{-1})	5.4
NO_3–N (mg l^{-1})	0.13	K^+ (mg l^{-1})	1.9
PO_4–P (mg l^{-1})	0.21	CO_3^{2-} (mg l^{-1})	31.8
SiO_2 (mg l^{-1})	16.8	Cl^- (mg l^{-1})	5.1
		SO_4^{2-} (mg l^{-1})	5.0
		Suspended matter (mg l^{-1})	21.7

Table 68 *Main morphometric features of Lake Cabora Bassa*

Length (E–W axis)	250 km	Mean depth	26 m
Max. width	38 km	Volume	62.25×10^9 m^3
Water level (above MSL)	326 m	Watershed area	200 000 km^2
Surface area	2739 km^2	Watershed area:volume ratio	0.31 m^2 m^{-3}

lakes (Kariba, Kafue and McIlwaine) and its own status is strongly dependent on the water quality of the upstream reservoirs.

The aquatic plants *Salvinia molesta*, *Pistia stratiotes* and *Azolla nilotica* were recorded in the watershed of Lake Cabora Bassa during preliminary surveys and *Eichhornia* was present in the Kafue River and the Hunyani River which join the Zambezi above the dam. Davies *et al.* (1975) predicted that *Eichhornia* would invade Lake Cabora Bassa and dominate over *Salvinia*. In January 1975, mats of floating macrophytes, transported by the Zambezi floods, accumulated near the dam wall. The mat was composed of *Eichhornia* and *Salvinia* in equal proportions of 40% and *Pistia* representing 20%. First *Pistia* suppressed *Salvinia* but *Eichhornia*, expanding rapidly by offset production, dominated *Pistia* in a few months and, in November, was the main component in all mats. However, from September onwards, the offset production of *Eichhornia* stopped and this limited considerably the colonization of the central parts of the lake. The combined effect of wind and wave action and the drawdown caused the accumulation on the shores and death of considerable amounts of macrophytes. The surviving mats showed signs of nutrient depletion and this prevented recolonization. Rapid rates of drawdown seem to have been decisive in the disappearance of the weed. Bond & Roberts (1978) discussed the numerous factors controlling the weed growth and the practical means of preventing it.

Lake Chilwa

Lake Chilwa is located in Malawi (15°0'–15°30' S and 35°30'–36°0' E) but its watershed (700 km^2) extends into Mozambique as well. Lake Chilwa originated in a depression which was tilted eastwards by the general uplifting of the escarpment of the Rift Valley. Terraces at 35, 24, 16 and 12 m above the present level indicate a wider extent of the lake in past periods. Lake Chilwa was then united with Lake Chiuta, presently located north of Chilwa and separated from it by a 16 m-high sand bar. It is likely that during this period of high levels, Chilwa was draining into the Indian Ocean via the Lugenda River.

Lake Chilwa is at present a closed basin separated from the valley of the Shire River in the west by the Zomba mountains and bordered in the south by the Mlanje Mountain. To the north, the lake is bordered by a 550 km^2 *Typha* swamp with *Typha domingensis* (Howard-Williams & Lenton, 1975) as the dominant species. The *Typha* swamp is surrounded by 570 km^2 of marshes and grasslands.

Lake Chilwa receives most of its inflow from rivers draining the mountains where the annual rainfall ranges from 760 to 1650 mm; rainfall does not exceed 600 mm in the depression. The wet season extends from November to April and the basic long-term average rainfall is 1170 mm y^{-1} (Morgan & Kalk, 1970). From 1965 to 1968, however, the precipitation decreased drastically (920, 1000 and 950 mm as basic average for these three years), which reduced the lake water income.

The lake itself plays an important role in the economy of Malawi because of its very prosperous fisheries (5000–9000 tons y^{-1}). However, the large fluctuations in water levels of this shallow water body endanger the fish stock and consequently a whole economic branch of the country. The Lake Chilwa coordinated Research Project of the University of Malawi was aimed at supplying the scientific elements for a better exploitation of the lake's resources. The project was initiated in 1966, during the first year of the drought and this was a rare occasion for a limnological team to follow the effect of lowering of water level on water, sediment and organisms and later to observe the new modifications caused by the refilling of the lake. The lake dried out completely for a few weeks in October 1968.

At its 'normal' extent, the lake surface area is 750 km^2. The lake basin being very flat, as much as 600 km^2 of littoral areas are flooded and dried every year. In the rainy season, the lake is about 5 m deep in the central area and 1–2 m deep in the littoral zone. The islands of the lake are of volcanic nature as are most of the rocks of the catchment area. Carbonate is frequent in the islands. Two hot salt springs are known near the lake and salt deposits of an additional spring have been found on an island (Garson, 1960).

The lake water belongs, as does most East-African water, to the sodium-carbonate type. It has a yellow-brown colour due to humic acids.

The phytoplankton includes Chlorophyta (Volvocales, desmids), diatoms and blue-green algae (Mwanza, 1972). Crustacea and Rotifera such as *Brachionus* sp. constitute the zooplankton of the lake.

Kirk (1967) reports 24 species of fish known in the Chilwa basin, of which only 12 species have been found in the lake: the cichlids *Tilapia shirana chilwae*, endemic to Lake Chilwa, *T. sparrmanii, T. melanopleura, Haplochromis callipterus*, and *Hemihaplochromis philander*; the clarids *Clarias mossambicus* and *C. theodorae*; the schilbeid *Pareutropius longifilis*; the characid *Alestes imberi*; and the cyprinids *Labeo cylindricus, Barbus paludinosus* and *B. trimaculatus*. Three species are abundant enough to be of economic importance: the cyprinid *Barbus paludinosus*, the cichlid *Tilapia shirana chilwae* and the clarid *Clarias* sp. *B. paludinosus* feeds mainly on zooplankton and occasionally on plant material. *T. shirana chilwae* feeds essentially on benthic diatoms and pieces of vascular plants. *C. mossambicus* feeds on the cyprinid *B. paludinosus* which rarely exceeds

10 cm, on insect larvae and fragments of vascular plants. Bourne (1973) established that the fingerlings of these three species are planktivorous to a large extent. *Clarias* and *Barbus* migrate into the swamp to breed in November–December and return two to three months later (Furse, 1972). *Tilapia shirana* breeds on sandy marginal areas of the lake. The predation on zooplankton by young fishes is then maximum from January to June.

The impact of water level lowering
The modifications of the composition of the lake water were investigated by Moss & Moss (1969) and by Morgan & Kalk (1970) during the 1966–68 drying phase. Although this period was marked by a significant decrease in rainfall, we note that during a cyclone, in March 1967, 200 mm rain fell on the area in 24 h.

The conductivity, which was 4210 μmhos cm^{-1} in June 1966, at the end of the rainy period, reached 12 000 μmhos cm^{-1} in December 1966, fell to 276 μmhos cm^{-1} after the cyclone and rose to 16 720 μmhos cm^{-1} in November 1967. Alkalinity, sodium, chloride, potassium and sulphates fluctuated as conductivity. The pH values increased during the dry period, decreased with the cyclone and rose again. Calcium and magnesium concentrations dropped during the dry period because of precipitation as carbonates. In the littoral mud uncovered by the level lowering, solutes concentrated and crystals of salt formed on the surface. The next rain stirred up the sediment of shallow areas, partly dissolving the newly formed crystals and releasing large amounts of phosphorus, nitrate and silicon. This explains why the salt concentration continued to increase during the initial rise in lake level. With the increases in salinity and alkalinity, the Chlorophyta disappeared and in the rainy period 1967–68, the phytoplankton was dominated by *Spirulina platensis* and *Anabaenopsis circularis*. On the newly covered mud, mats of *Oscillatoria* sp., *Nitzschia palea* and *Anomoeneis sphaerophora* developed. At the end of the rainy period, epipelic *Oscillatoria* were the only algae present.

In the zooplankton, Crustacea disappeared in late 1967. *Brachionus* was the main survivor. Lake Chilwa had a large population of molluscs. Vast amounts of *Lanistes* died during lake-level lowering but many survived in the wet mud, in particular *Aspatharia* sp.

As far as the fish are concerned, *Barbus paludinosus* was the first to disappear from the lake as the salinity increased to about 4000 μmhos cm^{-1}. In December 1966, at the peak of salinity, a mass mortality of *Tilapia shirana chilwae* was observed. It is related both to the modification of water composition and to winds which mixed the sediments causing turbidity and depletion of oxygen. *Clarias mossambicus* was the most tolerant species since specimens were still found in the lake at conductivity exceeding 10 000 μmhos cm^{-1}.

The refilling of the lake and recovery of the flora and fauna
From November 1968 to April 1969, the dry lake filled again to a depth of 2–3 m (Kalk, 1971). The runoff freshwater reaching the lake did not mix immediately with the saline lake water and a core of relatively saline water in the central part of the lake (2500 μmhos cm^{-1}) was surrounded by a belt of diluted water (250 μmhos cm^{-1}). Later, the two water bodies mixed with a uniform conductivity of 1200 μmhos cm^{-1}. During the following dry season, the conductivity reached 1800 μmhos cm^{-1} but dropped to very low values with the first rains (90 μmhos cm^{-1}) (Mwanza, 1972).

The recovery phase was characterized by a reappearance of Chlorophyta (Mwanza, 1972) although *Oscillatoria* still dominated the phytoplankton in 1975 (Kalk, 1979).

In the first eight months of lake filling, in 1969, the zooplankton was dominated by Rotifera which were later replaced by Crustacea. In her observations of 1975–76, Kalk found the zooplankton to be composed mainly of *Tropodiaptomus kraepelini, Diaphanosoma excisum, Daphnia barbata, Ceriodaphnia cornuta* and *Moina micrura. Mesocyclops leuckarti* and the rotifers *Brachionus calyciflorus* and *Keratella tropica* were occasionally present. Zooplankton development was partly limited by the availability of food sources: the turbidity of lake water did not favour the development of algae and the main plant food consisted of the detritus washed down from the swamps.

Quantitatively, the zooplankton biomass remained low three years after lake filling and recovery of the main fish species (1972). It is believed that fish predation was a major factor in operating a drastic selection on zooplankton species (e.g. elimination of large zooplankton). Moreover, the spawning behaviour of the three main fish species explains the seasonal fluctuations in zooplankton; in the hot dry period the fish, mainly composed of adult specimens, have a diversified diet and exert a low predation pressure on zooplankton which multiplies rapidly and builds up a high biomass. In April–June, the appearance in the lake of the juveniles born in the swamps decimates the zooplankton. The alternating feeding pattern of the fish allows the reconstitution of the zooplankton population and confers upon it a great stability. These observations by Kalk (1979) in 1975–76 demonstrate that seven years after the complete drying out of the lake a prey–predator system had been 'reinvented'.

The decrease of salinity in the lake was accompanied by the return of fish species which had found refuge in streams, lagoons and marshes. In addition, a stocking programme with the endemic *Tilapia* was initiated. All this resulted in a rapid resurrection of the fisheries: in 1976, the landings amounted to 20 000 tons. The decline and rise of Lake Chilwa fisheries have been described in detail in Morgan (1971).

McLachlan & McLachlan (1976), studying the development of the mud habitat during the recovery of Lake Chilwa, showed how wave erosion disrupts the dry and cracked mud of the newly flooded bottom and resedimentation of disrupted material allows the formation of a mud layer suitable for recolonization by benthic fauna. The tube-building animals stabilize the mud whereas mud-burrowing animals and mechanical mixing release nutrients and inorganic ions. During the recolonization period, the littoral area was occupied by a large population of *Chironomus transvaalensis* (McLachlan & McLachlan, 1976).

The very rapid biological recovery after the 1968 ecological catastrophe which destroyed a large part of the biological stock of the lake is mainly due to the role of sanctuaries played by the swamps and marshes (Howard-Williams & Lenton, 1975). The fauna seeks refuge in these areas which makes recolonization possible. It is probable that in the absence of swamps the reconstruction of the lake food chain would have taken much longer.

THE NIGER BASIN
The Niger River

The watershed of the Niger River extends over 1 125 000 km^2 between 4°22′ N at the delta in the Gulf of Guinea and the Tropic of Cancer in the Ahaggar Mountains. The Niger River springs at about 1000 m altitude in the Mounts of Loma (border area between Sierra Leone and Guinea) within 250 km of the Atlantic Ocean. It flows towards the northeast under the name of Tembo and 300 km from its source becomes the Joliba River. The course of the river is an impressive loop, with an eastward direction, beyond Timbuktu. The River, then called Isa, 'the black water', curves southwards through Niger and Nigeria and finally flows into the Gulf of Guinea by a large delta which is in fact a mangrove swamp stretching for 200 km along the coast and 250 km inland (Kirk-Greene, 1966). The river is 4183 km long, the tenth longest in the world and the third longest in Africa.

The present course of the Niger dates from the late Pleistocene. In the late Pliocene and Pleistocene, the present upper course of the Niger flowed westward into the Gulf of Senegal, whereas the middle and lower Niger drained the Ahaggar Mountains and flowed as today into the Gulf of Guinea. During the dry periods of the Pleistocene, the westward flow of the Upper Niger was interrupted by sand dune barriers and during the next wet period a vast lake covered the Araouane area and deposited large amounts of sediments. This lake was subsequently captured by the Middle Niger. This area between Segou and Timbuktu, characterized by a very small slope (3 m between these two towns which are nearly 600 km apart) and a very flat topography, is known as the 'internal delta of the Niger'. It is a swampy area where the sediments of the upper river and the Bani River are deposited causing the river to meander, split into several channels and form permanent and temporary lakes.

The hydraulic regime of the Niger has been described by Guilcher (1965). The Niger River has an average yield of 6000 m^3 s^{-1} or 192 × 10^9 m^3 y^{-1}. Its specific yield (172 m^3 km^{-2} y^{-1}) is identical to that of the Zambezi River and only half the specific yield of the Zaire. It has only three major tributaries: the Bani River in Mali, the Benue River and the Sokoto River in Niger. The river twice crosses the same pluviometric zones: upstream and downstream, the precipitations diminish to 200 mm in the northernmost part of the loop. Upstream, the Niger River and Bani River have low water in April–May and floods in September–October. Beyond Mopti, the swamp area delays the maximum which occurs only in February at Niamey. Beyond Niamey, the Niger goes back to a typical tropical regime with a peak in October which is superimposed on the February peak of Middle Niger. The Benue, with its 2500 m^3 s^{-1}, has again a regime with only one maximum.

The seasonal fluctuations in the Niger River discharge range from 500 m^3 s^{-1} in May–August to 7000 m^3 s^{-1} in September–October. The chemical features of the Niger River water are shown in Table 69.

Holden & Green (1960), studying the plankton of the Sokoto River, one of the main tributaries of the Niger, found that the number of zooplankton organisms does not exceed 20 000 m^{-3} which is very low when compared to the results of Rzoska, Brook & Prowse (1955) on the Blue Nile (108 000 m^{-3}) or any eutrophic lake of the temperate zone. The poor transparency of the water, its paucity in nutrients and the diluting effect of the flood seem to be responsible for this low planktonic productivity.

Lakes of the internal delta

The main morphometric features of these lakes are presented in Table 70.

Lake Lere

Lake Lere is located at 9°37′ N and 15°10′ E at the border of Cameroon and Tchad at an altitude of 231 m. Its surface area is 40 km^2; its depth is 8 m (max.) and 4.5 m (mean) and its volume 160×10^6 m^3. The Mayo Kebbi River, an affluent of the Bene is the inflow and outflow of the lake. Lévêque (1971) reports the following physico-chemical features (results in mg l^{-1}): Ca^{2+} 10; Mg^{2+} 2.6; Na$^+$ 2.3; K$^+$ 1.8; HCO$^-_3$ 27; pH 8; conductivity 89 μmhos cm^{-1}.

Kainji Reservoir

General

The Kainji Dam is located in Nigeria at 10°00′–10°55′ N to 4°20′–4°50′ E at 140 m altitude. The Kainji Dam was built during the 1964–68 period some 1000 km upstream of the Niger mouth. The reservoir is formed of three main areas: a deep and narrow basin near the dam, a wide

Table 69 *Chemical features of the Niger River (after Visser, 1973)*

Conductivity (μmhos cm^{-1})	70.0	Na$^+$ (mg l^{-1})	54.2
pH	7.1	K$^+$ (mg l^{-1})	5.2
Ca^{2+} (mg l^{-1})	6.4	Cl$^-$ (mg l^{-1})	2.6
Mg^{2+} (mg l^{-1})	2.7	SO$_4^{2-}$ (mg l^{-1})	traces

Table 70 *Main morphometric features of the lakes of the internal delta*

	Area (km^2)	Depth (m)	Volume (m^3)
Aougoundo	80–130	–	–
Do	120	9.5 (max.)	–
Faguibine	800	10.0 (max.)	–
Gakore	34	4.4 (max.)	150×10^6
Garo	60–120	9 (max.)	–
Haribonga	25	11.5 (max.)	290×10^6
Kobongo	8	9.2 (max.)	75×10^6
Korarou	80–170	–	–
Niangaye	400	6.5 (max.)	–
Tele	190	5	350×10^6

basin over the inundated Foge Island and a narrow riverine upper basin. Before the impoundment, the vegetation of the Island was cleared over an area of 535 km² and a detailed survey of the swamps of the Kainji basin was carried out by Imevbore & Bakare (1974). The main morphometric parameters are shown in Table 71.

Hydrological, physical and chemical features

The annual flow of the Niger into the basin of Lake Kainji can be divided into the 'white flood', rich in suspended material, originating in the Nigerian basin and estimated at 47×10^9 m³ y⁻¹ and the 'black flood', rich in humic material, coming from the Guinea equatorial rain forest and estimated at 33×10^9 m³ y⁻¹. It is clear that in dry years the white flood may be reduced by 50%, whereas the black flood is not reduced by more than 10%. The level fluctuations reach 10 m a year. The level is high from October to April. The cool and dry harmattan, a northerly wind, dominates from December to February whereas southerly winds are frequent during the rest of the year.

In 1969–1970, Henderson (1973) studied the thermal regime and circulation pattern of the lake. In August, the level is minimum and part of the central island is emergent so that the water flows in the eastern and western channels on both sides of the island. Then begins the white flood, which can be easily traced by turbidity measurements: most of the turbid water then flows in the western channel, protected from the south-easterly wind by the islands, whereas the eastern channel remains limpid. In September, the water level has already risen by 2 m and most of the water flows in the eastern channel. The water level goes on rising and general turbidity declines as, upstream, the black clear water dilutes the turbid flood. In December, the air temperature drops rapidly and reaches 15°C in January–February: the water temperature decreases then from 28–30°C to 22–25°C and complete mixing takes place. Since the mixed period corresponds to the peak of the black flood the whole lake is flushed again

Table 71 *Main morphometric features of Lake Kainji*

Max. length	137 km	Max. depth	60 m
Max. width	24 km	Mean depth	11 m
Surface area	1270 km²	Volume	14×10^9 m³
Watershed area	1.6×10^6 km²	Mean annual	
Watershed area:volume ratio	115 m² m⁻³	evaporation	2.5×10^9 m³

completely. In February, the air temperature rises and stratification begins and lasts until May. Oxygen disappears completely from hypolimnion. During the stratified period, the discharge of the dam is taken entirely from the hypolimnion which causes the rapid deepening of the thermocline and an 'artificial destratification' sometimes interrupted by temporary thermal gradients and partial deoxygenation. The water level fluctuations may reach 10 m.

The material in suspension consists mainly of kaolinitic colloids which reflect blue and green light. Even when the clay concentration of the water does not exceed 10 mg l^{-1}, the Secchi-disc readings are around only 50 cm. The rapid decrease in turbidity of the lake is not due to sedimentation but to replacement by black waters. This periodic renewal process restocks the lake in nutrients and reduces the amount of sediments deposited in the lake. The annual sediment load of the lake has been estimated to be 4.6×10^6 tons or 6.5×10^6 m^3, representing only 0.04% of the maximum volume of the lake. The suspended material contributes major ions and nutrients to the water and therefore significant correlations have been found between the concentration of suspended solids and concentration of nitrates, iron and organic matter (Imevbore & Adeniyi, 1977).

Hydrogen carbonate is the dominant anion (16 mg l^{-1}) and calcium and magnesium are the main cations (4.2 and 1.8 mg l^{-1}, respectively. Silica reached 30 mg l^{-1} and PO_4-P 4 mg l^{-1}, whereas nitrate rarely exceeds 110 µg l^{-1}.

Biological features

Data concerning phyto- and zooplankton in the Niger near Kainji prior to impoundment were reported by Visser (1973). The phytoplankton consisted mainly of blue-green algae and diatoms of the genera *Melosira* and *Nitzschia*; the zooplankton was essentially composed of *Moina dubia, Diaphanosoma excisum, Ceriodaphnia cornuta, Cyclops leuckarti, C. hyalinus* and *Diaptomus banforanus*. The low plankton biomasses were related to the paucity of nutrients in the water and to low transparency. Visser gives also a list of the Niger fishes near Kainji before closure of the dam: *Alestes leuciscus, Citharinus citharus, Synodontis gobroni, Gnathonemus niger* and *Polypterus senegalus* represented 28% of the ichthyological fauna.

Adeniji (1973) studied the seasonal fluctuations of the phytoplankton in the lake and showed that the total number of algae rises just before the beginning of the rainy season and peaks in July. During the dry period and beginning of the wet season, green algae and especially *Volvox* sp. are dominant; then they are replaced by diatoms (*Melosira* sp.) which decline in October when blue-greens (*Anabaena* and *Microcystis*) dominate. The zooplankton biomass also rises with the rain and peaks in April (13 organisms l^{-1}) with Cladocera as the dominant group. As in other impoundments it has been found that the plankton population in the lake is greater than that of the river before the closure of the dam (Visser, 1973).

After the closure of the dam the fish population consisted mainly of Citharenidae, Characidae, Centropomidae and Mochokidae; Cichlidae were rare. The most noticeable change which took place was a significant increase in predator fish in comparison with other species. The predator fish feed mainly on the small pelagic clupeids *Sierrathrissa leonensis* and *Pellonula afzeliusi*. *Citharinus* sp., very abundant in the river and in the lake just after impoundment, shows signs of decline, whereas the genus *Tilapia*, especially the species *T. galilaea*, increases rapidly in shallow water (Lelek & El-Zarka, 1973). The stabilization of the littoral zone and the development of vegetation is particularly favourable to Cichlidae and accounts for their spectacular increase along the shoreline. In 1968 they formed only 9% of the shallow-water fish families, but 79% in 1970; by mass, however, the Characidae dominate the catch (62%), followed by Cichlidae (19%) and Citharidae (11%). The fish community includes herbivores (*Alestes macrolepidotus*, *Distichodus* sp., *Synodontis schall*), detritivores (*Citharinus*, *Labeo* spp.), zooplankton-eaters (*Schilbe mystus*, *Eutropius niloticus*, *Chrysichthys* spp., *Alestes* spp., *Synodontis* spp.), and carnivores (*Lates niloticus*, *Hydrocynus* spp., *Bagrus* spp., *Clarias lazera*, *Heterobranchus* spp.).

In the northern part of the lake there is an increase in free-floating and semi-attached aquatic weeds (*Echinochloa stagnina*, *Polygonum senegalense*, *Jussiaea repens*, *Pistia stratiotes* and *Utricularia inflexa*). In other parts of the lake shallows, the distribution of weeds is limited. The undesirable *Salvinia auriculata* and *Eichhornia crassipes* have not been recorded in Nigeria. The evolution of the shoreline has been described by Halstead (1973).

LAKE CHAD

Lake Chad is located at 12°30′ N and 13°–15°30′ W at 282 m at the border of Cameroon, Chad, Niger and Nigeria. The Lake has been intensively studied by the Centre de Récherche Scientifique et Technique d'Outre-Mer (ORSTOM) based at N'Djamena (Fort-Lamy) and the Federal Nigerian Fisheries Services based at Malamfatori. The ORSTOM

Lake Chad

research was part of the IBP programme and was defined as 'Etude d'un lac tropical sous climat semi-aride (lac Tchad)'.

Origin, climate changes and water-level fluctuations

Lake Chad lies in the tectonically stable part of Africa, in a very flat sedimentary basin which has been a drainage area for surrounding regions for at least the whole Cenozoic period. The lake does not occupy, however, the lowest point of the basin (170 m) which is located north of Lake Chad and south of the Tibesti Mountains (Bodele Depression). During the Pleistocene, the lake's water level varied considerably; the study by Servant (1970) of the lacustrine formations of the basin and their flora of diatoms established that, between 5000 and 6000 B.P., Lake Chad covered an area 16 times that of the present lake.

The lake drainage area extends over 2.5×10^6 km², between 5 and 25° N and 7 and 25° E, an area periodically swept by the subtropical belts. The dry north-east wind, called the harmattan, blows from October to April, whereas the humid south-west wind brings rain during the second half of the year. The air temperature has a yearly annual value of 27°C (29–32°C during the warm season and 22–24°C in the cool period). The combined effect of high temperature and strong winds causes a very high evaporation: 2.05–2.25 m³ y⁻¹ (Carmouze, 1979). Rainfall occurs from July to September and varies from 500 mm in the southern part of the lake to 250 mm in the northern area. The observed long-term extreme variations at Bol are 125 and 565 mm.

Because of the flat topography of the basin, minor climatic modifications cause significant fluctuations in the lake surface. Since the lake was discovered in 1823 by Oudney, Denham and Clapperton, its limits and water level have changed considerably. During the last hundred years, three main 'sizes' of the lake were observed: 'Big Chad', with a mean depth of 4 m and surface area of 20 000–25 000 km², developed in the second half of last century; 'Normal Chad' (mean depth: 2.5 m and surface area 18 000 km²) was observed in the period 1960–72; 'Little Chad' (mean depth 1.5 m and surface area 12 000 km²) was typical of the periods 1908–1915 and 1972–1977.

General features and ecological zones

In its 'normal stage', the lake can be divided into two basins (north basin and south basin) on each side of a line extending from Baga Kawa to Baga Sola and representing a sill of shallow waters: la Grande Barrière. The eastern and northern parts of the lake are bordered by an erg, the dunes of which have a NE–SW orientation. The flooded lower parts of the dunes and their emergent crests form a complex archipelago elongated into the lake by reed islands. At level 281.5 m, the lake surface area is 20 000 km² with 4720 km² in the archipelago area, 6210 km² in the reed-island zone and 7900 km² of open water. La Grande Barrière covers ∼2000 km² (Table 72).

The northern basin contains three main ecological zones. The *northern reed islands* occupy the deepest part of the lake (4–7 m at 281.5 m above MSL) and have larger dimensions than in other parts of the lake. The *northern archipelago* is composed of sandy islands surrounded by a belt of submerged meadows. Depth varies between 3 and 6 m. The *northern open waters* (4200 km²) have a depth varying from 4 to 7 m. Identical units are found in the southern basin but depths are shallower (2–3 m in the *southern archipelago*, 1.5–3 m in the *southern reed islands* and 1.5–3.5 in the *southern open waters*) (Table 72).

The relatively dry year 1972 limited considerably the discharge of the lake inlets and, in April 1973, the lake was divided into three isolated zones: the northern basin, the southern open waters and the south-east archipelago, separated by emergent drying areas rapidly covered by a dense vegetation. From March to September 1973, the vegetation covered an area of 5000 km² in the southern basin. During the following, more rainy years, the southern basin, periodically fed by the Chari River, recovered a normal level in 1974, whereas the level continued to decrease in the northern basin even when the water level in the southern basin rose above the sill altitude. It appears that when very large areas are emergent in the northern basin, considerable amounts of water are needed to compensate for infiltration and evaporation.

At low levels, the vast areas of macrophytes considerably enhance the losses through evapotranspiration and exceptional discharges of the rivers are needed to re-establish previous levels (Lemoalle, 1979*b*). That is why in 1976, four years after the drought, the level in the northern basin was still a few metres lower than that in the southern basin.

Hydrology and water chemistry

The Chari–Logone River and El Beid River are the main inlets of Lake Chad, with respective yearly average discharges of 40×10^9 and 1.4×10^9 m³. The interannual discharge of the Chari River varies between 19.5×10^9 and 54.5×10^9 m³. The mean direct water input by rainfall is 6.3×10^9 m³ with variations from 2.7 to 8.7×10^9 m³. The total annual input of water varies therefore from 23×10^9 to 61.4×10^9 m³ with a mean

Table 72 Morphometric and physico-chemical features of Lake Chad at water level 281.5 m above MSL (from Carmouze et al., 1972)

Region	Total surface area (km²)	Water surface area (km² (%))	Depth (m)	Secchi disc (cm)	Conductivity (µhmos cm⁻¹)	Volume (10⁹ m³)	Yearly water renewal (%)
Northern reed islands	3560	2955 (83)	5–8	60–80	500–1500		
NE archipelago	2200	1144 (52)	4–6	60–80	200–1200		
Northern open waters	4200	4200 (100)	4–7	40–60	250–500		
NORTHERN BASIN	9960	8299 (83)				46.7	40
Eastern archipelago	1050	515 (49)	2.5–4	35–50	150–650		
SE archipelago	1470	911 (62)	2.5–4	20–50	70–200		
Southern open waters	3700	3330 (90)	2–4	10–50	50–120		
Southern reed islands	2650	2120 (80)	2–3	15–30	50–250		
Barrier zone	2000	1600 (80)	2–3	15–30	50–400		
SOUTHERN BASIN	10870	8476 (78)				25.3	85
TOTAL LAKE	20830	16775 (80)				72.0	

value of 48×10^9, 87% coming from rivers and 13% from rainfall (Carmouze, 1979).

The only obvious water loss is evaporation, since Lake Chad has no outlet. However, since the Chad water salt content is very low ($\sim 100\,\mu$mhos cm^{-1} at 'normal' level), it has long been assumed that infiltration and salt precipitation were the additional mechanisms accounting for the hydrochemical balance of the lake (Table 73). Carmouze (1979), on the basis of the sodium balance of the lake, determined that the losses by evaporation and infiltration represent, respectively, 92% and 8% of the total outputs. The separate water balance for each basin adds interesting new information: the southern basin losses are divided as follows: 50.9% through evaporation, 2.9% through infiltration and 46.7% through discharge into the northern basin; there the losses are 89% through evaporation and 11% through infiltration. Moreover, the northern basin contains twice as much water as the southern basin and has a higher water residence time (two years instead of six months). For any element under purely climato-geographical control, 8.2% of its annual input remains in the southern basin and 91.8% flows into the northern basin. These flow patterns generate differences of hydrochemical behaviour: the climato-geographical increase of salinity is 3.2 times greater in the northern basin than in the southern. When river input increases, concentrations drop rapidly in the southern basin; in contrast, any deficit of river input causes a greater salinity increase in the southern basin which underlines the instability of this basin.

Approximately 73% of the river's input flows into the northern basin which is consequently the main salt reservoir. The salt balance is regulated

Table 73 *Chemical features of Lake Chad (from Lemoalle, 1979b)*

Southern basin (Sept. 1972)	
Ca^{2+} (mg l^{-1})	10.0
Mg^{2+} (mg l^{-1})	4.8
Na^+ (mg l^{-1})	11.5
K^+ (mg l^{-1})	5.1
Cl^- (mg l^{-1})	1.0
CO_3^{2-} (mg l^{-1})	45.0
pH	8

Lake Chad

by infiltration and sedimentation. The infiltration is made possible by the existence of a 50 m-thick water-bearing layer under the lake and takes place mostly in the northern basin (89% in comparison with 11% in the southern basin). Arad & Kafri (1975) note that, although superficial waters converge towards Lake Chad, the Bodele Depression has phreatic levels lower than those of the Chad depression and consequently serves as a regional base level and a drainage area for the northern basin of the lake. Sedimentation is nearly equal in both basins; the sedimented material includes calcium carbonate and autochthonous smectites (nontronite, beidillite and magnesium montmorillonite). Molluscs and macrophytes also affect the calcium chemistry significantly.

Thermal structure and dissolved oxygen

Measurements by ORSTOM at Bol have shown that the water profile is thermally homogeneous in the morning, undergoes a superficial stratification in the afternoon and becomes homothermal again at the end of the day. Very often, the profile shows no stratification but a low gradient from surface to bottom indicating that turbulence is not high enough to transmit heat to deep layers. In general, however, the lake can be considered as tropical polymictic. The absolute temperatures vary from 17°C in January to 30–31°C from June to September (monthly averages). Comparing the annual averages, Lemoalle (1979b) noticed that water temperature is lower when the lake volume is small. This decrease in temperatures is related to perturbations in the movements of the subtropical belts. During the dry period (1973), an annual cycle of stratification appeared; it was caused by the appearance of vegetation at low levels which diminishes the effect of wind on the water surface.

For the 'normal-sized lake', the levels of oxygen are homogeneous in the morning and range from 80 to 100% saturation. Oxygen stratification develops at mid-day during the warm season and disappears in the evening. Such diurnal stratification vanishes in October.

In 1974–75, when the lake size and depth were reduced, vertical homogeneity was observed only in winter with concentrations near to zero. In summer, the epilimnion had very variable oxygen levels whereas the hypolimnion remained anoxic. During the 'Little Chad' periods, the deficit in oxygen was especially high in the archipelago which considerably affected the fish populations. In the northern open water, the lowering of the water level during the drought caused a simultaneous increase in oxygen-consuming resuspended material and oxygen-producing algae, resulting in a very variable oxygen regime.

Biological features

Compère (1967, 1974, 1975a,b, 1976, 1977), and Iltis & Compère (1974) have identified > 1000 species of algae in Lake Chad. The algal population is dominated qualitatively by Desmidiaceae and diatoms but Cyanophyta are sometimes abundant. During the 'Normal Chad' period, different assemblages were observed in the various ecological zones of the lake. The northern open waters were dominated for most of the year by Desmidiaceae (*Closterium aciculare*). *Pediastrum*, *Botryococcus* and *Microcystis* were also abundant. In the north-eastern archipelago, *Anabaena* and *Microcystis* predominated and *Closterium*, *Pediastrum* and *Botryococcus* were abundant. In the southern open waters, the diatoms *Melosira granulata* and *Surirella muelleri* formed the main assemblage, whereas *Microcystis* and *Anabaena* dominated in the eastern and south-eastern archipelago.

The algal wet biomass was relatively low in the southern basin (0.03–0.22 mg l^{-1} in open waters and 1.4 mg l^{-1} in the archipelago) and higher in the northern basin (0.7–1.6 mg l^{-1} in open waters and 1.4–2 mg l^{-1} in the archipelagoes) (Iltis, 1977). In February 1971, the algal biomass was estimated to be 40 800 tons fm. The drop in water level caused by the drought caused an increase in algal biomass which reached 76 800 tons in January 1972, 183 000 tons in April 1974 and 244 000 tons in February 1975 (Iltis, 1977). This increase was accompanied by a modification of the algal species. Euglenophyta became abundant and, in the north basin, centric diatoms and *Spirulina* (*Oscillatoria*) *platensis* became predominant. Léonard & Compère (1967) emphasized the nutritional value of this alga which is sold as 'algal cakes' at the market of Fort-Lamy.

The primary productivity is maximum in the northern basin (1.8–3.7 g C m^{-2} d^{-1} in the 'Normal Chad' phase; 3.4–13.3 g C m^{-2} d^{-1} in the 'Little Chad' period in 1975 (Lemoalle, 1979a), intermediate in the eastern archipelago (1.5 g C m^{-2} d^{-1}) and minimum in the southern basin (0.33–0.71 g C m^{-2} d^{-1} in winter and 0.7–2.7 g C m^{-2} d^{-1} in summer for periods of high levels).

In the 'Normal Chad' phase, 2400 km^2 were covered with macrophytes (*Vossia cuspidata* in the Chari delta, *Cyperus papyrus* in the south basin and *Typha australis* in the northern basin) with a total biomass of 7.2×10^6 tons dm of aerial parts and 13×10^6 tons dm of roots.

Zooplankton in Lake Chad includes some 20 species of Rotifera, eight species of Cladocera (mainly *Moina micrura*, *Diaphanosoma excisum*, *Bosmina longirostris*, *Daphnia barbata* and *Ceriodaphnia cornuta*) and four species of Copepoda (*Tropodiaptomus incognitus, Thermodiaptomus galebi, Thermocyclops neglectus* and *Mesocyclops leuckarti*). Quantitatively, rotifers play a minor role and the crustacean plankton represents 99% of the biomass (see Table 74) (Dussart & Gras, 1966; Gras, Iltis & Lévêque-Duwat, 1967; Rey & Saint Jean, 1968, 1969).

The zooplankton biomass has been measured by Robinson & Robinson (1971) in the northern open water, by Gras *et al.* (1967) in the south basin and by Carmouze *et al.* (1972). The south-eastern open water has the lowest annual average biomass (150 mg dm m^{-3}). Moreover, the seasonal minimal values of biomass are concomitant with the flood of the Chari River: it is likely that the turbidity of the lake water during floods is very unfavourable to zooplankton survival. Maximum biomass is observed during low-water and high-temperature periods (Lévêque, 1979). In contrast, only slight seasonal variations are recorded in the southern archipelago and reed islands and mean annual biomass is high (350 mg dm m^{-3}). The northern basin has a smaller biomass (240 mg dm m^{-3}) but is deeper. In the north and south basins, *Daphnia* and *Bosmina* dominate in winter whereas *Moina, Diaphanosoma* and *Ceriodaphnia* are preponderant in the warm period. The well-marked minimum of the diaptomids of the south basin does not exist in the north basin. The total zooplankton mass of Lake Chad was estimated at 14 000 tons dm (February, 1971) 60% of which is in the north basin. The dry mass averages are 11.5 kg ha^{-1} in the north basin, 7.5 kg ha^{-1} in the eastern archipelago and 2.5 kg ha^{-1} in the southern and south-eastern open waters.

Embryonic and post-embryonic development rates of the crustacean zooplankton have been measured (Gras & Saint Jean, 1969, 1976, 1978a,b). In both Cladocera and Copepoda, the embryonic development lasts from 1–1.5 days at 30°C to 1.9–2.3 days at 22°C. Rates of post-embryonic

Table 74 *Biomass and productivity of zooplankton, 1964–1965, eastern archipelago (according to Lévêque, 1979)*

	Biomass (mg m^{-3})	Productivity (mg m^{-3} dm)
Cladocera	125	14 110
Diaptomids	152	3504
Cyclopids	56	3580
Total zooplankton	333	21 194

development are more variable according to the species, diaptomids being slow-growing animals in comparison with Cladocera.

In the eastern archipelago, the annual production is 21.2 g dm m^{-3}, two-thirds of which is contributed by Cladocera. In 1971, the total lake production was estimated at 860 000 tons dm (Lévêque & Saint Jean, 1979).

Interesting interannual variations have been observed: during the period 1968–73, characterized by a rapid decrease in lake level, the duration of embryonic development diminished significantly because of the increase in algal concentrations which improved the feeding conditions. A similar chain of events (drought → lower water levels → increase of edible algae → reduction of zooplankton development time and increase in fertility) was observed in Lake Kinneret in 1973–75 (Serruya et al., 1980).

Various groups of benthic fauna have been investigated: oligochaetes – six species (Lauzanne, 1968); molluscs – 12 species (Lévêque, 1968, 1972); and insects – 200 species (Dejoux, 1969, 1976).

A clear seasonal variation has been observed in worms and insects, the maximal biomass occurring in winter (December–March). A general survey of benthic communities was carried out in 1970 (Carmouze et al., 1972). In the north basin, oligochaetes were represented mainly by Tubificidae whereas Alluroididae (*Alluroides tanganikae*) predominated in the south; both families were found in the archipelagoes. Molluscs were most abundant in the Grande Barrière, the northern open water (Table 75). *Melania tuberculata* and *Bellamya unicolor* were predominant in the north basin and *Cleopatra bulimoides* in the eastern archipelago. Both species were common in the south-east archipelago and the Grande Barrière.

The benthic biomass was ~71 000 tons dm (not including shells for molluscs – values of 1970). The lake being very shallow, the whole bottom is colonized by molluscs which contribute ~90% of the benthic biomass. From 1968 to 1972, the mollusc population diminished as a result of the lowering of the lake level. Reproduction was continuous throughout the year in prosobranch molluscs but occurred only in the cold season in *Corbicula*. The annual production of benthic molluscs was estimated to be 279 000 tons dm organic matter (Lévêque, 1979). Lévêque (1972) estimated that a mollusc population of the size of that of 1970 takes up a yearly amount of 700 000 tons calcium, i.e. four times the annual calcium input of the rivers.

The ichthyological fauna of the lake includes 140 species (Blache, 1964). In the south-eastern open water, *Schilbe* sp. are abundant as well as *Citharinus citharus*, *C. distichoides*, *Labeo coubie*, *Alestus* sp. and *Synodontis* sp. *Tilapia* sp. and *Heterotis niloticus* are common in the SE

Table 75 *Biomass and productivity of benthic fauna (according to Lévêque, 1979)*

	Worms Biomass	Insects Biomass	Molluscs Biomass	Productivity
North reed islands	2.1 kg ha^{-1}	1.4 kg ha^{-1}	0.2 kg ha^{-1}	1 kg ha^{-1} y^{-1}
Northern open water	8.0 kg ha^{-1}	2.1 kg ha^{-1}	64.2 kg ha^{-1}	353 kg ha^{-1} y^{-1}
NE archipelago	1.1 kg ha^{-1}	2.9 kg ha^{-1}	47.8 kg ha^{-1}	241 kg ha^{-1} y^{-1}
Grande Barrière	1.9 kg ha^{-1}	1.6 kg ha^{-1}	72.0 kg ha^{-1}	256 kg ha^{-1} y^{-1}
SE open water	1.5 kg ha^{-1}	0.1 kg ha^{-1}	38.6 kg ha^{-1}	114 kg ha^{-1} y^{-1}
South reed islands	2.6 kg ha^{-1}	0.1 kg ha^{-1}	11.8 kg ha^{-1}	338 kg ha^{-1} y^{-1}
SE and eastern archipelagos	0.8 kg ha^{-1}	0.6 kg ha^{-1}	10.6 kg ha^{-1}	80 kg ha^{-1} y^{-1}
Total biomass (dm)	5540 tons	2300 tons	63 280 tons	
Mean biomass	2.9 kg ha^{-1}	1.2 kg ha^{-1}	32.9 kg ha^{-1}	
Total productivity (dm)				279 000 tons
Mean productivity				145 kg ha^{-1} y^{-1}

archipelago. The northern basin is poorer in species since many of them (*Schilbe mystus, Synodontis* sp.) do not get over the Grande Barrière. This distribution indicates that, in the 'Normal Chad' period (1971–1972), the archipelagoes were mainly populated with planktivorous fishes whereas predators were more abundant in open waters (Quensiere, 1979).

The annual catch is about 100 000 tons. This abundance is partly due to the large flood plain surrounding the lake. As the water rises, the fish move upstream, when the flooding occurs they spread on vast inundated areas where food is abundant and shelter easy to find. Many species breed in the flood plain, then, as the water decreases, adults and juveniles move back into the river beds and into the lake. The Nile perch, *Lates niloticus*, is a noticeable exception since it does not migrate but lays its eggs in open water. In contrast, the zooplanktonivorous *Alestes baremose* gather at low water around the delta of the Chari River and migrate upstream of N'Djamena.

The drought and consequent lowering of water level in the early 1970s transformed the lake into three shallow water bodies. The reduction of the lake volume and the isolation of different water masses caused massive fish killing because of the high turbidity caused by resuspension of sediment and disappearance of oxygen. The return to normal conditions (flood of 1974) also had disastrous consequences for the fish population when the water got over the vegetation barrier which isolated the eastern archipelago, the dissolved oxygen was consumed at once by decomposing vegetation and many fishes which had survived the drought then died (Quensiere, 1979).

The sensitivity of the Chad ecosystem to minor climatic perturbations and its resulting instability is probably the main reason why this endorheic basin has almost no endemic species.

The Cladocera of Lake Chad feed basically on bacteria and algae of 4–30 μm size and detritus. *Anabaena flos-aquae* is consumed by *Tropodiaptomus incognitus* and *Thermocyclops neglectus*. The other copepods in their adult stages are essentially carnivorous.

From a trophic point of view, the fishes divide into three groups:
1. primary consumers feeding on algae (*Sarotherodon galilaeum*), on detritus (*Labeo senegalensis, Citharinus citharus*) and on macrophytes (*Alestes macrolepidotus*);
2. secondary consumers feeding on zooplankton (*Alestes baremose, Synodontis* sp.) and on benthos (*Synodontis schall, Heterotis niloticus*); and
3. terminal consumers feeding on fish (*Lates niloticus, Hydroxinus brevis*) and on aquatic and terrestrial insects.

Lauzanne (1975) has determined the quantitative pattern of energy transfer through the various levels of the food web. Globally, zooplankton and benthic production correspond to 6% and 1%, respectively, of the gross algal production.

THE VOLTA BASIN
General
The Volta system consists of four major headwaters which unite to form the Volta River. The whole Volta watershed has a surface area of 390 000 km² and extends over Dahomey, Ghana, Ivory Coast, Togo and Upper Volta. The main tributary is the Black Volta which arises at 300 m altitude in Upper Volta and drains an area of 150 000 km². The Oti River arises in Dahomey and drains 73 000 km². The Red and White Volta rise in Upper Volta (Welcomme, 1979).

The chemical features of the water reflect the sedimentary nature of the basin. Concentrations of calcium and sulphate are particularly high and strongly contrast with other African waters draining metamorphic Precambrian formations (Table 76).

Lake Volta
The Volta's River project materialized in 1962 with the beginning of the construction of the dam at Akosombo. The lake started to fill in 1964. In 1965, the Institute of Aquatic Biology was founded and, in 1968, the Volta Lake research project was launched by the Government of Ghana and by the UN Development Programme, in order to follow the

Table 76 *Chemical features of the Volta system after Blanc & Dajet (1957)*

	Black Volta	Red Volta	White Volta	Oti River
Na^+ (mg l^{-1})	3.9	8.9	8.9	3.2
K^+ (mg l^{-1})	0.25	2.9	2.2	0.2
Ca^{2+} (mg l^{-1})	380	220	280	300
Mg^{2+} (mg l^{-1})	11.9	13.4	12.4	11.6
CO_2 (mg l^{-1})	11.3	71.7	84	39
SO_4^{2-} (mg l^{-1})	72	120	108	96
SiO_2 (mg l^{-1})	125.5	148	150.6	132
Cl^- (mg l^{-1})	17.5	17.5	17.5	23.4
pH	6.4	6.5	7.6	6.4–6.7

physical, chemical and biological transformations occurring in Lake Volta, the largest man-made lake of Africa.

Lake Volta is located in Ghana, at some 100 km of the Volta mouth, at 6°15'–9°00' N and 1°00' W – 0°15' E. The lake level is at an altitude of 85 m above MSL.

The lake's main inlets are the Black Volta, the White Volta, the Oti and Afram Rivers. The Volta River system is ancient and flows on sedimentary rocks consisting of shales, mudstones, sandstones, arkoses, conglomerate, tillite and limestone (Obeng, 1973) covering an area of 400 000 km². The Volta basin is alternately under the influence of the harmattan, the northerly wind blowing from the Sahara desert, and the southerly monsoon. The boundary between these two systems reaches its northernmost position in August and its southernmost limit in January. In January–February, the climate, dominated by the harmattan, is dry and warm during the day and dry and cool during the night; in August, the monsoon brings prolonged rain. In intermediate periods, violent thunderstorms with heavy precipitation of short duration are frequent (De-Heer Amissah, 1969). The yearly amount of rainfall varies from 710 mm at Accra to 2180 mm in the Ahanta Nzima area. The temperature is high throughout the year. The mean monthly maxima range from 35°C in February to 29°C in July and the mean monthly minima from 22.8°C in February to 21°C in August.

From north to south the lake crosses the savanna belt and the tropical rain forest.

The main morphometric features of the lake are presented in Table 77. The very high value of the development of shoreline is caused by the dendritic form of the lake which can be divided into four main regions: 1) the main north–south water body; 2) the major 'arms' marking the lower reaches of the Volta River and its tributaries; 3) shallow littoral areas cleared of trees; and 4) the Gorge area from the Afram confluence to the dam.

The thermal structure and evolution have been studied by Biswas (1969*a*,

Table 77 *Morphometric features of Lake Volta*

Max. length	400 km	Max. depth	74 m
Max. width	24 km	Mean depth	19 m
Surface area	8300 km²	Volume	165×10^9 m³
Shoreline	4828 km		
Development of shoreline	15		

1973) and Viner (1970). The initial stage of filling was accompanied by a general cooling of the water. The characteristic temperature–oxygen profile which developed in November 1965 consisted of: 1) a superficial layer (0–5 m) with high diurnal variation; 2) an intermediate layer from 5 to 17 m, with small thermal variations; and 3) a deep layer almost without variation. A net thermocline and oxycline were present at 17–20 m; the difference in temperature in the thermocline was only 0.5°C but, between 27.5°C and 28°C, such a difference determined a significant density gradient. The Gorge area shows a certain thermal instability possibly caused by southerly winds funnelled in the Gorge and by internal seiches. Horizontal density flows have been observed in the Afram confluence and at Ajena and identified by their high oxygen values (Viner, 1969b). These density currents probably come from the cooling of superficial well-oxygenated waters and the subsequent sinking of the dense water. The Gorge area is sometimes nearly deoxygenated: the southerly winds blow the superficial well-oxygenated waters northwards causing a local upwelling of hypolimnic anoxic water. An inverse situation occurs during the 'harmattan' period when oxygenated waters from the lake accumulate in the Gorge and flush the deep anoxic water northward.

Biswas (1973) reports that in the Gorge, the temperature of the hypolimnion, which was 27°C in 1965 and 1966, rose to 28°C in 1967. Oxygen, which was completely absent from the hypolimnion in 1965–66, was at 15–20% saturation during the stratified period of 1967.

Viner (1970) underlines that Lake Volta is one of the rare tropical lakes of low altitude with a persistent thermocline. In fact, Biswas (1969b) observed that two periods of thermal instability occurred in 1966 and 1967. The first one in September was caused by the cool floods and the second one by the January–February harmattan. These two unstable periods were clearly observed in the southern part of the lake where warm and oxygenated water reaches the bottom. A similar pattern was observed in 1974 in the Afram arm, in the south-west part of the lake (Obeng-Asamoa, 1977).

The chemistry of the lake was investigated by Entz (1969), Viner (1970) and Obeng-Asamoa (1977).

Lake Volta water is a typical calcium–hydrogen carbonate water with low salt content (80–100 mg l^{-1}) (Table 78).

Viner (1970) showed that the electrical conductivity, which is around 60 μmhos cm^{-1} in the epilimnion, increases in the hypolimnion to about 100–120 μmhos cm^{-1}. Moreover, this author found that the hypolimnic conductivity was systematically higher at all measured stations in 1966.

The investigation of the nutrient chemistry in the Afram arm by Obeng-Asamoa (1977) revealed only traces of phosphorus and iron in the water column and concentrations of 0.02–0.04 mg l^{-1} ammonium at all depths, as well as levels of 0.01–0.03 mg l^{-1} nitrate. Conversely, silica was abundant and reached concentrations of 16–28 mg l^{-1}. A clear accumulation of CO_2 and NH_3 was noted in the hypolimnion between the unstable periods.

Biological features

At Ajena, the phytoplankton was examined weekly, almost without interruption, from May 1964 until March 1973 (Biswas, 1975). In 1965, the algal population was only one-quarter that of 1964 because of a drastic decrease in *Cryptomonas*. From 1965 to 1968, a constant increase in algal biomass was observed, followed by a steady decline from 1968 to 1973. From 1965 to 1966, the biomass doubled, mainly because of *Peridinium africanum*. A second doubling of phytoplankton in 1967 was due to *Synedra acus*, which was also responsible for a three-fold increase in 1969; the slight decrease in *Synedra* in 1969 initiated the phytoplankton decline in Lake Volta. *Lyngbya limnetica* replaced *Synedra* but the total population continued to decrease and reached its lowest in 1973. At Ajena, the algal density showed clear seasonal fluctuations including a peak in November and low values in August during the flood (period 1964–73).

In comparison with Ajena, the density of algae decreased progressively uplake, with the exception of *Anabaena* and *Melosira* which were more abundant in northern stations. It is interesting to note that *Melosira* was

Table 78 *Physico-chemical features of the Volta Lake water (Entz, 1969)*

Depth (m)	0	5	35
Temperature (°C)	30.3	28.5	27.2
pH	6.99	6.91	6.76
O_2 (%)	88.3	71.3	0.0
HCO_3^- (mg l^{-1})	65.6	66.4	73.5
CO_2 (mg l^{-1})	9.0	12.6	20.0
Ca^{2+} (mg l^{-1})	7.82	7.91	1.25
Mg^{2+} (mg l^{-1})	2.06	1.87	–
K^+ (mg l^{-1})	5.07	4.73	3.10
Na^+ (mg l^{-1})	3.76	3.49	3.15
Fe (mg l^{-1})	0.019	0.028	–
Cl^- (mg l^{-1})	1.6	1.6	–
SO_4^{2-} (mg l^{-1})	+	0	–
SiO_2 (mg l^{-1})	15.7	16.7	16.0
NH_3 (mg l^{-1})	0.01	0.02	1.01

The Volta basin

abundant before the impoundment. It seems that pre-dam conditions still prevailed in the upper reaches of the lake four years after the closure of the dam. In contrast, near the dam the blue-green algae, abundant in the pre-dam river, were replaced by diatoms and dinoflagellates during the few years after impoundment and reappeared after conditions stabilized.

Primary productivity was measured by Biswas (1978): it was minimum in the upper part of the lake (0.121 g C m^{-3} h^{-1}) and increased steadily towards the dam (0.478 g C m^{-3} h^{-1} at Ajena). Similar values were measured by Obeng-Asamoa (1977) in the Afram arm which supports a luxuriant growth of *Ceratophyllum demersum*. These results are in good agreement with previous measurements by Viner (1969*b*) who found production rates ranging from 0.8 to 5.2 g C m^{-2} d^{-1} in the different stations.

No outburst of aquatic weeds similar to those observed in Lake Kariba took place in Lake Volta; however, *Pistia stratiotes, Jussiaca repens, Ceratopteris cornuta, Nymphaea lotus* and other aquatic plants rapidly developed. *Pistia* was found to have the highest percentage of *Schistosoma*-carrying snails (Obeng, 1969).

The zooplankton population is especially abundant in the Afram arm and consists of rotifers (*Filinia, Asplanchna*), copepods and copepod nauplii, and *Moina* and *Bosmina* among the Cladocera (Rajagopal, 1969). Proszynska (1969) studied the zooplankton in 1964 and 1965 and observed several dramatic population crashes due probably to massive deaths of the plankton generally following the rains. In the Gorge area, the large and sudden variations in biomass and the absence of larval stages suggest that this area is very unfavourable to the crustacean plankton. The pesticides utilized for control of onchocerciasis ('river blindness' caused by the filarial worm *Onchocerca volvulus* and transmitted by flies belonging to the genus *Simulium*) in the Volta River basin may be related to these mass deaths, although Samman & Thomas (1978), studying the effect of aerial spraying of such pesticides in the White Volta basin, showed that the microcrustacean plankton was little affected.

With the closure of the dam, changes were observed in the fish population (Evans & Vanderpuye, 1973). Fish kills caused by the decrease in oxygen particularly affected *Chrysichthys* spp. Certain typically riverine species (*Labeo* and *Alestes nurse*) were reduced in the southern part of the lake and were replaced by *Tilapia galilaea*. The mormyrids, very abundant after the impoundment, disappeared nearly completely in a few years. The schilbeids *Eutropius niloticus* and *Physalia pellucida* expanded rapidly and became the dominant fish species in open water.

In 1970, the ichthyofauna included four main groups of fish. The most important were the semi-pelagic omnivores. This group was dominated by the clupeids *Cynothrissa mento* and *Pellonula afzeliusi* (75% of the total fish population by number and 20–25% by mass). This group also included the Characidae (*Alestes* spp.) and the previously mentioned Schilbeidae.

The second group comprised the herbivores feeding on aufwuchs and detritus. This group was dominated by *Tilapia* spp.; of secondary importance were *Citharinus, Labeo* and *Distichodus* spp. This group contributed 75% of the commercial catch but represented only 25% of the ichthyobiomass. The third group was composed of piscivores, mainly *Lates niloticus* and *Hydrocynus* spp. The fourth group consisted of benthic omnivores (*Chrysichthys* spp. and mormyrids). In 1969 the catch was 61 000 tons; it declined in following years to 35 000–40 000 tons (Welcomme, 1979).

As in Lake Kariba and Lake Kainji, a noticeable increase in piscivores took place following the expansion of the semi-pelagic fish.

The downstream freshwater clam (*Egeria radiata*) fisheries were affected by the reduction of the water flow in 1964–65. Since 1966, however, the continuous outflow from the dam has established suitable conditions for the development of the clam (Ewer, 1966).

Much attention has been devoted to the possible extension of tropical diseases. A survey by Paperna (1969) indicated that the bilharzia-bearing snail *Bulinus truncatus rohlfsi* had formed large colonies in certain habitats along the shores. The clearing of the forest and savanna vegetation enhances the development of aquatic plants which harbour the snails; since *Schistosoma haematobium* infections are common in fishermen's villages near the lake, there is little doubt that there will be a significant spread of the disease. An opposite trend seems to characterize the evolutions of onchocerciasis (Warmann, 1969) and trypanosomiasis (Kuzoe, 1973). In the first case, the submersion of the river rapids – main breeding sites of the fly *Simulium* – may be partly responsible for the observed decrease in fly density; it is however difficult to be certain since a heavy DDT treatment was applied in the lake area in 1962 and 1963. The tsetse fly *Glossina palpalis* is the main vector of trypanosomiasis in Ghana. Before the impoundment it was widespread in the Volta basin; it is now restricted to the southern parts of the lake since most breeding sites in the north have been drowned.

MADAGASCAR

It is generally accepted that the Mozambique Channel appeared in the Permian but did not definitively separate the Island of Madagascar from Africa. The Cretaceous was marked by intense rifting and volcanic eruptions but the presence of identical Cenozoic animals on both sides of the Channel leads to the assumption that a land connection was still active in the Upper Miocene. The isolation became complete some time in the Pliocene.

The flora of Madagascar reflects these successive steps of geological history. In pre-Upper Cretaceous time, when the island was connected with Africa and possibly with other Gondwanaland territories, Madagascar was covered by an ancient vegetation of pantropical affinities. This vegetation retreated under the pressure of the Angiosperms but it still represents 40% of the present flora and includes palaeoendemic families such as the Didiereaceae. The links with Africa in the Tertiary explain why about 27% of the Malagasy flora has African affinities (Koechlin, 1972).

The asymmetric geological structure of the island is reflected by the topography. The crystalline NE–SW backbone reaching nearly 3000 m in some places is limited in the east by the steep escarpment and the narrow eastern coastal plain, whereas the western side has a more gentle slope and is partly covered with sedimentary formations.

The island has several lakes of different origins: two large tectonic lakes, Lake Alaotra (220 km^2) in the central eastern zone and Lake Loza (156 km^2) on the north-west coast; and flood plain lakes, Lake Kinkony (139 km^2) in the western area, Lake Ihotry Morombe, a closed saline lake with a surface varying from 8 to 85 km^2, and Lake Tsimananpetsotsa (16–29 km^2), a very saline lake rich in calcium sulphate and magnesium sulphate and lacking any fish fauna.

Lake Ihotry has been studied by Mars & Richard-Vindard (1972). Located at 50 m altitude in the south-west areas of the island, the lake is separated from the sea coast by 30 km of ancient dunes. It is composed of two parts: a small permanent water body in the western area and a large eastern region inundated only at the end of the wet period. The permanent lake is fed by springs and by runoff water. During the dry period, the lake water becomes hypersaline but water holes in the underlying aquifer supply potable water.

The water chemistry of the lake shows very wide fluctuations. The chloride concentration reaches 21 g l^{-1} at low level but decreases down to 1.3 g l^{-1} at maximal level (Table 79). The salinity of Lake Ihotry is not caused by intrusion of sea water but results from the excess of evaporation losses over the water incomes.

These extreme variations of concentration considerably limit the number of fish species. Petit (1930) described the presence of the cichlid *Ptychochromis grandidieri* (*P. oligacanthus*) in Lake Ihotry but, in 1937, this species disappeared consequent to a dry period during which the lake water was more saline than the sea water.

In the 1950s, another cichlid *Tilapia melanopleura* (*T. rendalli*) was introduced into the lake and has survived since then in spite of periods of very high salinity (38‰ in August 1962 and 45‰ in July 1966). This species, however, does not survive in other biotopes with salinity exceeding 20‰. Mars & Richard-Vindard think that the relatively high level of calcium found in Lake Ihotry favours osmoregulation.

Many small crater lakes can be found in the volcanic zone. Lake Itasy (35 km²) was formed by a lava bar. Finally two artificial reservoirs should be mentioned. They are located in the central highlands east of Tananarive: Tsiazompaniri and Montasoa (32 and 20 km², respectively) (Kiener & Richard-Vindard, 1972). Brackish littoral lagoons and lakes are numerous along the east coast.

The fish fauna of these water bodies is characterized by the rarity of true freshwater fishes. Most of the species are euryhaline animals which have secondarily colonized continental waters. The Cichlidae and Mugilidae are the most important families. Then come the Cyprinidae which are represented by two species introduced into the island (*Carassius auratus* and *Cyprinus carpio*). The Anguillidae with four species are widely distributed in the island with the exception of the crater lake Tritriva, near Antsirabe. They reach 2 m long and 20 kg. Bagridae (5 species), Channidae with the euryhaline milk-fish (*Chanos chanos*), Gobiidae and Eleotridae represent the additional main families. *Salmo iridens*, the rainbow trout, and *Salmo trutta* were introduced in the 1920s and became acclimatized in high-altitude streams. The fish fauna includes some 230 euryhaline species belonging to 62 families and 32 endemic freshwater species belonging to eight families (Kiener & Richard-Vindard, 1972).

Bourrelly (1964) studied the phytoplankton of 44 samples from streams and rivers in east-central and southern Madagascar. Besides the diatoms, 82 species of algae were found. The Chlorophyta were most abundant (37

Table 79 *Chemical composition of Lake Ihotry water (Mars & Richard-Vindard, 1972) (results in g l⁻¹)*

	Ca^{2+}	Mg^{2+}	Na^+	K^+	Cl^-	SO_4^{2-}	HCO_3^{2-}
Maximum	1.90	1.25	12.90	–	21.0	7.90	0.11
Minimum	0.23	0.09	0.98	0.04	1.46	0.77	0.13

Madagascar

species), then the Cyanophyta (29 spp.) and the Rhodophyta (six species). Rapid waters are populated with *Nostoc verrucosum, Phormidium retzii, Nostochopsis lobatus, Cloniophora spicata, Sirodotia* sp., *Batrachospermum* sp. and *Tuomeya* sp. *Caloglossa* sp. inhabit the river mouths. Six among the rheophylic algae belong to the pantropical element. Slow-velocity streams are colonized by cosmopolitan Cyanophyta whereas Desmidiales and Chroococcales are common in outlets draining lakes, ponds and bogs. Other studies of freshwater plankton were carried out by Ramanankasina (1969, 1978) on a river east of Tananarive.

7
Middle East

SINAI PENINSULA
Bardawil Lagoon

The Bardawil Lagoon is located on the northern coast of Sinai at 31°02′–31°14′ N and 32°40′–33°35′ E. It is 78 km long and has a maximum width of 21 km; its surface area is ~650 km². The lagoon is separated from the sea by a sand bar a few metres high and a few hundred metres wide. Water depth does not exceed 3 m. In the past, the lagoon communicated with the sea through three openings, but in 1969 the middle one was obstructed by sand.

Salinity in the lagoon has periodical seasonal variations of about 15–20‰; they are caused by winter dilution of sea water when storm waves overspill the sand bar and by summer evaporation. At any given period, salinity shows an increasing gradient from north to south, that is with distance from the sea channels; Krumgalz, Hornung & Oren (1980) report values of 73.018‰ and 39.09‰ at a southern and northern station, respectively, in a 1967 survey. Salinity is also considerably affected by changes in hydrological conditions. In 1970–71, the silting of the channels caused a rapid increase in salinity which reached 75–90‰. After the cleaning of the eastern channel in August 1971, the salinity dropped back to 44–53‰ (Rot, 1972).

Various chemical parameters have been studied by Rot (1972) and Krumgalz et al. (1980).

pH varies very little (8.00–8.20) and the water is always well oxygenated and even oversaturated with oxygen. Concentrations of PO_4–P fluctuate from 0 to 12 μg l^{-1} and levels of total phosphorus from 15 to 38 μg l^{-1}. NO_3–N shows large seasonal variations (4.4–38.4 μg l^{-1}). The concentrations of SiO_3–Si range generally from 100 to 600 μg l^{-1} but may, exceptionally, exceed $1000 \mu g \, l^{-1}$. These nutrient levels are very similar to those of the Mediterranean Sea water.

Krumgalz et al. (1980) examined the question of the origin of the lagoon water. The constancy of the magnesium:salinity and calcium:salinity ratios and their identical values to those of the Mediterranean Sea water indicate that the brines of the Bardawil Lagoon originated from evaporation of sea water without mixing of non-marine brines. This was confirmed by laboratory experiments: during evaporation of sea water, the pattern of variation of the concentration of the calcium ion was identical to the behaviour of this ion in the lagoon.

Biological features

A distinctive feature of the Bardawil Lagoon is an intensive growth of the macrophyte *Ruppia maritima* on the bottom, between 0.5 to 1.5 m deep. A thick pellicule of epiphytic filamentous algae covers the growths of *Ruppia* (Por, 1969a).

A plankton survey, based on three collections (July and October 1967 and January 1969), indicated that the algal community is mainly composed of diatoms and dinoflagellates. Among the diatoms, *Rhabdonema adriaticum, Striatella unipunctata, Licmophora flabellata, Synedra hennedyana, S. gailloni, S. undulata* and *Campylostylus striatus* were particularly abundant. The dinoflagellates were dominated by *Ceratium furca* and *C. fusus. C. furca* is represented by a small-diameter form, characteristic of tropical environments. The Cladocera were represented only by *Bosmina coregoni maritima*, and this was 'the first time that this species was recorded in a marine or hypersaline environment in this area' (Kimor & Berdugo, 1969). We note that the genus *Bosmina* is well represented in Lake Kinneret.

The benthic diatoms were studied by Ehrlich (1975) who found a total of 147 taxa belonging to 140 species and 45 genera. The diversity is maximal near the inlets and decreases towards the inner and more saline part of the lagoon.

Four main groups of benthic diatoms are found. The species of the first group are widely distributed in the lagoon and comprise *Cocconeis bardawilensis, Amphora coffeaeformis, Nitzschia* sp. and *Mastogloia* sp. The members of the second group are found mostly in the northern and central part of the lagoon. *Nitzchia fusoides, N. granulata, Synedra tabulata* and *S. laevigata* are the most abundant. The species of group three are essentially found near the inlets, while the diatoms of group four belong to freshwater species and are mostly found along the northern sand barrier.

There is no hyperhaline species restricted to the most saline areas of the lagoon. New species have been described, such as *Cocconeis bardawilensis*, *Mastogloia sirbonensis*, *M. smithii* var. *heteroloculata* and *Nitzschia fusoides*.

The zoobenthos, studied by Por (1969a), consists of large populations of *Cardium edule* and the snails *Pirenella conica* and *P. caillaudi* in shallow water. The microbenthos includes portunide crabs (*Charybdis* sp.), penaeid shrimps, *Balanus amphitrite* and *Mytilus variabilis*. The meiobenthos is formed by *Heterolaophonte sigmoides*, *Robertsonia salsa*, *R. knoxi*, *Longipedia* and *Neocyclops salinarum*. The Ostracoda are mostly represented by *Cyprideis torosa* and *Aglaiocypris* sp.

A list of 41 species of fish found in Bardawil was established by Ben Tuvia & Herman (1972). The most common commercial fishes are the Mugilidae with *Mugil cephalus*, *Liza ramada*, *L. aurata*, *L. saliens*, *L. carinata* and *L. provensalis*, the Sparidae with *Sparus auratus*, and the Serranidae with *Dicentrarchus labrax* which feeds on small fishes and shrimps. Three other species are common: *Aphanius dispar*, *Atherina mochon* and *Pomatoschistus marmoratus*.

Among the 41 species observed, 23 belong to the Atlantico-Mediterranean group and 13 to the Red Sea fauna. This latter group, however, represents 30% of the Bardawil fauna while it represents only 11% of the fish fauna of the Mediterranean coast.

The Bitter Lakes

The Bitter Lake, divided into the Great and Little Bitter Lakes, is an old basin of the Red Sea which became hypersaline after its separation from the Gulf of Suez. With the opening of the Suez Canal in 1869, the ancient fauna of the lakes was mixed with the marine fauna of the Mediterranean and the Red Sea. Stratification developed with a deep saline layer overlaid by a sea-water layer. The slow mixing which took place between both layers caused a progressive decrease in salinity of the deep layer from 68‰ in 1869 to 53‰ in 1924 (Por, 1972). In 1960 the salinity of the whole water column fluctuated from 45 to 47.5‰, making the mixing possible.

Solar Lake

Solar Lake is a small pond (130 × 60 m), 5 m deep, located on the Sinai coast of the Gulf of Eilat at about 50 m from the sea. It was discovered in 1967 and described as being the site of heliothermal processes, which explains why its deep brines have a temperature of 50°C (Por, 1969b).

The pond occupies a depression in the igneous rocks which border the coast of the Gulf. The depression is surrounded by high cliffs except on the eastern side towards the sea from which it is separated by a gravel bar. The water level is 0.95 m above sea level at the end of the winter and 1.25 m below sea level in late summer.

In summer, the upper layer has a temperature of 32°C whereas the lower brines reach 54°C. The upper layer is submitted to intense evaporation and reaches a salinity twice that of sea water, while the deep layer has a chloride concentration of 90 g l^{-1}. The stability of the system is primarily due to the sheltered location of the pond and to the resulting chemical stratification. The deep brines, continuously heated by solar radiation and well isolated from the atmosphere by the upper lighter brines, have little heat losses until evaporation thins the protective upper layer. In September the profile is homothermal around 30°C and has a chloride level of 90 g l^{-1}. In winter, sea water spills over the gravel bar or seeps through the gravel bar and replenishes the pond. In December the surface temperature is around 20°C; the chloride concentration is ~ 34 g l^{-1} in the upper layers in comparison with 175 g l^{-1} in the deep water. The monimolimnion is permanently saturated with $CaSO_4$, and precipitation has caused the formation of a hard layer of gypsum on the pond bottom and slopes. The pH varies from 8.2–8.5 in upper layers to 7.1 at the bottom, where hydrogen sulphide reaches a concentration of 10 mg l^{-1} (Eckstein, 1970). A more detailed physico-chemical survey of the lake, covering a full year, has been carried out by Cohen, Krumbein, Goldberg & Shilo (1977*a*). It confirms that Solar Lake is of monomictic type with a very stable winter stratification, where thermal gradients reach 18°C m^{-1} and a summer homothermy and overturn.

The distribution and production of photosynthetic organisms have been studied by Cohen, Krumbein & Shilo (1977*a*), Hirsch (1978) and Potts (1979, 1980). During holomixis the photosynthetic organisms were evenly distributed in the water column and dominated by the blue-green alga *Aphanothece halophytica*, but *A. littoralis*, *Oscillatoria salina*, *Microcoleus* sp., *Spirulina labyrinthiformis* and *Spirulina* sp. were also present, together with the diatoms *Nitzschia thermalis*, *Amphora coffeaeformis* and *Navicula* sp. During stratification, the previously mentioned algae prevailed in the epilimnion. In the metalimnion a dense population of the sulphur bacterium *Chromatium violescens* developed whereas the upper hypolimnion was dominated by the green sulphur bacterium *Prosthecochloris* sp. Cyanophyta were in the hypolimnion; their density increased towards the bottom where they formed a benthic mat which received only 0.5% of the surface light intensity. During the mixed period, primary productivity was uniform with depth and did not exceed 100 mg C m^{-3} d^{-1}. During stratification, the peak of productivity was observed in the metalimnion and hypolimnion where values of 4960 mg C m^{-3} d^{-1} were measured while epilimnic production was very low. Solar Lake has a productivity pattern which contrasts very much with that of other monomictic lakes: the bulk of carbon fixation is done by prokaryotes in an anoxic, H_2S-bearing hypolimnion at temperatures ranging from 38 to 48°C. The annual productivity is 59 g m^{-2}.

Photosynthetic and non-photosynthetic bacteria were studied by Cohen, Krumbein & Shilo (1977b). Six different bacterial communities were observed during stratification. A maximal value of 1014 mg C m^{-3} d^{-1} was found for dark CO_2 incorporation; assuming that the CO_2 incorporation accounts for 6% of total bacterial carbon production, this latter parameter might be as high as 16 900 mg C m^{-3} d^{-1}, which is approximately three times more than the maximal photosynthetic carbon fixation.

Krumbein, Cohen & Shilo (1977) studied the bottom algal mat. The dominant algal communities at the surface sediments have been studied by Potts (1980). The following blue-green algae have been recorded in the shallow water mats: *Aphanocapsa* sp., *A. concharum, Aphanothece stagnina, Aphanothece* sp., *A. pallida, Dactylococcopsis* sp., *D. acicularis, Gloeothece, Johanesbaptistia* sp., *Lyngbya confervoides, Oscillatoria redekei* and *Pseudoanabaena catenata*. The primary productivity of the mat ranges from 5000 to 10 000 mg C m^{-2} d^{-1}. The accretion rates of the organic material to the bottom range from 0.5 to 5 mm y^{-1} only, since 99% of the organic production is remineralized. Cores, taken in the sediments, permitted the determination of the history of Solar Lake with precise dating based on counting of laminites and ^{14}C age determinations. The pond was originally a bay which was closed at about 3400 B.P. The intense evaporation caused the precipitation of carbonate and gypsum. At about 2490 B.P. a cyanobacterial mat replaced the chemical precipitation. At about 1935 B.P. a deepening of the central part of the lake occurred, transforming the shallow and uniform algal mat into a mat presenting characteristic variations with depth, as can be presently observed.

Wilbert & Kahan (1981) found 16 species of ciliates in the benthic algal mat located above the thermocline. Some of them are bacteriophagous, others are algae eaters and others are predators on other ciliates. A new species, *Condylostoma reichi*, and a new genus and species, *Uronychia transfuga*, have been described. The rest of the fauna includes 14 species (Zalcman & Por, 1975): six species of Insecta, one species of Platyhelminthes, one species of Nematoda, one species of Arachnida, and five species of Crustacea. This latter group includes *Artemia salina* living in epilimnic water, the copepods *Robertsonia salsa* and *Nitocra lacustris*, the Ostracoda *Cyprideis littoralis* and the Isopoda *Halophiloscia* sp. Most animals are herbivorous and feed on algae. *Artemia* feeds on bacteria whereas a few coleopterans are carnivorous and the spiders are scavengers. Most animals reproduce in winter and spread on the whole bottom. During the summer stratified period, benthic life is limited to the upper oxygenated sediments.

THE JORDAN VALLEY
General

The Jordan Valley originates in the Hermon mountains at 1500 m altitude and ends 200 km southward in the Dead Sea at 398 m below MSL. It occupies a tectonic depression, the history of which goes back to the Cretaceous when the Arabian plate started to move northward with respect to Africa. The northward displacement of Arabia and of Transjordania with respect to Cisjordania created the fracture line of the Syro-African rift which originates north of the Hula Valley and ends in the Zambezi area. The rift is presently occupied by various geographical units: the Jordan Valley, the Red Sea and the East-African depression. In addition to the lateral slide-rule displacement, vertical movements occurred but their intensity remained limited until the Pleistocene when the present graben was formed. In the Jordan Valley, the downfaulted area became the base level of drainage water and was nearly permanently occupied by lakes. From 70 000 to 20 000 B.P., the salty Lisan Lake covered the Jordan Valley from Hazeva, south of the Dead Sea, to Tiberias in the north. In this lake, evaporites, clays and marls were deposited. In the late Pleistocene, the level of Lake Lisan dropped down to a level lower than that of the present Dead Sea. Vertical movements probably downfaulted the Dead Sea area which became separated from the Kinneret basin, and the present hydrographic network came into existence: Lake Kinneret in the north, drained by the Jordan River which became the main inlet of the endorheic Dead Sea.

From a hydrological point of view, the Jordan River results from the junction of the Snir (Hatsbani) River, the Dan River and the Hermon (Banias) River (Fig. 25). Formerly the Jordan River used to flow into the swamp area of Lake Hula. In 1957, the artificial drainage of the Hula plain was completed and the river flow regulated. It now flows into two canals (western and eastern canals) which drain the plain on both sides and join into a single canal south of the Hula plain. Further south, this canal links

Fig. 25. The Jordan River and Arava Valley system

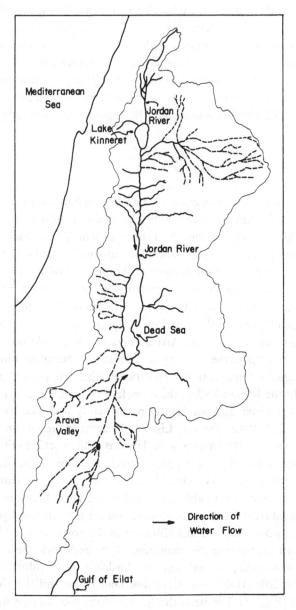

with the natural river bed of the river which digs its way through the basaltic massif of Korazim and reaches Lake Kinneret at 210 m below MSL. From Lake Kinneret, the Jordan river meanders among the Lisan sediments of the Jordan Valley before reaching the Dead Sea.

Lake Kinneret

Lake Kinneret is located at 210 m below MSL at 32°45′–32°53′ N and 35°30′–38′ E. It has a surface area of 168 km² and a volume of 4300×10^6 m³. Its mean and maximal depths are, respectively, 25 and 43 m. It is the only large natural freshwater lake of the Middle East.

In 1964, Lake Kinneret became the main reservoir of the National Water Carrier. An approximate amount of 400×10^6 m³ water is yearly pumped out of the lake and injected into the Carrier for water-supply purposes. Since the lake contributes 25% of the total needs of Israel, its water quality became a question of national importance. That is why the lake is permanently monitored by the Kinneret Limnological Laboratory, and concrete measures aimed at improving water quality are implemented by the Kinneret Water Authority, a local branch of the Water Commission. These circumstances have led to a thorough study of the lake which has been carried out on the basis of a 10-year master plan. The results of the investigation on Lake Kinneret have been reported in the Lake Kinneret Monograph (Serruya, 1978*c*).

Lake Kinneret is a warm monomictic lake with a winter holomixis at 13–16°C and a stratification period extending from June to December. Then the epilimnion reaches 28–30°C whereas the hypolimnion remains near to the winter temperature. The thermal pattern and its variations in time and space have been described by Serruya (1975). Very clear internal waves (seiches) have been observed from temperature data recorded by automatic instruments. For wind intensity greater than 12 m s⁻¹, the amplitude of the seiche is 5–10 m in peripheral stations; for winds lower than 12 m s⁻¹ the amplitude is only 3–4 m. In any case, the central part of the lake, located on the nodal point, is not affected by the seiche. The seiche has a period of 24 h. The characteristics of the system are such that, in the evening after the cessation of the wind, warm water is piling up in the south-west corner of the lake, 21 h later the accumulation of warm water takes place in the north-eastern part of the lake, and another 12 h later the situation has returned to its starting point. The direction and velocity of the seiche wave have been followed in the various stations around the lake: it moves counterclockwise at about 0.5 m s⁻¹, that is 2–3 times more rapidly than the water flow.

The general, wind-induced circulation of the upper layers is counter-clockwise, whatever the season or wind direction. The maximal speed recorded was 20.2 cm s^{-1} (hourly mean). The accumulation of upper water in one corner of the lake generates a deep return current of opposite direction to the current of the upper layers (S. Serruya, 1975).

The heat budget of the lake has been found to be 40 000 and 47 240 Kcal cm^{-2} y^{-1} by Stanhill & Neumann (1978) for the years 1965 and 1966, respectively.

The Jordan River draining the northern watershed is the main inlet of the lake. Smaller water supplies are ponded by the eastern and western watershed of the lake. The lake is dammed at its southern end and only small amounts of water are released through this outlet. The pumping station of the National Water Carrier is presently the main outflow of the lake. The hydrological balance of the lake is shown in Table 80.

The waters of the Jordan River belong to the calcium–hydrogen-carbonate type and have a chloride content of 20 mg l^{-1}, whereas the sublacustrine thermo-mineral springs are of the sodium-chloride type with chloride levels of 1800 to 2500 mg l^{-1}. The chloride supply of the Jordan River (11 000 tons y^{-1}) is approximately seven times smaller than that of the sublacustrine thermo-mineral springs (70 000 tons y^{-1}). The thermo-mineral springs of the western coast, which flowed into the lake in the past, were diverted in 1965, and the removal of pure water by evaporation has increased the salt concentration of the water. These various factors explain that the lake water belongs to the sodium-chloride type and has a chloride content of 250 mg l^{-1}.

The chemical features of the lake and Jordan River water are shown in Table 81. The hypolimnion of the lake is anoxic from May to December and sulphides reach a concentration of 6 mg l^{-1}.

Table 80 *Hydrological balance of Lake Kinneret (Mero in Serruya, 1978)*

Total net Jordan River inflow	558 × 10^6 m^3 (66.4%)	
Total net inflow from eastern and western watershed	122 × 10^6 m^3 (14.5%)	
Direct rainfall	70 × 10^6 m^3 (8.3%)	
Thermo-mineral springs on western coast	40 × 10^6 m^3 (4.8%)	
Sublacustrine thermo-mineral springs	50 × 10^6 m^3 (6.0%)	
TOTAL INFLOW		840 × 10^6 m^3
Evaporation	295 × 10^6 m^3	
Lake outflows (pumping + southern outlet)	545 × 10^6 m^3	
TOTAL OUTFLOW		840 × 10^6 m^3

Table 81 Chemical features of the Jordan River and Lake Kinneret water

	pH	Ca^{2+} (mg l^{-1})	Mg^{2+} (mg l^{-1})	Na^+ (mg l^{-1})	K^+ (mg l^{-1})	Cl^- (mg l^{-1})	SO_4^{2-} (mg l^{-1})	HCO_3^- (mg l^{-1})	Conductivity (μmhos cm^{-1})
Jordan River		56	11	15	1.6	20	30	98	
Lake Kinneret	8.1	40	33	125	5	250	48	77	1100

The level of nutrients is relatively low: total phosphorus rarely exceeds 20 μg l^{-1} in the epilimnion and PO$_4$–P is rarely higher than 10 μg l^{-1}. In spite of the long anoxic period, no massive increase in phosphate is observed in the hypolimnion as in many productive lakes. This is due to the fact that in Lake Kinneret the phosphorus deposited in the sediments is bound to calcium rather than to iron. In other words, a decrease in redox does not cause any significant release of phosphorus as would a decrease of pH. However, the water is well buffered and even at 40 m, pH rarely drops below 7.4. Any process which would destabilize the buffer system and cause a decrease in pH would release large amounts of phosphorus: the total amount of phosphorus varies from 50 to 100 tons in the lake water but amounts to 800 tons in the upper centimetre of the sediments. The levels of nitrate peak generally at 300 μg l^{-1}, and concentrations of ammonia are low in the epilimnion. In summer, ammonia levels increase in the hypolimnion and exceed 1 mg l^{-1}. About 60% of the nitrogen income of the lake is denitrified as demonstrated by nitrogen balances (Serruya, 1975) and microbiological research (Cavari, 1977). Denitrification and precipitation of phosphorus in association with calcium are the main mechanisms which keep the nutrients at low levels.

In spite of this self-controlled nutrient chemistry, Lake Kinneret has a yearly dense bloom of the dinoflagellate *Peridinium cinctum* fa *westii* which extends from February to June. The bloom has been studied by Pollingher (1978*b*), Berman & Pollingher (1974), Pollingher & Berman (1977), and Pollingher & Serruya (1976).

At the peak of the bloom, the algal biomass reaches very high values (300 g wm m^{-2}). However, *Peridinium* is a large (42–66 μm diameter) alga, dividing once every three days in optimal conditions. How, then, does a slow-growing species succeed in developing a nearly monoalgal bloom in a medium relatively poor in nutrients? The answer to this question lies in the complex life cycle of the alga which includes a six-month benthic stage, during which the *Peridinium* resting form (cyst) is in the sediments, and a six-month vegetative phase in the water. The mechanism which triggers the excystation process is not well known but resuspension of cysts and first-generation vegetative cells in shallow sediments by autumn storms always precedes the bloom. The new cells, excysted in or near the sediments, have very high levels of internal nitrogen and phosphorus (Serruya & Berman, 1975; Wynne, 1977). This initial storage confers to the *Peridinium* cells a relative independence from the medium and an enormous ecological advantage. The mobility of the vegetative cell, enlarging its field of nutrient supply, is a second non-negligible advantage over passive algal species. *Peridinium* dominance in Lake Kinneret is probably the result of a long

The Jordan Valley

process of natural selection. This explains why the *Peridinium* life cycle and the behaviour of its vegetative cells seem so closely 'adapted' to the meteorological–physico-chemical cycle of the lake.

Peridinium dominance does not exclude the development of many other algal species. Pollingher (1981) has numbered 222 species and distinguished four succession stages. The first stage starts in November with the destratification process and its associated injection of nutrients; small species belonging to the genera *Erkenia, Rhodomonas, Cryptomonas, Cyclotella, Crucigenia, Chodatella* and *Tetraedron* dominate and continue to increase during the second stage, which corresponds to the mixed period. During this period, the small-size algae reach their maximum and are accompanied by larger ones (*Melosira* spp., *Closterium aciculare* var. *subpronum, Coelastrum* spp., *Pediastrum* spp., *Microcystis* spp.) and by young cells of *Peridinium* emerging from the cyst. The third stage (March–June) is dominated by *Peridinium*. The bloom phenomenon occurs as a result of the sudden rise in the daily division rate of *Peridinium* cells, from 5–10% to 40% of the population, which occurs during the March windless period (Serruya, Serruya & Pollingher, 1978). *Peridinium* is accompanied by other dinoflagellates: *Glenodinium gymnodinium, Peridinium cunningtonii, P. inconspicuum* and *Ceratium hirundinella*. The end of the bloom, in June, coincides with stratification, nutrient depletion and high temperature. The fourth stage in the succession corresponds then to the summer and is dominated by species able to grow at low nutrient levels and high temperatures: coenobial blue-greens (*Microcystis* spp., *Radiocystis geminata*), coenobial green algae (*Pediastrum, Coelastrum, Scenedesmus*) and nanoplanktonic forms of *Tetraedron, Cosmarium laeve, Cyclotella* and *Chroococcus*. Most of the coenobial forms present in this period are the same ones which are able to take up organic compounds under experimental conditions (Pollingher & Berman, 1976). In the lake these species thrive probably on the decomposition products of the bloom. The complete list of algal species in the lake can be found in Pollingher (1978b).

The primary productivity of the lake has been investigated by Berman (1978). The average value for the period 1969–75 was 1700 mg C m^{-2} d^{-1}. Monthly averages ranged from 968 mg C m^{-2} d^{-1} (December) to 2419 mg C m^{-2} d^{-1} (April). The annual production is about 635 g C m^{-2}. Although the peak of primary production occurs during the *Peridinium* period, the nanoplanktonic species are far more active producers than *Peridinium*; in January–March 1972, the nanoplankton, representing 15% of the algal biomass, was responsible for 50% of the autotrophic carbon fixation during this period (Serruya *et al.*, 1980).

Peridinium is not grazed by zooplankton (Gophen, 1972). The herbivorous zooplankton, composed mainly of *Diaphanosoma brachyurum, Bosmina longirostris, B. longirostris* var. *cornuta, Ceriodaphnia reticulata* and *C. rigaudi*, feed mainly on nanoplanktonic algae as shown by the analysis of their gut contents (Serruya *et al.*, 1980). In contrast, the adult stages of *Mesocyclops leuckarti*, the dominant copepod, prey on the Cladocera *Ceriodaphnia* and *Diaphanosoma*. The complete list of zooplankton can be found in Gophen (1978).

The benthic fauna is described in detail in Serruya (1978a,b). The benthic copepods are of special interest. They include euryhaline species of harpacticoids and cyclopoids, but the widespread freshwater harpacticoids (Canthocamphidae) are lacking. Por (1968) considers *Nannopus palustris, Pseudobradya barroisi, N. palustris tiberiadis* and *Nitocra incerta* as the oldest forms of copepods in the lake, relicts of the Pliocene sea. The endemic blind prawn *Typhlocaris galilea* (Tsurnamal, 1978) is also a relict of the last marine transgression, three million years ago. The present fauna of Mollusca is composed of Bivalvia (*Unio semirugatus, U. terminalis* and *Corbicula fluminalis*) and Prosobranchia (*Theodoxus jordani, Pyrgula barroisi, Bithynia hawaderiana, Melanopsis praeniorsum* and *Melanoides tuberculata*); Pulmonata are completely lacking. A comparison with the fauna of Mollusca found in sediments of Early and Middle Pleistocene (Tchernov, 1975) shows that the lake fauna is an impoverished version of the Pleistocene fauna. In particular, in the Middle Pleistocene, the Jordan Valley was occupied by a freshwater lake (Lake Ubediyya) when Pulmonata were abundant. The post-Mendel development of the Lisan Lake, a salty water body, eliminated a large part of the freshwater fauna. The Lisan stage ended some 20 000 years ago and the salty lake was replaced by the present freshwater lake which was recolonized by the species that survived the Lisan stages in freshwater refuges. Conversely, a few marine species remained in salty niches within the freshwater lake; the brackish-water harpacticoids can survive on the lake sediments which are constantly fed by a saline seepage.

The fish fauna of the lake includes 24 species belonging to 20 genera and eight families. The families can be divided as follows: 1) primary freshwater fishes (Cyprinidae, Cobitidae, Clariidae); 2) secondary freshwater fishes (Cyprinodontidae, Poecilidae, Cichlidae, Blenniidae); and 3) euryhaline fishes introduced by man (Mugilidae) (Ben Tuvia, 1978). Cyprinids (ten species) and cichlids (seven species) are the dominant groups. Most of the lake species are benthic animals living in shallow water. However, the few pelagic species such as the endemic cyprinid *Mirogrex terraesanctae* contribute the bulk of the fish biomass ($\sim 75\%$, Serruya *et al.*, 1980).

The zoogeographical affinities are diverse: the cyprinodontid *Aphanius mento* is a relict of the last marine transgression. All the cichlids are of Ethiopian origin and include four endemic species (*Tristramella simonis simonis, T. simonis intermedia, T. sacra* and *Haplochromis flaviijosephi*). The other three species (*Sarotherodon galilaeus, S. aureus* and *Tilapia zillii*) are known in Israel, Jordan and Africa. The cyprinid *Tor canis* and the clarid *Clarias lazera* are also of Ethiopian origin. Most of the other species have Palaearctic affinities (Ben Tuvia, 1978).

Sarotherodon galilaeus feeds selectively on *Peridinium* (Spataru, 1976) whereas *Mirogrex terraesanctae* feeds essentially on zooplankton. *S. aureus*, a species from Lake Hula artificially introduced into the lake, has a mixed diet with an important zooplankton component (Spataru & Zorn, 1978). Moreover, nearly all the juvenile fishes feed on zooplankton. It follows that at least 80% of the fish biomass feeds on zooplankton, sometimes causing a dangerous pressure resulting in a severe diminution in zooplankton biomass and consequent summer blooms of small algae ordinarily grazed by zooplankton.

The carbon flow of the lake has been studied in detail (Serruya *et al.*, 1980).

The Dead Sea

The Dead Sea is a hypersaline lake, located in the Jordan Valley at nearly equal distance from Lake Kinneret and the Gulf of Eilat, between 31°04′ and 31°47′ N (Fig. 25).

The extreme salinity of the water and the desolate landscape which surrounds the lake conferred upon the Dead Sea a reputation of aridity and hopeless sterility, transmitted through the centuries by the travellers who visited the area (Nissenbaum, 1979). The modern studies of the Dead Sea started with the chemical analyses of the Dead Sea water by Lavoisier, Gay-Lussac, Klaproth and Gmelin (Nissenbaum, 1970). Then came, in 1848, the bathymetric survey of Lynch (1849), followed by the chemical survey of Lartet (1877). The industrial exploitation of the lake chemicals started in 1930 and was intensified by the Dead Sea Works Ltd after 1948. A comprehensive study of the Dead Sea was published in 1967 by Neev & Emery. The stratigraphic investigation permitted the determination of the climatological modifications which took place from the Upper Pleistocene until the present.

From 100 000 to 70 000 B.P. the climate was dry and salt was deposited in the Dead Sea basin. From 70 000 to 20 000 B.P. the climate was more humid and the deposits consisted of aragonite, calcite and gypsum (Lisan facies).

The last 20 000 years were characterized by a new dry episode with salt deposits interrupted by a few humid phases. During this period the water level varied by hundreds of metres; from 1800 until now it has varied by nearly 10 m (Klein, 1961).

At the beginning of the twentieth century the lake level was at −392 m and the lake comprised two basins, a deep Northern Basin separated from a shallow southern one by the Straits of Tisan located at −403 m. In recent years, the irrigation projects of the Jordan Valley utilized a large part of the water of the Jordan River, decreasing significantly the freshwater income of the Dead Sea. Finally, the level of the lake dropped below −403 m, the two basins became separated in 1976 and the water surface area decreased from 1000 to 750 km². In addition, water was pumped from the Northern Basin into the Southern Basin, which played the role of an evaporation pond for mineral extraction. The lake responded to these recent developments as to climatic modifications (Gat & Stiller, 1981).

Physical features of the Dead Sea
Before the lowering of the water level. From 1900 to 1930, the water level was at about −392 m; it dropped to about −395 m from 1930 to 1950 and reached −403 m in the 1970s. The Northern Basin reaches a depth of 330 m so that the floor of the Dead Sea is at approximately −730 m. In contrast, the Southern Basin was only a few metres deep (Hall & Neev, 1975).

For a level of −398 m, the lake has an area of 940 km² and a volume of 136×10^9 m³. It is 80 km long in the north–south axis and 17 km wide.

From the hydrological point of view, the Jordan River is the main inlet of the Dead Sea, which also receives minor contributions from freshwater springs on the eastern shore and runoff in winter. It also receives small amounts of salty water from shore seepage.

The surface temperature of the Dead Sea varies from 34–36°C in August to 19–23°C in December–January. The South Basin is colder in winter and warmer in summer than the North Basin. It follows that in any season the water of the South Basin is denser than that of the North Basin, in winter because of relatively low temperature, in summer because of high salinity.

The data on temperature, density and salinity of the water allow us to distinguish two main water masses. From 0 to 40 m, the Upper Water Mass represents the 'seasonal lake' with an epilimnion (surface member) and hypolimnion (intermediate or cold member). In August, the surface member has a uniform temperature of about 34°C and is separated by a sharp thermocline from the cold member, which is then at 23°C. In winter

The Jordan Valley

the cooling of the surface member causes homothermal conditions down to 40 m and mixing of the 'seasonal lake'. The Lower Water Mass begins at 40 m; temperature, density and salinity increase down to 100 m and remain constant to the bottom. The temperature of the deep layer is 21.7°C and its salinity 332 g l^{-1}, whatever the season or the locality. It is permanently devoid of oxygen. Neev & Emery (1967) interpret the Lower Water Mass as a fossil water body.

The higher rate of evaporation in the South than in the North Basin causes a more rapid lowering of water level in the South Basin. This generates a surface current from the North Basin along the western coast of the Strait into the South Basin. A deep return current flows northward to the North Basin. Because of its relatively high density, this water sinks down and mixes within the Upper Water Mass.

After the lowering of the water level. The previously described thermal structure and circulation have prevailed in the Dead Sea for at least 150 years (Assaf & Nissenbaum, 1977). During the last 15 years the income of freshwater of the Dead Sea has decreased by 50%. As a result, the salinity of the upper layers increased and the differences in density between the Upper Mass and the Lower Mass diminished. The weakening of the thermal stability became decisive around 1975. The salinity of the Southern Basin was 308 g l^{-1} in 1960 and 340 g l^{-1} in February 1975 (Beyth, 1976). These modifications considerably decreased the thermal stability of the Northern Basin, the halocline deepened and, in 1977, Beyth found homogeneous thermal and chemical profiles. Moreover, deposition of gypsum on the bottom of the North Basin was observed for the first time. Prior to this historic turnover, $CaSO_4$ could not be precipitated in the anoxic permanent hypolimnion because sulphate was rapidly reduced into sulphites. The presence of fresh gypsum on the bottom implies the presence of oxygen in deep water.

Chemical features

In 1960 the Dead Sea water (Northern Basin) had a TDS of 300 g l^{-1} in upper layers and 332 g l^{-1} in deep waters. The average ion content for the entire Dead Sea was 322 g l^{-1} (Neev & Emery, 1967). It now amounts to 338.5 g l^{-1}. The composition of the water is shown in Table 82. Magnesium, bromide, potassium, chloride and sodium are more concentrated in the South Basin. In contrast, sulphate and hydrogen carbonate are less concentrated because of active precipitation of gypsum and calcium carbonate.

Table 82 Average chemical composition of Dead Sea water

	Ca^{2+} (g l^{-1})	Mg^{2+} (g l^{-1})	Na^+ (g l^{-1})	K^+ (g l^{-1})	Cl^- (g l^{-1})	Br^- (g l^{-1})	SO_4^{2-} (g l^{-1})	HCO_3^- (g l^{-1})	pH	TDS (g l^{-1})
1960 (Neev & Emery, 1967)	16.86	40.65	39.15	7.26	212.40	5.12	0.47	0.22	5.7–6.2[a]	322.13
1977 (Beyth, 1978)	17.20	44.00	39.60	7.64	224.10	5.20	0.45	0.25	6.2	338.50

[a] The first value refers to pH of deep water, the second to surface water

The lower values of the ratios of ions to magnesium in the Lower Water Mass than in the Upper Water Mass indicate that the deep waters have lost, through precipitation, considerable amounts of chloride and sodium (probably as halite deposits), calcium, hydrogen carbonate and sulphate. The Lower Water Mass probably evolved during the last dry stage from 4300 B.P. to a few hundred years ago. In the anaerobic environment which developed after the formation of the Upper Water Mass some 1500 years ago (Neev & Emery, 1967), the sulphate of the gypsum deposits was reduced into sulphides which combined iron into FeS giving the black colour to bottom sediments. The released calcium was precipitated as calcium carbonate. This geochemical evolution of the Dead Sea water explains its percentage magnesium, sodium-chloride composition. The Dead Sea has a total of 44×10^9 tons of dissolved salts.

The origin of the Dead Sea salts is thoroughly discussed in Neev & Emery (1967). It is clear that the annual salt supply of the Jordan River and surrounding springs cannot account for the present concentration of the Dead Sea water. The current opinion is that the dissolved salts come from recycled older brines which originated in the salty water bodies that preceded the present Dead Sea.

Recently, Kafri & Arad (1979) have suggested that Mediterranean Sea water infiltration to the northern Rift Valley may account for the saline water of the Dead Sea area. According to these authors, the Mediterranean Sea water intrusion takes place through the Yisreel and Beersheva low topographic valleys which connect the Mediterranean and the Dead Sea Rift. In the valleys, there is a combination of a low water table and a low groundwater divide resulting in a shallow sea water–freshwater interface and allowing a flow of sea water eastward, controlled by the difference in elevation between the base levels.

The sediments of the Dead Sea are mainly composed of chemical precipitates: calcite, aragonite, halite and gypsum. In the past, this latter compound was found mostly in shallow sediments and not in the deep sediments of the North Basin for the previously mentioned reasons. The recent turnover of the Dead Sea and the renewed deep precipitation of gypsum initiate a new sedimentation cycle. Shallow sediments are laminated: white layers consisting of summer deposits of gypsum and calcium carbonate alternate with dark layers composed of detrital material carried by floods.

Biological features

For more than 2000 years, the Dead Sea has been described as a sterile lifeless water body and numerous legends made the Dead Sea a synonym of desolation and danger (Nissenbaum, 1979). Even in the mid-1880s, Barrois did not find any living organisms in the water. This background explains the great surprise of Lortet (1892) who found active bacterial populations in the Dead Sea mud. The legend of sterility was, however, more powerful than the scientific facts and little attention was given to this discovery. It was only after the publication of the research work of Volcani (1936, 1940, 1943a) that the existence of living organisms in the Dead Sea became known and accepted. The subject was reviewed by Nissenbaum (1975).

The algae are represented by the chlorophyte *Dunaliella* spp.; Kaplan & Friedman (1970) reported 40 000 cells ml^{-1} in surface water and showed that the number of cells rapidly decreased with depth.

Among the bacteria the halo-resistant organisms include three new species: *Flavobacterium haemophilum* (non-motile, short rods), *Pseudomonas halestorgus* (motile, uniflagellate rods), and *Chromobacterium maris mortui* (Gram-negative, motile rods). These organisms can grow in salt concentrations from 0.5 to 30%. Two types of halo-obligatory organisms were also found: *Halobacterium marismortui* (short non-motile Gram-negative rods) and *H. tarpanicum* (red, non-motile Gram-negative rods able to reduce nitrate). Both organisms can grow only in media with >15% NaCl. Gram-negative cocci requiring a salt concentration >6% were also found. Kaplan (1963) and Kaplan & Friedman (1970) reported surface concentrations of bacteria from 2.3×10^6 to 8.9×10^6 ml^{-1}. At 100 m, the concentrations dropped to $6-8 \times 10^5$ ml^{-1} and at 250 m they were only 2×10^4.

In the sediments, Volcani (1943b) isolated a cyst-forming *Dimastigamoeba* which could not grow below 6% salt in the medium, grew optimally at 15–18% salt, and could thrive at 33%. The blue-green algae *Microcystis*, *Phormidium*, *Nostoc* and *Aphanocapsa* were also isolated from the sediments but only *Aphanocapsa* was found to be halo-obligatory. The sediments also include numerous glucose-fermenting and protein-decomposing bacteria as well as sulphate reducers.

A common feature of most Dead Sea microorganisms is their low growth rate, confirmed by field experiments which demonstrated the very slow decomposition of organisms immersed *in situ*.

Osmoregulation of the bacteria and algae living in a concentrated medium is done by accumulation of solutes inside the cell. *Halobacterium* showed large amounts of KCl in the intracellular fluid (Ginzburg, Sachs & Ginzburg, 1971), while *Dunaliella* has high concentrations of intracellular, photosynthetic glycerol which can amount to 80% of the alga's dry mass

(Ben Amotz & Avron, 1972; Ben Amotz, 1973). Intracellular glycerol not only balances the osmotic pressure of the external salt concentration but also helps in maintaining enzymatic activity in a medium where water activity is low.

It is interesting to note that the two most abundant organisms of the Dead Sea, *Halobacterium* and *Dunaliella*, found different answers to the high salinity and unusually high Ca^{2+} and Mg^{2+}: *Halobacterium* became an obligate halophile by developing an internal osmotic pressure similar to that of the medium, whereas *Dunaliella* withstands high salinity by producing intracellular glycerol and modifying its concentration as a response to the changes of salt concentrations of the environment. Strict adaptation and dependence, against flexible regulation and opened options, are the two ways of evolution.

The relative abundance of these organisms in the Dead Sea is partly explained by the complete absence of zooplankton. Moreover, the possibility is not excluded that the glycerol, released from dead cells of *Dunaliella*, is a substantial food source for *Halobacterium* (Nissenbaum, 1975).

The Dead Sea is also known for its asphalt, which may appear as blocks up to 100 tons floating on the lake surface or in veins, cavity fillings and seepages, mostly on the western shore or associated with ozocerite, the mineral wax on the eastern shore. The asphalt seems related to the Upper Cretaceous oil shales of the western shores. Nissenbaum (1978) mentions that written evidence of the past leads him to think that asphalt emissions were more frequent then than in the nineteenth and twentieth centuries. He suggests that the asphalt is associated with diapiric activity and that the decrease in the frequency of asphalt emissions corresponds to a phase of tectonic relaxation.

The future of the Dead Sea

It is remarkable that the desolate lake which, during centuries, exerted a fascinating apprehension on men, is now an object of scientific, industrial and balnear attraction. The extraction of salts goes back to 1930. The production of potash consists of the evaporation of the brine until the formation of carnallite (KCl, $MgCl_2 . 6H_2O$). The bromides are concentrated in the liquid phase of the carnallite ponds and treated in a special plant. In the 1960s part of the Southern Basin was transformed into an evaporation pond, and when the Southern Basin became separated from the Northern Basin, water was pumped into the evaporation areas. With time, it became more and more difficult to evacuate the hypersaline brines released from the plants and the canals filled up with masses of precipitated salts. This excess of salts and lack of water to dissolve them reactivated old ideas concerning the supply of Mediterranean water to the Dead Sea. In the late 1970s, feasibility studies for the project now known as the Two Seas Project were completed. Two trajectories were envisaged for the Mediterranean Sea–Dead Sea canal: one, starting in the Haifa Bay, would carry the sea water into the Jordan River south of Lake Kinneret; the other, starting on the southern coast of Israel, would cross the Negev south of Beersheva and reach the Dead Sea directly. The second trajectory, which crosses areas poor in freshwater underground aquifers, seemed a better choice. An annual discharge of 1250×10^6 m^3 sea water would raise the Dead Sea level to -395 m in 10 years. Besides its 'diluting functions', the Two Seas Project would produce 100 MW electric energy. It would also allow the creation of several fish farms along the canal. In the last ten years, the increasing demand on freshwater for irrigation, domestic and industrial use has led to a progressive replacement of freshwater aquaculture by marine fish farming. The new fish farms would fit into this general scheme.

The recent developments of energy production in solar ponds (see pp. 434–5) have led many Israeli scientists to consider the Dead Sea shores as a particularly favourable area for the supplementation of large-scale solar basins. Other more far-fetched projects would utilize the totality of the Dead Sea area as a giant solar collector. The technical difficulties of such an enterprise will require years of research, but these might be worthwhile if the expected 5 MW km^{-2} represent a realistic evaluation.

A more unexpected aspect of the economic exploitation of the Dead Sea is to be seen in the industrial glycerol production from its main alga, *Dunaliella* sp. (Ben Amotz & Avron, 1980). At present, this alga is grown in artificial ponds; in the future framework of the Dead Sea development project this alga will be grown *in situ* in alga farms on the Dead Sea shores.

This rapid review indicates that in the future the Dead Sea area may become an important centre of production of energy, food and valuable industrial chemicals.

THE WATERS OF IRAQ

The Euphrates and Tigris are the only large rivers of the Middle East. The geomorphological, climatological and limnological information concerning these major waterways has been collated by Rzoska (1980) in a recent monograph from which much of the following information has been taken. The Euphrates and Tigris arise at about 2000 m in the High Plateau of eastern Turkey. The Euphrates, 2600 km long, flows westwards over about 500 km, crosses the Anti-Taurus, describes a west-oriented arc in Syria and enters the alluvial lowlands. The Tigris, 2000 km long, drains the regions situated south and south-west of Lake Van (Van Goln), then flows in a south-eastern direction and receives on its way five tributaries (the Khabur, Greater and Lesser Zab, the Adhyem and the Diyala Rivers), draining the Kurdistan and Zagros mountains which border the river on its eastern side. The Mesopotamian alluvial plain extends between the two rivers. The Euphrates and Tigris join at Qurna and form the Shatt al Arab which empties into the Persian Gulf after receiving the Karun River on its left side. Many historical facts show that Shatt al Arab is a recent formation; in particular Abadan was a sea port in the tenth century A.D. and the Karun River entered the sea directly. The Euphrates and Tigris have very unstable courses and their floods may cause large-scale disasters. Several reservoirs aimed at controlling floods and supplying irrigation waters have been constructed: the Samarra and Kut Barrages on the Tigris River, the Ramadi, Hindiya, Yao, Meshkab, Hafar and Akaika Dams and regulators on the Euphrates (Fig. 26).

Several temporary or permanent shallow water bodies called 'hors' result from the inundation of lowlands: the Habbaniya (430 km^2), Thartar (2700 km^2), Dibbis (1985 km^2) and Hammar (2500 km^2) are the most extensive. The Hammar forms part of the southern swamps, extending over 15 000 km^2, which constitute the lower course of the rivers.

The river discharge is very variable with maximal values in spring: the Tigris at Mosul carries up to 4667 m^3 s^{-1} in March but as little as 180 m^3 s^{-1} in September, and the Euphrates at Hit reaches its maximum in May (up to 3120 m^3 s^{-1}) and has its low water in autumn (232 m^3 s^{-1}). The annual discharges fluctuate from 45×10^9 to 63×10^9 m^3 for the Euphrates and from 46×10^9 to 93×10^9 m^3 for the Tigris (Al-Sahaf, 1975, quoted in Rzoska, 1980). The concentration of suspended matter shows a parallel variation: 2800 mg l^{-1} and 28–45 mg l^{-1} in spring and autumn, respectively, for the Tigris at Baghdad and 6920 mg l^{-1} and 46–86 mg l^{-1} in spring and autumn, respectively, for the Euphrates at Hit. These values concern the 1958–59 flood.

The salinity of the river water steeply increases downstream. In the Tigris, the electrical conductivity jumps from 400 μmhos cm^{-1} at Mosul to 900 μmhos cm^{-1} at Qurna; in the Euphrates it varies from 600 μmhos cm^{-1} at Qaim to 900 μmhos cm^{-1} at Qurna. Both rivers have very large seasonal variations with considerable dilution during the flood period In the Shatt al Arab, conductivity is always around 1000 μmhos cm^{-1} at Basra and increases near the Gulf. Among the lakes of the Mesopotamian plain, Lake Abu Dibbis has the highest salinity with 1969–72 averages of 6315 mg l^{-1} (Sahaf, 1975, quoted in Rzoska, 1980). In the upper course of the rivers, the waters are of calcium–magnesium hydrogen carbonate type; downstream sodium–magnesium and chloride–sulphate dominate. In the seas of the Miocene which occupied Mesopotamia, salt and gypsum were deposited and covered with more recent alluvial formations. Seepage of saline groundwater occurs in the lower valley or at the occasion of intensive irrigation and leaves its 'fingerprints' in the composition of the river water.

The pH of the river water ranges from 7.0 to 8.0. Saad & Antoine (1978b) found high concentrations of nitrate in autumn, phosphate levels exceeding 1000 µg l^{-1} in summer and SiO$_2$ concentrations varying between 7 and 10 mg l^{-1}. In general, it seems that inorganic nitrogen is often present in

Fig. 26. Shatt al Arab and the position of dams and reservoirs

very small amounts, whereas phosphorus is still at high levels. A similar feature has been found in African waters and might be related to denitrification.

Phytoplankton and primary productivity of the Tigris, Euphrates and Shatt al Arab have been studied by Kell & Saad (1975), Saad & Kell (1975), Saad & Antoine (1978a,b,c), Hug, Al-Saadi & Hameed (1978, 1981) and Al-Saadi, Rattan, Muhsin & Hameed (1979).

The study of Kell & Saad (1975) indicated the dominance in the rivers of diatoms, with *Diatoma elongatum, Synedra ulna, Bacillaria faxillifer, Navicula halophila, Nitzschia spectabilis* and the green algae *Pediastrum boryanum* and *Scenedesmus acuminatus. Cocconeis pediculus* and *Gomphonema acuminatum*, abundant in the Tigris, were not found in the Euphrates. In contrast, *Surirella ovata, Synedra tabulata* and *Volvox africanus* were not observed in the Tigris. Later studies confirmed the dominance of diatoms (*Cyclotella, Navicula, Nitzschia* and *Synedra*) and of green algae (*Chlamydomonas, Crucigenia, Pandorina, Pediastrum* and *Scenedesmus*). In the Shatt al Arab, diatoms are still preponderant but there is a high proportion of marine species. The primary productivity varies between 6 and 37 mg C m^{-3} h^{-1}.

The zooplankton of the Shatt al Arab includes the halophilic phyllopod *Artemia salina* and the Cladocera *Daphnia lumholtzi, D. magna, D. pulex, D. longispina, Ceriodaphnia reticulata, Moina rectirostris, M. dubia* and *Simocephalus exspinosus*. Among the Copepoda the most abundant are: *Cyclops vicinus, C. viridis, C. vernalis, C. bicuspidatus, C. leuckarti, C. crassus, C. albidus, C. agilis, C. bicolor* and *C. diaphanus*. The diaptomids comprise *Diaptomus vulgaris, D. blanci* and *D. chevreuxi*.

The fish fauna of Mesopotamia has been reviewed by Banister (1980). His list includes 84 species, among which are 60 cyprinids and 10 cobitids. This fauna is characterized by the nearly complete absence of non-ostariophysan fishes (the exception being *Mastacembelus mastacembelus*). The cyprinids are mainly represented by the genera *Acanthobrama, Alburnus, Barbus, Garra, Leuciscus* and *Varicorthinus*. This fauna is composed of European and Asiatic elements and has very few endemic species.

Lake Habbaniyah has a varied algal population including diatoms and green and blue-green (*Microcystis aeruginosa*) algae.

Lake Abu Dibbis, 120 km SW of Baghdad, receives its water from Lake Habbaniyah during the flood period. It is 75 km long and 31 km wide and has a maximal depth of 55 m. The average salinity is 7‰. Since the water cannot be used for irrigation, the lake has been stocked with a shrimp from

Al-Fao, on the Gulf. It appeared that the species involved was *Palaemon elegans*, a species common on the Atlantic coast from Norway to South Africa, the Mediterranean and the Black Sea, but it has not been recorded in the Indo-West Pacific area (Holthuis & Hassan, 1975).

THE WATERS OF IRAN

The uplift of the highlands of Iran took place in the Upper Miocene. Vast amounts of salt and gypsum were deposited and many water bodies became saline. The 'Upper Red Formation', a tertiary formation rich in salt and gypsum, covers large areas of the Kaviz, the Great Salt Desert in eastern Iran. Liassic sandstones and Lower Devonian Red Beds are additional sources of salt (Ruttner & Ruttner-Kolisko, 1973). The high mountains of Mesozoic limestone have abundant and fresh water but salt contamination occurs when the water reaches the alluvial fans and appears as natural springs or is tapped by 'ganats' (nearly horizontal underground galleries dug from the alluvial fans into the mountain). In these areas, because of the dust, even the rain water is loaded with salt: Ruttner-Kolisko (1966) reports a conductivity of 268 μmhos cm^{-1} in rain water of the Ozbakhuh region. The spring and ganat water is generally rich in chloride and sulphate.

Lake Niriz, the largest lake of the Iranian Warm Belt, has been investigated by Löffler (1953, 1956, 1959, 1961, 1981). It is located at 1525 m altitude and occupies an area of 1200 to 1800 km^2. Its depth is around 1 m. It has no outlet but receives freshwater from the surrounding mountains in winter–spring. The summer evaporation causes drastic variations in water level and in surface area.

Islands separate the lake into two basins: the northern Nargis basin and the southern Niriz basin. The latter basin has a concentration of 4 g l^{-1}. In Nargis, the concentration is much higher and varies in different locations from 18 to 36 g l^{-1}. A significant decrease in salinity is observed at the mouth of the Gomum River where the fish *Cyprinodon sophiae* sometimes appears in large numbers. The waters of Lake Niriz are heavily dominated by Cl$^-$ and Na$^+$; their pH is around 9.

The saline Nargis basin harbours *Artemia salina* and the Niriz basin has the following fauna: the cyclopids *Diaptomus salinus* and *Cletocamptus blanchardi*, ostrocods, the rotifers *Pedalia fennica oxyure* and *Brachionus mülleri*, the marine foraminifer *Nonion*, and the mollusc *Hydrobia* sp. nov., a genus generally present in freshwater bodies connected to the sea.

In the lake vicinity the halophytic water weeds *Ruppia maritima* v. *rostrata* and *Althenia filiformis* are found.

8
South-East Asia

INTRODUCTION

Humid tropical Asia was defined by UNESCO in 1956 as the region covering peninsular India, Burma, Thailand, the Indo-Chinese peninsula, the Malaysian peninsula, Indonesia, Philippines, New Guinea, Southern China and Sri Lanka, that is the area between 73 and 150° E and between 25° N and 10° S. This definition, mainly based on climate, concerns all the areas subject to the influence of the monsoon. Here we adopt this definition with the following modification: the northern boundary of the area is extended a few degrees northwards in order to include the water bodies of the Indus, Ganges, Brahmaputra and Yang-tse Kiang basins as well as the upper valleys of the Irrawaddy, Salween and Mekong Rivers.

All aspects of life in South-East Asia are dominated by the monsoon, the mechanism of which is briefly recalled. In winter the anticyclone prevailing over central Asia governs the air-flow pattern in the whole area. The northern parts of South-East Asia are then swept by cold and dry north-easterly winds, except for the northern regions of the Indian subcontinent which are protected by the Himalayan barrier. In areas located south of central Vietnam, the dominant winds also come from the north-east but they originate in the western part of the Pacific Ocean and not on a continental area. In summer, the land areas are heated rapidly in comparison with the ocean and become the site of a pressure low, well defined over south-west China and north Vietnam and culminating over Pakistan and north-west India. A secondary depression lies in the eastern Indian Ocean between the Equator and 5° S whereas the northern Pacific is an anticyclonic area. The wind pattern is as follows: south of 5–10° S the winds are easterly, north of this latitude to the Equator they become south-south-westerly and bring considerable humidity.

The rain is frequently accompanied by violent thunderstorms and is consequently rich in nitrate and ammonia. Richard (1960) estimated that the atmospheric nitrogen input was as high as 851 mg N m^{-2} y^{-1} in the Indo-Chinese peninsula. The major rivers (Indus, Ganges, Brahmaputra, Irrawaddy, Salween, Mekong, Yang-tse Kiang) drain the Himalayan and monsoon waters to the Indian Ocean and the China Sea.

With the exception of the mountainous regions, South-East Asia is characterized by the absence of a marked winter: south of 17°N the temperature is rather uniform throughout the year at around 27°C (Koteswaram, 1974). More specific climatological data will accompany the description of each region of South-East Asia.

Another characteristic of South-East Asia is its high population density: the monsoon area of South-East Asia (this excludes the Ganges and Indus plains) extends over 5% of the world's dry land but contains 25% of the earth's population. No more than 30% of the land is cultivated in spite of the great variety of soils. In contrast to tropical Africa and South America, where poor soils are mostly formed from weathered granite and metamorphic rocks, humid tropical Asia soils include very fertile soils from volcanic and alluvial origin. The favourable soils and the humid climate account for the extension of the rain forest which covers nearly 40% of the land. An ancient tradition and the pressure of increasing demography led to a flourishing fishing industry and pisciculture. It is estimated that 400 000 km^2 (i.e. 5% of the land) of fresh and brackish water are suitable for pisciculture. The mastering of the techniques of breeding of fish, shell-fish and edible molluscs has favoured the multiplication of artificial ponds and lakes which cover an area of 25 000 km^2, whereas natural lakes cover only $\sim$12 000 km^2 (Dussart, 1974). This long familiarity of the people of South-East Asia with ponds, lakes and fish makes this area especially interesting for limnological studies and the investigation of long-term human influence on aquatic ecosystems.

INDIA
General

India can be best described as a stable continent of Precambrian formations, which in its northward movement compressed the Tethys Sea where 15 000 m of sediments had accumulated since the Palaeozoic. In this geocollision, the maximal compression on the plastic sediment occurred in the area of contact of the Indian craton and Laurasia: this corresponds to the present Great Himalayas with its immense folds and impressive thrusts.

India

At either end, where the compression of the Indian block ceased, the chains bent, anticlockwise in Kashmir and clockwise in Assam (Krishman, 1975). Moreover, the present Indo-Gangetic plain with its thick alluvial filling is a trough created by the warping down of the Indian Continent under the Asian mass. It is also a 300 km-long drainage system where three powerful rivers, the Indus, the Ganges and the Brahmaputra, carry annually an amount of water equal to the yearly discharge of the equatorial Zaire.

According to Mani (1974) the uplift of the Himalayas conditioned to a large extent the present biogeographical composition of India. This new terrestrial link between peninsular India and Asia put the original and mature Gondwanalandian flora and fauna into contact with Asiatic as well as European and African elements. The obliteration of the Tethys Sea started in the north-east and gave birth to the Assam–North Burma area. This was the major and earlier route of faunistic migrations mainly from the east (China and Indo-China) to the west (Himalaya and peninsular India). Later, when the Tethys disappeared entirely the north-west bridge with Asia was established, allowing faunistic exchanges with Western Asia, Europe and Africa.

The access of new fauna and flora to the Indian continent led to the progressive retreat of the autochthonous elements to sanctuary areas, for example the southern part of the peninsula. A new cycle of impoverishment of both old and new flora and fauna started about 5000 years ago with intensive human colonization. Until then, climax forests covered extensive areas and their constitutive species had a large and continuous distribution; at present many of them are restricted to isolated relict pockets. Deforestation has not only caused the disappearance of tree species and their irreversible replacement by savanna or grassland in the best case, but has also impoverished the soil by erosion and demineralization resulting in vast expanses of infertile laterites, and has destroyed the natural habitats of an ever-retreating fauna. The final blow was the direct extermination of fauna by the hunting of the last 200 years. As a result, two species of birds and 15 species of mammals have disappeared from India and another 21 species of vertebrates have undergone an average 90% spatial regression. The case of *Antelope cervicapra* is a good example (Mukherjee, 1974); common all over peninsular India a hundred years ago, its present area of distribution is 4.6% of the former.

This ecological deterioration is not limited to the peninsula, the Himalayan slopes are being deforested at a dangerous rate (Eckholm, 1975). For example, in Nepal, the demographic pressure forces farmers to cultivate steeper slopes and destroy more trees for fodder and firewood.

The resulting landslides wash down considerable amounts of fertile soil into India and Bangladesh, and reduce the agricultural potential of Nepal. In the Eastern Hills as much as 38% of the area is composed of abandoned fields. When the firewood becomes scarce, the farmer utilizes domestic-animal dung for fuel and consequently decreases the fields' fertility. In Pakistan, only 3.4% of the area is presently covered with forests but overgrazing, demand of timber for industry and clearing for expansion of agriculture constantly reduce the forested area and enhance erosion. The subsequent heavy silt load of the Indus and its tributaries rapidly fills the existing reservoirs and makes it useless to build new ones.

Destruction of old fauna and flora by invasion of new elements, man-made ecological disturbance and especially deforestation exist in other places in the Warm Belt, for example the Andes and East Africa, but as emphasized by Mani (1974) 'the destruction of natural habitats by man (by deforestation) has been far more extensive and complete in India than perhaps anywhere else in the world'. The eradication of the forest cover of India has altered the hydrological regime of rivers and made conditions suboptimal for most species. India is what Amazonia might become if the demographico-economic pressure is strong enough to destroy the rain forest.

The Indus basin

The Indus basin extends over an area of 970 000 km² in West Pakistan, India and Afghanistan between 37° N and 23°27′ N.

The basin is drained by the Indus River and its tributaries. The Indus River rises in Tibet in the Kailas range of the Himalayas at about 5500 m altitude. A surprising feature is that the Indus system drains not only southern slopes of the Himalayas but also the northern ones in such a way that the topographical crest line does not correspond to the waterdivide which lies much farther northwards. According to Mani (1974), the Indus drainage system is more ancient than the Himalayas and the orogenic process did not change the course of the river notably; it only increased its erosion rate. The Brahmaputra River and the Sutlej River are also old drainage systems established before the Himalayan upheaval.

The basin lies in the subtropical area and has a very seasonal climate. In the mountains, the winter is cold (mean temperature of −9.4°C in January) and the summer mild (mean temperature of 17.8°C in July). In the plains, the winter average temperature varies from 15 to 21°C and, in summer, the average temperature ranges from 29 to 34°C. The Indus basin is semi-arid and the amount of rainfall in the different parts of the basin is very variable. In the Lower Indus in the vicinity of Sukkur, rainfall does not exceed

100 mm y^{-1} whereas, in the hills, precipitations are rarely below 1300 mm y^{-1}; in the plains most of the rains fall in summer while in the hills precipitation is distributed over the whole year. The Indus and its tributaries receive most of their water from the mountains; in summer, the snow melting at high altitudes and the summer rains of the lower valley generate strong floods, whereas in winter, the absence of rain in the lower valley and the snow in the upper valley reduce the river discharge to sometimes one hundredth of the summer maximum.

The Indus River

The Indus River is 2900 km long. After a 1000 km course in a north-westerly direction, it takes a right angle at 1200 m altitude and flows south-west. At Mithankot, it receives the Panjnad River composed of the Jhelum, Chenab, Ravi, Beas and Sutlej Rivers. Lower down, the Indus flows on its own sediments between embankments built to protect the plain from floods. In the lower valley at Kotri it divides into two branches, the Ochito and the Hydri Rivers, which both flow into the Arabian Sea.

The Indus and its tributaries carry annually an average amount of 208×10^9 m^3 water. This represents approximately two to three times the Nile discharge. The specific yield exceeds 200 000 m^3 km^{-1} y^{-1}, that is one order of magnitude higher than that of the Nile but less than one-fifth of the specific discharge of the Amazon. The regime is very seasonal and inundations of the lower valley are frequent. The scarcity of water in the plains led to the early development of irrigation canals, weirs and barrages (Sukkur Barrage on the Indus, Bhakra–Nangal project on the Sutlej, the Tamsa Barrage, the Guda Barrage and the Kotri Barrage on the Indus). Recent projects include the Mangla Dam on the Jhelum River (6.9×10^9 m^3) and the Tarbela Dam on the Indus (10.4×10^9 m^3). These water works allow the irrigation of an area seven times that of the irrigated lands of the whole Nile Valley. The utilization of the Indus water by India and Pakistan is subject to the terms of the Indus Water Treaty signed by the two countries in 1960.

Among the natural lakes of the Indus basin, the Kashmir lakes have been recently investigated by Zutshi, Kaul & Vass (1972), Zutshi (1975), Khan & Zutschi (1980*a,b*), Zutshi, Subla, Khan & Wanganeo (1980) and Vaas (1980). Other studies concern fish migration in the Indus (Islam & Talbot, 1968) and artificial reservoirs (Nazneen, 1980).

The natural lakes of Kashmir Himalaya

These are all high-altitude lakes (1580–4000 m) and belong to four different groups. The first group includes lakes situated between 3000 and

4000 m along the Pir Panjal Range. They are glacial lakes which originated during the third Himalayan glaciation. The second group is found between 2500 and 2000 m and is of tectonic origin. The third series, known as Valley lakes, is located around 1600 m in the Jhelum Valley and is probably of fluviatile origin. The lakes of the fourth group in Ladakh near Tibet are generally saline.

A high mountain lake: Lake Alipather. The high mountain lakes are generally small (10 km^2), shallow, situated above the tree line and fed by glaciers (Fig. 27). Macrophytes are completely lacking and conductivity is very low. They are ice-covered from October to May.

Lake Alipather is located at 3200 m and is typically dimictic. Its surface area is only 8 ha and maximum depth 6.5 m. It is a calcium–hydrogen-carbonate lake but concentrations (in mg l^{-1}) are very low: Ca^{2+} 2, Mg^{2+} traces, Cl$^-$ 3, HCO$_3^-$ 10, alkalinity 18 mg l^{-1} CaCO$_3$; pH is neutral. The watershed of the lake has no vegetation cover and there are no human settlements in the lake vicinity. The moraine deposits of the catchment area supply a glacial silt which limits water transparency. In summer, the lake is stratified with 13°C from 0 to 2 m; at 6 m, the temperature is 6°C. The

Fig. 27 A small Himalayan lake (by courtesy of Prof. H. Löffler, Institut für Zoologie, Vienna, Austria)

oxygen concentration is around 7 mg l^{-1} at all depths (data of August 1971).

The lake's algal assemblage includes mainly diatoms and Chlorophyta; *Cymbella ventricosa, Diatoma elongatum, Fragilaria crotonensis, Synedra ulna* are the most common diatoms, whereas the green algae include *Ankistrodesmus spiralis, Cosmarium* sp., *Pediastrum duplex* and *Staurastrum* sp. The Pyrrophyta are represented by the genus *Peridinium*. Pigmented *Daphnia* (probably *D. tibetana*) have been found in the deep layers of the lake. The biomass of plankton is small and the lake is very oligotrophic.

Lake Nilnag. Lake Nilnag is located at 2180 m altitude at 33°52' N and 74°42' E. It seems to be of recent tectonic origin. Its surface area is only 6.3 ha, its maximal and mean depths are 7 and 18 m, respectively, and its volume is 108×10^3 m^3. It is oval in shape and the greatest depth corresponds approximately to the central part of the lake.

Water transparency ranges from 1 to 2 m, indicating a permanent turbidity. The lake is ice-free in May and stratified in summer with temperatures of 22–24°C in the upper 2 m and 15–20°C in deep waters. Isothermy is achieved in November at 7°C. In winter, inverse stratification develops with 0°C at the surface and 4°C at the bottom. In spring, a second turnover occurs. The lake belongs to the typical dimictic category. The oxygen concentration is high at all depths at turnover times. In intermediate periods, the oxygen level decreases rapidly in deep water (1.9 mg l^{-1} in March 1975).

The lake water belongs to the calcium–hydrogen carbonate type with a relative ionic composition conforming to that of average inland water. Conductivity ranges from 230 to 370 μmhos cm^{-1}. The main chemical features are presented in Table 83.

Lake Nilnag is abundantly colonized by macrophytes: *Typha angustifolia, Phragmites communis, Cladium marcicus* and *Juncus lampocarus* are among the emergent plants. The common floating species are *Polygonum amphibium, Nymphoides peltata, Trapa natans, Nymphaea alba, Potamo-*

Table 83 *Chemical features of Lake Nilnag (from Khan & Zutshi, 1980) (results in mg l^{-1})*

Ca^{2+}	27–39	Na^+	2.7–4.6	SO_4^{2-}	6.5–8.5	Alkalinity	147–220
Mg^{2+}	5.0–7.1	K^+	0.4–1.0	Cl^-	5.0–7.0	(mg l^{-1} CaCO$_3$)	

geton natans and *Marsilea quadrifolia*. Among the submerged species are *Myriophyllum spicatum* and *Ceratophyllum demersum*. The aquatic vegetation is utilized as fodder for livestock, manure and food. The fruits of *T. natans* saved people from starvation in 1893 when severe floods destroyed the crops.

The phytoplankton is rich in blue-green algae (*Oscillatoria rubescens, Nostoc* sp. and *Spirulina* sp.) but also includes many Chlorophyta: *Chlamydomonas* sp., *Pediastrum duplex, Scenedesmus bijuga, Sphaerocystis schroeteri, Ankistrodesmus spiralis, Micrasterias radiata, Staurastrum* sp., *Sphaeroplea* sp., *Crucigenia rectangularis, Pachycladon* sp., *Selenastrum gracile, Closterium* sp., *Tetraedron minimum* and *Oocystis lacustris*. The diatoms are represented by *Fragilaria crotonensis, Gomphonema montanum, Navicula* sp., *Cymbella prostrata, Amphora* sp., *Achnantes* sp., *Synedra ulna, Diatoma elongata* and *Pinnularia* sp., and the Dinophyta include *Ceratium hirundinella*. The maximal growth period of the phytoplankton and the macrophytes extends from August to October.

The zooplankton is composed of the rotifers *Keratella tropica, K. cochlearis, Brachionus quadridentatus, B. angularis, Hexarthra major, Filinia terminalis, Asplanchna brightwelli, Pompholyx complanatus* and *Horaella brehmi*. The copepods comprise *Cyclops vicinus, C. ladakensis, Mesocyclops leuckarti, Thermocyclops crassus, Paracyclops fimbiatus* and *Acanthodiaptomus denticornis*. The Cladocera *Ceriodaphnia quadrangula, C. pulchella, Chydorus eurynotus, Alona rectangularis* and *Moina micrura* were also found.

During the ice-free period, the primary productivity was observed to vary from 120 to 562 mg C $m^{-2} d^{-1}$. However, because of the unfavourable winter conditions the annual production is estimated to be only 90–100 g C m^{-2}.

According to Khan & Zutshi (1980*a*), it seems that the development of human settlements and human activity in the lake watershed has increased the erosion and worsened the water quality. By its thermal dimictic structure Lake Nilnag resembles the high mountain lakes; however its trophic structure is closer to that of the turbid and mesotrophic valley lakes.

The valley lakes

Zutshi *et al.* (1980) investigated five lakes located at 1584 m altitude in the Jhelum Valley between 32°–34° N and 74°–75° E. Their main general features are presented in Table 84.

The valley lakes are not surrounded by forests. Tourism is developing rapidly as well as erosion and turbidity of the water. The climate is submediterranean with 750 mm y^{-1} precipitation falling mainly in winter–spring. Summers are dry and warm (28–30°C). In winter, the air temperature falls below 0°C, causing occasional freezing of the lake surface.

Lake Manasbal is deep enough to have a distinct thermocline from April to October: it is a typical warm monomictic lake. In the shallow lakes, a weak summer stratification develops without a clear thermocline. Trigam Lake has a pond-type thermal structure with rapid decrease in temperature below the surface. This lake has a high conductivity ($>400\,\mu$mhos cm^{-1}) whereas the other lakes are all below this value. The waters of Lakes Dal, Anchar, Naranbagh and Manasbal belong to the calcium–hydrogen-carbonate type, whereas in Trigam Lake water, sodium is dominant over calcium.

The valley lakes have a rich macrophyte vegetation, described in detail by Zutshi (1975). Among the submerged plants, *Myriophyllum spicatum* dominates in Lake Dal and Lake Naranbagh whereas in Lake Manasbal *Ceratophyllum demersum* is more abundant. Lake Trigam has a rich growth of *Typha* and *Phragmites*.

The phytoplankton of Lake Dal is dominated by *Euglena* sp., *Phacus* sp., *Merismopedia elegans* and *Microcystis aeruginosa*, but numerous diatoms are also present. In Lake Anchar, the diatoms dominate with *Cocconeis placentula, Fragilaria crotonensis, Navicula radiosa* and *Nitzschia acicularis* but *Euglena* sp. and Chlorophyta are found. The dominant species of Lake Naranbagh are *Fragilaria crotonensis, Staurastrum* sp. and *Ceratium hirundinella*. The algal community of Lake Manasbal is dominated by diatoms and especially *Fragilaria crotonensis, Navicula radiosa* and

Table 84 *General features of the valley lakes*

	Dal	Anchar	Narambach	Manasbal	Trigam
Area (ha)	1250	700	24	280	11
Max. depth (m)	6	3	5	12	2
Secchi disc (m)	0.4–5.0	0.2–1.8	1.2–4.0	2.5–5.5	0.2–0.6
Max. surface temperature (°C)	30	28	29	29	30
Stratification type	WM–PM	WM–PM	WM–PM	WM	P
Dominant cation	Ca^{2+}	Ca^{2+}	Ca^{2+}	Ca^{2+}	Na^+
Residence time (y)	0.29	0.29	1	1	1

WM = warm monomictic, PM = polymictic, P = pond type

Nitzschia acicularis but includes also the blue-greens *Merismopedia elegans, Microcystis aeruginosa* and *Oscillatoria* sp. Lake Trigam is especially rich in Chlorophyta with *Crucigenia crucifera* and *Staurastrum* sp. but also has an abundant flora of *Microcystis aeruginosa*.

As far as zooplankton is concerned, Dal Lake has a more abundant population of rotifers (*Keratella cochlearis*) than of crustacean plankton. In Lake Anchar, rotifers also dominate with *Keratella cochlearis, Lecane* sp. and *Trichocerca similis*. In Naranbagh, *Filinia longisete* and *Keratella cochlearis* show two peaks in spring and autumn. Lakes Manasbal and Trigam, although dominated by rotifers, also have copepods and cladocerans in higher amounts than the first three lakes.

Lake Trigam is the most productive of the five Valley lakes with an annual average of ~ 400 g C m^{-2} y^{-1} and maxima exceeding 3000 mg C m^{-2} d^{-1}. It also has the highest concentration of phosphate (40 μg l^{-1} PO$_4$-P) and is considered eutrophic. Next comes Lake Anchar, described as meso–eutrophic, and Lakes Dal (350–600 mg C m^{-2} d^{-1}) and Manasbal (240–475 mg C m^{-2} d^{-1}) classified in the mesotrophic range. The productivity of Naranbagh, studied in detail by Khan & Zutshi (1980*b*) shows very large seasonal differences (36 and 905 mg C m^{-2} d^{-1} in July and December 1976, respectively). The annual average is about 134 g C m^{-2} y^{-1}.

The Siwalik lakes

Two lakes of the Siwalik Himalayas (Jammu region, north of Lahore) have been studied by Zutshi *et al.* (1980). Lake Surinsar and Lake Mansar are located at 604 and 666 m altitude, respectively. Their main features are presented in Table 85. In contrast with the mountain and valley lakes, the Siwalik lakes are in the subtropical monsoon area: annual rainfall

Table 85 *Main features of Lake Surinsar and Lake Mansar*

	Surinsar	Mansar
Altitude (m)	604	666
Area (ha)	29	45
Maximum depth (m)	26	50
Secchi disc (m)	1.6–2.7	2–5
Temperature (°C)	31	30
Stratification type	WM	WM
Thermocline depth (m)	5–9	6–9

WM = warm monomictic

is as high as 1500 mm and the winter is dry and mild with temperature above freezing point. The lakes are typically warm monomictic. The water chemistry is dominated by calcium carbonate.

In Mansar Lake, Chlorophyta and Chrysophyta dominate in winter but Cyanophyta become very abundant in summer. In Surinsar, the winter forms belong mainly to Chlorophyta, and Chrysophyta dominate in spring and summer. Rotifers and copepods are the main zooplankton in Mansar whereas in Surinsar rotifers dominate throughout the year.

These lakes are very productive with annual averages of primary production exceeding 500 g C m^{-2} d^{-1} and maxima reaching 200 mg C m^{-2} d^{-1}.

In conclusion, we see a clear progression from the high mountain, oligotrophic lakes lacking macrophytic vegetation and having a typical temperate dimictic regime to valley lakes and Siwalik lakes showing increasing values of conductivity, nutrients, algal biomass and productivity. Moreover, these lakes are warm monomictic and the limit between di- and monomictism in this region seems to be around 2000 m. The human pressure increases as altitude decreases and signs of eutrophication, already visible, result from accelerated deforestation, erosion and silting. Vaas (1980) recommends a set of conservation measures including land-use control, diversion of effluents, dredging, harvesting of weeds and algae and nutrient removal from wastewater.

Artificial reservoirs

Bhakra Reservoir. The Bhakra lake has been created by damming the middle Sutlej River approximately 350 km north-west of Delhi at 515 m altitude. The dam was designed for irrigation and electric power and completed in 1963.

At maximal level storage, the surface area of the lake is 168 km^2 its maximal depth 155 m, its level 515 m above MSL and its total length 97 km. The permissible drawdown being 445.6 m, the lake live storage amounts to 7438×10^6 m^3. The Bhakra Dam drains an area of 56 876 km^2, 11% of which is covered with permanent glaciers. The dam is built downstream of a vast meander of the river and the resulting lake is 'S'-shaped.

The Sutlej River is fed both by rainfall in the catchment area and snow melting in summer. The lake starts filling in May and reaches its maximum in late September. The measured evaporation is $\sim 160 \times 10^6$ m^3 y^{-1}.

Sedimentation within the reservoir is the main problem of the Bhakra project. On the basis of observations carried out prior to the dam building, the suspended silt load was estimated at 35×10^6 tons y^{-1}. The actual measurements indicated values of 37×10^6 tons y^{-1} (1959–69 average). The silt is particularly thick (36 m) in the main curve of the river, some 26 km upstream from the dam. About 65% of the silt deposit is in the dead storage zone and 35% in the live storage zone. As much as 99% of the incoming silt is trapped in the lake. There is presently no silt outflow since the main site of silt deposit is too far from the dam. The water admitted into the irrigation canals is therefore clear and allows an unrestricted growth of weeds in the canals, which considerably diminishes the efficiency of the irrigation scheme. The problem was solved by 'drain the bed dry' operations consisting of pumping all the water out of the canal when it is not utilized.

Since the Sutlej River served as a cheap transportation system for Himalayan timber, special mechanical arrangements were made to carry the timber downstream of the dam.

The phytoplankton, dominated by Chlorophyta, is high in spring and autumn whereas zooplankton peaks in summer.

In the pre-dam period, the fish species *Labeo dero, L. dieochylus, Tor putitora, L. bata, Barbus sarana, Schyzothorax* and *Mystus seenghala* were most frequently found in the Sutlej River. The ecological changes introduced by the lake conditions caused the drastic decrease of most of these species with the exception of the detritus-feeding *L. dero* and *L. bata* which increased. The lake was stocked with *Cyprinus carpio, Labeo rohita, L. calabasu, Cirrihana mrigala* and *Catla catla* which have adapted fairly well to the new environment. From 10 tons in 1963, the fisheries reached 400 tons in 1973.

Kinjhar Reservoir. Kinjhar Reservoir is located 120 km north of Karachi at 24°47′ N and 68°2′ E. This reservoir of 80 km² and 3 m depth, supplied by a canal carrying water from the Indus River is the main source of domestic water of the city of Karachi. The monthly mean values of temperature vary from 18°C in January–February to 30°C in May–June. The reservoir water is very rich in nutrients (up to $1300 \mu g\, l^{-1}$ PO_4–P, up to $1500 \mu g\, l^{-1}$ NO_3–N and about 30 mg l^{-1} silica).

Nazneen (1980) studied the succession of algae in the reservoir. In spring the most abundant species were *Anabaenopsis elenkinii, A. milleri, Stigonema ocellatum, Spirogyra neglecta, S. subreticulata, Mougeotia thylespora, Cosmarium saprospeciosum, Gyrosigma attenuatum* and *G. scalproides*. In summer, *Microcystis aeruginosa* is heavily dominant but also abundant are *Lyngbya birgei, Anabaena* sp. and *Merismopedia* as well as the Chlorophyta *Spirogyra fuellobornei, S. hollandiae, S. mirabilis, S. quadrilaminata* and *Scenedesmus quadricauda*. In autumn, the algal population decreases and includes *Anabaena circinalis*, *A. spiroides*, *Lyngbya polysiphonae*, *L. saxicola*, *Microcystis ramosa*, *M. robusta*, *M. viridis*, *Tetrachloris merismopedioides*, *Navicula rhyncocephala* and *N. viridula*. The typical winter species are *Cymbella helvetica* and *Eunotia arcus*. *Melosira granulata* is present nearly all the year round.

The Brahmaputra River and basin

The 2580 km-long Brahmaputra River rises in Tibet at 30°31′ N and 82°10′ E south of Lake Konggyu Tso above 5000 m altitude in the Kailas range of the Himalaya. The Brahmaputra upper watershed is separated from that of the Indus and Sutlej Rivers by the Mariam La Pass (5150 m). As the Indus River, the Brahmaputra River is antecedent to the Himalayan relief.

The river, under its Tibetan name of Tsang Po, flows in southern Tibet for 1100 km with a course parallel to the Himalayan chain. On the left bank, it receives the Raja Tsang Po and the Kyi Chu which passes near Lhasa and, on the right bank, the Ngang Chu. For about 640 km from Lhatze Dzong (87°37′ E) the Tsang Po is wide and deep and represents the highest navigable inland waterway (3650 m altitude). At Tsela Dzong, it turns to the north-east and flows through gorges, falls and cascades. After a sharp hairpin turn, it runs southwards and emerges from the mountain under the name of Dihang River and receives two important tributaries, the Dihang and Luhit. Downstream of this point, the river is known as Brahmaputra. It crosses the Assam plain and receives the Subansin, Kameng and Manas on the right bank and the Buri, Dehing, Disang, Dikho, Jhansi and Dhansira on the left side. The Brahmaputra is then very broad (up to 10 km). Then the river turns southwards and flows into the alluvial plains of Bangladesh and joins the Ganges River at Goalundo.

The Brahmaputra basin is characterized by its high amount of rainfall (2125 mm as annual average rainfall for the whole basin).

From the source to Goalundo, the Brahmaputra watershed is 580 000 km^2. At Bahadurabad, the average annual peak discharge is about 64 300 m^3 s^{-1} and the average dry period yield 6510 m^3 s^{-1}. The annual discharge of the river is 614×10^9 m^3 (ECAFE, 1966). The specific yield is consequently 1.050×10^3 m^3 km^{-2} y^{-1}, very close to that of the Amazon (1137×10^3 km^{-2} y^{-1}).

It has been estimated that, through a tunnel 16 km long shortcutting the hairpin bend after Tsela Dzong, the power potential of the river is around 20×10^6 kW.

The Meghna River

The Barak River, the head-water of the Meghna, rises in the mountains of Assam at 94°10' E and 25°35' N at 2900 m altitude. It flows south-westerly for 250 km and emerges from the hills. Then, it runs west and divides into two rivers: the Surma and the Kushiraya, which join again at Markuli to form the Kalni River flowing south. The Kalni meets with the Ghorantia river and becomes the Meghna which flows into the Bay of Bengal.

The river is 950 km long and drains an area of 80 200 km². The Meghna watershed has a regime of very high precipitation: 5000 mm y^{-1} rainfall in the foothills of Assam and 12 500 mm y^{-1} at Cherrapunji at 1360 m altitude. For the whole basin, the average annual precipitation is 3500 mm. The average annual discharge at the Meghna at Bhairab Bazar is 3515 m³ s^{-1} and the total yearly discharge 111×10^9 m³. The specific discharge is 1375×10^3 m³ km^{-2} y^{-1} that is approximately 20% higher than that of the Amazon.

The Ganges basin

The Ganges River

The Ganges River or *Ganga* is 2527 km long. It drains the southern Himalayas and flows across the Hindustan Plain and into the Bay of Bengal (Fig. 28).

Two small rivers, the Alaknanda, rising at about 7800 m altitude near the Nanda Devi peak, and the Bhagirati, rising at about 6600 m altitude near Gangotri, form the head-waters of the Ganges. The river flows first westwards but, at Devaprayag, turns south and south-east. Near Farrukhabad, it receives the Ramganga River and, at Allahabad, the Yamuna River which also originates in the Himalayas, west of the sources of the Ganges. After the sacred town of Varanasi (Benares), the Gumti River, a left bank tributary, joins the Ganges. Further east, it receives the Gogra River and the Gandak River from the north and the Son River from the south. From the city of Patna, the Ganges becomes a wide and deep river which receives three additional tributaries from the north, the Buhri Bandak, the Baghmati and the Kosi. It then turns south and again south-east after the city of Farakka and receives the Mathabhanga River from the north. Then the river, called Padma, flows into Bangladesh and joins the Brahmaputra River at Goalundo. The combined river proceeds south-east and joins the Meghna River and the Bay of Bengal.

The delta of the Ganges extends from Farakka to the sea, covering an

India

area of 60 000 km². The Ganges has numerous spill channels running north–south from the river to the sea. The most important of these spill channels, the Bhagirathi–Hooghly, forms the western boundary of the delta.

The Ganges watershed extends over 967 000 km² and is one of the most densely populated of the world. The precipitation over the Ganges basin comes from the south-west monsoon and the cyclones originating in the Bay of Bengal. The average annual rainfall varies from 750 mm in the western part of the basin to about 2000 mm in the delta with a basin average of 1250 mm. The annual average flow (1934–62) is 11 610 m³ s^{-1} with seasonal variations ranging from 2460 m³ s^{-1} (January–May) to 31 000 m³ s^{-1} (July–September). The average annual discharge at Hardinge Bridge is 368×10^9 m³. The specific discharge is then 380×10^3 km^{-2} y^{-1}. An amount of 480×10^6 tons of silt is yearly transported to the sea via the Ganges River.

The Ganges' physico-chemical and biological features have been studied by Lakshminarayana (1965a,b,c) at Varanasi (Benares) (Fig. 29). The water level is maximal in July–August and minimal in March, the

Fig. 28. The Ganga (Ganges) River system (From Jhingran & Ghosh, 1978)

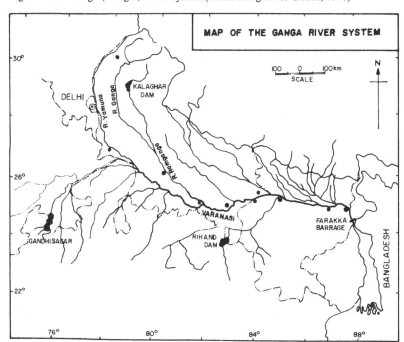

328 South-East Asia

amplitude of variation being as high as 15 m. The temperature of the water ranges from 19°C in December to 33°C in June–July (Fig. 29). The flood is accompanied by a fall in transparency as a result of the high silt content of the river. The content of dissolved oxygen which oscillates between 5 and 7 mg l^{-1} during most of the year falls to 2 mg l^{-1} in July–August. This drastic decline is related both to the increase in temperature and the considerable diminution of algal biomass and photosynthetic activity during the flood (Fig. 29).

The main chemical characteristics of the Ganges River at Varanasi are presented in Table 86.

The phytoplankton is heavily dominated by diatoms, mainly *Melosira*

Fig. 29. Physical, chemical and biological parameters of the Ganges River at Varanasi

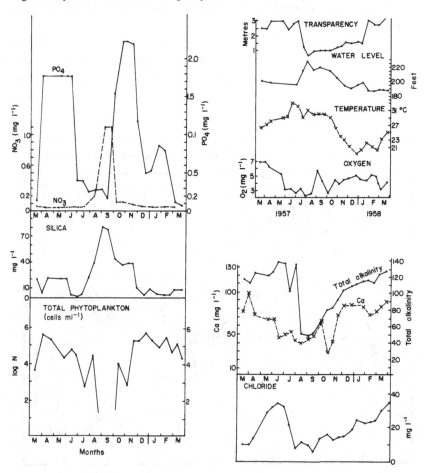

granulata, M. granulata var. *angustissima, Cyclotella meneghiniana, C. kuetzingiana, Fragilaria crotonensis, F. capucina* var. *lanceolata, Asterionella formosa, Synedra ulna, S. ulna* var. *aequalis, Navicula cryptocephala* var. *veneta.* The blue-greens form the second component of the algal assemblage with *Microcystis aeruginosa, Merismopedia tenuissima, M. minima, Gomphosphaeria lacustris, Anabaenopsis tanganikae* and *Anabaena* sp. The seasonal distribution is represented in Table 87.

The phytoplankton biomass is high all the year round with the exception of the flood period. The dramatic decline of algae during the rainy season is due to the dilution–flushing effect of the increasing amount of water on the concentration of algal cells and to the high turbidity which diminishes photosynthesis. The dilution effect of the flood seems to affect nearly all the chemical parameters, the concentrations of which drop steeply with the increasing discharge; for example, hydrogen carbonate alkalinity, high in summer, falls sharply in July. This could be interpreted as a hydrogen carbonate depletion resulting from the high algal activity of the previous months. However, the behaviour of hydrogen carbonates is very much similar to that of chlorides, conservative ions not involved in biological

Table 86 *Chemical features of the Ganges River at Varanasi (from Lakshminarayana, 1965a)*

pH	8.2–10	Cl^- (mg l^{-1})	6.4–36.4
CO_2 (mg l^{-1})	5.8–17.6	SO_4^{2-} (mg l^{-1})	1.5–3.3
Alkalinity (mg l^{-1})	44–130	NO_3 (mg l^{-1})	0.5–1.08
Ca^{2+} (mg l^{-1})	26–100	NH_4 (mg l^{-1})	0.01–0.44
Mg^{2+} (mg l^{-1})	24–167	PO_4 (mg l^{-1})	0.06–2.02
		Silica (mg l^{-1})	0.6–80.5

Table 87 *Seasonal distribution of algae in the Ganges River at Varanasi (from Lakshminarayana, 1965b) (number of cells per ml)*[a]

		Bacillariophyta	Gyanophyta	Chlorophyta	Euglenophyta
Summer	(Mar.–June)	888 717	105 675	55 545	2955
Rain period	(July–Oct.)	36 448	1 625	975	0
Winter	(Nov.–Feb.)	1 764 997	158 540	32 641	2500

[a] These numbers do not represent absolute numbers of cells of the river water since samples were taken with nets

synthesis but considerably diluted by the flood water. The dilution effect accounts also for the July–August minimum of phosphate (Fig. 29). The Ganges River receives considerable amounts of sewage throughout the year from Varanasi itself and from the surrounding country via the Assi River and the Barna River. The Ganges being a fast-flowing river, the algal development is more limited by strong currents than by nutrients. Signs of nutrient depletion are only occasionally observed (in March in the case of phosphate (Fig. 29) and in winter in the case of silica). It seems that should the current velocity be reduced by drought or human interference, the Ganges water would produce a much higher algal biomass. In highly polluted areas (Rajghat), Rai (1978) has reported the presence of the algal genera *Stigeoclonium* and *Schizomeris*.

Lakes of the northern bank of the Ganges

Lakes of the Mount Everest region. Löffler (1969) has investigated 24 lakes located south-west of Mount Everest in the watershed of Imja Khola, a tributary of the Kosi River. The lakes are located between 4500 and 5600 m, above the tree line, in the subnival belt. These lakes are all postglacial; among them, eleven may even be very recent, resulting from the progress of glaciers around 1850.

Most of the lakes are small (Tsola Tso, the largest, has a surface area of 0.55 km²) and their depth does not exceed 30 m. Nearly all these lakes dry up in spring and are refilled in summer. In Tsola Tso, the annual amplitude of water level exceeds 16 m and because of low air temperature no stable stratification is observed. In July the whole lake is homothermal around 8°C. In general, these lakes are typically cold monomictic. In lakes which are in contact with glaciers a frequent thermal anomaly occurs: water colder than 4°C loaded with glacial silt is overlain by water at 4°C creating a special type of meromictic stratification.

The conductivity of these lakes ranges from 7 to 43 microsiemens (μS). During Löffler's survey, concentrations of Cl^- were always < 0.4 mg l^{-1} and SO_4^{2-} varied between 0.1 and 4.3 mg l^{-1}. PO_4–P level was generally $< 1\,\mu$g l^{-1}. Turbidity was generally high due to morainic material and Secchi-disc readings were < 1 m.

The algal population was very thin and consisted mainly of *Cyclotella* sp., *Kephyrion* sp., *Chlamydomonas* sp. and *Ankistrodesmus* sp. The maximum density of any species was 2000 cells l^{-1}. Primary productivity, measured with a ^{14}C method, was as low as 3 mg C m^{-2} d^{-1}. The crustacean zooplankton consisted of *Arcodiaptomus jurisovichi*, *Daphnia tibetana* and *Daphnia* sp. The rotifers included *Hexarthra* cf. *bulgarica*,

Euchlanis sp. and *Polyarthra*, the maximum density never exceeding 50 organisms l^{-1}. Fish were completely absent. It is interesting to note that the pigmented *Daphnia tibetana* is found in clear lakes whereas the unpigmented *Daphnia* are observed in turbid water. The presence of the pigment seems to be an adaptation to high ultraviolet radiation. Nematodes dominated the benthic fauna followed by tardigrades. Oligochaetes were not found in all the lakes but harpacticoid copepods (genera *Maraenobiotus* and *Epactophanes*) were sometimes abundant. Ostracods were poorly represented. The chironomids were fairly abundant but molluscs, isopods and amphipods were totally absent.

Lakes and ponds of the Pokhara and Kathmandu Valleys (Nepal). In 1967–68, Hickel (1973*a,b*) investigated three lakes of the Pokhara Valley (900 m altitude) and two ponds in Kathmandu Valley (1400 m altitude). These two valleys are relatively flat areas of the Nepalese midlands which have developed in the basin of ancient lakes which have dried up. The morphometric features of the three lakes of the Pokhara Valley are shown in Table 88.

Phewa Tal is homothermal in winter at a temperature varying from 18 to 10°C. This high water temperature reflects the relatively high winter air temperature of the region (monthly mean temperature of 12–13°C in January). In May, a typical stratification develops with a surface temperature of 27°C and a hypolimnic temperature of 18°C. The pH is neutral during the homothermal period. During the stratified period, the pH is slightly alkaline in the upper layers and around 6 in the hypolimnion. The water is poor in electrolytes (conductivity 35–50 μS). In summer, there is a sharp decrease of oxygen in the hypolimnion.

The other two lakes were investigated only in winter. Their conductivity was lower than in Phewa Tal.

During the period of observation, the phytoplankton in Phewa Tal was

Table 88 *Morphometric features of three lakes of the Pokhara Valley*

	Surface area (km²)	Max. depth (m)	Mean depth (m)	Observations
Phewa Tal	5	23	6	Dammed for irrigation and power production
Begnas Tal	2.5	8	–	
Rupa Tal	1.4	4	–	

dominated by a diatom–desmid association. *Melosira islandica* fa *spiralis* dominated the winter algal assemblage. *Diatoma elongatum* was subdominant. The desmids reached a maximum in December with *Staurastrum phimus, S. leptodermus, S. protectum, Cosmarium moniliforme* and *C. pseudophaseolus*. In late December, the desmids decreased but rose again in March when *Melosira* declined. In winter, the Chlorococcales were abundant and diversified; they comprised mainly *Crucigenia tetrapedia. Ceratium hirundinella* and *Cryptomonas* were abundant especially in winter.

In Begnas Tal, *Melosira granulata* was dominant and *M. islandica* subdominant. Among the desmids, *Cosmarium* sp. and *Staurastrum* sp. were common. The blue-greens were represented mainly by *Microcystis aeruginosa*.

In Rupa Tal, *Ceratium hirundinella* was observed in large numbers although, in terms of biomass, *Melosira islandica* was dominant. *Cosmarium luetkemuelleri* was the most abundant desmid; the Chlorococcales were mainly represented by *Dictyosphaerium ehrenbergianum* and *Oocystis borgei*. The Cyanophyta were more abundant than in other lakes with *Anabaena* sp., *Microcystis aeruginosa* and *Spirulina gigantea*.

Most of the diatoms of these lakes belong to the temperate zone, for example *Melosira islandica* is found in oligotrophic lakes of alpine regions, but many of the Pokhara desmids are confined to tropical Asia.

The two ponds Tan Daha and Nag Daha, 150 m wide and with maximum depth of 6 and 4.5 m, respectively, are the largest water bodies of the Kathmandu Valley. They are not used for carp breeding.

The Kathmandu Valley receives an annual rainfall of 1700 mm; it has a cool and dry winter and a hot summer (10 and 25°C mean temperatures of January and July, respectively).

Tan Daha was found to be homothermal in January at 11°C. In June, the surface water was at 30°C and the bottom water at 20°C. The conductivity was 190 μS and Secchi visibility varied from 2 to 3.5 m. The phytoplankton was characterized by a moderate number of species, a small number of organisms and the dominance of nanoplankton over netplankton. During the cool season, the algae were mainly diatoms, especially *Navicula* sp.; in late spring, the diatoms declined to the benefit of green algae and all sorts of micro-algae such as *Chroococcus* sp. and *Monoraphidium convolutum*. In May, the small algae were replaced by *Ceratium hirundinella* and *Merismopedia*. In April–May, the zooplankton biomass composed of cladocerans and later of copepods exceeded the algal biomass.

Nag Daha was slightly acid, had a Secchi visibility of only 80 cm and was much richer in plankton ($83\,\mu$g C l^{-1}) than Tan Daha ($28\,\mu$g C l^{-1}). The zooplankton biomass, dominated by *Ceriodaphnia rigaudi* and *Daphnia longispina*, was also higher than the algal biomass which was especially rich in *Trachelomonas* sp., *Chrysococcus minutus, Ceratium hirundinella, Chroococcus, Monoraphidium convolutum* and *Merismopedia*.

From the quantitative point of view there is generally a decrease in plankton biomass during the monsoon, probably because of dilution and flushing. The subtropical-tropical desmids, abundant in the Pokhara Valley, were absent from the Kathmandu water bodies.

Lakes of Western Himalaya: Lake Nainital. Lake Nainital is located at 1937 m altitude at 29°24' N and 79°28' E in Western Himalaya. The lake is 1.4 km long and 0.4 km wide with a maximal depth of 24 m. A large portion of the sewage of the population of the watershed, including the 40 000 inhabitants of Nainital town is dumped into the lake. The air temperature varies from -5 to 30°C and the annual mean precipitation is 2271 mm. The lake was investigated by Pant, Sharma & Sharma (1980) in 1977–78. It belongs to the warm monomictic type with a winter homothermal profile at 7–8°C and a weak stratification in summer with 20°C in surface layers and 12°C in deep waters. The dissolved oxygen was about 6 mg l^{-1} at all depths in winter. In spring, it varied from 10 to 17 mg l^{-1} in the upper 10 m and ranged from 8 to 10 mg l^{-1} in the lower layers. In summer, the surface layers had 15–16 mg l^{-1} oxygen; at 10 m, the concentration was only 5 mg l^{-1} and 2 mg l^{-1} at 20 m.

From 1954 to 1975, the total hardness increased by 33% due to man-made deforestation and manipulation of the dolomitic catchment area. Similarly, ammoniacal nitrogen increased from $19\,\mu$g l^{-1} to $165\,\mu$g l^{-1} and NO$_3$–N from traces to $338\,\mu$g l^{-1} during the same period due to increasing sewage dumping. The level of PO$_4$–P of surface water varied from 6 to 7 μg l^{-1}.

The phytoplankton showed two distinct peaks: one in spring was composed of *Closteriopsis longissima, Chlorella vulgaris, Chlorococcum humicola, Closterium siamensis* and *Synedra ulna* and the second, in August, was dominated by *Chlamydomonas* and *Closterium acerosum*. In October, Cyanophyta were dominant with *Microcystis* sp. The small winter population was composed of diatoms. It is very interesting to note that *Ceratium hirundinella*, reported in the lake by Atkinson in 1892, has now completely disappeared in the newly established eutrophic conditions.

Rotifers like *Brachionus* and *Phillodina* dominate the zooplankton of Lake Nainital. Copepods form the bulk of the crustacean fraction of the zooplankton. Very numerous protozoans are also found (*Eutricha, Bursaria, Epistylis* and *Vorticella*).

The lake is a source of drinking water, produces hydroelectric power and attracts many tourists for angling. However, the famous game fish *Tor tor*, once abundant, has declined drastically. Moreover, the winter turnover is accompanied by mass fish killing caused by the recirculation of accumulated toxic substances.

Water bodies of the southern bank of the Ganges

The Yamuna River, the main tributary of the Ganges River, was investigated in 1954–55 upstream of Allahabad by Chakraborty, Roy & Singh (1958) and between 1958 and 1960 at Delhi by Rai (1974). Upstream of Delhi an irrigation barrage raises the water level of the Yamuna River. This allows the diversion of water into the West Yamuna and East Yamuna Canal systems which irrigate some 1800 km^2. The river then reaches the water works of Delhi which depends entirely on the river for its water supply. From Delhi and the countryside, the river receives considerable amounts of domestic sewage and agricultural drainage. At Okhla, 21 km downstream of Delhi, the river is dammed to permit irrigation. It is clear that this portion of the river is extremely loaded in inorganic and organic nutrients. At Delhi Intake Works the Yamuna discharge is about 33 m^3 s^{-1} in winter, but only 8 m^3 s^{-1} in summer. During the flood (August–October), the yield may reach 140 m^3 s^{-1}. In summer, the discharge of the river does not cover the city water requirements and ~10 m^3 s^{-1} are released from the Yamuna canals. The general bad quality of the water of the river and of the irrigation canals makes the water supply of the city of Delhi a difficult task.

Rai (1974) monitored eight stations and tried to characterize their degree of pollution by the biological assemblages and their quantitative variations. However, this latter parameter was not very useful since quantitative changes were essentially dependent upon fluctuations in discharge and turbidity, the highest biomass values being observed at low discharges. Conversely, floods caused nutrient dilution and poor plankton production. The prominent influence of hydrographic factors led, as in the Ganges River, to a bimodal distribution of algae with a minimum at monsoon period. Diatoms, as in the Ganges River, was by far the dominant group with green and blue-green algae as the second and third components.

Zooplankton was dominated by Rotifera although Protozoa were also abundant at polluted stations. In general, polluted stations were characterized by a systematic increase in blue-green algae, Rotifera and Protozoa if compared with cleaner stations.

Upstream of Allahabad, the general physico-chemical conditions are very similar: the water temperature fluctuates from 19°C in February to 32°C in June and the variations in the chemical parameters are mostly affected by the monsoon dilution. The phytoplankton peaks from April to August when it drops from 570 units* l^{-1} to 120 units l^{-1}. The bulk of the biomass consists of diatoms and Chlorophyta. From August to November, the small biomass is dominated by diatoms; blue-green algae represent the second most important group. From December to March, the Chlorophyta prevail, followed by blue-greens. The zooplankton has a similar April–July peak mainly composed of Rotifera and Entomostraca and reaching 180 units l^{-1}. From August to November, the biomass does not exceed 70 units l^{-1} and includes Rotifera and Protozoa. From December to March, the biomass slightly increases but remains dominated by Rotifera and Protozoa. The sharp decline in phyto- and zooplankton in August is caused by the flood dilution whereas the summer, pre-flood maximum is favoured by the quiescent conditions prevailing at this period. The concentration of nitrate and phosphate increases significantly during the flood probably because of the high levels of these nutrients in the upper valley.

Various water bodies located in the Ganges watershed have been investigated. George (1966) studied five fish tanks in the Delhi area in 1958–59. These are small and shallow water bodies (surface areas from 20 000 to 50 000 m² and depths around 2 m). Naini Lake, the largest of these tanks was made in 1948 for carp culture. The Naini water has a high hardness (> 100 mg l^{-1} $CaCO_3$) and a high sulphate content (85–163 mg l^{-1}). During the period of study, the phytoplankton showed two peaks, a small one in December dominated by the green alga *Pediastrum* and the diatoms *Synedra* and *Gomphonema* and a high maximum in July nearly entirely composed of the green alga *Tetraspora*. The zooplankton is dominated by Rotifera with high numbers of *Brachionus diversicornis* in May and June. It is interesting to note that, in spite of the general abundance of nutrients, the number of cells of algae in these tanks is relatively small: during the algal peak in Naini Lake there were only 88 000 cells l^{-1}. Mathew (1977) investigated the zooplankton of Govindgarh Lake near Rewa (24°24′ N and 81°15′ E) during two consecutive years (1968 and 1969). The quantitative and qualitative distribution of zooplankton was very similar during these two years: a January–February peak, mainly composed of Rotifera (150–300 units l^{-1}), followed by a July minimum and a minor October–November peak of Protozoa. The most frequent species of Rotifera were: *Pompholyx sulcata, Polyarthra vulgaris, Brachionus* sp.

*1 unit = 400 μm^2

and *Keratella tropica*. The Protozoa population included *Centropyxis, Difflugia, Euglypha* and *Didinium*. Misra & Singh (1968) investigated the periphyton developing in a temporary 2 m-deep pond in the botanical garden of Banaras Hindu University at Varanasi (Benares) and found that the most frequent subsurface species in October–November–December were *Amphora veneta, Cosmarium circulare, Characium ambiguum,* and *Gomphonema parvalum*. These were also the dominant species on the bottom although in smaller amount. *Pandorina, Volvox* and *Melosira* were restricted to the surface and *Botrydium* to the bottom.

Pichhola Lake, west of udaipur, was studied by Vyas (1968) in 1965–1966. The lake is 4 km long and 2.4 km wide and has a maximum depth of 5.8 m and a volume of 14×10^6 m^3. In complete contrast with what was observed in the Ganges River, the floods are accompanied by an increase of nutrient concentration (PO_4, NO_3, SiO_2 and dissolved organic matter). This is caused both by additional drainage and sewage water carried by the flood and by mixing of the bottom sediments. The blue-green algae, exceeding 2500 cells ml^{-1} in May–June with *Microcystis flos-aquae* and *Oscillatoria* sp. as dominant species and the green algae, reaching 1700 cells ml^{-1} with desmids as the main group, drop to 400 cells ml^{-1} in July with the onset of the flood. Then the diatoms peak with *Eunotia pseudolunaris, Synedra ulna* and *Nitzschia tryblionella* as the main species. In winter both diatoms and blue-greens remain very low whereas green algae fluctuate around 1500 cells ml^{-1}. As can be seen from Table 89 the total number of cells is nearly constant throughout the year.

The small and shallow Indrasagar Tank, also located in the vicinity of Udaipur has been investigated by Vyas & Kumar (1968). As in Pichhola Lake, the monsoon causes a substantial increase of nutrients. The seasonal succession of algae is very similar to that observed in Pichhola Lake: the blue-greens, exceeding 7000 cells ml^{-1} in May–June with *Microcystis flos-aquae* dominant, drop with the flood whereas diatoms develop until reaching 7000 cells ml^{-1} with *Navicula lanceolata, Nitzschia gracilis, Fragilaria brevistriata* and *Synedra ulna* as dominant species. Green algae develop also during the flood period and nearly reach 10 000 cells ml^{-1} with *Spirogyra, Oedogonium, Closterium, Stichococcus* and *Volvox* as the main genera represented. Green algae remain dominant also in winter as can be seen in Table 90.

India

Table 89 *Fluctuation of algae in Pichhola Lake (cells ml^{-1})*

	Bacillario-phyta	Cyano-phyta	Chloro-phyta	Total
Summer (Mar.–June)	5320	7714	3059	16 093
Rain period (July–Oct.)	7448	3724	3325	14 497
Winter (Nov.–Feb.)	3724	3458	6650	13 832

Table 90 *Fluctuation of algae in Indrasagar Tank (cells ml^{-1})*

	Chloro-phyta	Xantho-phyta	Chryso-phyta	Bacillario-phyta	Eugleno-phyta	Cyano-phyta
Summer (Mar.–June)	10 108	–	–	3458	–	21 413
Rain period (July–Oct.)	28 467	1995	3059	19 817	3059	13 699
Winter (Nov.–Feb.)	26 467	1596	–	5586	–	8113

Brackish water bodies of the eastern coast of India
Lake Chilka

Lake Chilka (Fig. 30A) is a shallow lagoon-like opening to the Bay of Bengal near Arakhakhud. It has a surface area of 1070 km^2. Its morphology is complicated by the presence of numerous islands between the lake mouth and the open lake (Fig. 30A). These islands are submerged only at maximal levels. On its north-east side, the lake receives the freshwater Daya River. The hydrography of the lake has been studied by Ramanadham, Reddy & Murty (1964).

Lake Chilka is a warm body where the annual fluctuations range from 24.4°C in February to 32.7°C in May and October. From July to September, the Daya River is in spate, the level of the lake is maximal (about 3 m). In the northern sector, the main currents are generated by the river inflow and directed towards the sea. They are strong enough to overcome tidal currents in the mouth. The salinity is then very low in the northern and central sectors (1.8–2.0‰) and in the mouth (3.5‰). It remains higher (10‰) in the southern sector where the water renewal is much slower. In post-monsoon period, the salinity remains low in the northern sectors, increases slightly in the central sector and in the mouth (5‰) but decreases in the southern sector (7–8‰) as a result of slow mixing. In February, the discharge of the Daya River is already negligible and there is no outflow from the lake into the sea. In May, the southerly wind is dominant and wind stress is higher than in other periods. The wind determines northward currents which cause the penetration of sea water into the mouth and into the lake where the level has dropped significantly as a result of evaporation. In the northern and central sectors, the salinity varies from 20 to 30‰ whereas sea water at 34‰ is found in the mouth. The southern sector remains at relatively low salinity (10‰). It follows that, in August, the lake is almost entirely 'fresh' and in May nearly completely brackish. This interesting regime confers on Lake Chilka original biological features which have been studied by Annandale & Kemp (1915), Devasundaram & Roy (1954) and Patnaik (1971).

The algal population was studied by Devasundaram & Roy in 1950–51 (published 1954) and during these two consecutive years it was maximal in June and consisted mainly of the diatoms *Asterionella japonica*, *Chaetoceros*, *Rhizosolenia* and *Coscinodiscus*, the blue-green *Anabaena* sp. and the green alga *Spirogyra* sp. *Asterionella japonica* constituted 40–50% of the total phytoplankton and *Chaetoceros*, 12–19%. *Anabaena*, completely lacking in 1950, contributed 14% of the total population in 1951. *Spirogyra* contributed 2–6%, *Coscinodiscus* 0.5–5% and *Rhizosolenia* 1–10%.

Fig. 30. A = Lake Chilka; B = Lake Pulicat

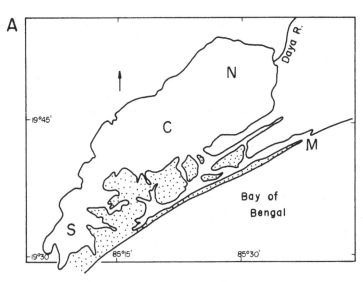

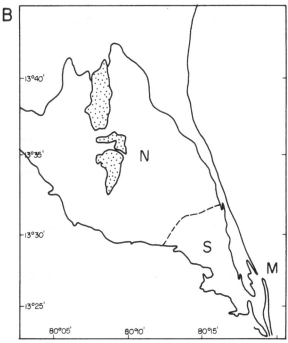

A later study by Patnaik & Sarkar (1976) indicates that diatoms form 67.3% of the algal assemblage, blue-greens 25.9%, dinoflagellates 3.5% and green algae 2.3%. These authors also observed two peaks in July and October.

The zooplankton was dominated by copepods which formed from 43 to 70% of the zooplankton population and consisted of *Acartia chilkaensis, A. centura, Acartiella major, A. minor, Paracalanus crassirostris, Pseudodiaptomus annandalei, P. binghami, P. hickmani, Oithona nana, O. brevicornis, Labidocera pavo* and *Parategastes sphaericus* var. *similis*. The veliger stage of gastropods contributed 5–32% and nauplius larvae 16–18% of the total zooplankton.

In absolute terms, the zooplankton is more abundant than the phytoplankton. However, it seems that its standing crop is too small to support the fish population of the lake which decreased by 24% in three years. The fish are dependent on the bottom fauna investigated by Patnaik in 1963–64. The Foraminifera dominated the benthos and were most abundant in the northern sector where they exceeded 200 000 organisms m^{-2} in March–April and dropped to 98 000 m^{-2} during the flood. *Polystomella, Spirillina* and *Rotalia* were the most common genera. Nematodes were the second most important group and were recorded throughout the year all over the lake. Copepoda was the third large group, recorded all year round in the whole lake. They were especially abundant in the outer channel. The most common genera were *Amphiascus, Pseudodiaptomus, Labidocera, Acartiella, Enhydrosoma, Nitaera* and *Canuella*. Gastropoda were also very abundant with maximal numbers in the northern and central sectors with *Stenothyra* as the main genus. Bivalves were also widely distributed in the lake especially from October to December.

In terms of biomass, the molluscs were dominant with values up to 35 g m^{-2} and represented from 74 to 93% of the biomass of the zoobenthos in the different sectors. Rajan (1971) studied the gut content of 23 species of fishes from Lake Chilka and showed that molluscs served as fish food in 78% of the examined animals.

The biomass of benthic fauna is maximal in October–December and minimal during the flood period. The mean annual standing crop varies from 111 to 183 kg ha^{-1} in the different sectors. Since the benthic fauna is abundantly consumed by the lake fishes, the annual productivity must be quite high which classifies Lake Chilka among the eutrophic groups of lakes.

Lake Pulicat

Lake Pulicat (Fig. 30B) is a large and shallow brackish water lagoon located on the east coast of India at 13°24′ N and 80°02′ E. Its surface area is 463 km² and its depth varies from 1 to 5 m. Its opening into the sea and connection to the inflowing freshwater Buckingham Canal are both located at the southern end of the lake. Its topographical and physical features have been studied by Chacko, Abraham & Andal (1953). During the three consecutive years 1968–69–70, a survey was conducted by Kaliyamurthy (1974) in order to define the fluctuation in major environmental parameters.

The lake can be divided into a large northern sector which includes a few islands, a smaller and narrower southern sector, and the mouth area.

During the previously mentioned survey, the water temperature was minimal in December (25°C) and showed two maxima in April–June and August–October. The monthly means reached 30.8°C in the southern sector and 32.8°C in the northern sector. Salinity dropped rapidly with the monsoon and peaked in May–June; variations were wider in the northern sector (0.5–51.5‰) than in the southern (7.3–40‰). Dissolved oxygen increased during monsoon and was minimal in the pre-monsoon period. The seasonal changes in pH were 8.3 to 8.7 in the southern sector and 7.4 to 8.8 in the northern sector.

The annual fish yield is about 1000 tons, 57–66% of which comes from the southern sector although its area is much smaller than that of the northern sector. Mullets, perches, clupeids, catfish and prawns constitute the bulk of the catch. The sandwhiting *Sillago sihama*, locally known as Kelangaan, represents 2–3% of the total catch (Krishnamurthy, 1969). The milkfish *Chanos chanos* is another valuable food fish and the migration of its larvae has been investigated by Rao (1970). The picturesque traditional fishery of the green crab (*Scylla serrata*) has been described by Thomas (1970). As in Lake Chilka, the benthic fauna is an important food item of the fish fauna. In the northern zone, the polychaetes were the predominant element whereas amphipods were the most frequent benthic animals in the south. Quantitatively, the northern sector had an average of 958 organisms m^{-2}, the southern sector 3767 m^{-2} and the mouth area only 385 m^{-2} (Krishnamurthy, 1970). Raj & Azariah (1967) reported the presence of the amphioxus *Branchiostoma lanceolatum* in Pulicat Lake. This was the first time this species was found in a tropical estuarine habitat.

The Godavari and Krishna watersheds

The Godavari River runs through forested mountains and in its lower course flows through the plains of the Andhra State and discharges into the Bay of Bengal. In its lower course, it is dammed at Dammagudam and Dowaleshwara where a lake of 352 km² has developed. This lake has been monitored for water quality and is reported by Pais Cuddou *et al.* (1973) to have low salinity and low hardness throughout the year.

Lake Gorewada and the Ambarasi Reservoir contribute to the water supply of the city of Nagpur (700 000 inhabitants). They have been investigated by Hussainy & Abdulappa (1973) and Krisnamoorthi, Gadkari & Abdulappa (1973), respectively.

Lake Gorewada is located 4 km north-west of Nagpur at 300 m altitude and has a surface area of 259 ha and a maximal depth of 5 m. At the time of investigation (2–3 March 1966), the lake temperature varied from 22°C (night) to 27.7°C (mid-day). pH varied from 7.9 to 8.3 on the bottom and from 8.4 to 8.6 in surface layers. Total dissolved solids varied from 100 to 120 mg l^{-1}; calcium and magnesium were the dominant cations with 52 and 40 mg l^{-1}, respectively. Phosphates, nitrates and nitrites were below detection limit. The main alga was *Ceratium hirundinella* which occurred in bloom concentrations. The productivity was 300 mg C m^{-2} d^{-1}. The crustacean zooplankton was dominant and represented by the copepods *Orthocyclops, Mesocyclops, Diaptomus*, the Cladocera *Bosmina, Alonella* and *Moina*, and the rotifers *Keratella cochlearis* and *Brachionus calyciflores*. Hussainy & Abdulappa assumed that the algae utilized nutrients released by zooplankton.

The Ambarasi Reservoir is located at 6 km from Nagpur and at 330 m altitude. Its surface area is 2.6 km^2 and it drains an area of 15.5 km^2; its maximal depth is 7 m. The reservoir supplies 440 m^3 d^{-1} with chlorination as the only treatment. The aquatic vegetation consists of partly submerged plants such as *Hydrilla, Chara* and *Hydrodictyon*. The seasonal variation of water temperature ranged from 20 to 33°C. The pH was always above 8. The bottom water was never anoxic although sometimes well under saturation in comparison with surface layers. The depletion was never sufficient to cause fish killing. The low BOD_5 (maximum of 10 mg l^{-1}) indicates the absence of organic pollution. As in Lake Gorewada, *Ceratium hirundinella* was the dominant alga. Occasionally blooms of *Microcystis* and *Spirulina* were observed. Crustacean zooplankton which easily escapes into the water supply system causes turbidity and bad taste and odour when it decomposes. Among the rotifers, *Keratella tropica* was especially abundant in summer.

The bottom fauna was composed of chironomid larvae, *Chironomus tendepediformis* and *C. tentans*, which are not affected by chlorination and are a source of trouble in water supply systems. A small number of tubificid worms, *Limnodrilus hoffmeisteri* have been recorded. Ten species of molluscs were observed but their biomass was small.

Sitaramaiah (1967) established the energy-flow diagram of *Tatayya gunta*, a pond situated at Tirupati town (79.5°E and 18.8°N).

In the Krishna watershed, Prasadam (1977) studied the development of plankton in washing ponds and irrigation tanks. In the former, very eutrophic water bodies, the blooms of phytoplankton somewhat limit the biomass of zooplankton mainly composed of Protozoa and Rotifera. In the irrigation tanks, high turbidity and rapid renewal of water maintain the algal biomass at low levels and the zooplankton, rich in crustacean species, reaches biomass values which exceed that of phytoplankton. However, in absolute terms, the zooplankton biomass of the washing ponds (300–2000 units l^{-1}) is much higher than that of the irrigation tanks (80 units l^{-1}).

The Krishna River has been dammed in its lower course and a large man-made lake has been formed. The Nagarjuna Sagar has a surface area of 286 km², a depth of 60 m and a volume of nearly 12×10^9 m³, with a live storage of 6.8×10^9 m³ (Ganapati, 1973).

South India
The Cauvery River

South India is mainly composed of the catchment area of the Cauvery River and the small watersheds of littoral streams on the eastern coast. The Cauvery River and its affluents rise near the western coast of India in the Western Ghats. The Cauvery River runs 764 km in a south-easterly direction and flows into the Bay of Bengal. In its upper course, the river crosses mountainous regions and passes through impressive falls: the 97 m Sivasamudram falls in Mysore State and the 21 m Hoginakal falls in Madras State which represent the upper limit of the Stanley Reservoir, a large man-made lake formed by damming the Cauvery River at the gorge of Mettur. Downstream of the Mettur Dam, the river flows in the plains where it is joined by the Bhavani, Amaravati and Moyar Rivers.

Numerous reservoirs and man-made lakes have been formed on the rivers to keep flood water for irrigation during the dry period. They were briefly described by Ganapati (1973) at the third Symposium devoted to man-made lakes.

The man-made lakes on the Cauvery River and its tributaries

The Krishnaraja Sagar. The Krishnaraja Sagar has been built about 20 km NW of Mysore City across the Cauvery River for hydro-electric and irrigational purposes. Its storage capacity is 1.2×10^9 m³.

Stanley Reservoir. Stanley Reservoir, resulting from the damming of the Cauvery River at Mettur in 1934, is located at 243 m altitude. Its surface area is 156 km² and its volume 2.7×10^9 m³. It has a maximal depth of 46 m and a mean depth of 18 m. The normal floods are evacuated through the 'Ellis Surplus', a 6 km channel built on the eastern side of the dam. The surplus channel is closed as soon as the level reaches 236 m. Limnological studies were performed in Stanley Reservoir by Sreenivasan (1966), Ganapati & Sreenivasan (1968), Ganapati (1969) and Ganapati & Sreenivasan (1970).

The reservoir is never really stratified. It is nearly homothermal in January–February and in October the difference in temperature between upper and bottom layers does not usually reach 4°C. The highest surface temperature recorded was 32°C in May 1962 and the lowest bottom temperature 25°C in February 1958. The circulation in the reservoir is probably not very active since, even in February, a severe oxygen deficit prevails in lower layers. During the wet summer, the oxygen-rich flood water penetrates into the deep layer and improves oxygen conditions. The Stanley Reservoir water is poor in electrolytes (conductivity ranges from 138 to 265 μmhos cm^{-1}) and alkaline (surface pH ranges from 8.2 to 9.1). Hydrogen carbonate (up to 180 mg l^{-1}) and chloride (10–17 mg l^{-1}) are the main anions and calcium the dominant cation (11–16 mg l^{-1}).

The primary production was found to be around 3.2 g C m^{-2} d^{-1} (oxygen method) for a net phytoplankton consisting mainly of *Oscillatoria* sp., *Nitzschia* sp., *Merismopedia* sp. and *Peridinium* sp. Ganapati & Sreenivasan (1970) worked out the oxygen budget of the reservoir and found that, as an average, the respiration of the lake community represented 48% of the total oxygen production.

The construction of the dam seriously limited the upstream migration of fish which however remains possible during the wet period through the 'Ellis Surplus'. The fish population consists mainly of carps (*Catla catla, Cirrhina cirrhosa, Labeo kontius, L. fimbriatus, L. calbasu, Barbus dubius, B. hexagonolopis*) and catfishes (*Mystus seenghala, M. aor, Wallagonia atu, Pangasius pangasius, Silondia*). *C. cirrhosa* is the typical indigenous carp whereas the Bengal carps *Catla catla* and *Mrigal* were introduced in 1934 and established satisfactorily in the reservoir. However, the total fish catch amounts only to 0.1% of the primary production instead of the generally recognized value of 1%. This discrepancy may be partly explained by the fact that it is impossible for the fish and fishbrood of the 'Ellis surplus' to reach the reservoir when the reservoir level is low and the surplus gates are closed. Then, the fish congregate in the rock pools of the surplus. There their existence is endangered by the scarcity of food and oxygen but also by noxious industrial wastes dumped into the surplus by the Mettur Chemical Industries. At low water levels, when dilution is minimum, this practice generates massive fish killing which doubtless contributes to the impoverishment of the reservoir fish stock.

Bhavanisagar Reservoir. Bhavanisagar Reservoir is formed by the impoundment of the Bhavani and the Boyar Rivers below their junction. At full capacity, the water level is 280 m, the mean depth 11.3 m and the volume 812×10^6 m^3. The two rivers originate in the Nilgiri Hills and drain lateritic soils which supply considerable amounts of silt during floods (Sreenivasan, 1964a).

Highest temperatures are generally reached in April (29–30°C in surface water and 28°C in bottom water) and lowest in July–August (24.2 and 23.1°C in surface and bottom water, respectively). The relatively low summer temperatures are due to the cold waters coming from the Nilgiri Hills. Bottom oxygen deficits are generally observed from March to May, sometimes even earlier. The electric conductivity is around 65 μmhos cm^{-1}. Thermal stratification is rarely observed because of the outflow of cool hypolimnic water through low level sluices. The pH is generally above 8. Although the silt is rich in apatite, dissolved inorganic phosphorus can hardly be detected, probably rapidly taken up by the relatively abundant *Microcystis*. Although the flushing time is short (4 months), the productivity of the reservoir and its algal stock are not affected since the water is removed from lower, oxygen-deficient layers and the carp population with the same species as in Stanley Reservoir is abundant.

Amaravathy Reservoir. The Amaravathy Reservoir is formed by impounding hill streams including the Amaravathy River. The rivers drain mountainous, uninhabited, forested areas. The surface area of the reservoir is 9 km^2 and its volume 115×10^6 m^3 at full capacity. The maximum and mean depths are, respectively, 35 and 14 m. The water level is at 358 m above MSL. Stratification is rarely observed in spite of the relatively great depth and absolute temperatures are close to those observed in Stanley and Bhavanisagar Reservoirs. In early summer, a severe oxygen deficit is observed in deep layers although the oxygen-poor hypolimnic water is continuously removed through low-level intakes.

Microcystis aeruginosa is present throughout the year. *Botryococcus* is the second most important species. *Brachionus, Daphnia* and *Cyclops* seem to be the dominant zooplankton genera especially in winter. Primary productivity is high: 1.3 (September) to 6.9 (April) g C m^{-2} d^{-1}. The indigenous fish population includes *Barbus dubius, B. tor, Labeo kontius* and *Macrones* sp. Introduced fishes comprise *Catla catla, Cirrhina cirrhosa, Labeo calbasu, L. fimbriatus, Cirrhina reba* and *Tilapia mossambica*.

Upland water bodies
  Ootacamund Lake (Ganapati, 1957). This small 'lake' has been formed by damming a mountain stream at 2317 m altitude in the Nilgiri mountains; its surface area is 0.28 km^2 and maximum depth, 9 m. The lowest water temperature recorded was 14.4°C at the bottom and the highest was 21.6°C at the surface. A distinct stratification was observed in the deepest part of the lake and was often accompanied by complete lack of oxygen and presence of sulphides in the hypolimnion, that is below 3 m. The lake is surrounded by a rich marginal vegetation and includes numerous hydrophytes such as *Potamogeton indicus, Vallisneria spiralis* and *Hydrilla verticellata* which are covered by epiphytic algae and zooplankton. In spring, the dinoflagellate *Ceratium hirundinella* produces a thick bloom and the only other alga found in significant amount belongs to the genus *Closterium*. The zooplankton is then dominated by the Cladocera *Daphnia, Ceriodaphnia, Bosmina, Simocephalus* and *Chydorus*, and by Rotifera. The Cladocera are generally found above or below the *Ceratium* layer; they are rarely mixed with the algae. The fish population includes indigenous species such as *Barbus*, loaches and minnows. The lake was stocked in 1827 with the carp *Carassius carassius* and the tench (*Tinca tinca*) and, in 1939, with *Cyprinus carpio*. These carps found excellent breeding conditions in littoral areas and constitute the dominant fishery of the lake.

Mukerti Reservoir. The Mukerti Reservoir is located at 2073 m above MSL in the Nilgiri mountains. It has a surface area of 3.7 km^2, a volume of 11×10^6 m^3 and a maximal depth of 21 m. In March, the reservoir was found to be homothermal around 18°C; 15°C seems to be the minimal temperature of the hypolimnion. The homothermy does not prevent anoxia developing below 18 m. The phytoplankton is dominated by *Cylindrocystis* sp. and *Peridinium* sp. but primary productivity is lower than in Ootacamund Lake. The oligotrophic conditions prevailing in this reservoir and the relatively small volume of water affected by anoxia allow the satisfactory development of the rainbow trout with which the lake is stocked.

A few other upland 'lakes'. The main characteristics of a few high-altitude water bodies are given in Table 91.
  Yercand Lake has its nutrients stripped by *Eichhornia crassipes*, and *Microcystis aeruginosa* occurs in mild blooms throughout the year. The

Table 91 *Main characteristics of a few high-altitude water bodies (Sreenivasan, 1964b)*

Name	Location (hills)	Nature	Altitude (m)	Area	Depth (m)	Temperature Winter (°C)	Temperature Summer	Observations
Yercand	Shevroy	Lake	1340	8.9 ha	5.5 max. 2.4 min.	19	24	Clean
Kodai	Palni	Reservoir	2285	39 ha	11.5 max. 3.0 min.	18	22	Clean
Mannavanur	Palni	Reservoir	–	20 ha	–	–	–	
Ooty	Nilgri	Lake	2500	34 ha	12	16	22	Polluted
Pykara	Nilgri	Reservoir	2130	16 km^2	35	–	23	Clean

range of hydrogen carbonate extends from 10 to 100 mg l^{-1} whereas the chloride, generally around 10 mg l^{-1}, never exceeds 50 mg l^{-1}.

Ooty Lake, studied by Ganapati (1957), and Sreenivasan (1964b) shows depletion of oxygen in lower layers even when stratification is not stable. It is heavily polluted by domestic sewage.

Sreenivasan (1964c) gives a detailed list of these water bodies. *Cyprinus carpio* varieties were introduced in all of them but got established only in the Pykara and the Mannavanur Reservoirs. In other reservoirs, the presence of *Gambusia affinis* seemed to be responsible for the failure of the German carp to breed since they bred and flourished after poisoning of the *Gambusia*. The fish production is 5.3 kg ha^{-1} in Kodai Lake, 32 kg ha^{-1} in Yercand Lake and 75 kg ha^{-1} in Ooty Lake; this represents 0.1% of the primary production in the first two lakes and 0.18% in the third.

Sathiar Reservoir. The Sathiar Reservoir was formed by the damming of the Sathiar Odai River, 30 km from Madurai City at 10°07′ N and 78°05′ E. The Sathiar project was completed in 1967. The surface area of the reservoir is 38 ha and its maximum depth 6 m. In January–February, the surface and bottom water are, respectively, at 28 and 26.5°C, without any clear stratification. In May, the corresponding temperatures are 32.8°C and 31.5°C but, then, the reservoir is only 2 m deep. In August, after the reservoir fills again the surface temperature is 31.5°C and the bottom water at 3 m depth drops back to 26°C. Then, the thermocline is well defined between 1.5 and 2 m. Below 2 or 3 m, the water is nearly always depleted of oxygen. Near the surface, the concentrations reach 14–15 mg l^{-1} in spring and the pre-monsoon period and drop to 6 mg l^{-1} during the flood. The pH ranges from 8.0 to 9.2 in the epilimnion and from 7.2 to 7.8 in the hypolimnion. Total phosphorus which varies from 0.12 to 0.20 mg l^{-1} from April to August rises rapidly after the flood (8 mg l^{-1}) and remains high until March of the next year. A similar pattern is followed by nitrates, organic nitrogen and dissolved organic matter (Kannan & Job, 1980a). Gross and net primary productivity increase from April to July when the lake is relatively shallow and decrease from August to November with the floods. The gross productivity ranges from 1.90 to 9.30 g C m^{-2} d^{-1} (Kannan & Job, 1980b). Saunders, Coffman, Michael & Krishnaswamy (1975) studied the extracellular release of carbon in three ponds near Madurai. In two of these ponds, rich in *Microcystis* and *Anabaena*, the released carbon amounted from 1.4 to 10% of the net photosynthesis.

Job & Kannan (1980), studying the detritus in the water column and in the sediments, showed that the detritus-bound energy of the water column is maximal in winter. The subsequent decrease of detrital energy in the water column is accompanied by an increment of detrital energy of the sediments, indicating that, even in tropical water bodies, productivity, sedimentation and bottom decomposition are cyclic processes. The cycles, however, are not generated by temperature variations but by rainfall seasonal distribution. This is very clear in coastal water bodies such as Lake Vembanad where salinity drops from 30‰ to 2‰ with the monsoon floods. This does not seem to be detrimental to the benthic fauna of this lake, Ansari (1974) reports values of 50 to 312 g m^{-2} with maxima of 418 g m^{-2}.

Western India
Ajwa Reservoir

The Ajwa Reservoir is the source of potable water for the City of Baroda. It has been formed by damming the river Surya at 25 m above MSL. It has a surface area of 145 ha, a maximal and mean depth of 9 and 3.4 m, respectively.

The water temperature varies from 17°C in January to 29–31°C in September–October. No durable stratification is observed. In spring or autumn, a labile thermocline develops in the afternoon and disappears at night. In consequence, the reservoir is oxygenated down to the bottom throughout the year. Jayagoudar (1980) thinks that these permanently oxidized sediments prevent the release of nutrients which explains their scarcity in the water column. The diatoms have the highest number of species (72%) but the biomass is dominated by *Pediastrum simplex*, and *Botryococcus braunii* is present at certain seasons. The phytoplankton is not abundant and its biomass is always lower than that of zooplankton which is mainly composed of rotifers. However, the reservoir is famous because of the presence of the Cladocera *Indialona ganapati* n.gen. et n.sp. Petkovski. The primary productivity varied from 0.15 to 2.90 g C m^{-2} d^{-1} in 1963 and from 0.15 to 0.59 g C m^{-2} d^{-1} in 1964 (Ganapati & Pathak, 1972).

WEST MALAYSIA
General

West Malaysia is located between the Equator and 5°N. Its latitudinal situation and its maritime exposure confer on the peninsula a warm and humid climate. There are two main seasons: the south-west monsoon from May to September and north-east monsoon from October to March. The mountain ranges being oriented north-north-west to south-

south-east, the east coast gets the maximum rainfall during the north-east monsoon; central and western Malaysia are sheltered from the north-east monsoon by the Malaysian mountains and from the south-west monsoon by those of Sumatra and get their maximum rainfall in the intermediary seasons between the monsoons. The western coast of Malaysia is very much affected by the 'sumatras', violent squalls with thunderstorms and rain developing at night in the Malacca straits from April to November. The annual rainfall generally fluctuates from 2000 to 3000 mm (Koteswaram, 1974). The mean monthly temperature is around 25°C throughout the year but the high humidity prevailing in summer makes the climate very oppressive in the lowlands.

The Malaysian peninsula is mainly composed of Palaeozoic rocks deposited in thick geosynclinal series. They emerged in the early Mesozoic and were metamorphized by massive intrusions of granites. In the late Jurassic, orogenic activity diminished, part of the area was submerged and came to final emergence only in the Tertiary and was subsequently submitted to intense erosion.

The rain forest extends over 93 000 km^2 (70% of the land) and contains a high percentage of large Dipterocarps. In 1965, the Dipterocarps represented 64% of the timber produced by the Malaysian forests.

The water bodies of West Malaysia

Besides the river system the inland waters cover an area of 6000 km^2, i.e. 4.4% of the area of the country. They include freshwater fish ponds (1.5 km^2), artificial lakes (5 km^2), salt or mangrove marshes (3000 km^2) and rice fields with fish culture (3000 km^2). These data are according to the Indo-Pacific Fisheries Council (1952). There are also numerous tin-mining pools utilized for fish production (Jothy, 1968). According to more recent data (Fernando & Furtado, 1975), the area of rice fields with fish culture has already reached 5000 km^2 and reservoirs extend over 200 km^2. The water bodies of the coastal areas have been much more thoroughly investigated than those of the interior (Johnson, 1967a,b).

The river waters of the western coast
The Gombak River

A detailed survey of the Gombak River located at 3°15′ N and 101°43′ E (Bishop, 1973) gives a global picture of the water chemistry and biology of the area. In the upper valley, covered with evergreen rain forest, the river water has low conductivity, a slightly acidic pH, a low alkalinity, low concentration of major cations and nutrients and high levels of silica

(Table 92). The Gombak waters are not as poor as those of Amazonia or southern Malaya but compare with the waters of southern Thailand and of rain-forest rivers of Central Africa. However, the Gombak waters are clear and not loaded with humic acids. A progressive enrichment takes place in the lower valley where the limestone bed rocks and the urban development supply major cations and nutrients, respectively.

These waters have a very poor phytoplankton population. The algal population is dominated by the epilithic flora. In the upper valley, in spite of the nutrient scarcity, the blue-green algae dominate with *Tolypothrix, Nostoc, Lyngbya, Chaemaesiphon, Phormidium*. In the lower valley, the epilithic flora is rich in diatoms with *Cymbella, Gomphonema* and *Frustulia*. The epipelic flora is abundant only in the middle and lower course of the river and comprises *Oscillatoria, Schizothrix, Anabaena* and *Nitzschia*. The root epiphyte community contains only a few diatoms such as *Achnantes* and *Cymbella*. In cascades exposed to high insolation *Spirogyra, Oedogonium* and *Rhizoclonium* develop in thick mats over the rocks. In summary, the taxa are more numerous in the epilithic community (47 as an average of the sampled stations) than in the epipelic (30 taxa) and epiphytic communities (eight taxa). Measurements of accumulation of ash-free dry mass on artificial substrates gave rates varying from 12 to 125 mg m^{-2} d^{-1}.

The benthic fauna in depositional areas showed mean biomass values of 639 mg dm m^{-2} in the lower nutrient-rich valley. In erosional areas, the

Table 92 *Chemical features of the Gombak River (Bishop, 1973)*

	Upper valley	Lower valley
pH	6.97	6.58
Conductivity (μmhos cm^{-1})	38.4	41.2
Ca^{2+} (mg l^{-1})	0.49	0.64
Mg^{2+} (mg l^{-1})	0.88	0.50
Na^+ (mg l^{-1})	2.11	2.59
K^+ (mg l^{-1})	1.27	2.62
Fe (mg l^{-1})	0.12	0.73
SiO_2 (mg l^{-1})	11.7	9.8
HCO_3^- (mg l^{-1})	17.35	17.93
Cl^- (mg l^{-1})	0.44	0.88
NO_3-N (mg l^{-1})	0.12	0.20
PO_4-P (μg l^{-1})	18	414
SO_4^{2-} (mg l^{-1})	0.90	0.13
BOD (mg l^{-1})	0.21	3.40

corresponding values were 714, 1320 and 3233 mg dm m^{-2}. In terms of number of organisms, the Diptera, mostly Chironomidae, were dominant especially in the upper and middle valley, followed by Coleoptera in the upper and middle course and by Annelida in the lower course. In terms of biomass Coleoptera were dominant (up to 79% of the total biomass) whereas Mollusca represented up to 57% of the biomass.

The ichthyological fauna includes 27 species belonging to 21 genera and 12 families. The Cyprinidae dominate with 30% of the species, the Channidae and Anabantidae account for another 26%. *Channa gachua*, *Clarias teijsmanni*, *Glytothorax platypogonoides* and *Tor soro* were found only in the upper valley. *Acrossocheilus deauratus*, *Mastacembelus maculatus*, *Silurichthys hasseltii*, *Glyptothorax major*, *Puntius binotatus* and *Macrones wyckii* are found all along the river. *Mastacembelus armatus*, *Mystacoleucus marginatus*, *Rasbora sumatrana*, *Clarias batracus*, *Osteochilus hasseltii*, *Channa striata*, *Hampala macrolepidota* and *Dermogenes pusillus* are mostly encountered in the middle and lower valley. *Betta pugnax* and *Poecilia reticulata* are more numerous in the upper course. Thirteen species were found to be omnivores such as *P. binotatus*, *G. major* and *G. platypogonoides*; among the 14 carnivorous species, *B. pugnax*, *S. hasseltii* and *D. pusillus* feed mostly on exogenous insects, *M. maculatus* and *P. reticulata* on endogenous invertebrates, whereas *M. wyckii* and *Channa* sp. eat mainly crustaceans and fish. *Osteochilus* is a detritivore. The food web, essentially based on exogenous material such as terrestrial plants and insects, is similar to that of Amazonian rivers and streams. The total yearly yield is 30 tons with *Macrones*, *Hampala*, *Clarias*, *Osteochilus*, *Rasbora*, *Puntius* and *Channa* as the main contributors.

Subang Lake

The Subang Lake, a reservoir on the Subang River, built in 1928 and located at 3°10′ N and 101°29′ E, was investigated by Arumugam & Furtado (1980). The reservoir has a surface area of 66.4 ha, a volume of 3.5×10^6 m^3 and a maximal depth of 8 m. In recent years, the reservoir has been polluted by effluents from oil palm and rubber industries and has experienced blooms of *Microcystis* and *Anabaena*. The chemical features of the reservoir are shown in Table 93.

In certain areas of the lake, a stable stratification takes place since the night mixing affects only the upper 3 m. This results in the formation of an anoxic hypolimnion in the deepest part of the lake. The 'destratification' of this lake occurs according to a very unusual process: after heavy rains, sublacustrine and underground seepage water accumulates in the lower part of the lake, pushing upwards the anoxic, nutrient-rich, hypolimnic water. Since the underground water is poor in salts and nutrients, the vertical profiles of most chemical elements show a marked maximum at the upper limit of the volume occupied by the underground water. When the rain subsides and the discharges of the sublacustrine spring diminishes, this maximum moves back downward, re-establishing the stratification. The 'destratification' process occurs in spring and autumn during the rainfall peaks.

Anabaena, Crucigenia, Staurastrum, Mallomonas, Peridinium, Phacus, Trachelomonas and *Euglena* are among the algal species most frequently encountered in the epilimnion. *Merismopedia* develops mainly in the anoxic, sulphide-rich hypolimnion. The epilimnic zooplankton comprises the protozoan *Difflugia*, the rotifers *Brachionus, Trichocera, Anuraeopsis* and *Polyarthra*, and several species of cladocerans and cyclopids. The metalimnion is rich in ciliates (*Microstomum* and *Chaetonotus*) and the hypolimnion in *Chaoborus* and *Microstomum*.

Nutrient budgets have shown that this naturally oligotrophic reservoir is artificially loaded with phosphate, ammonia and potassium which cause planktonic algal blooms and proliferation of *Salvinia molesta*.

Five freshwater species of red algae and one variety (*Ballia pinnulata, Batrachospermum beraense, B. cyclindrocellulare, B. tortuosum, Calaglossa ogasawaraensis* var. *latifolia* and *Tuomeya gibberosa*) were recently described by Kumano (1978) as new taxa; another three species (*Batrachospermum vagum, Hildenbrandia rivularis* and *Sirodotia delicatula*) were described for the first time from peninsular Malaysia. This finding is especially interesting if we take into account the scarcity of Rhodophyta in freshwater.

Table 93 *Range of chemical features of Lake Subang (Arumugam & Furtado, 1980)*

Temperature (°C)	24.6–32.8	K^+ (mg l^{-1})	0.4–3.6
pH	4.6–9.5	Ca^{2+} (mg l^{-1})	0.2–2.8
Conductivity (μmhos cm^{-1})	7–54	Mg^{2+} (mg l^{-1})	0.1–3.2
D.O. (mg l^{-1})	0.6–18.2	SiO_2 (mg l^{-1})	1.9–4.5
Alkalinity (mg l^{-1})	1.6–17.2	NH_3–N (mg l^{-1})	0–0.9
Cl^- (mg l^{-1})	1.6–8.5	NO_3–N (mg l^{-1})	0–0.3
SO_4^{2-} (mg l^{-1})	0–4.1	PO_4–P (mg l^{-1})	0–0.8
Na^+ (mg l^{-1})	1.0–4.2		

The River Pahang basin

The River Pahang rises in the central mountains and flows into the Sea of Borneo after draining a vast area through a very extended network of tributaries.

Tasek Bera is a shallow swamp-like lake with a dendritic shape. It is one of the sources of the River Pahang on its left side. Tasek Chini is another shallow lake of the Pahang watershed. These lakes have been studied by Mizuno & Mori (1970), Lim (1974), Lim & Furtado (1975) and Richardson & Jin (1975).

Tasek Bera is located in an undisturbed area and surrounded by rain forest. Many aquatic plants such as *Nymphoides* and *Utricularia* and the forest supply large amounts of organic detritus which decompose in the lake. The water is dark and belongs to the acid blackwater type of tropical areas. The lake is 4.8 m deep but the level rises up to about 3–4 m in the rainy season. The pH, measured by Mizuno & Mori in May 1968, was 4.7. The thick bottom detritus layer releases gases and does not harbour any benthic fauna. Conversely, *Chironomus* larvae, leeches, Cladocera and small shrimps are very abundant among the macrophytes. Tasek Bera has a high flushing rate which may contribute to its paucity in plankton. In contrast it has a rich fish fauna which includes the rare osteoglossid *Scleropages formosus*. These waters have a low primary productivity: in Tasek Chini where *Dinobryon* and *Staurastrum* are the dominant algae, the productivity is only 400 mg C m^{-2} d^{-1}.

Tasek Merah is an ancient swamp on the River Kuran dammed in 1965. Its surface area is 30 km^2 and maximum depth 10 m. Its pH is 5.5–5.8. The Cameron High Land Reservoir is also a man-made lake at 1300 m altitude. Its pH varies from 6.6 to 7.0.

Tasek Bera and Tasek Merah have a rich flora of desmids such as *Euastrum, Micrasterias, Desmidium, Hyalotheca, Staurastrum, Cosmarium, Xanthidium, Closterium* and of diatoms like *Eunotia* and *Frustulia*. A detailed list of algae for these water bodies is given in Mizuno & Mori (1970). The small shrimp *Caridina pareparensis* (Atyidae) was commonly found together with *Macrobachium sudanicum* (Palaemonidae) in the aquatic plants of Tasek Bera. Molluscs are absent in Tasek Bera and rare in Tasek Merah.

As in Amazonia, the food web of these lakes is not based on phytoplankton production. Mizuno & Mori have found that many fish of Tasek Bera feed on aquatic insects, young fishes and terrestrial insects. It is interesting to note that bottom-dwelling larvae were abundant in fish stomachs although they were not found in sediment samples.

Waters of southern Malaya

The waters of southern Malaya have been studied in detail by Johnson (1967a,b, 1968). This author studied blackwaters in tree country, open country, streams, drinking-water, reservoirs, ponds, polluted streams and hot springs. The oxygen concentration is usually low, not exceeding 5 mg l^{-1} at saturation because of constantly high temperature (25–28°C). Most of these waters are acid (pH of 3.6 to 5.4 in blackwaters and up to 6.2 in other types). The high concentrations of sulphate indicate that the low pH may be due to the presence of sulphuric acid. The alkalinity is extremely low (0–0.45 meq l^{-1}), the lowest values characterizing the tree country. The total salinity of the waters of South Malaya is variable; Johnson (1967a) reports values from 4.5 to 734 mg l^{-1}. Even the blackwaters do not form an homogeneous group and their salinity fluctuates from 9.3 to 702 mg l^{-1}. The highest salinities are observed in gelam-tree blackwaters (waters draining areas where the dominant tree is the gelam, *Melaleuca leucadendron*) and are caused by high levels of magnesium and sulphates. In blackwaters, the anion order is $SO_4^{2-} > Cl^- > HCO_3^-$ whereas it is $HCO_3^- > SO_4^{2-} > Cl^-$ or $SO_4^{2} > HCO_3^- > Cl^-$ in other water types. Chloride is however dominant in hot springs. Sodium is the dominant cation of Southern Malayan freshwaters and the most frequent order is $Na^+ > K^+ > Mg^{2+} > Ca^{2+}$. Calcium is especially low with an average concentration of 2 mg l^{-1} which compares with the calcium content of Amazonian water. Magnesium is also low, never exceeding 10 mg l^{-1}. The blackwaters are the exception with up to 79 mg l^{-1} Mg^{2+}, especially the gelam-type water (Table 94).

The Malayan blackwaters resemble those of Amazonia in their low pH, low alkalinity and high content of coloured humic components. They differ from them in their variable salinity and their high sulphate content. This ion is produced by the occasional oxidation of the large quantities of sulphides of the forest soils. In 1963, the soil of the Malacca area cracked, as a result of a prolonged drought, allowing the oxidation of the deep sulphide layer; when the rains drained this area the pH of the streams fell to values as low as 1.8 (Johnson, 1968).

This extreme chemistry has a filter effect on the flora and fauna. Algae are rare, though a few species are abundant in unshaded areas such as the red alga *Batrachospermum*, *Spirogyra* sp., *Hapalosiphon* sp., the diatoms *Desmogonium*, *Eunotia* and *Synedra* and the desmids *Cosmarium* and *Staurastrum*. Higher plants are rare although *Sphagnum* is abundant in the blackwaters of Klang. The productivity is consequently very low: 0.7 g C m^{-3} d^{-1} in an acid-water pond according to Fish (1960). Similar low productivity values have been measured by Richardson & Jin (1975) in unpolluted reservoirs of the State of Malacca (Asahan Reservoir and Ayer Keroh Reservoir). The vegetation fauna (Cladocera, vegetarian insects, annelids, rotifers, nematodes, protozoans) are scarce or absent. The prawn *Macrobrachium trompii* is abundant in certain areas and the crab *Potamon johorense* is found in dilute blackwaters. The low levels of calcium prevent the development of snail populations and explain the scarcity of such tropical diseases as schistosomiasis and lung fluke infection. Snail-mediated diseases are known in Singapore freshwaters where pollution has increased the calcium concentrations.

The most frequently encountered fish is *Rasbora einthovenii*. Also frequent are *R. cephalotaenia, Puntius pentazona johorensis, Hemirhamphodon pogonognathus, Betta picta* and *Sphaerichthys osphronemoides*. In gelam-type water, *Anabas testudineus* dominates. These animals feed mainly on exogenous vegetable detritus and insects and the fish productivity is not related to primary productivity.

In more alkaline water, productivity may be very high: Dunn (1967) estimated that the photosynthetic activity in a Malacca pond reached 353 mg C m^{-3} h^{-1} during a bloom of *Anabaenopsis philippinensis*. This activity was accompanied by daily pH variations of 2.5 units.

Fertilized fish ponds

The Chinese of Malaya have imported the management technique of fish ponds which have been in use for centuries in China. This technique consists of raising pigs, ducks and fish in a nearly closed system. A fish pond or a mining pool is associated with a pig enclosure located near the pond.

Table 94 *Chemical features of some southern Malaya waters (Johnson, 1967a) (upper and lower numbers are minimum and maximum values)*

	Alkalinity	Cl$^-$	SO$_4^{2-}$	Na$^+$	K$^+$	Mg^{2+}	Ca^{2+}	Salinity
	←			(meq l^{-1})			→	(mg l^{-1})
Blackwaters	0.00	0.08	0.10	0.091	0.015	0.015	0.00	9.28
	0.06	1.54	12.13	2.50	0.268	6.58	0.72	702.20
Other tree-	0.04	0.000	0.026	0.012	0.010	0.005	0.000	4.50
country streams	0.30	0.036	0.095	0.143	0.026	0.074	0.015	17.51
Open-country	0.08	0.064	0.001	0.077	0.004	0.053	0.000	11.75
streams	0.42	0.260	0.506	0.600	0.190	0.226	0.228	87.30
Ponds	0.20	0.042	0.058	0.075	0.016	0.005	0.038	26.82
	2.56	0.718	0.740	0.952	0.829	0.402	0.583	206.70

West Malaysia

The pig wastes, washed to the pond, serve as nutrients for algal blooms and proliferation of macrophytes. The algae are eaten by fish and ducks whereas the macrophytes such as the water hyacinth are harvested and processed into pig food. In such systems, the primary productivity reaches values of 10 g C m^{-2} d^{-1} with the blue-green algae *Anacystis* and *Oscillatoria* and the flagellates *Euglena, Mallomonas* and *Trachelomonas* as dominant species, whereas *Melosira, Ceratium* and *Anabaenopsis* are often found in mining pools.

The fishes of Malaya

More than 250 freshwater fish species are known in Malaya with 100 as common species (Johnson, 1967*b*). The Malayan freshwater fishes are distributed according to a north–south pattern in spite of the natural central north–south mountainous barrier which naturally isolates the western and eastern areas. The following species are characteristic of the northern area. *Chela laubuca, Danio (Brachydanio) tweediei*, endemic to northern Malaya, *D. regina, Rasbora borapatensis, Acrossocheilus hendersoni, Labiobarbus lineatus, L. siamensis, Puntius leicanthus, P. simus, Garra taeniata, Amblyceps mangois, Parasilurus cochichinensis* and *Anguilla bicolor bicolor*. Other species are found only in southern Malaya: *Rasbora caudimaculata, R. cephalotaenia, R. einthovenii, R. maculata, R. pauciperforata, R. taeniata, Osteochilus spilurus, Puntius pentazona johorensis, Homaloptera tweediei, Kryptopterus macrocephalus, Silurichthys hasseltii* and *Luciocephalus pulcher*.

Johnson (1967*b*) thinks that the north–south distribution of fish is mainly determined by the water chemistry: the northern species are hard-water species and southern species are acid-water tolerant. This author does not exclude a possible effect of the climate: the northern fishes with clear breeding seasonal cycle may be disturbed by the constant climatic conditions of southern Malaya. At any rate, the environmental factors have more influence on fish distribution than the geographical obstacles.

The various habitats encountered in Malaysia show a very different diversity: the blackwater fish fauna comprises only 25 species whereas the fish list of the less-acid tree country streams contains 109 species. The fauna of the ricelands includes 13 common species with a notable dominance of air-breathing animals such as *Trichogaste trichopterus, Trichopsis vittatus* and *Anabas testudineus*. The ricelands of Thailand have a much more diversified fauna which may be due to their more ancient existence in that country.

Cyprinidae clearly dominate the ichthyological fauna of Malaya. This is especially true of the acidic and unproductive blackwaters and tree-country habitat. Cyprinidae are less important in the highly productive riceland waters.

THE INDO-CHINESE PENINSULA
General

Similarly to the Malaysian peninsula, Indo-China is made of Precambrian and Palaeozoic formations and emerged in the early Mesozoic. Metamorphic rocks in Vietnam, Laos, Kampuchea and Thailand have been attributed to the Upper Precambrian, whereas marine Palaeozoic deposits are widespread. In Kampuchea, Laos and Vietnam, they are covered with the continental fluvio-lacustrine 'Indonesian formation' which range from the late Carboniferous to the Cretaceous with a few marine intercalations. The Indo-Chinese peninsula belongs to the stable continental areas and lies outside the South-Asiatic seismic belts.

The climate of Thailand is characterized by moderate precipitations and a hot and dry summer. The country is protected from the summer south-west monsoon by the Tennesarim mountains in Burma and from the winter north-east monsoon by the Vietnamese Ranges which exceed 2000 m altitude. The mean annual rainfall fluctuates from 1000 to 1500 mm. Laos, Kampuchea and Vietnam are somewhat protected from the north-east winter monsoon but the eastern coast of Vietnam is exposed to the east has heavy rainfall throughout the year. West of the Annam ranges, ~80% of the yearly precipitation occurs from May to October, as can be seen in Table 95.

The southern part of the Indo-Chinese peninsula experiences very small seasonal fluctuations: the mean monthly temperature varies from 26 to 30°C at Bangkok and from 28 to 31°C at Saigon. In contrast, the northern area has a cooler winter and, at Hanoi, the mean monthly temperature varies from 17°C in January to 30°C in May (Koteswaram, 1974).

Table 95 *Mean rainfall distribution in Lower Mekong basin (Committee for the Coordination of Investigations of the Lower Mekong Basin, 1970)*

Month	J	F	M	A	M	J	J	A	S	O	N	D	Yearly average
mm	8	15	40	77	198	241	269	292	299	165	54	14	1672
%	0.5	0.9	2.4	4.6	11.8	14.4	16.1	17.5	17.9	9.9	3.2	0.8	100

In 1968 the population of the Lower Mekong basin was 61 million with an average demographic density of 43 inhabitants km^{-2} which is relatively low.

The Mekong basin
The Mekong River

The drainage area of the Mekong River extends over 795 000 km². This includes nearly the whole Indo-Chinese peninsula (except the east coast of Vietnam and the small watershed of the Menam River reaching the Gulf of Thailand at Bangkok), a small part of Birman territory and $\sim$ 180 000 km² in China.

The Mekong (called Lan Tsang in China) rises in the Tibet–Sikiang plateau at about 33° N and 94° E and 5000 m altitude. It flows south-east for 425 km until Chamdo at 3000 m altitude and then runs south and enters the Yunnan province of China. It crosses Yunnan through 1000 km of steep gorges and high mountains. It then forms, successively, the borders between Burma and China, Burma and Laos, and Thailand and Laos. At Pak Tha it enters Laos and flows east through rapids. At Vientiane, the river enters the open plain; it is then at an elevation of 160 m above MSL at 1584 km from the sea. Downstream from Savannakhet the river cuts the Khemmarat sandstones with deep gorges and rapids. Next, the Mekong receives the Mune River from Thailand and the Se Bang Hieng from Laos. The Mekong flows 190 km in an open valley and cascades over the Khone Falls and enters the plains of Kampuchea. Below Kratie, the Mekong flows on its alluvial deposits. At Phnom-Penh, the Tonle Sap, draining the Great Lake, joins the Mekong. South of this confluence the river divides into the Mekong on the east and the Bassac on the west. Downstream, in its delta area, the river forms five branches before entering the South China Sea.

The Mekong watershed is divided into an upper and lower basin, the limit between them being the region of Chiang-Saen near the Burma–Laos–Thailand border. The lower basin covers 609 000 km².

The Mekong is 4350 km long with 1955 km from the sources to Chiang Saen and 2395 km in the lower basin. The Upper Mekong is fed by the Himalayan snow and the Lower Mekong by the monsoon rainfall. Its flood regime is very regular. The river begins to rise after the south-west monsoon sets in, reaches its maximum discharge in the period August–October and subsides until May. At Kratie, the minimum discharge is 1250 m³ s^{-1} whereas the average high-water discharge is 33 200 m³ s^{-1}. The average annual discharge is 14 800 m³ s^{-1}. At high water, there is extensive flooding in Kampuchea and the Vietnamese territories; flash floods in tributaries are then frequent and very destructive. The Great Lake has a powerful regulating influence on the discharge in the delta. From June, the lake begins to rise and stores 80×10^9 m^3 over a period of four months, 43% of which comes from the immediate drainage area and 57% from the Mekong itself. The lake level rises by 9 m and its surface area expands from 3000 to 10 000 km^2. In October, the water level in the Mekong decreases and the flow is reversed in the Tonle Sap from the lake to the river. A flood warning system is being developed in order to protect people and livestock living in vulnerable areas.

At Phnom-Penh, south of the Tonle Sap–Mekong confluence, the 14 years average discharge 1960–73 is 13 131 m^3 s^{-1} with a minimal discharge of 1250 m^3 s^{-1} and a maximal yield of 49 700 m^3 s^{-1}.

The yearly discharge of the Mekong is 475×10^9 m^3. Its specific yield is $\sim 700\,000$ m^3 km^{-2} y^{-1} which is lower than that of the Amazon and of the order of magnitude of that of the Ganges.

The water quality of the Mekong was studied by Kobayashi (1969). The results obtained at Nong Khai near Vientiane are shown in Table 96. The Mekong water belongs to the calcium–hydrogen-carbonate type. Its salinity is low and Na$^+$, K$^+$ and Cl$^-$ are the least abundant ions. A significant dilution occurs at the flood period.

Water is plentiful in the Indo-Chinese peninsula but the seasonal uneven distribution of the rain and the high temperatures make irrigation necessary part of the year. In 1951, the ECAFE (Economic Commission for Asia and the Far East) suggested the study of the Lower Mekong basin. In 1957, the Committee for the Coordination of Investigations of the Lower Mekong Basin was created to prepare water resources development projects. Progress in data collection and basin planning has been

Table 96 *Chemical features of the Mekong at Nong Khai (Kobayashi, 1969)*

	15 Sep. 1956	15 Apr. 1957		15 Sep. 1956	15 Apr. 1957
Ca^{2+} (mg l^{-1})	23.7	37.0	Cl$^-$ (mg l^{-1})	2.5	10.2
Mg^{2+} (mg l^{-1})	3.2	6.9	SiO$_2$ (mg l^{-1})	16.0	15.9
Na$^+$ (mg l^{-1})	3.3	11.6	Dissolved solids (mg l^{-1})	100.0	177.0
K$^+$ (mg l^{-1})	1.2	1.6	Suspended solids		
HCO$_3^-$ (mg l^{-1})	89.6	134.3	(mg l^{-1})	470.0	68.5
SO$_4^{2-}$ (mg l^{-1})	4.8	25.0	pH	6.8	7.0
			Hardness (mg l^{-1} CaCO$_3$)	72.2	121.0

presented in two reports, in 1957 and 1970. The later report gives a list of potential projects which includes 17 mainstem and delta projects and 87 possible tributary projects.

The mainstem projects are intended to store the Mekong flow in order to allow the utilization of the water for irrigation and hydroelectric power, even during the dry season. This will be achieved by a system of reservoirs which will also have a moderating effect on floods. Five projects, High Pak Beng, High Luang Prabang, Pamong, Stung Treng and Tonle Sap, have large storage capacity and 12 projects have a storage capacity allowing only a short-term flow regulation (Fig. 31). It is not planned to carry out all the

Fig. 31. The Mekong watershed and main storage reservoirs

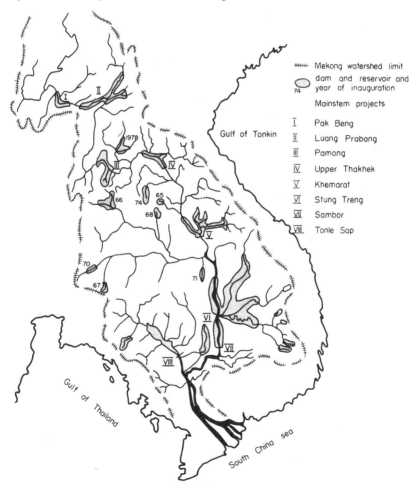

mainstem projects. Several are alternative solutions. The seven main projects would permit a gross storage capacity of 170×10^9 m^3. The execution of a certain number of tributary projects was scheduled for the period 1971–80: for example the Nam Ngum Dam and its 30 000 kW powerplant were inaugurated in December 1971. Other projects, including 15 mainstem multipurpose projects (some of these projects are exclusive alternatives), are scheduled for the period 1981–2000. On Fig. 31, the year of inauguration of the tributary projects already completed has been noted (Committee for the Coordination of Investigations of the Lower Mekong Basin, Annual Report, 1978). Bardach (1972) analysed the impact of these projects on fisheries, agriculture and soil conservation.

The Great Lake of Kampuchea

Located between 12 and 14°N, the region of the Great Lake is bordered in the south-west by the Cardamones and Elephant mountains which protect it somewhat from the south-west monsoon. The Great Lake of Kampuchea lies in a very flat plain at 10 m altitude above MSL. At low water, the Great Lake is 110 km long and has a maximum width of 30 km and covers an area of ~ 3000 km^2. Its depth is then 0.7 m and its volume 2×10^9 m^3. The Great Lake has a drainage area of 90 000 km^2. The outlet of the lake, the Tonle Sap meets the Mekong River near Phnom-Penh.

In April, the water level is low in the lake and the rivers. The flood plain of the lake is dry and cultivated. In the Mekong the current is very slow, the solid yield very small and the water green and transparent. The Lower Mekong and the Tonle Sap are then affected by the tide. In early May, the tributaries of the Great Lake and in particular the Stung Sen, a left-side affluent reaching the Tonle Sap south of the lake, start rising. The coarse suspended material of the Stung Sen is deposited in the Tonle Sap delta. Although at this period the Tonle Sap flows south, the Stung Sen swift waters flow northwards and reach the lake. During the meantime the current velocity increases in the Mekong but there is no significant variation in water level. In June, the Mekong flood starts; its waters are red and turbid. The Lower Mekong cannot evacuate the water and the level starts to rise. When the level of the river rises by 1 m, the current of the Tonle Sap is reversed but the Mekong waters do not yet flow into the lake. High waters in the Stung Sen cause the deposition of sediments in the delta and the water level rises in the lake. Finally, the water level rises in the Mekong which exceeds 10 m at the flood peak. The Mekong water spreads into the Tonle Sap which begins to function as an inlet of the Great Lake. The lake level rises by more than 8 m and the lake extends over 10 000 km^2

and has an approximate volume of 100×10^9 m^3. In September, the Mekong current decreases and in November its level drops down. As a result the Tonle Sap flows to the Mekong and drains the Great Lake plain. However, the hydrological events are not exactly synchronous in the Mekong and in the local tributaries of the Great Lake. Dussart (1960) reports that in October 1959, although the Mekong level had been decreasing for a few days, the Great Lake level was still rising because of the high discharge of the lake's local tributaries.

A volume of 37×10^9 m^3 is carried from the Mekong to the Great Lake via the Tonle Sap during the flood whereas 80×10^9 m^3 are drained to the sea from the lake area. The lake area supplies to the Mekong a net discharge equivalent to about 10% of its total discharge.

In its lower course, the Mekong carries 250–300 mg l^{-1} suspended solids (average concentration from June to October) corresponding to the silt fraction. These concentrations are low when compared to the 1500 mg l^{-1} of the Nile and the 1950 mg l^{-1} of the Ganges. The sediments carried by the Mekong flood into the Tonle Sap are partly deposited some 20 km before reaching the lake. As a result, a delta has formed, with the Tonle Sap dividing into various arms separated by sand banks. Only the fine sediments reach the lake and the lake-bottom mud consists of a fluid clayey material with organic debris. This is clearly demonstrated in the cores taken by Dussart (1960): the cores taken in the delta consist of compact clay or fine sands; those sampled in the lake are made of 1 m of inconsistent clayey ooze lying on a more compact clayey formation. In the upper sediments, 99.4% belong to the fine phase ($< 50 \mu$m). The shells collected in this ooze belong to the following species: *Corbicula fluminea, C. moreletiana, Melania rudicostis, M. scabra* and *Paludina* sp. The relative fluidity of the bottom sediments of the lake explains that the combined effect of lowering levels and strong waves and currents, due to the winter northeast monsoon, resuspends easily the bottom mud which then leaves the lake during the flood recession. Dussart (1960) estimates that, on an average yearly basis, 5670×10^3 tons of sediments are carried by the Tonle Sap to the Great Lake whereas 4450×10^3 tons are taken back via the same river. The Great Lake sediment filling is therefore due to the solid input of the local tributaries and not so much to the Mekong. The flood plain of the Tonle Sap is partly covered with inundated forest which limits the erosion. The human migration towards the Great Lake which took place in the first half of this century led to the destruction of forested areas and then replacement by rice fields. The soils, previously fixed by the forest, were washed into the lake. This process has caused a sudden acceleration of the lake filling.

The flora and fauna of the hydrographic complex formed by the lake, the Tonle Sap and the Mekong was studied by Blache (1951) and Brehm (1951). In the Mekong, the plankton was studied during the low-water period and early flood. The diatoms heavily dominate the phytoplankton (82–97% of the total number of algae) with *Melosira spiralis* as the main species. Other frequent genera are: *Attheya* (*A. zachariasi*), *Rhizosolenia*, *Stephanodiscus*, *Tabellaria*, *Fragilaria*, *Synedra*, *Gyrosigma*, *Nitzschia* and *Surirella*. The green algae (1.9–14.5% of total number) are especially abundant from March to May with *Pediastrum boryanum* as the dominant species and the genera *Actinastrum*, *Micractinium*, *Schroederia*, *Treubaria*, *Golenkinia* and *Coelastrum*. The blue-green algae are always present in small numbers with *Lyngbya*, *Anabaena* and *Oscillatoria*. In quantitative terms, the number of algal cells varies from 2300 at low water to 400 in the early flood. The zooplankton is dominated by rotifers with *Brachionus capsuliflorus*, *Schizocerca diversicornis* and *Pedalion* sp. Among the Cladocera, the Bosminidae are most abundant. The copepods are not abundant: the Cyclopidae are represented by *Microcyclops varicans*, *Mesocyclops leuckarti* and *Thermocyclops hyalinus*. Among the Diaptomidae, a new species, *Eodiaptomus blachei*, was described; the maximal concentrations do not reach 50 individuals l^{-1}.

In the Tonle Sap, two periods of maximal current occur: in July–August towards the lake and in November–January towards the Mekong. The intervening periods are characterized by a weak water current which favours the phytoplankton development. The first peak is concomitant with the onset of the flood and the second peak with the beginning of the recession. As in the Mekong the phytoplankton is dominated by the diatoms which form 28–93% of the algal cells. The species *Melosira granulata*, *M. granulata* var. *angustissima*, *M. spiralis* and *Attheya zachariasi* are dominant. The genera *Surirella*, *Synedra*, *Pleurosigma*, *Fragilaria* and *Navicula* are also found. During the low waters, the blue-greens are abundant and form about 60% of the algal community from February to April with *Anabaena*, *Lyngbya* and *Microcystis* as the main genera. In April–May, *Microcystis* forms thick blooms upstream of Phnom Penh. In October, the genera *Eudorina*, *Gonium* and *Volvox* appear in the Tonle Sap: they originate from the inundated zones and are carried into the river by the recession current. The green algae *Pediastrum simplex*, *P. duplex* and *P. boryanum* are especially common in January–February. The maximal density of the green algae occurs in May–June with *P. boryanum* and the genera *Actinastrum*, *Micractinium*, *Schroederia*, *Treubaria*, *Scenedesmus* and *Coelastrum*.

The zooplankton is low in January but increases progressively to reach a peak in May at low waters; then, it drops rapidly and nearly disappears in July–August; a new peak occurs in October. Nauplii and copepods are abundant in periods of low current; the diaptomids are represented by *Alladiaptomus* Raoi n. var. *membranigera* and *Pseudodiaptomus beieri* n. sp. and the Cyclopidae by: *Mesocyclops leuckarti, Macrocyclops distinctus, Eucyclops serrulatus, Paracyclops affinis* and *Thermocyclops hyalinus*. The Cladocera, abundant at the decline of the flood, are dominated by *Bosmina longirostris* and *Bosminopsis deitersi; Ceriodaphnia rigaudi* and *Diaphanosoma sarsi* are also found. The rotifers, especially abundant in October, include *Brachionus capsuliflorus, B. falcatus, B. angularis, B. forficula, B. plicatilis, B. leydigi, Keratella cochlearis, K. valga, Polyarthra* sp., *Trichocerca cylindrica, Schizocerca diversicornis, Tetramastyx opoliensis, Pedalion* (very common) and *Ascomorpha ecaudis*.

The stagnant waters of the inundated forest are particularly rich in blue-green algae (*Microcystis* dominant from November to January and *Lyngbya* and *Anabaena* abundant in September to October). Large areas covered with *Eichhornia* and *Nelumbium* have no submerged vegetation and their algal flora is composed of Desmidiales with *Micrasterias* as dominant species and nearly pure cultures of *Peridinium*. These inundated areas serve as algal reservoirs for the Great Lake which is characterized by very thick blooms of *Microcystis*. However, other groups are represented such as the Desmidiales (*Micrasterias, Cosmarium, Spirogyra*) and Euglenophyta with *Euglena, Trachelomonas* and *Strombomonas*.

The hydraulic complex of Southern Indo-China is composed of a diatom-dominated river and a blue-green-dominated lake which alternatively affects the Tonle Sap which connects the two systems. This mechanism has considerable impact on fish ecology.

Nam Pong Reservoir

The Nam Pong Reservoir (site 1 of Fig. 32) is situated in north-eastern Thailand; it was mainly designed for electricity production and completed in 1966. Its maximal and mean depths are, respectively, 19.5 and 15.5 m. Its algal population is dominated by *Microcystis incerta*, but also includes *Aphanocapsa delicatissima* and *Synedra ulna;* the zooplankton is mainly composed of *Bosminopsis detersi* and *Cyclops* sp. The freshwater shrimp *Macrobrachium lar* (Palaemonidae) is abundant in the reservoir. The molluscs are rare; Mizuno & Mori (1970) report the presence of *Gyraulus chinensis convexiusculus* and *Sinotaia ciliata basicarinata*. Among the fish, *Cyclocheilichthys apogon* feeds essentially on chironomids and zooplankton, *Puntius orphoides* feeds on detritus, diatoms, Copepoda, Cladocera and statoblasts of Bryozoa, *Osteochilus hasseltii* is the only phytoplankton-eater observed, *Ompok bimaculatus* feeds on young fish, *Clarias batracus* on aquatic insects and *Mystus vittatus* on aquatic insects and shrimps.

Nong Raharn Reservoir

Nong Raharn Reservoir (site 2 of Fig. 32) is situated also in north-eastern Thailand. Nong Raharn is a natural lake converted into a 55 km² reservoir of 1–2 m depth. The bottom is covered with *Vallisneria* sp. The phytoplankton comprises *Eudorina elegans, Volvox aureus, Ceratium*

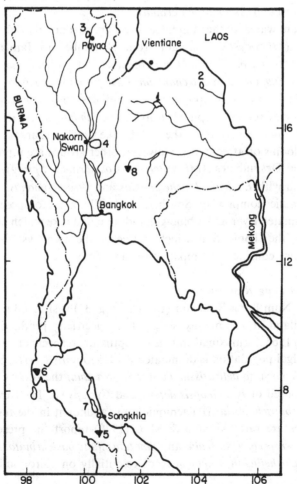

Fig. 32. The hydrography of Thailand and the south-western Mekong tributaries (from Geisler *et al.*, 1979)

hirundinella, Oscillatoria limnetica, Lyngbya sp. and *Oedogonium* according to the report of Mizuno & Mori (1970). These authors also indicate the abundance of *Bosmina longirostris* in the zooplankton of the reservoir. The shrimp *Caridina wyckii* (Atyidae) was commonly found. Molluscs seem limited to the snails *Pila polita, Bithynia funiculata* and *Sinotaia ciliata basicarinata*.

The fish of the Lower Mekong
The fish population of the Lower Mekong basin is dominated by the Cyprinidae and Siluridae but other families such as the Ophicephalidae, Anabantidae and Notopteridae also contribute to the prosperous fishery of the area. The number of known species is high: 206 species belonging to 55 genera of Cyprinidae and 100 species from 34 genera of catfishes in Thailand (Smith, 1945). In Kampuchia, Bardach (1959) counted 171 species of fish from 30 families. These numbers are however much below the number of species found in Amazonia and in the Great Lakes of Africa for instance. Fernando & Furtado (1975) claim that the absence of natural old lakes in South-East Asia has limited the evolution and speciation of lake fish species and invertebrates with the exception of cyprinids in Lake Lanao.

Fishery resources have long been exploited in the area; fish presently provides from 50 to 75% of the animal protein in the people's diet. It is estimated that the fish production of the basin is $\sim 300\,000$ tons^{-1}. The abundance of fish is related to the flood pattern of the Mekong; Chevey (1940) reports that, in June, thousands of fingerlings 0.5 cm long, born in the Mekong, penetrate into the Tonle Sap, then into the inundated areas. They are followed all along by a very high concentration of adequate planktonic food which favours a very high rate of survival. Moreover, the sheltered areas of the flood plain offer ideal conditions for the development of young fishes. Original techniques are used by fishermen such as the barrier nets and have been described by Goossens (1951); the fish which have developed during the flood in the inundated areas follow the receding flow and reach back the Mekong. However, the fish are not passively carried by the river current and their back migration is affected by the lunar cycle: a massive migration occurs between the sixth and seventh days of the first quarter of the moon and the full moon. The fishermen block the streams by various techniques and capture very large amounts of fish composed of *Puntius* sp., *Puntioplites proctozysron, Cryptopterus* sp., *Thynnichthys thynnoides, Cirrhinus jullieni, Dangila cuvieri, Spilopleura, Cirrhinus auratus, Wallago attu, Ophicephalus micropeltes* and *Notopterus notopterus*.

A list of 105 species observed in the fishery from the inundated area north of Phnom-Penh has been established by Goossens (1951). Pantulu and Bardach (unpublished report) have reported 150 species near the Pa Mong dam site (quoted in Pantulu, 1973). The distribution of the fish was as follows: Cyprinidae 54%, catfishes (Siluridae, Clariidae, Schibeidae, Bagridae, Sisoridae and Akysidae) 19%, Ophicephalidae 8%, Notopteridae, Clupeidae, Anabantidae and other minor groups 19%. At the onset of the flood, most of the animals operate a 'lateral migration' to the inundated areas. Fish with low oxygen requirements like catfishes and Anabantidae remain for longer periods in the periphery of the flood plain in spite of low oxygen and pH conditions. The more exacting cyprinids occupy the central zone of inundation where they spawn and move back into the river with the receding flood (Le Van Dang, 1970). If the flood recedes too early, numerous eggs are desiccated and an unusually long inundation period causes mass mortality of fingerlings by lack of oxygen due to the decomposition of organic matter in stagnant water. High fecundity and adaptation to extreme conditions of pH and dissolved oxygen characterize the fish fauna of the Mekong.

Another 27 species of marine and estuarine fishes are found in the Mekong. *Hilsa kanagurta* is an anadromous species migrating several hundred kilometres upstream, whereas *Anguilla japonica* and *A. mauritiana* migrate seawards for spawning.

Part of the fish is consumed as fresh fish and part is processed by sun drying, smoking, fermentation, salting or made into fish sauce or paste. In the past, *Pangasius* sp. represented 30% of the salted fish (Lafont, 1951). With time, this technique declined because more fish was transported as fresh fish, but also because of the rarefaction of the fish. Blanc (1959) who studied the species *P. pangasius* noticed that the fish do not spawn in the flood plain where mature animals are never found. The local fishermen think that it spawns in deep areas of the Mekong between Kratie and Stung Treng. *Cryptopterus apogon* is the main species used as smoked fish but *Notopterus, Dangila siamensis* and *Xenentodon canciloides* are also utilized.

A general decline in fish production is observed in the Lower Mekong basin. In Kampuchea where data are available the catch which amounted to 147 000 tons in 1939 was only 123 000 tons in 1965. The major decline has been observed in the Great Lake where 70 000 tons were caught in 1940 and only 35 000 tons in 1967; overfishing, reduction of fish nursing areas by destruction of the inundated forest, increase of sedimentation are among the causes of deterioration of the fishery.

The construction of dams and formation of reservoirs considerably

The Indo-Chinese peninsula

modified the fish fauna and the fishery. After impoundment, the plankton and benthos of the Nam Pong Reservoir became more abundant. The number of fish species decreased from 74 before closing to 51 but the fish catch passed from negligible values to 1300 tons in 1967 (Pantulu, 1973) and to 2140 tons from May 1977 to May 1978 (Committee for the Coordination of Investigations of the Lower Mekong Basin, Annual Report, 1978). Migrating species such as *Cirrhinus jullieni* and *Bagarius bagarius* disappeared as well as the sedentary bottom-dwellers *Labeo bicolor, Wallagonia attu* and *Cryptopterus* sp. The contribution of the Cyprinidae to the catch (by mass) dropped from 56% to 24% whereas the Ophicephalidae now form 35% of the catch instead of 16.5% prior to impoundment.

A similar evolution of the fishery has taken place in other reservoirs of Thailand, such as Lam Dom Noi, completed in 1971 and Lam Pao, completed in 1968 where the 1977–78 fish catches were, respectively, 3160 and 1980 tons (Annual Report, 1978). It is clear that the Mekong fish, well adapted to alternatively lentic and lotic conditions, have generally benefited from the creation of impoundments. However, the dams represent a serious obstacle to the long-range migrating species and especially the big animals *Pangasius* sp., *P. sulchi* and *Pangasianodon gigas*. Pantulu (1973) suggests that spawning of these species be induced below the dam sites and the larvae be reared until the age at which they naturally migrate downstream. Similarly a solution should be found to the reproduction of the freshwater prawn *Macrobranchium rosenbergii*; this prawn lives in freshwater but migrates to brackish water to spawn. The regulation of the Mekong will modify the salinity of the Mekong estuaries. During 1978 the Interim Committee of the Lower Mekong increased the capacity of the prawn hatchery at Vung Than (Vietnam) and built rearing ponds on a 5 ha area to grow juveniles which will be subsequently introduced into large water bodies. The project also includes the creation of five farms for prawn production in the Mekong delta. The Committee has also recommended to convert non-arable lands in irrigated areas into fish ponds. Two such projects are being carried out, one at Tha Ngone, Laos and one in Thailand.

The Mae Nam and Mae Klong water systems
The Mae Nam River

The Mae Nam River is a north–south fluviatile system draining western Thailand and flowing into the Gulf of Siam in the Bangkok area through two branches, the Mae Nam Chao Phya and the Mae Nam

Nakoru Sri. South-western Thailand is drained by the Mae Klong River formed of the Kwai Yai and the Khai Noi Rivers. The Mae Klong also flows into the Gulf of Siam, west of the Mae Nam.

The water of these rivers is yellow-brown or red-brown and generally turbid because of lateritic soils. From the report of Mizuno & Mori (1970) the pH varies from 7.4 to 8.8 along the river.

These authors also reported an unusual abundance of molluscs in the Kwai River including mussels such as *Ensidens scobinata, Corbicula noetlingi, C. siamensis* and *C. petiti* and the snails *Pachydrobia siamensis* and *Mekongia pongensis*. A fish survey carried out by using poison indicated that in the Kwai rivers the migrating species *Cirrhinus jullieni* was the most abundant species at the time of the survey (June 1968). In the species examined for gut contents, *Asteochilus hasseltii* is, here too, the only phytoplankton-eater; *Mystus nemurus* and *Ompok bimaculatus* feed on small shrimps, *Mystus vittatus* and *Cyclocheilichthys* on aquatic insects, *Puntius gonionotus* on bivalves and plant detritus and *Pteropangasius cultratus* on terrestrial insects.

Khan Payao Lake

The Khan Payao Lake (site 3 of Fig. 32) is a small natural lake located in Northern Thailand; Khan Payao was dammed and now has a surface area of 21 km^2. The water is brown and muddy and the southern part of the lake is covered with aquatic plants. Its maximum depth is 25 m and pH ranges from 6.4 to 9.2. Its phytoplankton is heavily dominated by *Microcystis aeruginosa* and, in the zooplankton, *Moina dubia* and *Thermocyclops* sp. are abundant. The shrimp *Macrobrachium lar* is also common. Mizuno & Mori (1970) have found seven species of mussels and two species of snails.

Bum Borapet Reservoir

The Bum (or Bung) Borapet Reservoir (site 4 of Fig. 32) is the oldest reservoir in Thailand. It was built in 1930 to drain the Borapet Swamp at the junction of the Mae Nam Nan and Mae Nam Ping near Nakorn Swan. The reservoir has a surface area of 208 km^2 and a depth varying from 2 to 4 m. During the survey by Mizuno & Mori (June 1968) the pH varied from 7.3 to 7.7 at different stations.

The reservoir can be divided into three sections. In the inflow area, the water flows among many small islands. Large amounts of organic detritus explain the relatively low pH (6.9–7.6). The outflow area has various large flat islands surrounded by macrophytes. The pH varies from 7.6 to 8.0. The central area is free of vegetation and its pH ranges from 7.9 to 8.3. The calcium concentration varies from 13.6 to 24 mg l^{-1} and conductivity from 135 to 162 μmhos cm^{-1}.

The algal community is dominated by the blue-green *Sphaerozosma vertebratum* but also includes many desmids (*Micrasterias, Xanthidium Cosmarium, Closterium*), and diatoms like *Synedra ulna*. This flora resembles that of Tasek Bera in Malaysia. The zooplankton is rich in *Bosminopsis deitersi, Bosmina longirostris, Heliodiaptomus kikuchii* and *Keratella valga*. The shrimp *Macrobrachium rosenbergii* is abundant and reaches a length of 32 cm.

The bottom fauna of the reservoir, studied by Junk (1975) was essentially composed of Bivalvia with *Corbicula, Hyriopsis, Union, Ensidens, Philobryconcha, Scaphula* and *Limnoperna* as the main genera. Their biomass amounted to 222 g m^{-2} dry mass and represented up to 99% of the total benthic biomass. The Ephemeroptera and Trichoptera with *Eatogenia* sp. and *Dipseudopsis* sp. were second in importance with maximal biomass values of 0.95 and 0.7 g m^{-2} dry mass, respectively. The fish *Tetraodon* and *Cyclocheilichthys apogon* feed on the smaller forms and especially *Scaphula pinna* which has soft shells at pH 7.6. The biomass of Mollusca in Bum Borapet compares with that of Lake Chad and is far higher than the biomass of this group in the electrolyte-poor water of Amazonia.

In their fish survey, Mizuno & Mori caught 29 species of fish; a detailed list is published in their paper of 1970. Many species were found to feed on insects (*Cyclocheilichthys apogon, C. enoplos, Cryptopterus cryptopterus, Pangasius larnaudii, P. siamensis, Pteropangasius cultratus, Mystus cavacius, M. nemurus*).

The waters of southern Thailand

Geisler, Schmidt & Sookvibul (1979) have studied the water and fish populations of two streams of southern Thailand (sites 5 and 6 of Fig. 32) and compared them with these of a right-side tributary of the Mekong on the Korat plateau (site 8 of Fig. 32). The rivers Bori Pat (site 5) and Lam Pi (site 6) have very low salt content (electrical conductivity 22–52 μmhos cm^{-1}) with sodium (4.1 mg l^{-1}) and chloride (3.6 mg l^{-1}) as dominant ions whereas site 8, located in the Mekong basin on the limestone Korat plateau, has a conductivity in the range 400–700 μmhos cm^{-1} with calcium (44 mg l^{-1}) and sulphate (10.5 mg l^{-1}) as principal ions. In sites 5 and 6, the fish population was dominated by cyprinids in number of individuals and in biomass. At site 8, $\sim$50% of the individuals were catfishes but, because of their small size, their contribution to the biomass was small. The fish standing stocks at sites 5, 6 and 8 were, respectively, 118, 186 and 81 kg ha^{-1} which is in good agreement with the average biomass values (180 kg ha^{-1}) found by Bishop (1973) in the Gombok River (Malaysia). The total number of species observed (33) is in good agreement with the 27 species of the Gombok River.

No relation was observed between the standing crop of fish and the total mineral content of the water. The low content of Ca^{2+} and Mg^{2+} at sites 5 and 6 in comparison with site 8 is not a disadvantage in fish production since many tropical fish are soft-water forms; such is the case with the genera *Rasbora* and *Puntius* and the species *Betta taeniata* and *Sphaerichthys osphromenoides*.

Lake Songkhla (site 7, Fig. 32) is a lagoon-lake located in southern Thailand at 100°4′ E and 7°5′ N. Its surface area is about 1040 km^2 and its depth does not exceed 2 m. As a former lagoon separated from the sea by sand dunes, its saline waters have been gradually replaced by freshwater. It is formed of several successive basins, and the southern one is a breeding area for young fish, shrimps and prawns. *Macrobrachium rosenbergii* moves in the lakes according to salinity, and *Mugil* sp., *Lates calcarifer* and other brackish-water fish and crustaceans are abundant. The salinity of the main basin in 1965 and 1966 varied between 0 g l^{-1} in the wet season and 15 g l^{-1} in the dry season. Limpadanai & Brahamananda (1978) reported that the salinity increased recently up to 32 g l^{-1} as a combined result of irrigation drainage and the dredging of the lake channel at the outlet for navigation purposes.

INDONESIA
General

Indonesia is an arc of islands separating the Indian Ocean plate from the Asiatic continental plate. The Java trench is a subduction zone where the excess of oceanic crust, ceaselessly produced along the Indian Ocean ridge, descends beneath the island and is reincorporated into the mantle. The subduction zone is characterized by a high seismic activity and a strong negative gravity anomaly, this latter feature being caused by the unusual accumulation of sialic material. The Indonesian islands are then part of the Asiatic plate and are separated from Indo-China by the very shallow south-western part of the South China Sea (< 100 m). Conversely, the Java trench south of Java Island is more than 7000 m deep.

In the early Cenozoic, all Malaysia, Sumatra, Borneo and most of the area inbetween were emergent. A mid-Cenozoic transgression covered parts of this area and, in the late Cenozoic, folding and thrusting intensely affected the Timor–Celebes sedimentation basin. The erosion material accumulated in other subsiding basins was affected by a Pliocene folding phase during which western Sumatra and southern Java were uplifted. The Pleistocene sea level was lower than the present one and drowned valleys can be seen on the bathymetric maps of the South China Sea and Java Sea.

Indonesia consists of more than 3000 islands forming a curved belt of 5500 km long and 1100 km wide. The total land surface is nearly 2×10^6 km^2, 75% of which comprise sparsely populated areas such as Sumatra, Kalimantau (Borneo) and Irian Java (New Guinea). The densely populated islands such as Java and Madura represent only 7% of the total area. In 1971, the total population of Indonesia was 119 million, 64% of which was in Java (demographic density of 565 inhabitants km^{-2}).

The climate is tropical without significant seasonal difference and precipitations are abundant, generally exceeding 2000 mm y^{-1}. The water balance is positive and many rivers with catchment areas up to 80 000 km^2 drain the superficial runoff.

Early studies of the lakes of Indonesia

From 1915 to 1922, Van Oye studied the algae of Indonesia. In Java, he distinguished the low-altitude water bodies, dominated by blue-green algae, the lakes of intermediate altitude (200–800 m), especially rich in desmids and those of higher altitude with Chlorophyta. In Sumatra, he noted especially the presence of *Dinobryon, Euglena, Peridinium volzi, P. javanicum, P. gutwinski, P. raciborskii, Ceratium hirundinella* and numerous Protozoa (*Arcella vulgaris, Difflugia* spp., *Nebela, Lacrymaria olor, Colpoda cuculus, Stentor, Strombidium, Vorticella campanula*) (Van Oye, 1922).

In 1928–29, the German Sunda Expedition, led by Ruttner, visited a certain number of lakes in Sumatra, Java and Bali. Numerous measurements and sample collection and analysis, made during and after the Expedition were subsequently published, among them Ruttner (1931, 1937, 1952), Steenis & Ruttner (1932) and Thienemann (1957). The morphological features of the lakes studied by the Sunda Expedition are shown in Table 97. Ruttner (1937) compared the difference in density generated by thermal stratification in a tropical and temperate lake (Ranu Klindungan and Altau, respectively). A difference in temperature of 5.5°C between epi- and hypolimnion in the tropical lake developed a stability of 190 kg m^{-2} whereas a difference of 7°C in the temperate lake led to a thermal stability of only 60 kg m^{-2}. This is indeed a major difference

Table 97 *Morphological features of the lakes studied by the Sunda expedition (from Ruttner, 1952)*

	Location	Altitude (m above MSL)	Area (km²)	Depth (m)	pH (Ruttner, 1928–29)	pH (Green et al., 1974)	Conductivity (μmhos cm⁻¹) (Green et al., 1974)
SUMATRA							
Ranau	5° S	540	126	229	8.5		
Singkarak	0°40' S	362	108	269	8.6		
Manindjau	0°20' S	465	98	169	8.4		
Danau di Atas	1°5' S	1531	12	44	8.0		
Toba	2°3' N	905	1130	529	—		
Southern basin			438	433	8.4		
Northern basin			586	529	—		
Poma basin			80	87	8.3		
Pangururan basin			26	97	7.9		
JAVA							
Lamongan	8° S	240	0.34	29	8.2		
Pakis	8° S	205	0.45	156	8.4		
Bedali	8° S	150	0.12	11	8.5		
Klindungan	7°40' S	10	1.90	134	8.1–8.9		
Pasir	7°40' S	1290	0.28	24	8.6		
Ngebel	7°50'S	730	1.48	44	8.0		
Tjigombong	6°40' S	500	0.30	17	7.5		
Sindanglaja	6°40' S	1050	0.02	19	8.0		
BALI							
Bratan	8°15' S	1231	3.8	23	6.8	7.1	39
Batur	8°15' S	1031	15.9	88	8.5	8.7	1650
Buyan	8°14' S	1214	3.9	30	—	7.6	220
Tamblingan	8°15' S	1214	1.3	28	—	7.8	170

between warm and cold water due to the higher difference in density per degree Centigrade (Celsius) in the higher range than in the lower range of temperature. The implications of such a difference are discussed in Part III. Ruttner studied the plankton of all the listed lakes and his qualitative and quantitative results are summarized in Table 98.

Recent studies of the lakes of Java and Bali

In 1974, a few of the lakes visited by the Sunda Expedition were visited again and studied by Green, Corbett, Watts & Oey (1976, 1978).

Ranu Lamongan (Java)

The crater lake is bordered by a floating mat of *Eichhornia crassipes* with *Ilomoea aquatica* and *Jussiaea repens*. *Ceratophyllum demersum* and *Hydrilla verticillata* cover the lake bottom in various places. The lake is fed by two inlet streams on its eastern side and has an outlet which is regulated by a sluice. The lake water is used for irrigation of the crops which have replaced the original forests.

The visit of Green *et al.* (1976) coincided with 'the' or 'an' overturn of the lake. The homogeneous thermal profile was at 27°C. Stratification developed during the following days. The turnover was accompanied by a drastic decrease in oxygen at all depths, probably due to the oxidation of hypolimnic reduced substances. Concentrations as low as 1 mg l^{-1} in the whole water column caused a massive fish kill. Oxygen levels increased in the epilimnion after the stratification was re-established. The hypolimnion was generally devoid of oxygen and contained hydrogen sulphide and sulphur bacteria as already observed by Ruttner. Vertical displacements of the thermocline and oxycline suggested the occurrence of internal seiches with a 4 h period and an amplitude of 3 m.

Diatoms, especially *Nitzschia* and *Synedra*, were very abundant as during Ruttner's survey but *Anabaenopsis raciborskii* was rare, apparently replaced by *Anabaena sphaerica* and *Lyngbya limnetica*. *Melosira granulata* was much more abundant than in 1928 and *Aphanotece microsphaera*, absent in 1928, was common. The cyclopids were the dominant zooplankton with, as in 1928, *Mesocyclops hyalinus* as the main species. Among the rotifers, *Brachionus caudatus* dominated and multiplied rapidly during the turnover. As the population increased, a dwarf form of species developed.

Eichhornia mats harboured a rich and diversified community; the plant roots constituted the main food of the water beetle *Hydrophilus bilineatus*, the prawn *Macrobrachium sintangense* and the fish *Barbus gonionotus*.

Table 98 The plankton of the lakes studied by the Sunda Expedition: qualitative and quantitative composition (from Rutner, 1952)

Island	Lake	Layer	Algae and hypolimnic organisms (% total plankton)	Algal biomass (g m^{-2})	Zooplankton (% total plankton)	Zooplankton biomass (g m^{-2})
SUMATRA						
South	Ranau	EPI	Bacillariophyta (33): *Synedra rumpens* var. *neogena* and *Synedra ulna* Desmidiales (4.9): *Staurastrum variodirectum*		Crustacea (49): *Daphnia carinata*	
		HYPO	Chlorophyta (6.1): *Oocystis crassa* sulphur bacteria (4.2): *Thiopedia rosea*			
Central	Singkarak		Cyanophyta (6.6) Bacillariophyta (53): *Synedra rumpens* var. *neogena*, *Denticula pelagica*	0–100 m 68.6	Crustaceae (12.9): *Mesocyclops leuckarti* and *M. hyalinus*	0–100 m 12.1
	Danua Manindjau		Chlorophyta (22): *Oocystis crassa* Cyanophyta (15): *Aphanocapsa delicatissima*	0–100 m 80.0	*Diaphanosoma Sarsi* Crustaceae (54.1): *Simocephalus serrulatus* and *Latonopsis australis*	0–100 m 115.2
	Danau di Atas		Chlorophyta (19): *Oocystis crassa* var. *Marssonii* Cyanophyta (17); Desmidiales (9.3): *Cosmarium atomus* Xanthophyta (36): *Chlorogibba irregularis*	0–40 m 28.7	Crustacea (22): *Eucyclops prasinus Diaphanosoma penarmatum*	0–40 m 8.2
North	Toba Northern Basin		Chlorophyta (9.9): *Oocystis crassa* Bacillariophyta (56): *Denticula pelagica* Chlorophyta (10)	0–100 m 17.0	Cyclopids (47.3) *Diaptomus Doriai* (35.1)	7.4
	Southern Basin		Bacillariophyta (27): *Denticula pelagica* and *Synedra rumpens* var. *neogena*	0–425 m 10.6	Cyclopids (14.7) Diaptomids (14.2): *Diaptomus* Insecta (21.6): Larvae of *Corethra*	12.6

Region	Lake		Species	Depth	Value	Additional Species	Value
	Porsea Basin		Bacillariophyta (20): *Denticula pelagica* Chlorophyta (15)	0–75 m	5.8	Cyclopids *Diaptomus Doriai* Larvae of *Corethra*	8.8
	Pangururan Basin		Bacillariophyta (6.7): *Melosira granulata* var. *valida* and *M. granulata* var. *curvata* Chlorophyta (17)		4.7	Cyclops, *Diaptomus Doriai*, *Ceriodaphnia dubia*, *Diaphanosoma*	11.8
BALI	Danau Bratan		Dinophyta (6.7): *Peridinium pygmaeum*, *P. Baliense*, *P. Willei*, *P. Volzi* Bacillariophyta (24): *Melosira granulata* Desmidiales (53): *Staurastrum excavatum* var. *planctonicum*, *S. perundulatum*	0–20 m	33.0	Crustacea (11.6): *Mesocyclops hyalinus* and *Moina latidens*	4.9
	Danau Batur		Bacillariophyta (36): *Nitzschia amphibia* Cyanophyta (4.6): *Anabaenopsis Raciborskii* Dinophyta (4.6): *Peridinium munusculum*	0–20 m	109.2	Crustacea (1.5): *Mesocyclops leuckarti*	2.5
EASTERN JAVA	Ranu Lamongan	EPI HYPO	*Anabaenopsis Raciborskii* (24) *Nitzschia acicularis* (12) Sulphur bacteria (11): *Achromatium mobile*	0–25 m	61.3 13.1	*Mesocyclops hyalinus* (11)	0–25 m 18.7
	Ranu Pakis		*Dactylococcopsis fascicularis* (18) *Lyngbya limnetica* fa *minor* (614) *Cymbella turgida* (27) *Synedra rumpens* var. *scotica* and *S. rumpens* var. *neogena* (18) *Peridinium inconspicuum* (3)	0–100 m	102.9	Rotifera (3.9) Crustacea (5.9) with *Mesocyclops hyalinus* Ciliates (2.5)	0–100 m 14.1

Table 98 (cont.)

Island	Lake	Layer	Algae and hypolimnic organisms (% total plankton)	Algal biomass (g m^{-2})	Zooplankton (% total plankton)	Zooplankton biomass (g m^{-2})
	Ranu Bedali		*Dactylococcopsis fascicularis* *Anabaenopsis Raciborskii* *Peridinium Wildemani* *Cymbella Ruttneri* *Lyngbya limnetica*		*Brachionus angularis* *Mesocyclops hyalinus* *Mesocyclops leuckarti*	
	Ranu Klindungan 26 XI 1928	EPI	Cyanophyta (9.7): *Dactylococcopsis fascicularis, Anabaenopsis Raciborskii, Lyngbya limnetica*	0–30 m 24.6		
			Dinophyta (10): *Peridinium Gutwinski, P. Elpatiewskyi, P. quadridens, P. Wildemani, P. inconspicuum*		Crustacea (3.8) Insecta (38) mainly larvae; *Corethra*	43.8
		HYPO	Sulphur bacteria (26.7) mainly *Thiopedia rosea* Ciliates (2.6)	34.5		
	Ranu Klindungan 28 VI 1929		Dinophyta (65) mainly *Peridinium Gutwinski* forming 48% of total plankton	85.1	Rotifera (2.6) Crustacea (2.0) Insecta (29.0)	43.2
CENTRAL JAVA	Telaga Pasir	EPI	Cyanophyta (20): *Dactylococcopsis fascicularis* and *Anabaenopsis Raciborskii* Chrysophyta (6.6): *Chromulina ovalis* Bacillariophyta (3.7) Desmidiales (21): *Cosmarium bioculatum, C. adoxum, Staurastrum perundulatum*	0–15 m	Rotifera (3.1) 84.7	6.6
		HYPO	Sulphur bacteria: *Chromatium* sp. (19), *Lamprocystis* (11), *Thiopedia rosea* (2.9)	53.4		

	Telaga Ngebel	Cyanophyta (36): *Lyngbya limnetica* fa *minor* Cryptophyta (8.2): *Cryptomonas* sp. Bacillariophyta (10.5): *Synedra rumpens* Chlorophyta (19.4): *Botryococcus Braunii*	0–25 m 29.9	Crustacea (11.9): *Mesocyclops decipiens, M. leuckarti, Ceriodaphnia cornuta, C. dubia*	6.3
WESTERN JAVA	Tjigombong (man-made)	Dinophyta (83): *Peridinium Gutwinski* (80), *P. quadridens, P. wildemani, P. inconspicuum, Ceratium hirundinella*	0–115 m 113.6	**Rotifera** (1.7) **Crustacea** (5.5): *Mesocyclops decipiens, M. leuckarti, Ceriodaphnia cornuta*	0–115 m 17.3
	Sindanglaja	Dinophyta (72): *Ceratium hirundinella* and *Peridinium Gutwinski* Bacillariophyta (4.3): *Synedra rumpens*	0–15 m 45.4	**Insecta** (5.6) **Rotifera** (50) **Crustacea** (18.0)	0–15 m 1.3

EPI = epilimnion; HYPO = hypolimnion

The fish population contains local species and introduced fishes. The endogenous species include: the cyprinids *Rasbora lateristriata* and *Barbus microps*, which feeds on higher plants, sponges, snails and ostracods; the clariid *Clarias batrachus* which eats copepods, *Chaoborus* larvae, *Macrobrachium* and ostracods; the ophicephalid *Ophicephalus gachua*, feeding on prawns, snails, fishes and zooplankton; and the synbranchid *Monopterus albus*, feeding on insect larvae, snails, prawns and fish. The introduced species are: the cyprinid *Barbus gonionotus*, an herbivorous fish brought to the lake in 1949; the poecilid *Poecilia reticulata*, feeding mainly on zooplankton; the cyprinodont *Aplocheilus pandax*, eating insect larvae, copepods and prawns; and the cichlid *Tilapia mossambica*, feeding on zooplankton in its early stages and on diatoms later on.

The fisheries include the prawn fishery mainly carried out by women and the open-water fishery concerning *Tilapia, Barbus gonionotus* and *Clarias*.

In spite of the forest clearance and the introduction of exotic fishes which has taken place since the Sunda Expedition, Green *et al.* (1976) conclude that very little change can be observed in the lake chemistry and biology. This is probably due to the abundance of food resulting from the presence of *Eichhornia* mats which also constitute excellent nursery grounds. In Ranu Pakis, another nearby crater lake which has no *Eichhornia* belt, *Tilapia mossambica* is also the main food-fish but the survival of the fry is poorer than in Ranu Lamongan and the lake has to be stocked periodically.

The lakes of Bali

Green *et al.* (1978) investigated the three crater lakes Danau Bratan, Danau Buyan and Danau Tamblingan, located in a large caldera in the Northern central part of the island, and Danau Batur located in the caldera of the active volcano Gunung Batur, east of the previously mentioned lakes. In 1928–29, only Danau Bratan and Danau Batur were visited by the Sunda Expedition.

Danau Bratan

This relatively shallow lake is mixed by strong winds most of the time. Temporary stratification accompanied by a decrease in oxygen in deep layers occurs on calm days. In 1974, the water conductivity was 39 μmhos cm^{-1} instead of the 18 measured by Ruttner in 1929. Significant changes were also observed in the composition of the algal community: the plankton was no longer dominated by desmids (*Staurastrum* was not even found) but by the blue-green *Lyngbya limnetica* with spherical green algae and *Melosira granulata* as subdominant species. The replacement of desmids by blue-greens may indicate that the lake is progressing towards a more eutrophic state. This is consistent with the development of agriculture, cattle rearing and tourism around the lake. In 1974, the zooplankton biomass was very low and the marginal fauna very sparse. The main food chain is a detritus pathway based on brown debris coming from exogenous plant material falling on the lake: this material is the basic food of chironomid larvae, may-fly larvae, Cyprinidae, *Poecilia* and *Xiphophorus*. In contrast, *Clarias* feed principally on larvae of aquatic insects and cyclopids.

Danau Buyan

The lake has an abundant macrophyte vegetation on its southern shore which represents a major fishing area. The lake is not clearly stratified but the temperature difference of 1°C observed between the surface and 29 m depth causes enough stability to reduce significantly the oxygen concentration of deep layers.

The phytoplankton is more abundant than in Danau Bratan, especially the desmids; spherical green algae and Cyanophyta such as *Aphanocapsa delicatissima, Lyngbya limnetica* and *Dactylococcopsis fascicularis* are also present. *Poecilia reticulata* was the dominant fish.

Danau Tamblingan

The lake, clearly stratified (22°C at the surface and 20.5°C at 28 m depth) was devoid of oxygen below 15 m. It has a richer desmid flora than Danau Buyan and its algal assemblage resembles that described by Ruttner in 1931 for Danau Bratan. It is probable that, in 1929, the three lakes had a very similar algal flora which was kept unchanged in Danau Tamblingan, nearly unchanged in Danau Buyan and was strongly modified in Danau Bratan. This corresponds to the degree of human pressure on the three lakes: non-existent in Tamblingan, mild in Buyan, considerable in Bratan. *Poecilia reticulata* was, here too, the dominant fish.

Danau Batur

This lake has a much higher conductivity and a higher pH than the other lakes. In 1974, it was not stratified but the decrease in temperature with depth (23.2°C at the surface and 22.7°C at 30 m) was accompanied by a gradual decrease in oxygen (from 7.4 mg l^{-1} in upper layers to 3.2 mg l^{-1} at 30 m). The algal flora was dominated by the blue-green algae *Lyngbya limnetica, Aphanocapsa delicatissima* and *Dactylococcopsis fascicularis*, with the diatoms and especially *Nitzschia amphibia* as the second most important group. Zooplankton was scarce. *Xiphophorus maculatus* was as abundant as *Poecilia reticulata*. The chemical and biological features of the lake in 1974 show no major difference with the description of Ruttner.

Danau Buyan and Danau Tamblingan are the most productive lakes with more vigorous fisheries than Danau Bratan and Danau Batur.

Fish ponds

In Indonesia, nearly all types of water bodies are used to grow fish: brackish water ponds along the sea coast where *Chanos chanos, Sarotherodon mossambicum* and prawns are grown, paddy fields where carp culture is carried out, and freshwater ponds. This latter type of pond is often found in the vicinity of human settlements and domestic raw sewage is used as fertilizer. Extreme conditions of carp raising are reported by Vaas (1954): in Bandung City (Java), carp are cultivated in cages immersed in a flowing open sewer. Such carp feed directly on the particulate fraction of the sewage and on an abundant fauna of insect larvae, oligochaetes and crustaceans. Near Djakjarta, the ponds are fed by a mixture of sewage water from the city and drainage water from rice fields. The main species raised in freshwater ponds are the carp *Cyprinus carpio* (feeding on filamentous algae, crustacean zooplankton, insect larvae and worms), *Sarotherodon mossambicum, Helostoma temmincki* and *Osteochilus hassletii* (feeding mostly on phytoplankton and periphyton), and *Puntius javanicus* (eating preferentially submerged plants). This latter fish is often cultivated together with the common carp which utilize the natural animal food not eaten by *Puntius*. Carp ponds are heavily fertilized with human sewage. It does not seem that infectious diseases result from this very common practice in Indonesian fish farming.

Vaas & Sachlan (1955) studied the development of the plankton community in a pond and a newly dug pit near Bogor. The fertilized pond was dominated by diatoms and rotifers whereas in the *Nitella furcata* belt bordering the oligotrophic pit, 66 species of desmids were found.

Ardiwinata (1957) describes the fish culture in paddy fields. This practice began in West Java about a century ago in the vicinity of religious schools from where it spread through the students returning to their places of origin. In 1955, more than 9000 tons of fish for consumption was cropped from paddy fields of Sumatra, Java, Celebes, Bali and Lombock. Formerly, fish were grown as a 'second crop' after rice had been harvested. Later, rice was planted more than once a year and the time for fish culture became shorter (1-3 months instead of 6 months) and the technique became known as the 'interval method'. Simultaneously, there was a shift in the aim from production of fish for consumption to production of fry.

Blue-green algae appear first after the field flooding and are later replaced by Desmidiales. The benthic organisms living on decomposing straw and their rich associated bacterial populations comprise tubificids, pulmonate snails and a very wide variety of aquatic insects.

Presently, the common carp *Cyprinus carpio* is the main species raised in paddy fields because of its rapid growth rate and its good adaptation to variable food sources and concentrations of oxygen. *Puntius javanicus* is mostly raised for fry; in Sumatra, in the fields near Lake Ranau, it is grown together with *Osteocheilus* until of suitable age for consumption. *Helostoma* is raised in West Java. In 1943, *Sarotherodon mossambicum* was introduced and its success led the Javanese farmers to spawn this cichlid in paddy fields.

THE PHILIPPINES
General

The Philippines is an archipelago of more than 7000 islands located between 4 and 21° N and 118° and 127° E. Temperature fluctuates around 28°C and precipitation is 2500 mm y^{-1} with distinct dry and wet periods. The area is yearly swept by 21 typhoons (average of the last 90 years), causing considerable damage.

In the Philippines, 61.1% of the animal protein produced comes from fish, molluscs and crustaceans. The annual production of fishery is 746 000 tons (1967 data), 47% from rivers and bays and 8.5% from fish ponds. The freshwater fishery resources are distributed in more than one million hectares of lakes, rivers, swamps, paddy fields and fish ponds. The development of fish ponds is hampered by severe problems of man-made pollution. Both the importance of aquatic organisms in the local diet and the new problems caused by rapid urbanization and industrialization underline the necessity of large-scale freshwater investigations.

Lake Lanao

Lake Lanao is located in Northern Mindanao at 8° N and 120° E at 701 m altitude. Lake Lanao was formed by a lava flow which dammed the streams flowing south-west. According to recent dating this event probably occurred in the late Tertiary which gives the lake an age of 3.6 to 5.5×10^6 years (Frey, quoted by Lewis, 1978c).

The mean annual rainfall is 2873 mm around the lake. The period

December–April is cool and dry and peak precipitation takes place in June. The air temperature (mean monthly values) varies from 21.5°C in January to 23.7°C in May.

Most of the lake watershed extends on its eastern side and culminates at 2815 m in the south-east corner. More than 50% of the watershed is forested and the nutrient load carried by the streams draining the watershed is very small.

The lake is drained northward through the Agus River, a fast-flowing river with a drop of 700 m over 36 km. The lake is biologically isolated from marine organisms by the 57 m high Maria-Cristina Falls on the Agus River. The lake is relatively stable (mean annual variation of 0.8 m with a maximal value of 2.09 m); in contrast, the discharge of the Agus River shows large fluctuations (233 m^3 s^{-1} in December 1955 and 12.8 m^3 s^{-1} in May 1958). The mean discharge from the lake is 104.7 m^3 s^{-1} or 3.3×10^9 m^3 y^{-1}. The main morphological features of the lake are summarized in Table 99.

The lake bottom is covered by accumulated sediments which mask topographic irregularities. The southern part is the deepest. The southern shore is very steep and a depth of 100 m is reached at about 200 m offshore. The low value of the watershed area: lake volume ratio reflects the wet climate of the Philippines. For comparison, the subalpine lakes have a ratio of 0.10.

Lewis (1973b) studied the circulation pattern of Lake Lanao. The lake is homothermal from December to March around 25°C. The turnover in late December is accompanied by a decrease in oxygen in the whole water column; in 1970, from 4.3 mg O_2 l^{-1} after homothermy it rose to 6.4 mg l^{-1} in February. Lewis distinguishes three types of thermoclines: the 'breeze thermoclines' which establish between 5 and 20 m as a result of a few days of sunny weather with moderate winds, the 'squall thermoclines' developing at 20–30 m after strong squalls which homogenize a thick water column, and the 'storm thermoclines' generated by typhoons at 40–50 m depth. Any temporary mixing, causing the homogenization of water layers of different chemical properties without disappearance of the hypolimnion

Table 99 *Morphological features of Lake Lanao (according to Frey, 1969)*

Altitude	701 m	Volume	21.5×10^9 m^3
Surface area	357 km^2	Replacement time	6.5 y
Mean depth	60.3 m	Watershed area	1680 km^2
Max. depth	112 m	Watershed area:lake volume ratio	0.078 m^2 m^{-3}

The Philippines

is called 'atelomixis' by Lewis. This process allows periodical reinjection of nutrients of deeper layers into the trophogenic zone. Modifications of the storm thermocline during calm weather were attributed to the existence of internal seiches; stable thermal inversions observed on the bottom were related to heat retention by the lowermost layer of the water column which was then stable and isolated from the rest of the water by its relative high salt content resulting from the release of dissolved substances from the mud.

The annual energy budget was found to be 7250 cal cm^{-2} in 1970 and 4500 cal cm^{-2} in 1971.

Lake Lanao has a low conductivity (105 μmhos cm^{-1}) with Ca^{2+} (4.5 mg l^{-1}) and Na^+ (5 mg l^{-1}) as the major cations; alkalinity is 51 mg l^{-1}. The concentrations of PO_4–P and NO_3–N in the epilimnion are, respectively, 29 and 9 μg l^{-1} and silicate averages 9.2 mg l^{-1}. Nutrient availability is maximal in September–October and in April–July. Lewis (1978b) explains the succession of algal groups as a function of these two factors. The period September–October, characterized by high light availability and low nutrient levels due to low turbulence, is dominated by blue-green algae. In November, during the precirculation period, radiation levels are lower but the frequency of atelomictic events increases and, with it, the availability of nutrients; these new conditions are more favourable to the Chlorophyta which replace the blue-greens. The circulation period with its constantly high nutrient flux and very low radiation levels is the period of diatoms and cryptomonads; the identical conditions which prevail in July–September during the storm period cause the reappearance of the same two groups. The extreme conditions of intense radiation and nearly complete absence of nutrients allow a short dominance of the dinoflagellates. It therefore seems that the nutrient and light climate prevailing at different periods determines, not only the presence or absence of growth pulses, but also the identity of the algal groups most likely to develop under such conditions.

The phytoplankton community of Lake Lanao includes 12 species of Cyanophyta (Cya), four species of Euglenophyta (E), 45 species of Chlorophyta (Chl), four species of Bacillariophyta (B), three species of Dinophyta (D) and two species of Cryptophyta (Cry) (Lewis, 1978c).

The 10 largest biomass contributors, ranked in order of importance, are as follows: *Nitzschia baccata* (B), *Dictyosphaerium pulchellum* (Chl), *Oocystis submarina* (Chl), *Lyngbya limnetica* (Cya), *Cryptomonas marssonii* (Cry), *Anabaena sphaerica* var. *tenuis* (Cya), *Aphanothece nidulans* (Cya), *Sphaerocystis schroeteri* (Chi), *Chodatella subsalsa* (Chl) and *Coelastrum cambricum* (Chl).

Lewis (1978a) studied the spatial distribution of the algae. During calm weather, the algal biomass shows a clear peak in the upper euphotic zone and so does primary productivity. Then, motile species and species with buoyancy control occupy the top layer, a few metres above the other species. In windy weather, the algal biomass is uniformly distributed in the whole mixed layer and no species has any privileged position.

Lewis (1974) studied the primary productivity of Lake Lanao with the ^{14}C technique. He did not find any marked diurnal rhythm in photosynthetic efficiency, i.e. for a given day, the production per unit of radiation was constant. He found experimentally that the threshold of photosynthetically active radiation causing light inhibition is 0.10 cal cm^{-2} min^{-1} in calm weather and that in Lake Lanao, light-inhibited production accounted for 83% of the total productivity.

Lake Lanao has a productivity range of 400–5000 mg C m^{-2} d^{-1} with an average value of 1700 mg C m^{-2} d^{-1}. This exceeds the productivity of many eutrophic lakes of the temperate zone. The naturally high productivity of Lake Lanao derives from a certain number of factors, some of them intrinsic to Lake Lanao, others typical of a large number of tropical lakes.

One of the main factors is the excellent light penetration due to the very low concentrations of dissolved substances and suspended matter which characterize Lake Lanao. The algal biomass is in fact the main suspended matter in the lake water. Its low value, fluctuating from 5 to 60 g wm m^{-2} with a yearly average of 24 g m^{-2}, indicates a high specific productivity and thence a very active population. This activity is maintained through an adequate supply of nutrients made possible by a rapid nutrient regeneration and by a thorough wind-mixing of the epilimnion. This description of Lake Lanao is very close to the conditions reported for Lake Kinneret in summer by Serruya et al. (1980). In this latter lake, a small biomass (20 g m^{-2}) produces 1400 mg C m^{-2} d^{-1} during several consecutive months without any external nutrient supply.

The mean annual value of respiration in the euphotic zone of Lake Lanao is 1076 mg C m^{-2} d^{-1}, 80% of which is respired by the phytoplankton and 20% by the heterotrophs. Since the ^{14}C primary productivity values are net values the gross productivity of Lake Lanao is 2776 mg C m^{-2} d^{-1}.

In his structural and compositional analysis of the phytoplankton communities, Lewis (1978c) found a greater species diversity between lakes in South-East Asia than between alpine lakes. Studying the overlapping between the phytoplankton genera of Lake Lanao and those of other tropical and temperate lakes, he found that, despite the different types of lakes compared and the geographical separation of these lakes, 80% of

Lanao genera are shared with other tropical lakes of Indonesia and South America. Only 45% of Lanao genera are shared with temperate lakes of Europe and North America. This suggests the existence of a pantropical algal assemblage dominated by Chlorophyta. In Lake Lanao, of 70 observed algal species, 44 belong to the Chlorophyta. This proportion is maintained among the dominant species: as mentioned above, of the 10 largest biomass contributors, five are Chlorophyta. Moreover, when unpolluted tropical lakes are compared with unpolluted temperate lakes, the average biomass of the former group is twice that of the latter.

The zooplankton of Lake Lanao is dominated by *Chaoborus* and copepods. Among this latter group, the cyclopid *Thermocyclops hyalinus* and the calanoid *Tropodiaptomus gigantiviger*. The copepods have populations of 250 individuals l^{-1} in July–September against 50 individuals l^{-1} in December–January. Rotifers of the genera *Conochiloides, Hexarthra, Keratella, Tetramastix* and *Polyarthra* are especially abundant in June–July. The Cladocera *Bosminopsis, Moina* and *Diaphanosoma* are not abundant.

Lake Lanao is famous for its population of endemic cyprinids discovered by Herre (1924, 1933). It consists of 13 species of the genus *Barbodes*, which is synonymous with *Puntius* which, according to Myers (1960), can hardly be distinguished from the genus *Barbus*. Five additional species belonging to four genera complete the known endemic fish fauna of the lake: *Spratellicypris* (1), *Mandibularca* (1), *Ospatulus* (2) and *Cephalakompsus* (1). Three non-endemic predators introduced by man are present: *Channa striata, Anguilla celebensis* and *Micropterus salmoides*. Myers (1960) thinks that the Lanao cyprinids have entered the Philippines from North Borneo through the Palawan–Calamianes–Mindoro pathway and through the Sulu–Mindanao route. Cyprinids are absent from the rest of the Philippines and from Celebes. The three dominant genera of cyprinids in North Borneo are the only ones found in Mindanao: *Barbus, Rasbora* and *Nematabramis*. Since *Barbus binotatus* is the commonest cyprinid species of the Sunda Islands and is widespread in Thailand and Malaysia, Myers concludes that this species is the ancestor of the 18 endemic species of Lake Lanao. The specializations developed by these species, especially the jaw modifications of *Mandibularca*, transcend the family limits of some 2000 species of cyprinids in the world. This type of extreme modification has been called 'supralimital specialization' by Myers.

Lake Mainit

Lake Mainit is located in north-eastern Mindanao at 9°30′ N and 125°30′ E at 27 m above MSL. The origin and age of Lake Mainit are unclear. Lewis (1973a) thinks the lake has a tectonic origin and that its steep west shore is a fault scarp.

Heavy rainfall (400–700 mm per month) and relatively low temperatures (down to 25.7°C, monthly average value) prevail from November to March. The rest of the year is dry (100–200 mm per month rainfall) and warmer (up to 27.8°C, monthly average value). The mean annual rainfall is 3180 mm and mean air temperature 26.8°C. Wet south-easterly winds dominate in November–December and the northern wet and cool monsoon prevails in January–February. The cloud cover, more than any variation of radiation, diminishes the temperature in the cool period.

The two main inlets of the lake drain the northern part of the watershed. The lake is drained by the Tubay River at its southern end.

The main morphological features of the lake are summarized in Table 100. It is worthwhile noting the extremely low value of the watershed area:lake volume ratio which explains the low concentrations of suspended solids in the lake water and its consequent high transparency. There is evidence that, in the past, the lake occupied a wider area and drained a larger watershed.

The lake fishes were studied by Manacop (1937). The lake was also visited by the Wallacea Expedition. Lewis visited the lake in August and November 1971.

The temperature profiles made on these two dates were very similar with the exception of the superficial layer. On both occasions, the maximal thermal gradient occurred between 30 and 40 m and the deep portions of the profiles were identical with a minimum temperature of 26.02°C at 200 m. This indicates that no deep mixing took place between the two visits. Oxygen concentration, steady from 0 to 10 m, declined gradually down to 30 m; below this depth it was negligible.

The Lake Mainit water has respective concentrations of Na^+ and Ca^{2+} of 6.0–6.5 and 5.5 mg l^{-1}. Both cations but especially sodium increase in concentration with depth. Lewis thinks the dominant sodium is of marine origin and transported by typhoons. This, however, does not

Table 100 *Morphological features of Lake Mainit*

Altitude	27 m	Volume	18×10^9 m^3
Surface area	141 km^2	Replacement time	15 y
Mean depth	128 m	Watershed area	313 km^2
Max. depth	223 m	Watershed area:lake	
Depth of cryptodepression	198 m	volume ratio	0.017 m^2 m^{-3}

explain the increase with depth of this element. Since the lake bottom is located far below sea level and not far from the sea shore, a slow seepage of seawater is not excluded.

In August 1971, the plankton was dominated by blue-green algae (68.2% of the biomass) with *Anabaena* sp., *Lyngbya limnetica* and *Aphanothece* sp., followed by Dinophyta (14.8% of biomass) with *Gymnodinium* sp. and *Peridinium* sp., Cryptophyta (9.4%) with *Cryptomonas marssonii*, Chlorophyta (5.8%) with *Coelastrum cambricum* and *Oocystis submarina*, and Bacillariophyta (1%). In November 1971, the contribution of the various groups was more homogeneous: the blue-greens dominated by *Anabaena* sp. represented only 27.9% of the biomass; the Cryptophyta with *Cryptomonas marssonii*, 26.9%; the Bacillariophyta with *Synedra rumpens* and *Melosia granulata*, 16.6%; the Chlorophyta with *Gloeocystis planctonica*, *Coelastrum cambricum* and *Dimorphococcus* sp., 15.2%; and Dinophyta 12.2%. The absolute values of the biomass were 28 g wm m^{-2} in August and 18 g wm m^{-2} in November. Blue-greens, represented by six genera, contributed much of the biomass in contrast with the less abundant but more diversified Chlorophyta (13 genera).

^{14}C primary production was 2 g C m^{-2} d^{-1} in August and 1 g C m^{-2} d^{-1} in November. As with Lake Lanao, Lake Mainit is a naturally productive lake in spite of its remarkable transparence. In August, 1% of the incident light reached a depth of 13 m.

In August, the zooplankton was dominated by Protozoa (55.6% of the biomass), followed by the copepods (38.5%) represented by one species similar to the *Thermocyclops* of Lake Lanao, Rotifera and the Cladocera *Chydorus barroisi* and *Diaphanosoma sarsi* (0.2%). In November, the contribution of the Protozoa dropped to 15% and that of copepods exceeded 80%. It seems that, at certain periods, the grazing pressure of young fishes on crustaceans and rotifers favours the predominance of the small-size Protozoa. The absolute biomass values were 12 g m^{-2} in August and 21 g m^{-2} in November.

The collected fishes included the endemic *Ophieleotris agilis* and *Glossogobius girus* which were abundant. The rest of the catch comprised *Trichogaster trichopterus* (Anabantidae), *Ophiocephalus striatus* (Syngnathidae) and a few specimens of the endemic *Solenophallus thessa*. Among the cyprinids, *Puntius (Barbus) binotatus* was present. There is no clear answer to the question why this species did not speciate as in Lake Lanao. A possible reason is that competition of euryhaline fish made the process of specialization much more difficult. The fishermen report that a catfish and an eel are very abundant in certain seasons.

Laguna de Bay

Laguna de Bay is one of the largest lakes in South-East Asia. It is located in Luzon Island at 14° N and 122° E at 125 m above sea level. It has a surface area of 900 km² and a mean depth of 2.4 m.

The major inlets of the lake are the outlet of the Calinaya Power Dam and the Pagsanjan River. On June 7 1961, the discharge into the lake was 25 m³ s⁻¹, a typical value for the dry season. The lake is drained by the Pasig River into the Manila Bay. Sea water intrusions via the Pasig River may occur at high tides. More than 120 industries discharge wastes into this river which is devoid of fish especially in summer (Lesaca, 1974).

The lake has an average oxygen concentration of 7.3 mg l⁻¹ with a range of 3.4 to 10.5. Its total alkalinity is around 114 mg l⁻¹ and its pH 7.9. The chloride level is generally very low (3 mg l⁻¹) but may reach several hundred mg l⁻¹ where there are sea water intrusions.

The algal assemblage of Laguna de Bay is dominated by the blue-green algae with *Oscillatoria* and *Phormidium*. *Melosira granulata* is the most frequently found diatom, whereas Chlorophyta are represented by the common *Sphaerella* and an abundant population of desmids including seven genera.

The zooplankton is dominated by rotifers (*Synchaeta*, *Anuraea*, *Filinia*) but also includes copepods, cladocerans and zooflagellates. Chironomid larvae, annelids and snails are the main elements of the mud-bottom fauna whereas clams dominate on sandy areas. An average number of 541 chironomid larvae m⁻², 1976 snails m⁻² and 60 annelids m⁻² are reported in the 1962 Report on freshwater fisheries investigations of FAO.

The most abundant native species of fish are *Glossogobius giurus* (Gobiidae) and *Therapon plumbeus* (Theraponidae). *Chanos chanos* (Chanidae), *Clarias batrachus* (Clariidae), *Trichogaster pectoralis* (Anabantidae) and *Atherina ondrachtensis* (Atherinidae) are also present. *Cyprinus carpio* (Cyprinidae) and *Tilapia* (sym. *Sarotherodon*) *mossambica* (Cichlidae) have been artificially introduced.

Duck rearing was practised on a large scale to produce a local delicacy, the embryonated duck egg. The birds fed on snails, mainly *Corbicula manillensis*, *Vivipara angularis* and *Melania lateritia*. One of the distinctive features of Laguna de Bay was its large-scale snail and clam dredge fishery. The continuous dredging of the bottom mud released enormous amounts of nutrients in the water and was probably responsible for the unusually high production of Laguna de Bay. In October 1961, the catch of snails, shrimps and clams by dredge fishing was 18 548 tons, i.e. > 126 billion animals. An additional approximate catch of 1000 tons of fish, mainly composed of therapon, was obtained from August to December 1961 from baklad fishery only.

During the last 10 years, heavy agricultural and industrial pollution have endangered the fisheries of Laguna de Bay. As previously mentioned, the Pasig River is heavily polluted and the snails on which was based the duck farming of the Pasig River have disappeared, as well as most of the river fish population. At high tides, the water of the Pasig River and its pollutants are carried into the lake; the man-made activities around the lake and deforestation in the watershed have accelerated the rate of sedimentation which will turn the lake into a marsh choked with vegetation. As a consequence, the 1962 FAO report warns that Laguna de Bay cannot be considered in the future as a permanent producer of fish and other aquatic organisms, as a source for water supply or as a transportation route. Similar pollution problems are experienced in the Tinajeros River System, North of Manila.

PAPUA NEW GUINEA

Geologically, Papua New Guinea consists of a southern core of old Palaeozoic formations on the northern side of which post-Palaeozoic series have been added by the Miocene orogenesis. A resulting mountainous backbone running west-north-west to east-south-east culminates at 5000 m bordering the lowland southern area. Little is known of the limnology of the Island and the studies carried out within the framework of the Purari River System Project have added valuable information to our knowledge.

The Purari River System Project

In early 1975, it was decided to examine the feasibility of the construction of a dam at the site of Wabo on the Purari River. Environmental studies accompanied the project. The Purari River has a catchment area of 33 000 km²; the mountainous part of the watershed consists of granodiorite, shales, gabbro and limestones while greywacke shale, tuff and lava dominate in the lowlands (Petr, 1976). The precipitations are high over the whole watershed, ranging from nearly 2000 mm y^{-1} in the highlands to 9630 mm y^{-1} at Wabo, the dam site, located at 61 m altitude and 3255 mm y^{-1} at Baimuru in the delta at 5 m altitude. The highlands are more populated than the lowlands, densely covered by the rain forest. The Purari River has an average annual discharge of 1525 m³

s^{-1}. The salinity is low (54 mg l^{-1}) and the water belongs to the calcium hydrogen carbonate type (Table 101).

The Wabo Lake will have a surface area of 260 km and a maximum volume of $14 \times 10^9 \, m^3$. It is planned to have a drawdown of 25 m and a retention time of 136 days (Petr, 1975).

In order to assess the impact of the dam on the fish fauna a detailed study of this fauna was carried out in the river and in the delta (Haines, 1979). It is interesting to note that the fish fauna is much richer in the delta than in the rivers probably because, as in Australia, most species are of marine origin. A few species, however, are found both in the rivers and in the delta. This is the case for *Toxotes chatareus*, *Scutengraulis scratchleyi* and *Parambassis gulliveri*. Other species such as *Kurtus gulliveri* and *Cinetodus frogatti* are found only in deltaic areas in the Purari System but are present in other Papuan freshwater bodies, such as Lake Murray on the Fly River. The diversity of the deltaic fauna is related to the high number of habitats of the mangroves of the Gulf of Papua which are among the largest in the world. The mangroves offer good conditions for prawn and fish nurseries and feeding grounds. They also determine the very special type of food web of the system. The scarcity of phytoplankton results in very low plankton primary productivity. The floating *Azolla* is the main aquatic producer. However, the bulk of the organic material of the system comes from terrestrial and mangrove sources. It serves as a substrate for numerous microorganisms and small invertebrates. Those, together with detritus, are consumed by a flourishing prawn community, mainly carid prawns in the river and penacid prawns in the delta. The fish fauna has nearly no herbivores or plankton-feeders and is mainly composed of carnivores, omnivores, and detritivores. The carnivores feed on prawns, fish, crabs, molluscs and insects in various combinations; the omnivores eat a mixed diet including the previous elements and plant material; and the detritivores are essentially mud-eaters. Quantitatively, prawns are the main item of fish food in the mangrove area whereas this role is played by insects

Table 101 *Chemical features of the Purari River and the Sirunumu Impoundment (results in mg l^-; (‰meq))*

	pH	Ca^{2+}	Mg^{2+}	Na^+	K^+	Cl^-	SO_4^{2-}	HCO_3^-	Salinity
Purari R.	7.8	7.0	3.1	0.5	0.45	6.0	0	37.2	
20 Jan. 1975		(0.35)	(0.258)	(0.022)	(0.012)	(0.169)	0		
Sirunumu I.	6.6	1.5	1.7	1.3	1.35	6.0	0	9.2	21.05
3 Feb. 1975		(0.075)	(0.142)	(0.057)	(0.035)	(0.109)	0	(0.150)	

Papua New Guinea

and plant material in the river. The absence of true freshwater fishes has led to the development of prawns which fill the niche of detritivore fish and to the overwhelming dominance of ariid catfishes which have filled all kinds of trophic niches in the rivers and the delta. This type of food web is very similar to that of Amazonia and tropical Africa.

The Sirinumu impoundment

The Sirinumu impoundment is located on the Laloki River near Port Moresby at the upper limit of the tropical forest at 538 m altitude. Magnesium and hydrogen carbonate are the dominant ions. Salinity which was 109 mg l^{-1} in 1971 was only 21 mg l^{-1} in 1975. This decrease in concentration occurred after the dam was raised and the storage capacity of the impoundment increased.

The Fly River system

Sorentino (1979) reports a survey carried out on freshwater and marine fish of Papua New Guinea aimed at determining their mercury content. Abnormally high concentrations were found only in the barramundi (*Lates calcarifer*) caught in the Fly River watershed. The mercury originates from geological formations and not from human activities.

Lake Wisdom (Long Island, Papua New Guinea)

Lake Wisdom fills the central caldera of Long Island and is located at 5°20′ S and 147°6′ E. It has been studied by Ball & Glucksman (1978). The lake came into existence around 1750 on the occasion of a major volcanic eruption (Ball & Johnson, 1976). A secondary volcanic cone, the Motmot, forms an island in the southern part of the lake. Lake Wisdom is surrounded by steep walls and the crater rim is located from 30 to 260 m above the lake level. The morphometric data of the lake are summarized in Table 102.

The average annual rainfall is 2800 mm and the net annual hydric balance (rainfall − evaporation) is 1250 mm. Although the lake bottom is 170 m below sea level, there is no contact with the sea since the lake contains freshwater down to the bottom.

The lake surface is nearly always at 28°C and, in November 1974, when the lake was surveyed there was a thermocline between 10 and 20 m; below and down to 60 m, where the measurements were carried out, the temperature was 27°C. This stratification does not affect the oxygen content of the hypolimnion where 100% saturation has been observed down to at least 60 m. The presence of *Melanoides tuberculata* and chironomid larvae below 300 m depth indicates that these deep layers are also well oxygenated.

The presence of oxygen in deep layers suggests some sort of mixing; however a regular turnover of the lake caused by seasonal thermal variation is unlikely to happen since the maximum variation of monthly mean temperature is 1.1 °C. Ball & Glucksman (1978) think that convection currents generated by the Motmot volcano are, together with strong winds, the driving forces of Lake Wisdom's mixing.

Although the lake has presently no outlet and the volcanic activity of the Motmot (the last eruption took place in 1973) contributes large amounts of dissolved elements, the salinity of the lake is only 1230 mg l^{-1}. The chemical features of the lake are shown in Table 103.

The lake water is dominated by Na^+–Cl^-–SO_4^{2-} and its composition is strongly affected by the volcano: the lower pH and higher content of sulphates measured in its vicinity suggests that eruptions are accompanied by emissions of sulphuric acid.

The lake has a low phytoplankton biomass and a complete absence of vascular plants; the zooplankton consists of two Cladocera (*Latanopsis australis* and *Diaphanosoma excisum*) and the notonectid *Anisops nasuta*. Copepods are absent. The sponge *Spongilla alba* is abundant in shallow water. Four species of Mollusca have been collected: *Melanoides tuberculata* from the surface down to 350 m, *Amerianna papyracea* and *Gyraudus convexiusculus* in shallow water, and *Thiara scabra* from the surface down to 63 m. There is a small number of species of aquatic insects including Hemiptera, Odonata and larvae of chironomids. Numerous birds and one or more crocodiles complete the list of the lake fauna. Fishes are completely

Table 102 *Morphometric features of Lake Wisdom and Lake Dakataua*

	Lake Wisdom	Lake Dakataua
Altitude (approx.) above MSL (m)	190	76 ± 12
Max. length (km)	13.4	8; 7.5[a]
Max. width (km)	10.6	5.3; 3.9[a]
Surface area (km²)	95	48
Max. depth (m)	360	120
Mean depth (m)	~220	69
Volume (10⁹ m³)	21	3.3
Shoreline development (including islands)	1.656	2.18
Watershed area (km²) (excluding lake surface)	30	47

[a] The first number refers to the west arm of the lake, the second to the eastern arm

Table 103 *Chemical features of Lakes Wisdom and Dakataua*

Depth	Ca^{2+} (mg l^{-1})	Mg^{2+} (mg l^{-1})	Na^+ (mg l^{-1})	K^+ (mg l^{-1})	HCO_3^- (mg l^{-1})	Cl^- (mg l^{-1})	SO_4^{2-} (mg l^{-1})	SiO_2 (mg l^{-1})	pH	Salinity (mg l^{-1})
				Lake Wisdom						
1 m	98	28	263	19	138	274	336	55	7.6	1226
360 m	100	28	290	19	145	278	356	75	7.8	1230
Near Motmot surface	106	42	265	20	93	198	461	110	6.9	1360
				Lake Dakataua *(west arm)*						
0 m	12	15.8	35	2.6	102	40	44	30	7.80	315
106 m	14.6	14.2	38	2.7	109	40	46	60	7.01	325

absent from the lake. The small number of species and the very simple food web of the lake seem to be related to its young age and its isolation from other lakes.

Lake Dakataua

Lake Dakataua is a crater lake located at the northern-most point of the Willaumez Peninsula (50° S, 150° E) in the island of New Britain. The lake is surrounded by rain forest. The area receives 4270 mm y^{-1} precipitation, mostly during the north-west monsoon. The lake has been studied by Ball & Glucksman (1980).

The Caldera occupied by the lake was formed more than 1000 years ago. The Caldera and the lake have a horseshoe shape: the two arms of the lake are separated by an andesitic volcano (Mount Makalia) the last eruption of which occurred a hundred years ago.

The lake has no outlets although four springs, outside of the crater, are possibly fed by the lake. The morphometric data of the lake are presented in Table 102.

In October–November when it was surveyed, the lake was stratified. The surface temperature varied from 30.8 to 31.9°C. The thermocline was around 25 m and at 50 m the temperature was 28.5°C. In contrast with Lake Wisdom the hypolimnion is anoxic below 70 m.

The lake water belongs to the $Na^+-Mg^{2+}-HCO_3^-$ type. The low salinity (Table 103) reflects the high amounts of precipitation received by the lake and the long inactivity of the volcano.

Higher plants such as *Najas tenuifolia* and *Chara* sp. are abundant in shallow water down to 8 m. Seven species of Cladocera have been found: *Biapertura karua, Bosmina meridionalis, Ceriodaphnia cornuta, Diaphanosoma excisum* (the most abundant species), *Dunhevedia crassa, Kurzia longirostris* and *Latanopsis australis. Mesocyclops hyalinus* was also common. The sponges *Spongilla alba* and *Eunapius carteri* were found in shallow water. Among the molluscs, *Melanoides pallens, Gyraulus convexiusculus* and *Physastra* sp. were not found deeper than 5 m whereas *Melanoides tuberculata* and *Thiara scabra* were recorded down to 20 m. The insects are more numerous than in Lake Wisdom. The vertebrates include frogs, crocodiles and abundant birds but fish are totally absent.

Most of the species existing in Lake Wisdom are also found in Lake Dakataua but, except for chironomid larvae, Lake Dakataua has a more diverse biota probably due to its greater age and higher possibilities of colonization. However, the lack of oxygen below 70 m in Lake Dakataua, limits the habitats of aerobic fauna.

9
Australia

Introduction

Australia is an old Gondwanalandian continent. The thoroughly eroded Precambrian shield of central and western Australia contrasts with eastern Australia, mainly composed of Palaeozoic formations folded during the hercynian orogenesis.

The topography reflects the geological history. The Precambrian area, extending over two-thirds of the continent, is a plateau of 150–600 m altitude rising to 800–1200 m near the northern and western coasts as a result of post-hercynian epeirogenic movements. The eastern Palaeozoic area, the Eastern Highlands, running from northern Queensland along the east coast to southern Victoria, is a mountainous area culminating in Mount Kosciusko (2229 m) in New South Wales.

The summer northern monsoon brings moisture to the northern and eastern coastal areas and the winter monsoon reaches only the south-western and south-eastern corners of the mainland. A thin belt on the north and east coast receives yearly between 1000 and 2000 mm rainfall, whereas in central Australia, the yearly precipitation ranges from 100 to 200 mm.

The relatively abundant precipitation falling on the Eastern Highlands is drained by the Murray–Darling river system, the only large permanent one in Australia. It drains a watershed of 1×10^6 km^2 but, because of relatively high evaporation, the specific yield is as small as 13 800 m^3 km^{-2} y^{-1} or 14 mm (the specific yield of the Mekong is 700 000 m^3 km^{-2} y^{-1} or 700 mm).

In central Australia there is practically no runoff. On four-fifths of the continent, the runoff is < 10 mm and there is no groundwater. West of the central lowlands, a dense system of impermanent creeks (Diamintina, Finke, Coopers Creek) forms large internal deltas when the creeks reach

397

the plain and divide into multiple arms. Much of the water is lost and does not reach Lake Eyre, the draining area of this 'Channel country'. In a few limited areas the runoff exceeds 1500 mm, for example in the Snowy Mountains in South-eastern Australia (L'vovitch, 1979).

Variability from year to year in rainfall and streamflow is a typical Australian feature. It is maximum in western Australia, but even for the Murray River at Eston (drainage area 342 000 km^2), extreme annual runoff values of 121 mm and 4 mm were recorded in 35 years of observation.

Victoria
General

Williams (1966) gave the frequency distribution of 77 lakes and 41 reservoirs as a function of concentration of dissolved solids. In 38 reservoirs and 11 lakes, this concentration was < 500 mg l^{-1}. In three reservoirs and 27 lakes, it ranged between 500 and 3000 mg l^{-1} and, in 39 lakes, it varied from 300 to > 100 000 mg l^{-1}. According to the same author, the total range of salinity is 20–780 mg l^{-1} in reservoirs and 32–343 700 mg l^{-1} in lakes. In exorheic areas (to the ocean or via River Murray), lakes have generally lower concentrations than in the arheic north-west area of the western basaltic plain, where salt lakes are numerous (for example Lake Tyrrell where salt is mined commercially). Variations in salinity with time are considerable. In Lake Corangamite, seasonal variations exceed 30 000 mg l^{-1} and from 1875 to 1956, the concentration of sodium chloride increased by 114 000 mg l^{-1}.

Considering only the natural lakes, Williams (1966) estimated that 49% of the Victoria lakes were fresh and 51% saline. This is most interesting since the State of Victoria is one of the best watered of Australia: most areas receive > 250 mm y^{-1} and the south-east corner receives 1500 mm y^{-1}. However, in many areas the annual evaporation exceeds the precipitation; the runoff is then only seasonal and lakes are shallow with a very variable water level. Many lakes dry out completely in dry years. Water levels are maximal in winter and low in late summer. The surface area of most Victoria lakes ranges from 0.4 to 6.4 km^2 (Williams, 1964a) and many have no outlet.

The volcanic lakes of the Western District

The volcanic plains of the Western District extend west of Melbourne. Pleistocene-to-recent vulcanism covered this area with basaltic lava where many lakes can be found. The morphological features of a few volcanic lakes are shown in Table 104.

Australia

Only the deep lakes are permanent; their surface temperature varies from 10°C in July to 24°C in January. They are stratified in summer but their winter temperature exceeds 4°C; they are isothermal in winter and classified as warm monomictic lakes.

Lake Bullenmeri (Fig. 33) is isothermal from July to September at 10°C and stratified in summer. Complete disappearance of oxygen in the deepest part of the hypolimnion is observed from April to July. A similar pattern

Table 104 *Morphological features of the lakes of the Western District*

Lake	Surface area (km^2)	Max. depth (m)
Corangamite	234	4.4 (1956)
Werowrap	2.5	35.0
	0.25	1.5
Beeac	6.0	–
Bullenmeri	–	66.0
Gnotuk	2.1	18.5
Purrumbete	5.5	45.0

Fig. 33. Regional distribution of some lakes of Australia

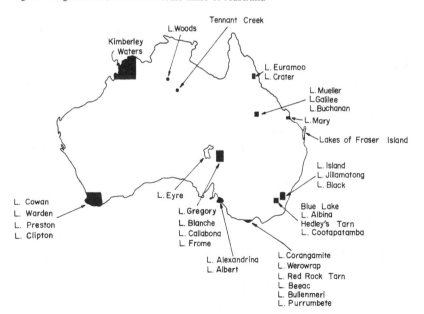

exists in Lake Purrumbete. In the shallower Lake Gnotuk, the isothermal period is longer (April–September) and hypolimnic oxygen depletion takes place from January to April (Timms, 1976). The absolute rates of hypolimnion deoxygenation are 0.012, 0.046 and 0.087 mg O_2 cm^{-2} d^{-1} in Lakes Gnotuk, Purrumbete and Bullenmeri, respectively.

A permanent meromixis has been described in West Basin lake, a sheltered small volcanic lake (Timms, 1972). Timms thinks the meromixis was established in 1968 when, after a prolonged drought, rainfall covered the saline lake water. The difference in salinity between mixolimnion and monimolimnion is 40‰. Shallow lakes are warmer in summer but are not stratified.

All the lakes are alkaline with pH > 8.0. The chemical composition of a few volcanic lakes is shown in Table 105. It is clear from Table 105 that the lake water of the Western District belongs to the sodium-chloride type and that the pattern of ionic prevalence is not much affected by salinity (Williams, 1967; Williams & Buckney, 1976a). In fact, as shown by Williams (1964a), there is an increase in the percentages of sodium and chloride and simultaneous decline in other ions with increasing salinity, probably due to the precipitation of the most insoluble salts.

The lakes of the Red Rock volcanic cone are slightly different. According to Bayly (1969) they belong to the sodium-chloro–carbonate water type; the carbonates originate from the local volcanic material and the water is very alkaline (pH 9.8).

The seasonal fluctuations in salinity were studied by Timms (1976) for the period 1969–76 in Lakes Gnotuk, Bullenmeri and Purrumbete. They clearly depend on variations of water levels which are caused by the timing and amount of rain. Precipitations are very variable from year to year, and so are the lake levels and the salinity. Besides the absolute high salinities, the extreme fluctuations of the physico-chemical environment constitute a nearly insurmountable obstacle to many living species.

As far as algae are concerned, blue-greens seem to be the dominant group. During measurements of primary productivity carried out by Hammer, Walker & Williams (1973) in 1970, the blue-green *Nodularia spumigena* was the main algal species in January in Lake Corangamite where the salinity was 25 g TDS l^{-1}. The primary productivity was then 1619 mg C m^{-2} d^{-1}. In Red Rock Tarn (15–25 g TDS l^{-1}), another lake investigated by the mentioned authors, a unialgal bloom of *Anabaena spiroides* var. *crassa* developed in January–February and gave very high values of productivity (10 356–17 281 mg C m^{-2} d^{-1}). In May, the bloom declined and was replaced by flagellates and diatoms, mainly *Nitzschia* sp.,

Table 105 *Chemical features of lakes of the Western District*

Lake	Na$^+$	K$^+$	Ca^{2+}	Mg^{2+}	Cl$^-$	SO$_4^{2-}$	HCO$_3$	Salinity (g l^{-1})	Source
	(in % of equivalent sums of cations or anions)								
Corangamite	86	1	<1	13	95	<1	4	37.9	
Werowrap	95	4	<1	1	65	<1	35	38.4	
Beeac (Sep. 1970)	93	<1	<1	6	98	2	<1	55.5	
Beeac (Mar. 1971)	95	1	4	<1	96	3	<1	317.8	
Bullenmeri (1969–72)	78.8	5	15.3	1	90.5	0.03	9.4	8.0	Timms (1976)
Gnotuk	78.1	1.7	1	19.2	98.5	–	1.5	53.5	Bayly & Williams (1966a)

and productivity dropped to very low values (61 mg C m^{-2} d^{-1}). In their study of Lake Werowrap, Paterson & Walker (1974) noted the disappearance of *Anabaena* from the lake in 1969 during a significant increase in salinity (from 23 to 56 g l^{-1} from September 1968 to May 1970), which was also accompanied by a decline in periphyton growth. *Anabaena* was replaced by *Gymnodinium aeruginosum* and *Chroococcus* sp. The productivity was high in March 1970 (2900 mg C m^{-2} d^{-1}) and low in August-September (369–1339 mg C m^{-2} d^{-1}), as measured by Hammer *et al.* (1973). The 1970 total productivity was, however, as high as 435 g C m^{-2} y^{-1}. On the bottom of the lake, the maximal area of which is 22 ha and maximal depth 1.5 m, the benthic fauna is dominated by the chironomid *Tanytarsus barbitarsis*, which occurs in large numbers (up to 150 000 individuals m^{-2}), and also includes the rotifer *Brachionus plicatilis*, the beetle *Necterosoma penicillatum* and a thin population of nematodes. Paterson & Walker (1974) found that the yearly mean standing crop of *T. barbitarsis* in Lake Werowrap is 538 kg wm ha^{-1} and productivity 664 kg dm ha^{-1} y^{-1}, a very high value in comparison with other inland water bodies. This is probably related to the simplicity of the food chain in Lake Werowrap: the chironomids which feed on benthic bacteria and small algae, and a beetle preying on chironomids. The absence of a multi-predation system gives the chironomid a decisive advantage.

Other Victoria lakes

Lake Tali Karng is located at 900 m altitude in Gippsland. It is one of the few natural deep lakes which is not of volcanic origin. It was formed by a landslide, probably 1500 years B.P. (Timms, 1974a). Its surface area is 16 ha and its volume 3.4×10^6 m^3. Its maximum and mean depths are, respectively, 51 and 21 m; however, the lake has no outlet and the water level varies by several metres. The lake is thermally stratified in summer. The lake water is not saline: 32.5 mg l^{-1} salinity (Williams, 1964b), 55 mg l^{-1} (Timms, 1974a).

Four species of zooplankton have been reported by Williams (1964b): *Calamoecia ampulla, Macrocyclops albidus, Ceriodaphnia* sp., and a hydracarinid mite. In 1973, Timms also found *Eubosmina meridionalis* and *Eucyclops* sp. The littoral fauna is restricted to *Micronecta robusta* and tadpoles of *Littoria phyllochroa* in *Chara* meadows developing in shallow areas and *Bulinus harnesii*. Among the benthic invertebrates, *Colubotelson* sp., *Ablabesmyia* sp. and *Procladius* sp. are dominant. Biomass is high in the littoral (up to 16 g m^{-2} wm), very low between 20 and 30 m, and intermediate in the deep area (2–3 g m^{-2} wm). Such a distribution has been

Australia

observed in Lake Kinneret (Serruya, 1978b), where the minimum has been attributed to the detrimental effect of the internal waves in the thermoclinal area on the fauna. It seems that, in Tali Karng, the artificially introduced *Salmo trutta* has eliminated *Galaxias coxi*. The sediments include large amounts of woody detritus which are probably an important source in the diet of the benthic animals.

The Centre Lake series lies north of the volcanic plains on quaternary sediments and was studied by Bayly & Williams (1966a). These lakes have a salinity reaching or exceeding 290 g l^{-1} with the ionic order Na$^+$ > Mg^{2+} > K$^+$ ≃ Ca^{2+} and Cl$^-$ > SO$_4^{2-}$ > HCO$_3^-$.

Lake Mulwala is a 45 km^2 man-made lake formed in 1939 on the Murray River which, in this region, forms the border between Victoria and New South Wales. This lake has been investigated for potential eutrophication problems due to the growth of the urban centre of Albury. Hart, McGregor & Perriman (1976) have determined that its non-calcareous, clayey sediments contain from 0.04 to 0.12% total phosphorus and that 50% of the sediment phosphorus reserve is theoretically exchangeable with the water.

New South Wales

New South Wales is geographically divided into two regions: the eastern highlands with precipitations ranging from 500 to 2000 mm, the higher value being typical of the Kosciusko plateau, and the western semi-arid lowlands, crossed by the Darling River and where yearly rainfall never exceeds 250 mm. The lakes and reservoirs of the north-eastern part of the State have been studied by Timms (1970) and those of southern and western New South Wales by Williams, Walker & Brand (1970).

North-eastern New South Wales lakes

In north-eastern New South Wales, Timms sampled 103 water bodies, 51 of which had a volume smaller than 10^5 m^3 and only three a volume greater than 50 × 10^6 m^3. This indicates the scarcity of large water bodies in this region. The waters are characterized by low salinities: of 103 water samples, 84 were in the salinity range 0–225 mg l^{-1}. In 44% of the samples the cationic sequence was Na$^+$ > Ca^{2+} > Mg^{2+} > K$^+$ and in 34% Na$^+$ > Mg^{2+} > Ca^{2+} > K$^+$; in 43%, the anionic sequence was Cl$^-$ > HCO$_3^-$ > SO$_4^{2-}$ and in 51%, HCO$_3^-$ > Cl$^-$ > SO$_4^{2-}$. The water chemistry will be discussed at length in a more general chapter (see pp. 436–43). Forty-three species of Entomostracean zooplankton have been observed in these water bodies. Temperature rapidly varying with altitude determines the distribution of various species: *Calamoecia tasmanica, Gladiofereus*

404 Australia

spinosus and *Ceriodaphnia cornuta* are confined to areas below 150 m altitude. *Daphnia lumholtzi* and *Diaphanosoma excisum* are found only below 650 m, whereas *Boeckella fluvialis, Calamoecia lucasi, Thermocyclops decipiens, Ceriodaphnia quadrangulata* and *Moina micrura* are not found above 1100 m, which is the lower limit for *Boeckella montana*. *Boeckella triarticulata, B. minuta, Mesocyclops leuckarti, Microcyclops sydneyensis* and *Daphnia carinata* do not live above 1280 m. *Bosmina meridionalis* is the only species found at all altitudes. Water chemistry also affects the appearance of species: for example, *Calamoecia tasmanica* is restricted to dilute, acid waters with predominance of sodium chloride.

South-eastern New South Wales Lakes

The subalpine lakes of the Kosciusko plateau have an extremely low salinity, 2.4–3.0 mg l^{-1} TDS or 4.6–5.3 μmhos cm^{-1} conductivity, and the most frequent ionic sequence is Na$^+$ > Ca^{2+} ≥ Mg^{2+} > K$^+$; the three main anions were alternatively dominant. This applies to Lakes Cootapatamba, Albina, Hedley's Tarn, Blue Lake and Club Lake. All these lakes are slightly acid and are very close in their concentration and composition to rain water. Two reservoirs: Jindabyne reservoir, 1000 m altitude, and Sponar's lake, 1525 m altitude, are ~10 times more concentrated than the natural lakes and both have magnesium–hydrogen-carbonate waters.

The lakes of the Monaro peneplain have very variable salinity and composition. In Island Lake (salinity = 88 mg l^{-1}) calcium and chloride are dominant but, in Lake Jillamatong (salinity = 21 244 mg l^{-1}), sodium and chloride are strongly dominant. In lakes of low salinity, hydrogen carbonate is always relatively high and sometimes dominant (Table 106).

The western rivers the Darling and Paroo, have an average salinity of 273 mg l^{-1} and the following ionic sequence: Na$^+$ > Ca^{2+} ≥ Mg^{2+} > K$^+$ and HCO$_3^-$ > Cl$^-$ > SO$_4^{2-}$.

The western lakes, those located in the valley of the Darling River, have a low salinity (156–835 mg l^{-1}) for this arid climate and have the same ionic sequence as the rivers. Other lakes, further from the river, are very saline or dry and the salt deposits are mainly composed of sodium chloride.

Lake Burley Griffin

The artificial Lake Burley Griffin was planned and built in 1963 as an ornament to the city of Canberra and to regulate the floods of the Molonglo River. It has a surface area of 720 ha, a mean depth of 4 m and a volume of 28 × 10^6 m^3. An amount of sediment of 250 000 m^3 enters the lake annually. Bed load traps have been built to remove coarse sand and

Table 106 *Chemical features of lakes of south-eastern New South Wales (from Williams et al., 1970) (results in mg l^{-1})*

	Na$^+$	K$^+$	Ca^{2+}	Mg^{2+}	Cl$^-$	HCO$_3^-$	SO$_4^{2-}$	Salinity
Kosciusko plateau lakes (average)	0.48	0.12	0.13	0.07	0.41	0.79	0.56	2.56
Rain water	0.39	0.08	0.26	0.12	0.36	1.40	0.43	3.04
Monaro peneplain lakes:								
Island lake	4.0	5.0	16.0	2.2	8.1	6.7	0	88
Anon. lake	54.0	14.0	58.0	16.0	95.0	256.0	3.8	497
Black lake	240.0	18.8	125.0	107.0	556.5	671.6	0	1719
Lake Cootralantra	450.0	10.0	82.0	18.0	493.0	809.0	0	1862
Beard's lake	740.0	20.0	92.0	11.0	874.0	1010	0	2747
Lake Jillamatong	7600	336.0	68.0	14.0	9803	865.0	2554	21 244

gravel. A copper–lead–zinc mine upstream was closed in 1962 to avoid pollution. The lake salinity is in the range 140–190 mg l^{-1}. Catfish, carp and trout are the main fishes of the lake (Minty, 1973).

Queensland

The difference in altitude between the coastal ridge and the central plain, together with the abundant precipitations falling in the mountainous area, determine the occurrence of the Great Australian Artesian Basin and the consequent presence of springs feeding rivers and more or less permanent lakes and water-holes.

North-eastern and central Queensland

This area was investigated by Bayly & Williams (1972) who distinguish four main regions:

Central eastern Queensland

Lake Galilee and Lake Buchanan, probably of tectonic origin, are both fed by underground water. Lake Galilee, which drains into the Thomson River, has a low salinity (600–700 mg l^{-1}). Although the water is of the sodium-chloride type, the other ions are not negligible; in particular, hydrogen carbonate and sulphate equivalents represent 25% of total anions. In contrast, Lake Buchanan, which drains an endorheic basin, is saline (salinity of 87 600 mg l^{-1}) and sodium chloride is massively dominant with hydrogen carbonate and sulphate equivalent forming only 0.1% of total anions. It is clear that the different flow pattern of these two lakes has channelled the water chemistry along two different pathways: the opened Lake Galilee keeps the ionic composition of the underground water relatively rich in hydrogen carbonate and to a lesser extent in sulphate, whereas in the closed Lake Buchanan the precipitation of carbonate has transformed the water into a nearly pure sodium chloride solution.

The Atherton Tablelands

This basaltic plateau, located south-west of Cairns, contains several deep crater lakes of low salinity: Lake Barrine (146 m deep), Lake Eacham (110 m deep), Lake Euramoo (165 m deep), and Crater Lake (70 m deep), are the largest. These waters have a salinity < 100 mg l^{-1} and are characterized by the dominance of divalent cations, especially magnesium, and of hydrogen carbonates, and by the absence of sulphates. In contrast, the thermo-mineral Innot springs are sodium-chloride waters.

The coastal lakes near Cape Bedford

These lakes resulted from the formation of an impervious layer of organic origin in the coastal sand dunes. The sand is poor in soluble substances; the lake water has the approximate composition of rain water, slightly concentrated by evaporation, and salinities do not exceed 60 mg l^{-1}.

The lakes near Rockhampton

Volcanic lakes in this area, such as Lake Mary, belong to the sodium-chloride type but magnesium and hydrogen carbonates are also major components.

South-eastern Queensland

The investigations of Bayly (1964) and Bayly, Ebsworth & Wan (1975) in this area concern mostly the lakes of the Fraser Island and the coastal sand dunes of the mainland. They have the same origin as the lakes near Cape Bedford. As in this latter area, the lakes lie in siliceous sands and do not receive the soluble products of rock weathering. They have therefore low salinities (30–40 mg l^{-1}) and an ionic composition very similar to that of sea water. The water is acid (pH = 4.1–6.0) and coloured by humic acids.

Phytoplankton is scarce in these dark waters and the most frequently found species belong to the desmids, especially *Staurastrum* sp., *Cosmarium* sp., *Micrasterias* sp., but there are other groups such as the Chlorococcales (*Botryococcus braunii*), Dinophyceae (*Peridinium* sp.) and Chrysophyceae (*Dinobryon* sp.). The zooplankton is abundant, with the calanoid *Calamoecia tasmanica* as the dominant species; the cyclopids *Eucyclops nichollsi* and *Macrocyclops albidus* are common in shallow waters.

The most common fish on Fraser Island is *Hypseleotris klunzingeri*, which seems limited to eastern Australia. *Melanotaenia* n. sp., *Rhadinocentrus ornatus* and *Craterocephalus fluviatilis* are also frequently found. Terrestrial insects, pollen, aquatic insects and crustaceans form the main components of the fish diet. Lake Wabby, with seven fish species, has an unusually rich fish fauna since the other lakes of this region have a maximum of three species.

Western Queensland

Lake Moondara is a man-made lake located 20 km downstream from the city of Mount Isa. It is located in a basaltic area with very variable precipitation (300–500 mm y^{-1}). The mean salinity is 177 mg l^{-1}. The lake waters are dominated by calcium, magnesium and hydrogen carbonates. The wastewater of a treatment plant is discharged into the lake inlet where concentrations of phosphorus of 10 mg l^{-1} have been measured. In the lake, the concentrations vary between 0.1 and 0.5 mg l^{-1} at the inlet mouth down to 0.01 mg l^{-1} near the dam. It is assumed that this rapid decrease is due to the active nutrient uptake of submerged macrophytes such as *Hydrilla verticillata* and the floating *Sabrinia modesta*.

Northern Territory

The Northern Territory is characterized by large temporary lakes such as Lakes Amadeus, Meale, Hopkins and MacDonald. These lakes have their bottom covered by a salt layer 1–2.5 cm thick and fill only in rainy years. Permanent water-holes occur, for example, in the McDonnel ranges: in this desertous area, they represent sanctuaries for certain aquatic animals which might have been widespread in previous more rainy periods. The water bodies of this area have been studied by Williams & Siebert (1963) and Williams (1967). It is worth noting that this arid area is provided with relatively low salinity water. In an analysis of samples taken from 1954 to 1962 by the authors, the salinity of most investigated water-holes ranged from 77 to 770 mg l^{-1}. As in Western Queensland, the ionic composition of these waters is dominated by calcium, magnesium and hydrogen carbonate. It seems that both the composition and the low salinity of the water-holes are explained by the fact that they are fed by an underground water body which prevents evaporation. In certain cases, however, underground water is saline. This explains the unusually high salinity of Glen Helen (3000 mg l^{-1}) which is partly fed by waters passing through salt-rich formations. In such waters sodium and chloride are the dominant ions.

Western Australia
South-western Western Australia

South-western Western Australia is, like Victoria, a well-watered area with high salinities. Nearly all the water bodies sampled by Williams & Buckney (1976*b*) are saline (5–298‰). In most of them, sodium and chloride are the dominant ions but magnesium represents more than 20% of the total cation equivalents in 56% of the cases. Calcium, sulphates and hydrogen carbonates being negligible, it is clear that this type of water results from the precipitation of calcium carbonate and gypsum and can be described as a Na^+–Mg^{2+}–Cl^- residual water.

Not only standing waters are saline but also tributaries. Morrissy (1979)

Australia

describes the presence of a summer halocline in pools of a non-tidal river south of Perth on the western coast. In summer, saline flow (with salinity exceeding 5000 mg l^{-1}) originating from inland fertilized agricultural catchments, reaches the river and the pools. When the saline flow ceases in January, the water of small perennial freshwater tributaries (salinity of 200 mg l^{-1}) accumulates on the residual waters of the pools: the resulting halocline and pycnocline remain stable during the dry season until April. In one of the pools investigated by Morrissy, the density of the water in March was 999 g l^{-1} at 1 m and 1002.7 g l^{-1} at 7.5 m. In January–February, thermal stratification (26°C in top layers and 17°C at 7.5 m) was an additional factor of stability. The summer stagnation in the pools is accompanied by a hypolimnic anoxia, which considerably limits the habitat of *Cherax tenuimanus*, a large detritivorous crayfish which was common in Western Australian streams. The man-made eutrophication which led to summer stagnation of these streams has caused a widespread decrease of the biomass of this species.

Northern Western Australia

The rivers, pools and lakes sampled by Williams & Buckney (1976*b*) in the Kimberley region all have very low salinities: 13–50 mg l^{-1} is the common range although various pools display salinities of 200–300 mg l^{-1}. These waters are generally of the magnesium-hydrogen carbonate type with sodium and chloride as the second most common ions.

Permanent freshwater bodies are also found in the western and southern coastal areas of Western Australia. They frequently consist of pools along the courses of rivers or spring-fed water-holes. Here also salinities are very low, generally below 200 mg l^{-1}.

The central part of southern Western Australia has been called the Salinaland. The pools associated with the rivers become saline and even dry out when the river flow diminishes. Concentrations of TDS reach 80‰ and may drop to 4‰ during rainy seasons (Williams, 1967).

Wood (1924) reports a gradual increase in salinity with time: a small dam near Formby had a salinity of 571 mg l^{-1} in August 1913 and 28 571 mg l^{-1} in November 1916. This rapid deterioration of the water resulted from vegetation clearance from the watershed between 1909 and 1913 for agricultural purposes.

South Australia
The south-eastern region

The south-eastern region of South Australia, bordering with Victoria, has recently been investigated by Bayly & Williams (1964, 1966*b*),

Timms (1974b) and Williams & Buckney (1976b). The volcanic area near the town of Mount Gambier includes several well-known lakes, such as Lake Leake, Lake Edward, Browne Lake, Valley Lake, Blue Lake and Leg of Mutton Lake. The morphometric features of several of these lakes are shown in Table 107. Their chemical features are summarized in Table 108.

These lakes, although belonging to the same area, can be divided into three groups: (1) Lake Leake and Leg of Mutton Lake, where sodium and chloride dominate, sulphate and hydrogen carbonates are low and magnesium relatively high (Fig. 34); (2) Browne Lake, Valley Lake and

Table 107 *Morphometric features of several lakes of south-eastern South Australia*

	Valley Lake	Lake Leake	Lake Edward	Blue Lake
Surface (ha)	28	65	29	–
Volume (10^6 m^3)	2.75	2.01	1.22	–
Max. depth (m)	16	4.5	7	79
Mean depth (m)	9.7	3.2	4.2	–

Fig. 34. Lake Nazdab, South Australia. This lake is typical of many throughout Australia in that it is temporary, saline, dominated by Na$^-$ and Cl$^-$ and has a fauna and flora comprising *Parartemia* (Anostraca), *Coxiella* (Gastropoda), *Calamoecia salina* (Calanoid) and the alga *Dunaliella salina* (by courtesy of Prof. Williams, University of Adelaide, Australia; photograph taken on 10 February 1982)

Table 108 Chemical features of several lakes of south-eastern South Australia (results in mg l^{-1} (% of equivalents))

Date	Lake	Ca^{2+}	Mg^{2+}	Na^+	K^+	Cl^-	SO_4^{2-}	HCO_3^- (m eq)	Source
Sep. '65	Lake Leake	30 (4)	92 (20)	650 (74)	27 (2)	1016 (75)	149 (8)	6.39 (17)	Bayly & Williams (1966b)
Sep. '65	Lake Edward	38 (6)	127 (35)	370 (53)	66 (6)	528 (49)	720 (50)	0.41 (1)	Bayly & Williams (1966b)
Mar. '64	Browne Lake	8 (1)	73 (16)	667 (75)	123 (8)	920 (67)	—	15.98 (41)	Bayly & Williams (1964)
Mar. '64	Valley Lake	19 (9)	43 (33)	138 (55)	15 (4)	250 (65)	—	5.06 (47)	Bayly & Williams (1964)
Mar. '64	Blue Lake	36 (28)	21 (27)	63 (43)	4 (2)	116 (51)	—	3.35 (53)	Bayly & Williams (1964)
Sep. '65	Leg of Mutton Lake	43 (7)	97 (25)	480 (65)	39 (3)	978 (86)	33 (2)	3.85 (12)	Bayly & Williams (1966b)

Blue Lake, where sodium is dominant followed by magnesium; chloride and hydrogen carbonate are nearly equal, and sulphate and calcium are nonexistent or low; and (3) Lake Edward, where sodium is dominant immediately followed by magnesium, with sulphate dominant over chloride.

It seems difficult to accept the interpretation of Bayly & Williams (1966b) concerning the 'catching effect' of conifer plantations around Lake Edward. According to this interpretation, 'the vegetation on the catchment area of Lake Edward, being a more efficient filtering mechanism, catches proportionately greater amounts of drifting salts and dust particles from terrestrial sources than does the vegetation of the catchment area of Lake Leake and that, as the chemical components of these particles are different from that of rain, this accounts for the disparity in the proportionate and subsequent distribution of ions to each lake'.

It seems reasonable that even lakes which are geographically very close, such as Lakes Edward and Leake, may have natural or man-made differences in their watersheds. Bayly & Williams mention, for example, the utilization of zinc sulphate in the conifer plantations around Lake Edward. This may certainly affect the water composition of a million m^3 water body. However, the low content of hydrogen carbonate and calcium suggests a previous precipitation of $CaCO_3$ and probably of a small amount of gypsum. These processes lead to the formation of a Na^+–Mg^{2+}–SO_4^{2-}–Cl^- residual water. Such precipitations of $CaCO_3$ are a common phenomenon easily induced by an increase in temperature for example (Eugster & Hardie, 1978).

Blue Lake has a scarce algal population whereas Browne Lake is rich in *Anabaenopsis circularis* and in zooplankton, especially *Mesocyclops leuckarti* and *Calamoecia gibbosa*. *Alona* sp., almost absent from Browne Lake, has been observed in the other lakes. Timms (1974b) has investigated the benthic fauna of Lakes Valley, Leake and Edward. Valley Lake is especially rich in oligochaetes and chironomids, Lake Leake benthos is dominated by chironomids, and Lake Edward has an abundant and diversified fauna of insects. The mean biomasses for Lakes Leake, Valley and Edward are, respectively, 23.3 g m^{-2}, 16.8 g m^{-2} and 8.7 g m^{-2}. The biomass is high and diversity very low when compared with lakes in North America and Europe. Timms relates the low diversity to the shallowness and small size of the lake which limit the types of habitats.

In the same region the Beachport–Robe series of lakes, located near the south-eastern coast of South Australia, is highly saline with salinities varying from 50 to 325‰. These waters are dominated by sodium and chloride and depleted in calcium and hydrogen carbonate, and may be

Australia

considered as Na^+–Mg^{2+}–Cl^-–SO_4^{2-} brines. The largest lakes of this area are Lake Eliza (38 km²) and Lake St. Clair (22 km²).

Northwest of the Beachport–Robe area, another series of water bodies includes freshwater lakes, a marine embayment and inland hypersaline lakes. The first category comprises Lake Alexandrina (salinity = 313 mg l^{-1}) and Lake Albert (salinity = 599 mg l^{-1}). Their waters are dominated by Na^+–Mg^{2+} and Cl^-–CO_3^-. The 'Coorong' represents the second type and, in its ionic composition, is very similar to sea water. The salinities are in the range 25–50‰. The inland saline lakes have a similar ionic composition but salinities reach 280‰.

The lakes of the York Peninsula, west of the town of Adelaide, are all saline, with salinities ranging from 26 to 359‰. They are of the Na^+–Mg^{2+}–Cl^-–SO_4^{2-} type with extremely low levels of calcium and hydrogen carbonate.

Central South Australia

Central South Australia contains many large impermanent lakes, the largest being Lakes Eyre, Fiome, Gregory, Blanche and Callabona. Some 5000 years ago, these lakes might have been permanent; presently they fill only sporadically in especially rainy years. Lake Eyre, located at 15 m below MSL, is for most of the time a depression covered with a salt crust. It was filled in 1950 and again from March 1974 to April 1975. Salinity was 20‰ in March 1974 and 7‰ in July 1974; then it climbed gradually and reached 40‰ in April 1975. The plankton has been investigated by Bayly (1976).

During the winter low-salinity period, the plankton was dominated by *Moina mongolica*, found for the first time in Australia, and *Apocyclops dengizicus*. *Boeckella triarticulata* was present. In September 1974, together with a very abundant population of *M. mongolica*, a few specimens of *Daphniopsis pusilla* were sampled. The scarcity of the latter species is possibly due to the selective predation of fish, *Craterocephalus evresii* and *Nematalosa erebi*. The cyclopids include *Microcyclops platypus* and *Apocyclops dengizicus*. The endemic ostracod *Diacypris* was also observed as well as the rotifers *Brachionus plicatilis* and *Hexarthra fennica*. The most common phytoplankton genera were *Nodularia, Anabaenopsis, Glaucocystopsis, Oscillatoria, Microcystis, Chodatella, Peridinium, Navicula* and *Gyrosigma*.

The chemistry of Australian waters

A salient feature of Australian waters is the high frequency of the sodium-chloride type contrasting with the general calcium-carbonate composition of most inland waters. This is true for certain very dilute waters such as the dune lakes of Queensland and hypersaline waters such as Lake Beeac, Victoria (Table 109). This characteristic is generally explained by the 'cyclic origin' of the sodium chloride, that is the continentward transport of marine salt by rain. Doubtless, large amounts of salt are yearly transported by this process, and Hutton & Leslie (1958) showed that, in Victoria, there is a rapid decrease in chloride with increasing distance from the ocean. However, if the rain composition was the dominant factor regulating the lake water's ionic patterns, then the lakes of Tasmania, being necessarily close to the sea, should exclusively belong to the sodium-chloride type. Whereas Cl^- is the dominant anion very closely followed by HCO_3^-; Mg^{2+} and Ca^{2+} are the leading cations. Na^+ representing only 18–40% of the cation-equivalent. Only in saline water is the sodium-chloride type again dominant. The features of Tasmanian waters (low salinity and Mg^{2+}–Na^+, Cl^-–HCO_3^- sequence) result from the weathering of dolerite granite and Palaeozoic rocks which are the major formations of the island; these well-leached silicates release only small amounts of salts but these are enriched in magnesium because of the ferromagnesian character of these rocks, and hydrogen carbonate is also produced during the weathering of silicate minerals. The influence of the bedrock on composition of water in Victoria and New South Wales has been studied by Reinson (1976). Therefore, rainfall has a determining effect on the ionic pattern of Australian waters in coastal lakes located on siliceous sand which contributes mainly silica. In other areas, the ionic pattern of freshwater is significantly affected by the composition of the local bedrock. Thus, the ionic sequence of the freshwater Lake Euramoo (Queensland): $Mg^{2+} > Na^+ > Ca^{2+} > K^+$ and $HCO_3^- > Cl^- > SO_4^{2-}$, reflects the composition of its volcanic substrate. The dilute magnesium hydrogen carbonate water of Lake Woods (Northern Territory) results from the leaching of magnesium-rich Precambrian rocks.

Superimposed on precipitation and rock-weathering effect, variations in temperature and evaporation and biotic effects are additional factors which deeply modify the composition of lake water. Increases in temperature, evaporation and active photosynthesis are three very distinct factors leading to precipitation of calcium carbonate. If the initial water was rich in hydrogen carbonate, then the residual water would be poor in calcium and would have moderate amounts of hydrogen carbonates and in certain cases, sulphate. This type of evolution would lead to an increase in salinity and ionic patterns similar to that of Lake Leake (South Australia) and Lake Werowrap (Victoria). If the water was not initially rich in bicarbonate, then the precipitation of calcium carbonate would deplete the solution in hydrogen carbonate and further evaporation would lead to precipitation of gypsum, which has been found in the salt deposits of many lakes in central South Australia. The residual water would be poor in calcium sulphate and hydrogen carbonate and would turn into a $Na^+-Mg^{2+}-Cl^-$ brine such as Lake Cowan (Western Australia) and Lake Corangamite (Victoria). The final stage is a nearly pure solution of sodium chloride such as Lake Beeac, Victoria (318‰) or a magnesium-chloride brine such as Lake MacDonnell, South Australia (347‰).

Thus, Australian waters show all the stages of evolution of natural water from purely rain water, to rain water transformed by dissolved substances originating from rock weathering, to residual water modified by precipitation of insoluble salts. In this latter category again all the stages are found from moderately saline and chemically diversified water, to chemically simplified brines and salt deposits.

We note that saline waters seem to be more numerous in the southern part of Australia although it receives more precipitation. The pattern of water supply to the lakes affects the salinity; lakes fed entirely by seasonal superficial runoff are generally saltier than lakes fed by underground water. This is a factor which probably explains the low salinities of many water bodies of the Northern Territory. However, it seems that there has been a constant salinization process since European colonization.

The problem of gradual salinization of the Murray River has been reviewed by Selby (1981). Three main factors are responsible for the observed increase in salinity. The first one is of a geological nature and concerns the slow seepage of sea water trapped in marine sediments deposited in Victoria during the Tertiary era. The other two factors are of man-made origin. Locks and weirs built on the river can raise water levels up to 3 m. The increased pressure on the saline groundwater favours the saline flux into the river downstream. Finally, irrigation has been a major factor of salinization from 1880 until the present. Although several measures have been taken to minimize salinization (pumping of saline groundwater away to evaporation basins, improved land drainage of irrigated areas), new irrigation projects are presently limited on the ground that they might increase dangerously the salt load of the river.

Biological features of inland waters of Australia

The phytoplankton has been described in various lakes of different salinities. It seems that blue-green algae are dominant in saline lakes and that in freshwater acid lakes, desmids are the leading group. Algal diversity

Table 109 *Chemical features of some Australian lakes*

	Salinity (mg l⁻¹)	Ca²⁺	Mg²⁺	Na⁺	K⁺	Cl⁻	SO₄²⁻	CO₃²⁻ + HCO₃⁻	pH	Sources
			←	———	% of meq l⁻¹	———	→			
QUEENSLAND										
L. Galilee	694	4	5	85	5	75	6	19	8.0	Bayly & Williams (1972)
L. Buchanan	87 600	5	4	90	1	100	0	0.1	8.6	Bayly & Williams (1972)
L. Euramoo (volc.)	31	25	35	28	12	50	0	65	6.8	Bayly & Williams (1972)
L. Mary (volc.)	106	6	20	76	4	63	4	33	7.5	Bayly & Williams (1972)
Dune lakes										
Fraser Is. lakes	39	4	16	78	2	82	16	2	–	Bayly (1964)
L. Moondarra	210	27	25	43	5	30	23	47	8.0	Farrell *et al.* (1979)
NORTHERN TERRITORY										
Finke River	196	43	15	32	10	38	15	46	–	Williams & Siebert (1963)
Lake Woods	46	4	75	10	11	7	19	74	–	Williams & Buckney (1976*b*)

WESTERN AUSTRALIA										
South-west										
L. Warden	24 000	2	16.6	79.8	1.6	95.9	3.4	0.4	9.5	Williams & Buckney (1976b)
L. Cowan	298 000	0.4	32.5	66.3	0.8	96.7	3.3	0	—	Williams & Buckney (1976b)
North										
Lennard Riverpool	210	20	54	23	3	6	5	83	—	Williams & Buckney (1976b)
SOUTH AUSTRALIA										
South-east										
L. Leake	2848	4	20	74	2	75	8	17	—	Bayly & Williams (1966b)
L. Edward	697	6	35	53	6	49	50	1	—	Bayly & Williams (1966b)
Blue Lake	340	28	27	43	2	51	0	50	—	Bayly & Williams (1966b)
VICTORIA										
Corangamite	27 900	<1	13	86	1	95	<1	4	—	Williams (1964a)
Werowrap	38 400	<1	1	95	4	65	<1	35	—	Williams (1964a)
Beeac	317 800	4	<1	95	1	96	3	<1	—	Williams (1964a)

decreases with salinity, and unialgal blooms reach very high rates of productivity as in Red Rock Tarn, for example. However, in Lake Werowrap, where the phytoplankton was studied by Walker (1973), the 1970 increase in salinity was accompanied by the disappearance of *Anabaena spiroides* and its replacement by a bloom of *Gymnodinium aeruginosum* during which peak density reached 158 000 cells ml^{-1}. In Lake Werowrap, the cell size of this dinoflagellate varied between 18 and 35 μm. The similarity of the algal blooms of Lake Werowrap and Lake Kinneret (Israel) and its implications are discussed in a later chapter (pp. 495-7). We note that the dinoflagellates and especially the genus *Gymnodinium* are present in a number of saline Victoria lakes.

In the zooplankton, the calanoid copepods and among them the genera *Calamoecia* and *Boeckella* occupy a special place. In the freshwater lakes, such as sand-dune lakes, *Calamoecia tasmanica* is frequently found. *C. clitellata* and *C. salina* are typical of salt lakes in the salinity range 16 to 131‰. These species are frequently bright red. *Boeckella triarticulata* is a very common species in Australian waters. It is known in a salinity range of 4.3‰ to 22.3‰ (Bayly & Williams, 1966a). *B. fluvialis* is widespread in New South Wales and Southern Queensland. *B. minuta* is found in New South Wales, southern Queensland and Victoria. A very common cyclopid copepod is *Halicyclops* n. sp. With the exception of *H. sinensis* occurring in the Yangtse-kiang River and other freshwaters of China, all the species of this genus are marine. *Mesocyclops leuckarti* is a common species in south-eastern Australia. The genus *Microcyclops* (*Metacyclops*) with various halophilic species contains the endemic species *M. arnaudi* which has been found in the salinity range 6-93‰. *Daphnia carinata*, *D. lumholtzi*, *Moina micrura* and *Bosmina meridionalis* are among the most frequently observed Cladocera. Special mention should be made of *Moina mongolica* found only in Lake Eyre.

The wide salinity range of Australian waters induced considerable physiological adaptation from many groups of animals. Certain animals of marine origin have colonized saline inland waters. This is the case with *Calamoecia salina* and *C. clitellata*, although no such colonization by the Centropagidae is known in saline athalassic water in the northern hemisphere. Similarly decapod crustaceans are originally marine organisms. One species, *Halicarcinus lacustris*, has invaded the Oceanian freshwater and is limited to south-east Australia, Lord Howe Island, Norfolk Island and the North Island of New Zealand. Crayfish (Parastacidae) are known in Oceania, Madagascar and South America. This distribution suggests that their invasion of inland water occurred before the

Cretaceous period of separation of Australia from the continent. The other families of decapod crustaceans (Atyidae, Palaemonidae and Potamonidae) have originated outside Australia. *Macrobrachium*, the Atyidae and the Potamonidae came from the north and are limited to eastern Australia. *Palaemonetes*, which came from Africa, is found only in the south-west corner of Australia (Bishop, 1967).

In contrast to these marine invaders, there exist animals which have an incredibly wide range of tolerance to salinity but have no marine representatives: this is the case of *Artemia salina*, which lives in a salinity range of 3–300‰, but there are no marine species in the family to which *Artemia* belongs (Anostraca). Bayly (1967) thinks that these animals have not been specially selected for salinity tolerance but for environmental impermanence; consequently they are able to surmount a very large range of ecological variations but are not fitted to the extreme stability of the marine environment.

The fish fauna of the Murray–Darling River system has been studied by Lake (1967), who gives a list of 19 native species and eight other species introduced during the second half of the last century. The native species are all endemic, and except for the lamprey *Mordacia mordax* they are Teleostean fishes belonging to the following orders: Clupeiformes, Galaxiiformes, Cypriniformes, Mugiliformes and Perciformes. The following native fishes are of commercial importance. The catfish *Tandanus tandanus* builds a nest in shallow water. Any lowering of the water level which exposes the nest before spawning compels the fish to abandon the nest and build a new one. The young fish feed on zooplankton and the adults on benthic fauna. The golden perch, *Plectropletes ambiguus*, spawns as a response to river floods. When floods are delayed, the ovaries remain mature for approximately two months. If floods do not occur, the ovaries regress. The young fish feed on zooplankton and the adults on large crustaceans, even shrimps. The Murray cod, *Macculochella macquariensis*, has the same feeding pattern as the golden perch: zooplankton in younger stages and shrimps, crayfish, fish and bivalve molluscs as adults. The silver perch, *Bidyanus bidyanus*, feeds on phyto- and zooplankton when young. Adult fish also have a mixed diet including filamentous algae, ostracods and shrimps. The western carp gudgeon, *Carassiops klunzingeri*, is very common and its small size makes it an easy prey for the Murray cod. *Macquaria australasica*, *Macculochella mitchelli*, *Madigania unicolor* and *Gadopsis marmoratus* are also commercial fishes.

Chessman & Williams (1974), studying the fish fauna of saline lakes in Victoria, found 14 native species and five exotic ones in lakes with a salinity range of 3–31‰. The most common species are *Galaxias maculatus, Retropinna semoni, Craterocephalus eyresii* and *Philypnodon grandiceps*, but only *G. maculatus* in Lake Corangamite and *C. eyresii* in Long Lake were recorded in lakes with salinities comparable with that of sea water. In water bodies with salinities higher than 31‰, true inland-water fish were not found. *Taeniomembras microstomus* recorded in waters with salinities of 3.3–82.4% is a coastal species. This lack of penetration of inland fish into saline non-marine waters strongly contrasts with the successful colonization of these water bodies by invertebrates. The impermanence of salt lakes and the lack of freshwater fish ancestors capable of elaborating mechanisms of osmotic regulation are the factors suggested by Chessman & Williams to explain their results. The massive fish killing which occurred in Lake Eyre in July 1975 (Ruello, 1976) and which mainly concerned *Nematalosa erebi* and *Craterocephalus eyresii* corroborated the previous observations. The dead fish were in good physical condition with large fat deposits in the abdomen which excludes disease or parasites as a possible cause of mortality. Ruello estimated that 40 million dead fish were left on the shores. He relates the fish killing to the increase in salinity of the water in 1975. Although in July 1975 the surface water had a salinity of 35‰ and many fish of both species were still alive, much higher salinity was measured on the bottom. Ruello envisages that the mixing of this deep saline water at the occasion of an internal seiche was the direct reason for the event.

The self-maintaining species of introduced freshwater fish are: *Salmo trutta, S. gairdneri, Cyprinus carpio, Carassius carassius, C. auratus, Tinca tinca, Rutilis rutilis, Perca fluviatilis* and *Gambusia affinis*. Although these species have been transported from Europe to the Southern Hemisphere, they continue in Australia to breed in the same season as in their original environment: for example, *Salmo trutta* which in Europe breeds in autumn–winter (September–February) still breeds in autumn–winter in Australia, that is from March to July. It follows that the process of gonad maturation is triggered by external mechanisms. When these are delayed for six months, so is the spawning cycle (Weatherley & Lake, 1967).

Interactions between the native and the introduced populations have been reviewed by Weatherley & Lake (1967). Tilzey (1976), working on the catchment area of Lake Eucumbene, New South Wales, has shown that *Galaxias coxii* and *G. oledus* were found in only four and one, respectively, of 27 streams sampled. In contrast, the introduced salmonids occurred in all but the stream where *G. oledus* was recorded. This seems to indicate that the survival of galaxiid populations, especially those restricted to small habitats, might be endangered by the well-established trout.

PART III

The dynamic processes in lakes of the Warm Belt

10
Mechanisms and patterns of circulation

The main factors affecting hydromechanical events in tropical lakes

Physically speaking, the water of the lakes of the Warm Belt is essentially characterized by the small seasonal variations in its temperature. In general, this temperature is high but the previous statement remains true for the colder water of tropical lakes at high altitude. There are numerous examples of warm tropical lakes with a seasonal variation of only 2–3°C (24 to 26°C for Lake Victoria) but the epilimnion of Lake Titicaca also shows a 3°C difference between summer and winter in a totally different range of temperature (12 to 15°C). This, of course, reflects the small changes in solar radiation and air temperature at low latitudes. It follows that, although the radiation rates are very high in the Warm Belt (140 to 220 kcal cm^{-2} y^{-1} in comparison with 80–100 kcal cm^{-2} y^{-1} at 40° latitude) the fact that they are constant with time turns them into a poor driving force of limnological events. Precipitations and winds, more seasonal, are likely to play a more decisive role. Fig. 35 shows the seasonal variations of the three meteorological parameters. The little seasonality of air temperature between 20°N and 20°S explains the limited seasonal variations in lake water temperature described by Talling (1966). It also accounts for the low value of the annual heat budget of tropical lakes (Table 110) in comparison with temperate lakes. Among the different terms of the budget, the evaporative heat loss (LE) is high. In Lake Titicaca, LE is of the same order of magnitude as the net radiation flux and relatively constant during the year (~ 300 cal cm^{-2} d^{-1}) (Richerson et al., 1977).

As pointed out by Lewis (1973b) the minimum heat content of warm lakes varies little from year to year whereas its peak value is affected by brief and non-periodic weather modifications, which explains the large

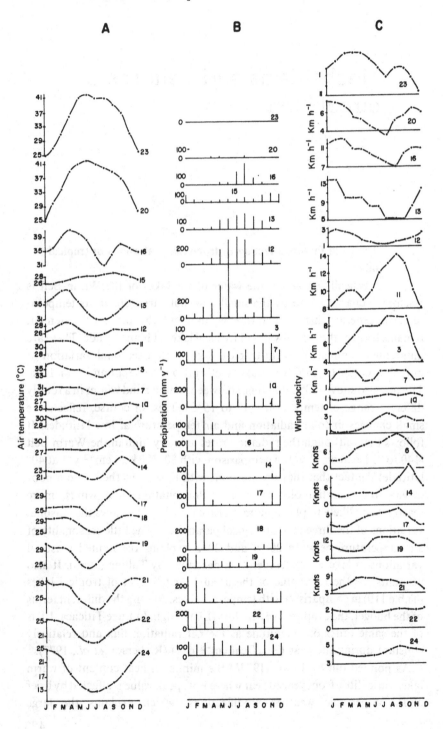

differences often observed in the heat budgets of consecutive years. In the temperate zone, the maximal heat storage occurs after a progressive heat accumulation due to an increase in insolation. In the tropics the peak value of heat storage may result from a unique meteorological event.

It is interesting to note that the maximal value of heat rate loss does not necessarily occur during the yearly temperature minimum. In Lake Lanao, Lewis (1973b) reports rates of 450 cal cm^{-2} d^{-1} during the winter minimum and 650 cal cm^{-2} d^{-1} during brief periods in summer. In the Bishoftu lakes, Wood et al. (1976) note that maximal rates of heat losses were observed when high solar inputs coinciding with low relative humidity, the evaporative heat losses were very high.

Conversely, precipitations have a clear seasonal distribution even at the Equator. In consequence, the rising of water levels of lakes and rivers by several metres is a common phenomenon in the tropics. Besides a considerable impact on lake water dilution and nutrient supply, the seasonality of precipitation deeply affects the biological events such as algal blooms, zooplankton reproduction and fish spawning. Winds also display

Table 110 *Heat budgets of various lakes of the Warm Belt and the temperate zone*

Lakes	Latitude	Altitude (m)	Heat budget (cal cm^{-2})	Source
Klindungan	7°40′ S	10	3400	Ruttner (1952)
Lanoa	8° N	720	4500–7250	Lewis (1973b)
Guija	14°13′ N	426	5410	Hutchinson (1957)
Amatitlan	14°25′ N	1189	8510	Hutchinson (1957)
Atitlan	14°40′ N	1555	22 100	Hutchinson (1957)
Titicaca	16°00′ S	4000	18 900	Richerson et al. (1975)
Kinneret	32°42′ N	−209	40 000–47 240	Stanhill & Neumann (1978)
Geneva	42°34′ N	263	32 300	Hutchinson (1957)
Michigan	44°00′ N	176	52 400	Hutchinson (1957)
Baikal	53°00′ N	453	65 500	Hutchinson (1957)

Fig. 35. Seasonal variation in air temperature (A), precipitation (B) and wind velocity (C) at different latitudes of the Warm Belt. Data from Arakawa (1969), Gentilli (1971), Griffiths (1972), Schwerdtfegar (1976) and Takahashi & Arakawa (1981). Meteorological stations: la Belem 1°28′ S, 1 Guayacil 2°12′ S, 3 Todwar 3°07′ N, 6 Tabora 5°02′ S, 7 Eala 0°03′ N, 11 Abidjan 5°15′ N, 12 San Fernando 7°53′ N, 13 Malakal 9°33′ N, 14 Mbeya Tanza 8°56′ S, 15 Curacao 12°12′ N, 16 Abfcher 13°51′ N, 17 Kazama Zam 10°13′ S, 18 Darwin 12°28′ S, 19 Santa Cruz 17°47′ S, 20 Faya Largeau 18°00′ N, 21 Normanton 17°39′ S, 22 Tres Lagoas 20°47′ S, 23 Wadi Halfa 21°50′ N, 24 Alegrete 29°46′ S

a seasonal pattern. Very generally, the wind speed is maximum during the dry season which maximizes evaporation. The resulting heat losses and the general turbulence caused by the wind generate or enhance water mixing at this period. Protection from the wind may lead to a nearly permanent stratification: for example, the shallow Lake Tupe, Amazonia, in spite of a 1.5°C difference between the epi- and hypolimnions during the dry, low-water period does not have a frequent turnover only because it is protected from the wind. In contrast, the much deeper Lake Albert is the site of an intense circulation during the stormy period and even later stratification is always weak because of its exposure to frequent and especially strong windstorms. Finally, the winds, being the main cause of water motions, determine the large-scale circulation which has been studied very little in warm lakes.

The idea that winds and rains have a greater impact on lake limnology than net radiation and air temperature has been already expressed by various authors such as Beadle (1974) but recent studies have brought new arguments to this claim.

Stratification and mixing
The vertical thermal gradient and thermal stability

The vertical profiles of warm lakes generally display a very small difference in temperature when compared to temperate lakes. It is minimal near the Equator (1°C in Lake Victoria and Lake Albert) and increases with latitude (3.5°C at Lake Nasser, Egypt, 22°N). Since, in the range of temperatures we are concerned with, the variation of density with temperature is high, even a small gradient may confer a reasonable stability. However, as stressed by Lewis (1973b), warm lakes have generally a much lower stability than temperate lakes: Lake Albert has an epilimnion to hypolimnion density difference of 0.3431 g l^{-1} for a thermal difference of 27–28°C, in comparison with 1.869 g l^{-1} for Lake Geneva with its 4–20°C thermal difference between bottom and surface.

Tropical lakes at high altitude offer an interesting case: having both a low temperature and a small thermal gradient they have a much lower difference in density than the lowland lakes of same latitude and than the temperate lakes. They are the lakes of lowest stability. Lake Titicaca (16° S, 3800 m altitude) has a density difference of 0.3720 g l^{-1} in contrast to 2.5696 g l^{-1} for Lake Kariba (17° S, 485 m altitude).

Conversely, the combined effect of high absolute temperature and large thermal gradients observed between 20 and 30° latitude gives the lakes of this belt a maximal stability: Lake Nasser, Egypt, 22° N and Lake

Mechanisms and patterns of circulation

Kinneret, Israel, 32° N have respective density differences of 4.67 and 3.75 g l^{-1}. Towards the Equator from these latitudes, the stability decreases because of the reduction of the thermal gradients; towards the poles it diminishes because of the lower temperature of the water. This interesting property should be considered when artificial reservoirs are built since prolonged water stagnation is damaging to water quality and fisheries.

To illustrate the preceding considerations, the differences in density between the epi- and hypolimnions of a certain number of lakes have been plotted against the latitudes of these lakes (Fig. 36). The differences in density considered here are those generated by the differences in temperature and do not take into account the influence of the salinity of these lakes. This is justified by the fact that, in freshwater, the contribution of mineral content to the water density is small in comparison with temperature. Considering that the lakes plotted on the graph spread over a wide range of

Fig. 36. Difference in density between the epi- and hypolimnions of lakes at various latitudes. The values used in the graph were calculated according to temperature differences and were not corrected for content of dissolved solids. Note the scarcity of lakes in the 15–25°C belt. Lake Titicaca has a low difference in density for its latitude because of its high altitude and consequent weak stratification.
A = Alber (Af); Bt = Batur (SE Asia); B = Bunyoni (Af); E = Edward (Af); Gi = Guri (SA); Kj = Kainji (Af); Ka = Kariba (Af); Ki = Kinneret (As); Lg = Lagartijo (SA); L = Lanao (SE Asia); Mu = Mucubaji (SA); Nl = Nainital (As); N = Nasser (Af); Na = Nilnag (As); Ng = Nkugute (Af); O = Ootacamund (As); P = Pawlo (Af); Pt = Perth (Aust); Ph = Phewa Tal (As); S = Sathiar (As); Tb = Tamblingan (SE Asia); Td = Tan Daha (As); T = Tanganyika (Af); Tc = Titicaca (SA); Tso = Tsola Tso (As); V = Victoria (Af); Vo = Volta (Af); W = Wisdom (SE Asia)

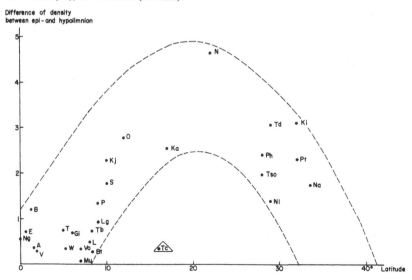

altitudes, depths and sizes, and have very different exposure to winds, it is remarkable to observe the general trend of increase in stability with latitude, the peak of stability near 20–30° latitude and its further decrease at higher altitudes. The lakes of high altitudes are systematically below the graph of lowland lakes for the reason previously mentioned.

The factors affecting the vertical gradient

The 'warming speed' of the surface layer in springtime has a considerable bearing on the development of the vertical gradient and thickness of epilimnion. A strong and rapid warming of the surface layer causes the formation of a shallow thermocline and leads to an apparent paradox: the warmer the surface layer, the more the subsurface layer is prevented from warming. Since the speed of spring warming increases with increasing annual variations of the surface temperature, the shallowest thermocline and stronger vertical gradients are expected in areas of high annual temperature range i.e. in areas of high latitudes (Dietrich, Kalle, Krauss & Siedler, 1980). In areas of low latitudes, the annual temperature range being much smaller, the speed of seasonal warming is slow, the resulting thermocline is weaker and more easily deepened by wind. This has been confirmed by Munk & Anderson (1948) in the ocean. A second factor affecting the thickness of epilimnion in the Warm Belt is the fact that, according to the drift current theory of Eckman (1905), the thickness of the drift current layer is reciprocal to the square root of the line of the latitude in such a way that $D = 7.6\ W/\sqrt{\sin \varphi}$, where D is the thickness of the drift current layer in metres, W the wind velocity and φ the latitude. For example, a wind of $7\ m\ s^{-1}$ would generate a drift current layer of 104 m at 15° latitude and only 63 m at 45° latitude. Since this theory has been developed for the ocean, it can be taken into account only in large and deep lakes or at least in lakes whose depth is greater than the drift current layer.

This explains that, in general, warm lakes have a much thicker epilimnion than temperate lakes. In Lake Lanao, Philippines, 8° N and maximal depth 112 m, the thermocline is located at 40 m below the surface. In Lake Geneva, 45° N and 310 m maximal depth, the thermocline is not deeper than 15 m in spite of the favourable exposure of the lake to strong winds.

When a thick epilimnion has developed in a warm lake, two different types of behaviour may occur: (1) the amount of mechanical energy received daily by the lake surface is large enough to maintain the epilimnic volume thoroughly mixed. This is the case of Lake Kinneret, Israel, 32° N

Mechanisms and patterns of circulation

and 42 m deep where the daily western wind (15 km h) generates daily storms on the lake. (2) The daily wind is not sufficient to mix the whole hypolimnion daily. The deepest part of the epilimnion is mixed only on the occasions of major storms. The resulting uneven circulation leads to the formation of several thermoclines, the depths of which correspond to the intensity of the most frequent winds of the considered area. Thus, Lewis on Lake Lanao described the 'breeze thermocline' destroyed by nearly any slight wind, the 'squall thermocline' developing at 20–30 m and eroded by squalls of medium intensity and duration, and the 'storm thermocline' eroded only during violent and long storms. If two consecutive storms are separated by a long period of time a certain chemical stratification takes place but will disappear with the next storm in the absence of complete mixing. This mixing of chemically non-homogeneous layers above the deep thermocline was called 'stelomixis' by Lewis (1973*b*). The formation of multithermoclines depends essentially on the regional intensity and frequency of winds and on the lake exposure. It has a very general occurrence in the Warm Belt, especially in areas of irregular winds.

Other factors tend to perpetuate stratification or at least to delay the mixing process. In the tropics, the diurnal air temperature range is greater than the seasonal range. Therefore, it is likely that the shallow water of tropical lakes cools at night and sinks into deeper layers. Talling (1966) explained the stratification in Lake Albert through this mechanism. Frequently, warm lakes are fed by rivers originating at high altitudes and carrying cold water. A density current is created and stratification strengthened from below as in Lagartijo Reservoir, Venezuela, 9° N, El Prado Reservoir in the Caribbean area and Madden Lake, Panama. In the Mwadingusha Reservoir, in the Zaire basin, the weak stratification which develops during the wet season is generated by the density current caused by the cold water of the Lufina River.

In Lake Subang, West Malaysia, 3° N, 8 m depth, the night mixing affects only the upper 3 m; the lake is stratified and the hypolimnion anoxic. With the rains, the increased discharge of the sublacustrine springs fills the hypolimnion and pushes the anoxic water upwards. This original process of hypolimnion renewal occurs twice a year with the rain peaks.

Salinity gradient is frequently an impossible barrier to mixing. Lakes Tanganyika and Kivu are classical examples. On a smaller scale, a similar stratification is described in a pool of a non-tidal river near Perth, Western Australia. In summer, the pool receives saline drainage from surrounding fields. The drainage flow stops in January; then the freshwater coming from the streams overlies the saline layer and warms up to 26°C whereas the saline layer remains at 17°C. The resulting pycnocline remains stable until April when it is flushed by the floods.

A few types of lake and their circulation patterns

The more data are collected on warm lakes, the clearer it becomes that their circulation pattern is highly affected (1) by morphometric parameters (size and natural depth configuration; (2) by their location and exposure to winds; and (3) by aperiodic meteorological phenomena.

Thus, the traditional classification of lakes based on the seasonal alternation of temperate regions is inadequate and there is not yet enough information to propose a better one. Therefore, in the following pages we review various problems associated with a certain number of lake types. This review does not pretend to exhaust either the problems or the lake types.

Crater lakes

Crater lakes are generally characterized by a small surface area, a relatively great depth and a sheltered position. These features indicate that such lakes cannot be much affected by the wind and are generally stratified. Green et al. (1976) visiting the Ranu Lamongan, Indonesia, 8°S, 29 m deep, found the lake completely mixed, probably as a result of very bad weather; however, stratification was re-established within a few days. Green et al. report that these short overturns are accompanied by massive fish kills due to the sudden decrease in oxygen, caused by the oxidation of the hypolimnic sulphides dispersed in the whole lake volume which indicates that such overturns are not frequent. In Danau Buyan, Indonesia, 8°14′S, 30 m deep, 1°C difference between surface and bottom develops enough stability to reduce significantly the oxygen concentration of deep layers.

Lake Nkugute, East Africa, 0°, 58 m deep is assumed by Beadle (1966) to be permanently stratified in spite of a maximal thermal gradient of 2.7°C, which can be expected from its sheltered position. However, it is difficult to understand its invariable hypolimnic temperature (21.9°C) while the epilimnion varies between 22 and 24°C, without assuming some cooling mechanism such as night cooling or evaporative heat losses. It is probable that Lake Nkugute undergoes occasional deep mixing, since fish kills have also been observed.

High-altitude lakes

With the exception of Lake Titicaca, tropical lakes of high altitude were formed by glaciation during the Pleistocene or, to a much lesser

Mechanisms and patterns of circulation 431

extent, by vulcanism. Since in the Pleistocene the snow line was around 3000 m, most of these lakes lie above this altitude (Löffler, 1968). Tropical high-altitude lakes are especially numerous in the Andes, where they have been uplifted during post-Pleistocene epirogenic movements, and in Central America but they are also found in East Africa, Indonesia and New Guinea.

Löffler (1972) distinguished the 'Paramos' type located in the wet tropical high mountain zone where the diurnal variation of temperature is of an order of a few degrees Centigrade and the 'Puna' type located in the dry tropical high mountain area characterized by a very large range of diurnal fluctuation in temperature.

A clear relationship exists between the temperatures of maximum depth of these lakes and their altitudes (Löffler, 1968): 10–13°C for lakes between 3000 and 4000 m and 5–10°C for lakes between 4000 and 5000 m.

Lakes of the 'Paramos' type, with their constant and low temperature, have very little stability and consequently frequent overturns, especially the shallow lakes. In spite of their equally low and nearly constant mean temperature throughout the year, the 'Puna' lakes have a more distinct pattern of stratification. This is due to the night cooling of superficial water and large evaporative losses, especially during the dry season. Here again, the degree of mixing of the lake depends on its depth and exposure to wind. In July 1973, Lake Titicaca (281 m depth) mixed down to 100 m and, without any deeper mixing, stratification was re-established in September 1973. In spite of its low difference in density between the epi- and hypolimnions (0.372 g l^{-1}) Lake Titicaca does not overturn completely, probably because of the minor role of the winds, average wind speed ranging from 0.89 to 1.3 m s^{-1} (Richerson et al., 1977). In fact, it is expected that convective vertical mixing by cooling should have a depth limit. The sinking of cool heavy water and rising of warmer and lighter water in neighbouring convection cells lead to an unavoidable lateral heat exchange between cells and thus to gradual mixing. Thus the penetration of cool water ceases at a given intermediate depth. Unfortunately, the process of convective cooling has been little studied.

Shallow large lakes

Lake Chad, Lake George and the Great Lake of Kampuchea offer good examples of this type of lake. They are characterized by a diurnal thermal pattern and as far as Lake Chad and the Great Lake of Kampuchea are concerned by very large fluctuations in water levels.

Homothermal in the morning, these lakes undergo a mid-afternoon stratification which is destroyed during the night.

Lake George, located at the Equator, shows very constant hydrological features which probably explain its permanent bloom of algae. Lake Chad (12° N) has a very fluctuating water level which depends upon the variable discharge of the Chari River and on a clearer thermal seasonality, but the daily water mixing maintains a dense algal population here also. The Great Lake of Kampuchea (12° N) shows extreme seasonal variations of water level (8 m) and surface area (3000–10 000 km^2). The existence of a very large and shallow flood plain is in this case one of the main factors of the lake productivity.

Large deep lakes

Lakes Malawi, Tanganyika and Kivu represent large deep lakes. They all have a 'seasonal' thermocline and a deep 'permanent' thermocline below which the water is always anoxic. In the past, the general opinion was that the deep water was excluded from circulation. However, R.S.A. Beauchamp (1939) noted that the uniform vertical chemical composition of Lake Tanganyika was evidence of deep mixing. More recently, the measurement of the rate of methane oxidation at the boundary between anoxic and oxygenated water was carried out in Lake Tanganyika and Lake Kivu (Hecky et al., 1978; Rudd, 1980). The rates were very close in both lakes in spite of a much weaker methane gradient below the interface in Lake Tanganyika than in Lake Kivu. This observation implies a higher rate of deep-water transport up to the interface in Tanganyika than in Kivu. Information obtained from isotope investigation (Gonfiantini et al., 1979) indicates that in Lake Malawi about 20% of the hypolimnic water is yearly recirculated, whereas the rate of circulation is much slower in Lake Tanganyika.

Thus, in the present state of knowledge, the recirculation of the deep water of these lakes has become a fact but the mechanisms remain to be clarified. In Lakes Tanganyika and Malawi, accumulation of warm water in the northern areas of the lakes accompanied by upwelling at the southern ends has been observed during the cool windy season as well as long period ($\sim$1 month) internal waves during the stratification. The current field of these motions is characterized by opposite directions across the thermocline or internal fronts producing increased vertical shear in the deep thermocline which should then mix with adjacent layers and deepen. Eccles (1974) accounts for the fact that the permanent thermocline does not deepen by the existence of cool density currents generated by cooling in the

Mechanisms and patterns of circulation 433

southern shallow areas of Lakes Tanganyika and Malawi and sinking into the hypolimnion.

A special mention should be made of the microstructure observed in Lake Kivu and which possibly exists also in Lakes Tanganyika and Malawi. Newman (1976) has described the existence of some 150 layers in the hypolimnion of Lake Kivu which he relates to the process of 'double diffusion convection'. A 'thermohaline' or 'diffusive' system consists of a stratified water mass with 'a linear stable salinity gradient initially at constant temperature and heated strongly from below' (Turner, 1973). This applies well to Lake Kivu which has a deep source of heat and salt. Heat, having a greater rate of molecular diffusivity than salt, is the driving force of the system and creates instability in the deep salty layer. The convective motion, however, does not spread through the whole water column but a succession of convecting layers develops progressively from the bottom upwards with a density discontinuity between layers in the following way. When heating has made the bottom layer unstable, the top of this layer rises and mixes with a certain amount of the fluid located just above it. A new layer is formed with salinity and temperature different from those of the initial bottom layer. Then, as heat is transferred upwards quicker than salt, a new instability area develops at the upper boundary of the newly formed layer, with the resulting formation of a new layer on top of the first. According to laboratory experiments, this mixing process is at first regularly organized in finger-like structures corresponding to upward and downward water motions of the lower and upper layers, respectively.

This hypolimnion structure is by no means characteristic of tropical lakes and has also been observed in Lake Constance (Kroebel, 1982). It simply emphasizes the fact that the more we learn about the physical structure of lakes, the more we realize that the two-layer models are a necessary but rough approximation to reality.

A new question arises concerning the effect of microlayering of the hypolimnion on stability. Has Lake Kivu a slower hypolimnic recirculation because of the existence of the microlayers? We cannot answer this question by a simple comparison with the other two lakes since we do not know at present whether they also have some sort of microstructure.

It is clear that the layered structure implies horizontal motions. For example a river inflow penetrating into a layered hypolimnion generates an advective transfer compensated by another horizontal return current. The persistence of these weakly stratified layers in the hypolimnion during long periods of time proves that the forces driving the deep-water motions are rather moderate. These advective motions should be investigated thoroughly since, in the absence of violent disturbances able to recirculate large portions of the hypolimnion, they account for the slow, long-term recirculation of the hypolimnion.

As far as we know at present, it is clear that the wind is the major driving force of circulation of the deep large lakes. It directly mixes the epilimnion but this mixing effect, at least in Lakes Tanganyika and Malawi, is not powerful enough to create homothermal conditions in the 'seasonal lake' and directly affect the deep thermocline. Deep circulation, including the 'permanent hypolimnion', is developed indirectly by the wind-induced motions in the epilimnion, since the deep thermocline is deflected from its position at rest and generates thereby a field of motion affecting the whole hypolimnion. The considerable effect of the wind in the elongated basins of Lakes Malawi and Tanganyika generates movements concerning very large volumes of water. As already mentioned, this effect is not limited to the 'seasonal lake' but also concerns the 'permanent hypolimnion' with resulting turbulence and water exchange in the deep thermocline. This slow but continuous wind-generated recirculation of deep water is characteristic of tropical deep lakes and is opposed to the thermal and seasonal mixing in temperate deep lakes. The cycle of the latter is annual whereas that of the former extends over several years.

Solar lakes and solar ponds

Several examples of anomalous stratification have been described in the literature. A certain saline lake in Hungary shows an increase in temperature with depth (Kalecsinski, 1901); a solar pond with deep warm water has been described on the shores of the Red Sea (Eckstein, 1970); whereas Lake Mahega, Uganda, has an intermediate warm layer (Melack & Kilham, 1972).

These natural occurrences and the recent urgent need for cheaper sources of energy have led R. Block, Israel, to imagine a solar collector of a new type: the solar pond. A solar pond is a shallow black-bottomed water body where the sun-heated water at the bottom is prevented from mixing by a strong saline gradient. The relatively cool (near the surrounding air temperature) upper water plays the role of an insulating layer allowing the deep saline, non-convecting layer to reach a temperature of 90–100°C. The hot water may then be used directly for heating or for conversion to power through heat exchangers and heat engines.

The operation of such ponds raises several problems reviewed by Tabor (1963) and Tabor & Matz (1965). The main problem consists of keeping the halocline by preventing mixing and diffusion. Mixing can be prevented by increasing the pond depth, but then less energy reaches the bottom. At present, the mixing is reduced by using nets of floating wind breaks.

The salt solution generally utilized is $MgCl_2$ and for a pond with a density of $1 g l^{-1}$ at the top and $300 g l^{-1}$ at the bottom (depth 1.3 m) and with a difference in temperature of 50°C, the diffusion rate is $50 g m^{-2} d^{-1}$. Periodically, a concentrated solution is injected into the deep layer whereas the top layer is washed with freshwater.

The extraction of the heat is another serious technical problem, partly solved by 'decanting' a layer of warm water at the bottom, passing it over external heat exchangers and reinjecting it into the pond.

The heating rate of the bottom water being a direct function of the light transmission through the water, it is essential to keep the pond clean from dust and biological growth.

Most of these technical difficulties have been overcome and on 16 December 1979 a 150 kW solar pond–thermal power station was inaugurated at Ein Boqeq on the shores of the Dead Sea, Israel. The heat of the pond was turned into electricity in a low-temperature, organic-vapour turbine (Tabor, 1980). A 5 MW plant is now being planned for completion in 1982.

The advantages of the solar-pond solution over the 'intensive collectors' are obvious. First, it allows storage of solar energy and therefore is not sensitive to diurnal variations in solar radiation. The pond keeps supplying power even after a few consecutive cloudy days. Deeper ponds are being studied with the aim of providing seasonal storage. A second advantage consists of the relatively low cost of building; large areas (several hectares for power production) and a cheap nearby source of salt are however required. A third point concerns the cost of solar-pond electricity: at present and expected fuel prices, solar-pond power competes with electricity generated by oil-powered plants and is much cheaper than diesel-generator power.

We note that developing countries are essentially located in the Warm Belt and consequently have ample 'reserves' of solar radiation; solar ponds may then represent a very attractive solution for meeting the future energy needs of these countries. Since the cost of solar-pond power stations is minimum for units of 20–40 MW, developing countries will be able to increase their power capacity by 'small steps' (Tabor, 1980) thus avoiding the considerable investments required by the implementation of nuclear- or oil-powered plants, the costs of which become optimal only for several hundred MW. New developments have led to the idea of floating ponds in lakes or in the sea. A new scheme under study in Israel is aimed at transforming part of the Dead Sea into a solar lake for power production and desalinization (Assaf, 1976).

11
Water chemistry

The mineral content of the water of tropical lakes covers a remarkably wide range from 2.4 mg l^{-1} in the lakes of the Kosciusko Plateau, New South Wales, Australia, to 415 000 mg l^{-1} in the brines of Lake Mahega, Uganda.

In Table 111, we have considered four groups of water of increasing salinities illustrated by examples taken from the different areas of the Warm Belt. Although dilute and saline waters can be found in each of these regional units, highly saline waters are mostly found in East Africa, Ethiopia, the Middle East and Australia, whereas dilute waters are common in South America, Western Africa and South-East Asia. The yearly amount and seasonal distribution of the precipitation are obviously decisive factors in the resulting salinity of lakes. However, we note that, in Australia, well-watered areas such as the south-west corner of Western Australia have saline waters. Similarly, although many regions in East Africa are drier than their latitude would suggest, the highlands bordering the Rift receive abundant rainfall which is drained into the Rift lakes. Both areas have in common an intense evaporation which, in topographically low areas, leads to the formation of endorheic basins or, as in the case of the unique large river system of Australia (Murray–Darling system) to incredibly small runoff (see p. 397). Storm and runoff water is often infiltrated whilst part of the infiltrated water is pumped back and evaporated by capillary processes during the dry season. Interstitial solutions as well as lake water and even river water become oversaturated and allow precipitation of calcium carbonate (streams draining Mount Meru in Tanzania), trona (Lake Magadi), northupite and thenardite (Lake Mahega). Such basins undergo endless cycles of dissolution, infiltration of brines and recrystallization (see Lake Magadi, pp. 193–4).

Table 111 Chemical features of some lakes of the Warm Belt. The lakes are classified according to increasing salinity

	Conductivity (μmhos^{-1})	Salinity (mg l^{-1})	pH	Alkalinity	Chemical dominance
I Very low salinity	<100				
South America					
Tota	75				
Mucubaji	12.6				
Guri	9		6.5	3.2 mg l^{-1}	Ca–HCO$_3$
Brokopondo	6.8		4.5–5.1		Ca
Castanho	20–50		6–7	3.7–31.7 mg l^{-1}	Ca–HCO$_3$
Waters of Middle Parana	60–200				Na–HCO$_3$
Central America					
Gatun	30		7.1	90 mg l^{-1}	
High mountain lakes					
Africa					
Chad	100		4.7		
Nabugado	25		7.8		K, Na
Victoria	97				Na
Nkugute	86				
Tumba	24–32		4.5–5.0		
Kariba	73	58	7.5–8.5		Ca–HCO$_3$

Table 111 (cont.)

	Conductivity (μmhos cm^{-1})	Salinity (mg l^{-1})	pH	Alkalinity	Chemical dominance
South-East Asia					
Mt Everest Lakes	7–43				Na–Cl
Southern Thailand lakes	22–52				Na–Cl
Subang Lake	7–54			1.6–17.2 mg l^{-1}	Na–SO$_4$
Southern Malaya lakes		4.6	3.6–5.4	0–0.45 meq l^{-1}	
Lakes of Purari River		54	7.8		Ca–HCO$_3$
L. Lanao	105			51 mg l^{-1}	
D. Bratan	39				
Australia					
Kosciusko pl. lakes	4.6–5.3	2.4–3.0			Na–Cl
Queensland sand dune lakes		30–40			Na–Cl
Fraser Island lakes		39			Na–Cl
II Low salinity	**<500**				
Central America					
Peten Itza	485		8.3	76 mg l^{-1}	Ca=Mg–SO$_4$
Nicaragua		151	7.0	82 mg l^{-1}	Na=Ca–HCO$_3$
Izabal	210–465		7.9	70 mg l^{-1}	Ca–HCO$_3$=SO$_4$
Atitlan	420		8.3	150 mg l^{-1}	Na–HCO$_3$=SO$_4$
Patzcuaro		428			Na–HCO$_3$

Africa					
Kyoga	320		7.7		Ca–HCO$_3$
Volta		80–100	6.7–7.0	65–70 mg l^{-1}	Ca–HCO$_3$
George	200		8.0	2 meq l^{-1}	Ca=Mg.HCO$_3$
Bunyoni	258				
Fort Portal group	276–535				Na–HCO$_3$
Naivasha	311–353				Ca–HCO$_3$
Tana	200–240				Na-Mg.HCO$_3$
Tanganyika		420	8.1	70.6 mg l^{-1}	Ca–Na–HCO$_3$
Malawi	210				
South-East Asia					
Nilnag	230–370			147–220 mg l^{-1}	Ca–HCO$_3$
Borapet	135–162	315	6.9–7.5		Na–Mg–HCO$_3$
Dakataua		100–177	7.8	90–134 mg l^{-1}	Ca–HCO$_3$
Great Lake of Kampuchea			6.8–7.0		
Australia					
Mary Q		106	7.5		Na–Cl
Moondarra Q		210	8.0		Na–HCO$_3$
Burley Griffin NSW		140–190			
III Medium salinity	<3000				
South America					
Valencia		980	8.2	2.1 meq l^{-1}	Na–Cl–SO$_4$
Titicaca		788			
Central America					
Managua	830	747	8.5	500 mg l^{-1}	Na–HCO$_3$
Amatitlan			7.7	152 mg l^{-1}	Na–Cl
Africa					
Edward	735	720	9.0		Na–K–HCO$_3$
Albert	785		9.1	7.3 meq l^{-1}	Na–K–CO$_3$
Kitangiri		404–432	8.0–8.9	6.65 meq l^{-1}	
Rudolf	2800		9.5–10	21.7 meq l^{-1}	Na–CO$_3$
Awasa	1050				Na–CO$_3$

Table 111 (cont.)

	Conductivity (μmhos cm^{-1})	Salinity (mg l^{-1})	pH	Alkalinity	Chemical dominance
South-East Asia					
Wisdom		1230	7.6		NaCl-SO$_4$
Australia					
Black Lake NSW		1719		672 mg l^{-1}	Na-Cl-HCO$_3$
Cootralantra NSW		1862		809 mg l^{-1}	Na-HCO$_3$-Cl
IV High salinity	> 3000				
South America					
Poopo		41 000		4.7 meq l^{-1}	Na-Cl
Africa					
Mahega	16 000–160 000	193 000–415 000			
Katwe	94 000				Na-SO$_4$-Cl
Manyara	160 000			806 meq l^{-1}	Na-HCO$_3$
Magadi	340 000		10.5	1180 meq l^{-1}	Na-Cl
Natron	162 000			2600 meq l^{-1}	
Nakuru	29 500			200 meq l^{-1}	Na-HCO$_3$
Shala	30 000			210 meq l^{-1}	Na-HCO$_3$
Abiata					Na-HCO$_3$
Australia					
Beeac (Vic.)		55 500–317 800			
Corangamite (Vic.)		27 900			
Jillamatong (NSW)		21 244			
Lake west of Adelaide (SA)		359 000			
Cowan (WA)		298 000			
Buchanan (Qld)		87 600			

Water chemistry 441

Relief also plays an important role in the salinity of rivers and lakes. Gibbs (1967), by multiple regression analysis of the 74 river water samples in Amazonia, demonstrated that relief explained 85% of the 92% variability in salinity accounted for and that 86% of the total dissolved salts yearly carried by the Amazon was supplied by only 12% of its watershed in the high Andean area.

The nature of the watershed rocks not only affects the rate of solubilization and the amounts of soluble substances reaching the lakes but also partly explains the composition of their water.

A large portion of the Warm Belt is composed of outcropping metamorphic Precambrian rocks (Amazonia, Congo basin, part of the Zambezi, central and western Australia), volcanic rocks (African Rift Valley, Zambezi basin, Indonesia) and sedimentary formations (Malaysia, Indo-China). In fact, not only the nature of the rocks should be considered but also their age and the period of time during which they have been subjected to the weathering process. In Amazonia, the weathering of granites and gneiss took place during very long geological periods; the consequent erosion and dissolution of soluble substances by heavy equatorial rain have left mostly insoluble compounds. The waters washing such rocks are extremely poor in minerals, especially calcium, sulphate and hydrogen carbonates, relatively richer in sodium and potassium, and have very low pH. The very low salinity of Amazonian waters characterizes most of the river-lakes associated with the large river. Extreme conditions of pH (as low as 3.5) and salinity (79, 82, 93‰) are found in 'blackwaters', the waters rich in humic acids and draining the white sands (see p. 65). A comparable situation is observed in rain-forest waters in other parts of the world. In the Zaire basin, acid blackwaters are common (see p. 213); moreover, the total dissolved solids of the Zaire River do not exceed 30 mg l^{-1} and Lake Tumba has a conductivity of 24–32 μmhos cm^{-1}. In West Malaysia, the waters also have a low conductivity and low alkalinity, but are less acid than in Amazonia and in Zaire. The water of the Gombak River has a conductivity of 28 μmhos cm^{-1} and belongs to the $NaHCO_3$ type but its pH is much higher (6.5–7.0) because of the absence of humic acid (Bishop, 1973). In Southern Malaya we again find acidic (pH 3.6–5.4) blackwaters with low salinity (minimal value 4.6 mg l^{-1}), low alkalinity and a high content of humic acid. These waters have very low levels of calcium (0–2 mg l^{-1}) and are sodium-dominated. However, blackwaters here have a larger range of salinities (up to 702 mg l^{-1}; Johnson, 1967a) and may be very rich in sulphates.

To these extremely dilute and acidic waters we can oppose the saline and generally alkaline waters of the African Rift, the Middle East and Australia.

Very generally, the weathering reactions in tropical Africa consist of the breaking down of sodium aluminosilicates by carbonic acid. Many of these reactions are incongruent, that is they cause the precipitation of another silicate as in the following transformation of albite (Eugster & Hardie, 1978):

$$\underset{\text{albite}}{NaAlSi_3O_8} + \tfrac{11}{2} H_2O \rightarrow \tfrac{1}{2} \underset{\text{kaolinite}}{Al_2Si_2O_5(OH)_4} + Na^+ + HCO_3^- + 2 H_4SiO_4$$

The cations (here sodium) and the silicic acid go into solution, hydrogen carbonate is formed and the pH of the solution rises. It follows that the rocks become progressively enriched in aluminium and the solutions in cations and silicic acid. This explains why, in tropical Africa and also in South America, most waters belong to the sodium-hydrogen carbonate type and not to the calcium-hydrogen carbonate group which characterizes the waters of Europe, North America and Asia. This is a basic feature of these waters since it is found in dilute waters such as those of Lake Nabugado and Victoria where it does not result from secondary transformations.

Local composition of rocks may accentuate or reduce this characteristic. In East Africa, the Virunga and Elgon volcanic deposits are especially rich in Na^+ and K^+ and so are the waters of the streams and lakes draining these mountains (Viner, 1975a). In contrast, in the region of Fort Portal, the carbonatites of the Kalyango volcano have a very low content of silica and alumina and very high levels of calcium (36%), carbonate (15%) and phosphorus (3.3%) (von Knorring & DuBois, 1961).

The works of Talling & Talling (1965), Arad & Morton (1969) and Jones et al. (1977) have shown that the concentrated and alkaline brines of East Africa are generated by evaporative concentration of the previously described type of water, accompanied by successive precipitation of the less-soluble salts.

Calcium and magnesium carbonates are the first salts which are removed by precipitation. The residual brine is depleted in calcium and magnesium but still rich in hydrogen carbonate and carbonate. Further concentration leads to precipitation of trona ($Na_2CO_3.NaHCO_3.2H_2O$) as in Lake Magadi and thermonatrite ($Na_2CO_3.H_2O$) leaving a final solution of sodium chloride.

Other pathways of evolution exist: in water less rich in hydrogen carbonate than the East-African waters, the precipitation of calcium

Water chemistry

carbonate utilizes all the alkaline reserve and further evaporation leads to precipitation of gypsum as in the lakes of central South Australia. The residual brine, dominated by Na^+, Mg^{2+} and Cl^-, corresponds to the waters of Lake Cowan (Western Australia) and Lake Corangamite (Victoria). More unusual geochemical evolution seems to cause the precipitation of northupite ($Na_2CO_3.MgCO_3.NaCl$) and thenardite (Na_2SO_4) in Lake Mahega, Uganda (Melack & Kilham, 1972; Kilham & Melack, 1972).

Human influence may possibly accelerate the salinization of freshwater lakes. This has been documented in Australia (see p. 415) where it seems that the early European colonization, by intensive deforestation, considerably increased evaporation. In Mesopotamia, thousands of years of irrigation of saline subsoils have caused a general salinization of soils which, together with the withdrawal of the rivers to a new course, caused the disappearance of great and advanced civilizations.

12
Biological diversity

Bacteria

It is a common observation that in the tropics organic matter is mineralized more rapidly than at higher latitudes. Relatively little is known about the main agents of mineralization, such as bacteria, protozoa, fungi, insects, etc., which intervene along the degradation chain. Moreover, in certain ecosystems where bacteria develop abundantly on allochthonous matter, they may be considered as primary producer-equivalents and constitute then a basic link in the food web.

The Amazonian fluvio-terrestrial biotope is a privileged field for bacterial activity, the organic substrates being quantitatively abundant and qualitatively varied. Rai (1979) and Rai & Hill (1981a) observed the periodic fluctuations in the number of bacteria in Lago Tupé. The high number of bacteria observed at high water, accompanied by a low glucose uptake rate, is related to the abundance of allochthonous amino acids carried by the Rio Negro into the lake; in contrast, at low water, a relatively smaller bacterial population, living on the sugars of dead cells of diatoms and *Peridinium*, has a glucose uptake rate 100 times higher than the preceding population. Similar observations were made by Richey *et al.* (1980). During two cruises of the R.V. Alpha Helix, two transects of, respectively, 2000 and 3400 km along the Amazon were carried out for assessment of carbon oxidation and transport. At rising water, the particulate organic carbon (POC) content of the water varies from 20.3 g m^{-3} upriver to 8.2 g m^{-3} downstream, whereas at high water the corresponding values are 3.7 and 1.5 g m^{-3}. Conversely, the dissolved organic carbon (DOC) content of the water is rather uniform along the river and varies only from 4.2 g m^{-3} at rising water to 6.5 g m^{-3} at high water. The respiration rate at rising water is two orders of magnitude

Biological diversity 445

higher than at high water (26 and 0.2 mg m^{-3} h^{-1}). These results indicate that at rising water labile organic substances originating from inundated forest, flood plains, lakes and floating macrophytes are drained into the river and immediately metabolized by bacteria. The low metabolizing rates at high water are due to both the decrease in the organic substrate and its dilution by the flood. Richey *et al.* (1980) hypothesize that peaks of oxidation occur at rising water and receding water periods and minimal values characterize the flood maximum. In consequence, a nearly constant rate of POC degradation takes place throughout the year, and the carbon export from the river to the ocean is constant over the hydrologic cycle. The study by Wissmar, Richey, Stallard & Edmond (1980) of the bacterial densities of the Amazon water, the ^{14}C-acetate rate constants for uptake, and the high values of the particulate organic C:N ratio (C:N = 20) found by Wissmar *et al.* confirm that the bacteria of the river are influenced by terrestrial carbon. These findings emphasize the predominant role of terrestrial plants over planktonic algae in the carbon load of the Amazon.

The rapid mineralization of organic matter in the Amazonian system is emphasized by the work of Leenheer (1980) on the origin of the humic substances in the waters of the Amazon. The soluble humic substances of the Amazonian blackwaters account for only 50% of the soluble organic substances of these waters, the remaining fraction being composed of colourless organic acids. DOC fractionation shows that organic solutes typical of plant materials, such as carbohydrates, amino acids, chlorophyll and polyphenols, are absent from blackwaters, and X-ray scattering gives small molecular dimensions to humic solutes which thus consist of relatively small molecules of aromatic nuclei, quinones or free radical groups. The absence of other undecomposed plant residues, commonly found in blackwaters of temperate regions, is characteristic of the rapid bacterial mineralization of these products in tropical soils and water.

Methane-oxidizing bacteria are typically found above the perennial thermocline of Lakes Kivu and Tanganyika, and the rates of methane oxidation have been measured by Jannasch (1975) and Rudd (1980). The results of Hecky & Fee (1981) indicate that in Lake Tanganyika it is difficult to account for the high fish production solely from the autotrophic primary production. The authors envisage the eventuality of additional sources of energy derived from the oxidation of reduced substances stored in the permanent hypolimnion. However, the annual net production of methane-oxidizing bacteria is less than 5% of the phytoplanktonic production and cannot bridge the gap between the energy required by the food web and the known sources of energy production. The small contribution of the methane-oxidizing bacteria and probably of the sulphur-oxidizing bacteria is related to the fact that these bacteria are restricted to a very small area on the top of the permanent hypolimnion. The results of Gonfiantini et al. (1979) indicate that in Lake Malawi 20% of the water of the 'permanent' hypolimnion is annually recirculated. Although they also note that this 'mixing' rate is slower in Lake Tanganyika than in Lake Malawi, the recirculation of only 10% of the permanent hypolimnion would cause a phosphorus flux of ~ 50 mg m^{-3} y^{-1}* to the upper seasonal lake which would classify it among eutrophic lakes. Assuming a C:P atomic ratio of between 50:1 and 100:1 in the permanent hypolimnion, the recirculation of 10% of this deep carbon would supply 45 000–90 000 × 10^9 g C y^{-1} to the upper lake, whereas the autotrophic production (1 g C m^{-2} d^{-1}) supplies only 12 000 × 10^9 g C y^{-1}. The resulting bacterial activity would explain the high respiration rates found by Hecky & Fee (1981), the presence of Protozoa which feed on bacteria, such as *Vorticella* and *Halteria*, and the existence of a long food chain including zooplankton-eating fish (sardines) and piscivorous fish. Such a food chain is energy-consuming and could not rely only on autotrophic productivity.

Photosynthetic sulphur bacteria are frequently found in the stratified lakes of Indonesia (Ruttner, 1952). They serve as food to zooplankton which is probably abundant since many fish species are zooplankton-eaters. Sulphur bacteria are also abundant in the metalimnion (20 m depth) of Lake Kinneret during the stratified period. The main species found, *Chlorobium phaeobacteroides*, has been studied in detail by Bergstein, Henis & Cavari (1978).

Comparison between conditions leading to optimal growth in experiments and natural conditions shows that these bacteria bloom in the lake (densities of 10^7 cells ml^{-1}) at light intensities two orders of magnitude lower than required by optimal photosynthesis. As the growth rate of *Chlorobium* is significantly increased by the presence of glucose and acetate in the medium and their active uptake by the cells, it is clear that such myxotrophy gives this species a distinct ecological advantage in the deep and relatively dark metalimnion. The total carbon fixation of these bacteria is small (0.25 μg C mg protein h^{-1}) and represents only 0.8% of the yearly total carbon fixed by planktonic algae (Cavari, personal communication). In spite of this modest quantitative contribution, the photosynthetic bacteria play a major role in the feeding of the nauplii and copepodite larvae of

*The calculation is done on a basis of 5 μg atomic phosphorus l^{-1} in deep water (Degens, Deuser et al., 1971).

Biological diversity 447

Mesocyclops leuckarti. During the bacterial bloom, the number of these organisms in the metalimnion is at least one order of magnitude higher than in the epilimnion, and it has been shown experimentally (Gophen, Cavari & Berman, 1974) that the herbivorous zooplankton can feed actively on bacteria.

Kilham (1981) found high concentrations of bacteria (10^7–10^8 cells ml^{-1}) in three alkaline, saline lakes in Kenya (Lakes Bogoria, Elmenteita and Sonachi). Since these lakes receive little or no domestic sewage, this author relates the abundance of bacteria to the abundance of dissolved organic matter resulting from the high rates of algal productivity. Such environments are also characterized by rich populations of ciliates feeding on bacteria and zooplankton such as rotifers and the copepod *Paradiaptomus africanus* preying on ciliates.

Phytoplankton

Most of the investigations of the phytoplankton of the water bodies from the Warm Belt (old and more recent) are based on samples taken with a net, and the results are lists of algae without quantitative data. In most cases the nanoplanktonic species were neglected.

The number of species recorded from the same water body depends on the sampling method (net or water samples), on the number of sampling stations and on the time span of the investigation: Thomasson (1955) described 38 species in one net sample of Lake Victoria; Van Meel (1954) compiled a list of 351 species based on the work of Woloszynska (1914), Bachman (1933) and Ostenfeld (1908, 1909); Talling (1966), counting the dominant forms, quoted 24 species (Table 112); Uherkovich (1976), studying samples from Rio Negro over a one-year period, described 204 species; later, Uherkovich & Rai (1979), studying samples taken during a three-month period, recorded only 107 species; West (1907) gave a list of 240 species of algae in Lake Tanganyika (sampling also in the bays) and 85 species as pelagic algae; and Hecky *et al.* (1978), using water samples, gave a list of 103 species. In the large African lakes the location of the sampling station plays an important role; bays are richer in species and sometimes contain dominant forms different from those in the open waters. Lind (1968), sampling in Lake Naivasha, described it as a Cyanophyta lake in Loydien Bay and a *Melosira* lake in Crescent Bay. Pollingher (1978*b*) gave a list of 222 species in Lake Kinneret, covering the whole lake (the pelagic zone, the shallow areas and the littoral zone) and a list of 52 species as the most common ones. Ruttner (1952) was the first to use water samples and Utermohl's quantitative method in his study of the phytoplankton from

Table 112 Number of algal species found in some warm lakes by various authors

Water body	Cyano-phyta	Bacillario-phyta	Chloro-phyta	Desmi-diales	Pyrrho-phyta	Crypto-phyta	Eugleno-phyta	Chryso-phyta	Total taxa recorded	Method of sampling	Duration	Sources
Lake Victoria (Af)	39	75	117	110	7		2		351	net	many sampling stns	Van Meel (1954) (after Woloszynska, Bachmann, Ostenfeld)
	27		7		4				38	net	1 sample	Thomasson (1955)
	4	5	7	7		1			24			Talling (1966)
Lake Edward (Af)	33	185	33	7					258	net		Van Meel (Damas' samples)
Lake Albert (Af)	6	14	24	4	2		1		48	net		Van Meel (West's samples)
Lake Kyoga (Af)	8		7	1					16	net		Van Meel (after Bachmann)
Lake Kivu (Af)	15	132	10	7					164			Van Meel (Damas' samples)
Lake Naivasha (Af)	15	32	27	23			3		100			Van Meel (after Bachmann and Rich)
Lake Tanganyika (Af)	52	102	43	40	4	5	1	13	240	net	~1 year	Van Meel (after West)
	22	18	35		9				103	water samples		Hecky et al. (1978)
Lake Malawi (Af)	28	167	124	52	5		7		331	net		Van Meel
Lake Bangweulu (Af)	7	2	17	13	1		2	1	82	net		Thomasson (1965)
Lake Kariba (Af)	15	20	38	9	5		4		95	net		Thomasson (1980)
Lake Edku (Af)	43	20	14	14					86	net	5 months	Elster & Vollenweider (1961)
Lake Mombolo (Af)	33	52	4		1	1		1	92	net (100μm)	1 year	Iltis (1971)
Lake Kinneret (As)	36	45	90	14	15	7	10	5	222	water samples 7 stns	15 years	Pollingher (1981)

Location										Total	Sample type	Duration	Reference
Lake Gatun (MA)	10	8	24	2	5	3			1	53	water samples	1 year	Pollingher (1978b)
	11	8	12	41	8	1				81			Gliwicz (1976)
Lake Sacnab (MA)	18	24	28	46	5				3	124	net (100 μm)	8 months	Deevey, Brenner et al. (1980)
Lake Valencia (CA)	16	4	12			1				33	water samples	1 month	Lewis & Weibezahn (1976)
Guanapito Reservoir (CA)	8	5	7		3	1				21			Lewis & Weibezahn (1976)
Lagartijo Reservoir (CA)	13	5	23	8		2	5			59	water samples	6 months	Lewis & Weibezahn (1976)
Lake Castanho (CA)	19	14	58	51	1		58		9	204	net	1 year	Uherkovich & Schmidt (1974)
Lake Lanao (As)	12	4	39	5	3	2	4		1	70	water samples	1 year	Lewis (1978a, b, c)
Lake Mainit (As)	8	4	20		2	2				36			Lewis (1973a)
Aswan-Nubia (Af)	7	5	13	3			1			29	net	1 sample	Elster & Vollenweider (1961)
River Nile (Af)	8	15	13	4			1			41	net		Elster & Vollenweider (1961)
River Jordan (As)	16	42	61	8		1	11		4	143	water samples	2 years	Pollingher (1978a)
River Shatt al Arab (As)	19	178	20		6		3			226	net + water samples	1 month	Kell & Saad (1975)
Rio Negro (SA)	7	33	57	86	5		6		6	204	net	~1 year	Uherkovich (1976)
only Rio Negro (SA)	7	15	29	44	6	2	4		4	107	net	3 months	Uherkovich & Rai (1979)
Rio Negro + tributaries (SA)	32	91	63	130	12	2	6		9	350	net	3 months	Uherkovich & Rai (1979)
Rio Tapajos (SA)	13	22	42	70	2		2		3	154	net	~1 year	Uherkovich (1976)
River Ganges (As)										86	net	1 year	Lakshminarayana (1965a, b)

⟵————— Bacillariophyta and Cyanophyta dominant —————⟶

KEY: Af = Africa; As = Asia; CA = Central America; SA = South America

450   Biological diversity

warm lakes (Sunda Expedition), but the investigation was based only on two surveys and most of the lakes were small crater lakes or reservoirs. In recent years, the quantitative method has been used in the study of the phytoplankton of Lake Victoria (Talling, 1966), Lake Kinneret (Pollingher, 1969), Lake Mainit (Lewis, 1973a), Lake Lanao (Lewis, 1977), Lake George (Ganf, 1974c), Lakes Gatun and Madden (Gliwicz, 1976), and the rivers Shatt al Arab (Kell & Saad, 1975) and Jordan (Pollingher, 1978a), and others.

Table 112 comprises the number of species belonging to different algal groups as they appear in the lists given by different authors. The comparison of the total number of species found in warm lakes (except the extreme shallow or saline lakes) with the number of species recorded in some lakes in temperate regions: Lake Zurich 133 species (Pavoni, 1963), Lake Mälaren 174 species (Willen, 1973), Lake Erie 250 species (Vollenweider, Munawar & Stadelmann, 1974), where water samples were used, shows that the number of algal species in warm lakes is not lower than in lakes in temperate zones. This situation is different from that observed in zooplankton, but agrees with the results described by Patrick & collaborators (1966). Comparing the flora and fauna of the rivers and streams from the Peruvian head-waters of the Amazon with those of the Guadalupe and the Potomac River in the temperate zone, Patrick concluded that in tropical rivers the ranges of numbers of species of algae, Protozoa and insects are similar to those found in the rivers from the temperate zone.

In the warm lakes the Chlorophyta and Bacillariophyta are represented by the highest number of species. The African lakes comprise many species of Bacillariophyta and Cyanophyta; South-American water bodies comprise more Desmidiales and Euglenophyta. In contrast to the conclusion of Ruttner, the Chrysophyta are present in the Warm Belt lakes (Lake Tanganyika, Lake Kinneret, Lake Castanho) and sometimes play an important role.

Lind (1968) described some small high-altitude lakes in Kenya as *Dinobryon* lakes (Lake Ruiru, 1960 m altitude and Lake Molo, 2493 m) accompanied by *Peridinium*. *Dinobryon* is also common in high mountain lakes in Central America (Löffler, 1972). The phytoplankton of these lakes is similar to that of lakes in temperate regions. The phytoplankton of the warm freshwater bodies is composed mostly of cosmopolitan forms, accompanied by some endemic species and tropical forms; some cosmopolitan species develop in higher numbers than in the temperate zone.

Table 113 is an attempt to summarize the mentioned endemic, tropical and common forms as they were quoted in the revised literature.

Biological diversity 451

Table 113 *Distribution of common algal species in warm lakes; endemic species are mentioned*

CYANOPHYTA		
Anabaena		
nodularoides	Indonesian lakes (As)	
recta	Indonesian lakes (As)	
Anabaenopsis		
arnoldii	Nakuru (Af), Rudolf (Af), Elmenteita (Af), Java (As)	
circinalis	Nakuru (Af), Baringo (Af), Boron (A), Lakes in the Andes (SA)	
circularis	Lakes in Africa and Asia	
magna	Nakuru (Af), Brackish pond Peru (SA)	Evans (1962)
philippinensis	South Malayan waters (As), George (Af)	
raciborskii	Lakes in Java (As), India (As), Valencia (SA)	
var. *lyngbyoides*	Indonesian lakes (As)	
tanganikae	African lakes, Nile (Af), Ganges (As)	
Dactylococcopsis		
fascicularis	Indonesian lakes (As), Lanao (As), Mainit (As), Valencia (SA)	
Spirulina		
laxissima	Lakes in Africa, Asia, Australia	
platensis	Lakes in Africa, Asia, South America, Central America	
BACILLARIOPHYTA		
Coscinodiscus		
rudolfi	Edward (Af)	
Cyclotella		
stelligera		
var. *tenuis*	Toba (As), Sindanglaja (As)	
Cymatopleura		
nyassae	Nyassa (Af), Victoria (Af)	
Cymbella		
ruttneri	Toba (As)	Littoral & fossil
Denticula		
pelagica	Indonesian lakes (As)	Endemic
van heurcki	Indonesian lakes (As)	Endemic
Fragilaria		
africana	Edward (Af), Tanganyika (Af)	Kufferath (1956a,b)
Gomphonema		
africanum	Tanganyika (Af), Edward (Af), Albert (Af)	African endemic
Melosira		
agassizii	Victoria (Af), Edward (Af), Kivu (Af), Lanao (As)	Tropical form
goetzeana	African lakes	Kufferath (1956a,b)
granulata	(Af), (As), (CA), (SA)	Very common
var. *valida*	Toba (As)	Plankton & fossil
var. *jonensis*	Nyassa (Af), Toba (As)	Fossil & tropical
ikapoensis	Ikapo (Af), Indonesian lakes (As)	Tropical & littoral
nyassensis		
var. *victoriae*	Victoria (Af), Albert (Af), Nyassa (Af)	Endemic
pyxis	Malombe (Af), Ikapo (Af), Nyassa (Af), Tanganyika (Af), Victoria (Af)	

452 Biological diversity

Table 113 (cont.)

Navicula		
nyassensis	Edward (Af), Tanganyika (Af), Kivu (Af), Victoria (Af)	
zanoni	Victoria (Af)	
Nitzschia		
bacata	Indonesian lakes (As), Lanao (As), Edward (Af), Kivu (Af)	
confinis	Kivu (Af)	Kufferath (1956a,b)
fonticola var. pelagia	Toba (As)	
lanatula	Edward (Af), Kivu (Af), Tanganyika (Af), Victoria (Af)	
striolata	Java (As)	Endemic
Rhizosolenia		
victoriae	Victoria (Af)	
Rhopalodia		
gracilis	Tanganyika (Af), Kivu (Af), Rudolf (Af), Edward (Af), Victoria (Af)	
hirundiformis	Tanganyika (Af), Nyassa (Af), Victoria (Af)	
Stephanodiscus		
damasi	Edward (Af), Kivu (Af)	Kufferath (1956a,b)
Surirella		
englerii	Victoria (Af)	
nyassae	Nyassa (Af), Malomba (Af), Victoria (Af)	
Synedra		
rumpens var. neogena	Indonesian lakes (As), Mainit (As)	Tropical form
ulna	Albert (Af), Edward (Af), George (Af), Kivu (Af), Naivasha (Af), Nyassa (Af), Tanganyika (Af), Victoria (Af), Kinneret (ME)	Very common
Tepsinoe		
musica	Shatt al Arab (ME) Pond Israel (ME)	Bourrelly (1973)
DINOPHYTA (PYRRHOPHYTA)		
Ceratium		
brachyceros	Victoria (Af)	Tropical form
Gymnodinium		
varians	Java lakes (As), Kafue R. (Af)	Evans (1962)
Peridinium		
africanum	Victoria (Af), Nyassa (Af), Tanganyika (Af)	
baliense	Indo–Malaysia Islands (As), Rio Janauperi (Amazonia) (SA)	Uherkovich & Rai (1979)
gatunense	Madagascar (Af), Gatun (CA)	
gutwinskii	Indonesian lakes (As), Madagascar (Af)	Tropical form
wildemani	Indo-Malaysia Islands (As)	
EUGLENOPHYTA		
Euglena		
spirogyra var. magnifica	Castanho (SA)	Endemic: Uherkovich & Schmidt (1974)
Lepocinclis		
mayali	Castanho (SA), Lakes in Egypt (Af)	Tropical–Subtropical

Biological diversity 453

Table 113 (cont.)

Phacus		
myersi		
var. *myersi* f.		
maior	Castanho (SA)	Tropical form
Strombomonas		
ensifera		
var. *brasiliensis*	Castanho (SA)	Tropical form
Trachelomonas		
acanthophora		
var. *minor*	Castanho (SA), Africa	Endemic: Uherkovich & Schmidt (1974); Evans (1962)
armata	Castanho (SA), Lakes in Australia	
var. *duplex australica*	Castanho (SA), Lakes in Australia	
var. *rectangularis*	Castanho (SA)	
dastuguei	Castanho (SA)	Tropical form
kellogii		
var. *nana*	Castanho (SA)	Endemic: Uherkovich & Schmidt (1974)
megalacantha	Castanho (SA)	Endemic; Uherkovich & Schmidt (1974)
var. *crenulatocollis*	Castanho (SA)	Endemic; Uherkovich & Schmidt (1974)
CHLOROPHYTA		
Allorgiella		
tobaica	Toba (As)	
Ankistrodesmus		
braunii	Toba (As)	
elakatotrichoides	Manindjan (As)	
malaiicus	Toba (As)	
pyrenogerus		
var. *gelifactus*	East Java lakes (As), Toba (As)	
Desmatractum		
nyansae	Victoria (Af)	
Gloeocystis		
ilapoae	Ikapo (Af), Victoria (Af)	
mirabilis	Bratan (As)	
Oocystis		
crassa		
var. *compacta*	Toba (As)	
var. *smithii*	Ranau (As)	
cruciata	Sumatra lakes (As), Toba (As)	
duplex	Sumatra lakes (As), Toba (As)	
gelatinosa	Manindjan (As)	
natans		
var. *simplex*	Lamongan (As)	
tobae	Toba (As)	
Schroederiella		
africana	Castanho (SA), African lakes	

454   Biological diversity

Table 113 (cont.)

Selenodictyum		
brasiliense	Castanho (SA)	Endemic
Sphaerocystis		
shroeteri		
var. *major*	Danau di Atas (As)	
Tetraedron		
trigonum		
var.		
scrobiculatum	Java lakes (As), Toba (As)	
victoriae	Victoria (Af)	
	DESMIDIALES	
Cosmarium		
adoxum	Pasir (As)	
bioculatum		
var.		
minutissimum	Klindungan (As), Pasir (As), Bratan (As)	
dorsirotundatum	Bratan (As)	
neglectum	Bratan (As)	
tethophoroides	Singkarak (As)	
Spondylosium		
nitens		
var. *triangulare*	Danau di Atas (As)	
Staurastrum		
asteryas	Toba (As), Bratan (As)	
excavatum		
var. *minimum*	Bratan (As), Madagascar (Af), (A)	
var.		
planctonicum	Bratan (As), Toba (As)	
floriferum	Danau di Atas (As), Bratan (As)	
var. *compactum*	Pakis (As)	
formosum	Bratan (As)	
gutwinskii	Bratan (As)	
javanicum	Danau di Atas (As)	
var. *maximum*	Danau di Atas (As), Toba (As)	
limneticum	Toba (As)	
longebrachiatum	Bratan (As)	
orthospinum	Bratan (As)	
variodirectum	Ranau (As)	

KEY: A = Australia; Af = Africa; As = Asia; CA = Central America; ME = Middle East; SA = South America

Cyanophyta

Among the 277 species of Cyanophyta recorded in the water bodies of the Sunda Expedition, 41 (14%) were tropical and subtropical forms (Geitler & Ruttner, 1936). The coenobial Cyanophyta belonging to the genera *Microcystis, Aphanocapsa* and *Aphanothece* and species of *Anabaena* are common and sometimes the dominant forms in warm water bodies. *Dactylococcopsis fascicularis* Lemm. fa *solitaris* Geitler is the most common species in the water bodies of the Sunda Expedition, in the Philippines and in Lake Valencia (Venezuela). During most of the year a large part of the offshore plankton of Lake Victoria is composed of colonial coccoid Cyanophyta (Talling, 1965a, 1966). Species of the genera *Anabaenopsis* and *Spirulina* (*S. platensis* and *S. laxissima*) are the dominant forms in saline and alkaline water bodies. The subtropical *Raphidiopsis curvata* is common in the Nile basin. Cyanophyta as a group are common in Africa and Asia and less frequent in South-American water bodies.

Bacillariophyta

Among the 645 taxa of diatoms recorded by Hustedt (1938) in the water bodies of the Sunda Expedition, 279 (44%) are warm water forms. Kufferath (1956a), studying the diatoms in cores and mud from Lake Tanganyika, has described 56 species comprising 13 African endemic forms.

The cosmopolitan *Melosira granulata* and its variety *M. granulata* var. *angustissima* are very common in the Nile, Amazonia and Ganges basins, the Murray River (Australia) and in many lakes of the Warm Belt. It is worth noting that Hecky *et al.* (1978), in their survey of the phytoplankton of Lake Tanganyika, have not recorded *M. granulata*. Some forms of *Melosira* are endemic (Table 113). The genus *Nitzschia* also has many endemic forms. Some cosmopolitan forms such as *Nitzschia amphibia* and *Synedra ulna* are very common in warm lakes.

Chlorophyta

Most of the green algae are cosmopolitan forms; species belonging to the genera *Sphaerocystis, Scenedesmus, Coelastrum, Pediastrum, Oocystis, Dictyosphaerium* and *Lagerheimia* are very common. *Botryococcus braunii* has a high development in all water bodies in the Warm Belt at all altitudes. Bourrelly (in Ruttner, 1952), studying the Chlorophyta of the Sunda Expedition, has described a new genus, *Allorgiella*, with the species *A. tobaica* present only in Lake Toba, a new variety of *Sphaerocystis shroeteri* and *Lagerheimia longisetta*, three new species of *Ankistrodesmus* and six forms of *Oocystis* (Table 113) (Ruttner, 1952). Uherkovich & Schmidt (1974) mentioned the presence of a tropical form, *Schroederiella africana*, in Lake Castanho and an endemic form, *Selenodictyum brasiliense*.

Desmidiales

Among the Desmidiales, the number of tropical endemic species is high. There are also many Desmidiales with a pantropical distribution which are present in all warm regions, but others are restricted to a smaller area (Bourrelly, 1966). The genus *Streptonema* was found only in India and the Sunda Islands, the genus *Apriscottia* in Brazil, and *Allorgeia* in Africa. *Phytomatodocis* is present only in warm regions, but *P. alternans* was found only in South America. *Micrasterias foliacea*, a warm-water form which was not found in Europe, is present in Terra Nova and Canada (Bourrelly, 1966). Among the 280 taxa of Desmidiales described by Krieger (1932) from the lakes of the Sunda Expedition, 80 (29%) are tropical forms. Only 39 species were counted by Ruttner (1952), and among them 21 (54%) are tropical forms; five new species of *Cosmarium* and four of *Staurastrum* were described (Table 113). Uherkovich & Schmidt (1974) have found 51 taxa of Desmidiales in Lake Castanho, and Uherkovich & Rai (1979) have found 130 taxa in Rio Negro and its tributaries. Forster (1969) has described 409 taxa in the Santaren region, and Thomasson (1971) has described 1000 taxa in the Amazonian basin. The Desmidiales from Brazil were studied also by Grönblad (1945) and Forster (1963, 1964, 1969, 1972, 1974).

The Desmidiales are generally less common in African lakes. They are not mentioned by Hecky *et al.* (1978) in Lake Tanganyika, but 40 species appear in the list of West (1907). A similar situation was found in Lake Victoria: Talling (1966) quoted seven desmids, and in the list compiled by Van Meel (1954) there were 110 species. Grönblad *et al.* (1964), studying one sample from Lake Victoria, gave a list of 21 desmids. Among them one new species (*Staurastrum lindae*), a new variety and a new form were described. Thomasson (1965) has recorded 52 desmids in Lake Bangweulu. The desmids in Africa are common in the shallow swampy water bodies. Bourrelly (1957), studying the floating meadows of the inundation zone of the Niger River, has recorded 213 taxa (excluding diatoms) comprising 162 (75%) desmids; 40 forms were new taxa. Among the 213 species, 69 (32%) were pantropical forms. Most of them were endemic Desmidiales. The highest number of tropical species belonged to the genus *Staurastrum*, followed by *Xantidium, Micrasterias, Arthrodesmus, Pleurotenium* and *Cosmarium*.

Grönblad *et al.* (1958) have recorded 200 taxa of desmids in Lake Ambadi, a shallow widened stretch of the Bahr el Gazar (a tributary of the White Nile) with a pH of 6.9. Among the 200 desmids, 18 new species and 25 new varieties were described. The desmids disappear by the time the Bahr el Gazar reaches Lake No. Brook (1954) recorded only 19 species of desmids in the White Nile.

Grönblad (1962) has recorded 89 species of epiphytic desmids on *Ceratophyllum* and other submerged plants in the Ghazal River.

Euglenophyta

The number of Euglenophyta species quoted in warm lakes is very low. Only in Lake Castanho did Uherkovich & Schmidt (1974) find 54 taxa; among them seven were tropical forms and five were endemic species. Four species of *Trachelomonas* were counted by Lewis (1978c) in Lake Lanao (Philippines). Thomasson (1960) has recorded six species of Euglenophyta in Lake Bangweulu.

Dinophyta (Pyrrhophyta)

The dinoflagellates are scarce in large warm lakes. In the literature, five tropical species of *Peridinium*, one of *Ceratium* and one of *Gymnodinium* are quoted (Table 113). Water blooms of dinoflagellates are described in tropical reservoirs, small crater lakes and high mountain water bodies. Ruttner (1952) described a water bloom of *Peridinium gutwinskii* in Tjigombong Reservoir (81% of the biomass) and Klindungan crater lake (47% of the biomass). Both are located in Java. The only large warm lake in which a bloom of *P. cinctum* fa *westii* develops every year seems to be Lake Kinneret (Israel) (Pollingher & Serruya, 1976). *P. cinctum* is common in high-altitude small water bodies in Kenya (Lind, 1968). It is found in Lake Tigoni at 2250 m altitude and in Sasumua Reservoir, where it is accompanied by *Ceratium hirundinella*. Löffler (1972) has found *P. willei* and *P. volzii* (very common species in the temperate regions) in high mountain lakes in Central America at 3800 m altitude. Gessner & Hammer (1967) have found the small *P. pusillum* and the large *P. willei* as the dominant phytoplankton in Lake Mucubaji, located at 3560 m altitude (Venezuela).

It is worth noting that *Ceratium hirundinella* was not recorded in the Amazon basin but is very common in small water bodies in Asia and was quoted in Lake Lanao (Philippines). Ruttner (1952) found it only in two small reservoirs in the lakes of the Sunda Expedition. *C. hirundinella* accompanied *P. gutwinskii* in Tjigombong and in Sindanglaja Reservoirs.

Bourrelly (1961), studying the algae from small freshwater bodies in Ivory Coast, has described two new species: *Dinococcus africanum* and *Stylodinium africanum*, and a new variety, *Cystodinium closterium* var. *crassa*, epiphytic on filamentous algae.

Chrysophyta and Cryptophyta

Due to the use of net samples in most studies of the warm lakes, the Chrysophyta and Cryptophyta were not recorded. Hecky *et al.* (1978), working with water samples in Lake Tanganyika, have found 13 species of Chrysophyta and five species of Cryptophyta, which contributed 13% and 9%, respectively, to the total algal biomass in October–November. The nanoplanktonic *Erkenia subaequiciliata* and *Rhodomonas minuta* var. *nanoplanctica*, which were described by Skuja (1948) as cold stenotherm forms, were recorded in Lake Tanganyika (Hecky et al., 1978) and Lake Kinneret (Pollingher, 1978b). *R. minuta* was also found (Lewis, 1973a) in Lake Mainit and (Lewis, 1978c) in Lake Lanao (Philippines).

Phytoplankton associations

The various regions of the Warm Belt are characterized by typical algal associations. In Middle America, the association Chlorophyta (with frequent desmids)–Pyrrhophyta is common (Lakes Gatun and Amatitlan), but the association Bacillariophyta (dominated by *Melosira*)–Pyrrhophyta is also found (Lake Izabal).

In South America, the blackwaters are characterized by blooms of *Peridinium* spp. (Rai, personal communication). In Africa, the association Bacillariophyta–Cyanophyta is common in the Nile system, whereas the delta lakes of Egypt have a frequent diatom–dinoflagellate assemblage.

The East-African and Ethiopian lakes have a strong Cyanophyta component with *Spirulina platensis* as the dominant species. Most lakes are not dominated by Chlorophyta; Lake Chilwa is a rare exception. In South-East Asia, Lake Lanao and various lakes of Indonesia have a Chlorophyta–Bacillariophyta association. In Indo-China, the Bacillariophyta dominate in the Mekong, the Cyanophyta in the Great Lake of Kampuchea, whereas the flood plain is rich in desmids and *Peridinium* spp.

In Australia, the freshwater acid lakes are dominated by desmids. The saline lakes of Australia and Africa are characterized by a paucity of algal species. The phytoplankton community is usually dominated by a single species: *Spirulina platensis* in East-African and Ethiopian lakes, *Nodularia spumigena* in Lake Carangamite (Australia), *Dunaliella salina* in Pink Lake (Australia), *Anabaena spiroides* in Red Rock Tarn and Lake Werowrap (Australia). In Lake Werowrap, the *Anabaena* bloom was replaced by a *Gymnodinium* sp. bloom when the salinity increased (Walker, 1973).

Seasonal succession

The succession of algal assemblages in lakes responds essentially to availability of nutrients and light, temperature, grazing and sinking rates.

In the Warm Belt, incident light is hardly ever limiting; however, local

Biological diversity

conditions in the water may limit light penetration (presence of humic acids, suspended organic and inorganic matter). These local conditions may vary seasonally due to hydrographic or internal limnological phenomena (rainfall, river discharge, algal biomass, mixing). Nutrient availability is as variable in tropical lakes as in temperate ones. The factors which govern nutrient availability are various. Nutrient supply by rivers is negligible for large lakes such as Tanganyika (Hecky & Fee, 1981) or Lake Lanao (Lewis, 1978b). It is of considerable importance for the river-lakes of Amazonia, where the river operates 'nutrient injections' at the flood period, not only by its own nutrient load but also by flooding large areas covered by vegetation. It seems that in many lakes of the Warm Belt, phosphorus is relatively more abundant than nitrogen (Lake Victoria – Talling, 1966; Lake Lanao – Lewis, 1978b; Lake Titicaca – Richardson et al., 1977). In many cases, wind-induced turbulence eroding the thermocline is the main process which recirculates nutrients from the hypolimnion into the euphotic zone. In the case of Lake Tanganyika, the hypolimnion is the only source of nutrient supply. The simultaneous occurrence of the various phases of the annual cycle of stratification and the stages of algal succession is striking. The mixing period in August–September is followed by a jump in biomass from 200 to 920 mg m^{-3} wm, whereas maximal stratification is accompanied by minimal values of biomass (50 mg m^{-3} wm); in the intermediate period the biomass oscillates between 100 and 200 mg m^{-3} (Hecky & Kling, 1981). The hypolimnic water of Lake Tanganyika has an atomic N:P ratio of 5.6, that is, considerably depleted in nitrogen (Coulter, 1977). If this is the source of nutrients, it is easy to understand that nitrogen limitation occurs soon after mixing. This is probably why *Anabaena*, a nitrogen-fixing species, is abundant in October–November.

Similar situations of nutrient supply associated with the annual circulation pattern and with wind-induced mixing are well documented in Lake Victoria (Talling, 1966), Lake Lanao (Lewis, 1974) and Lake Kinneret (Serruya et al., 1978). In Lake Kinneret, the timing of the dinoflagellate bloom development is a function of the duration and intensity of the mixing period in the lake (Pollingher & Zemel, 1981). The wind regime in Lake Victoria is responsible for the cyclic appearance of *Melosira nyassensis* var. *victoriae* (Talling, 1965b).

The Lagartijo Reservoir (Venezuela) represents an extreme case of dependence on the wind for nutrient supply. The Lagartijo River channels into the hypolimnion its cold water and associated nutrients and thus maintains a permanent stratification. Algal growth is strictly dependent on turbulence, and the plankton of the reservoir is dominated by nitrogen-fixing species of Cyanophyta.

In the Warm Belt the seasonal and even diurnal changes of temperature are too mild to cause direct modifications of the algal assemblages, but these changes affect the metabolism of the zooplankton (enhanced grazing and respiration rates). This imposes a permanent pressure on algal cells throughout the year. Finally, high temperature maintains high bacterial activity and respiration losses, resulting in rapid regeneration of nutrients.

The combined effect of these factors creates seasonal variations of the limnoclimate, which are different for each lake, and the resulting seasonal succession of algae varies from lake to lake. In Lake Tanganyika, the low-biomass period is dominated by Chlorophyta, the intermediate period by Chrysophyta and Bacillariophyta, and the high-biomass period by Chlorophyta and Bacillariophyta in a first stage and by *Anabaena* in a later one. Although Lake Victoria shows a similar general sequence of events related, as in Lake Tanganyika, to the turbulence and circulation patterns, the algae dominating at each period are different, at species level, from those of Lake Tanganyika. This is because the different morphometry of Lake Victoria has combined the factors regulating algal growth into a different limnoclimate.

Nutrients and underwater light were chosen by Lewis (1978b) as the two main factors causing the seasonal shifts in algae. In a situation with high nutrient levels and low light intensity, diatoms and cryptomonads dominate. When nutrient levels progressively decrease and light intensity increases, the previous groups are successively replaced by Chlorophyta, Cyanophyta and Dinophyta.

In Lake Kinneret, a warm monomictic lake with low concentrations of available phosphorus, four stages of algal succession were described (Pollingher, 1981) as a function of mixing, nutrient availability, algal competition and temperature. The algal succession in the lake starts with thermal and chemical destratification and ends with stratification. In many shallow and saline lakes, algal seasonal succession is nearly absent.

Biomass, chlorophyll and primary production

Data on the phytoplankton biomass in water bodies of the Warm Belt are scarce (Table 114). The concentrations of chlorophyll do not always give a real picture of the algal standing crop since its concentration varies in different taxonomic groups. Low phytoplankton biomass and chlorophyll have been recorded in the large lakes of the Warm Belt (Lake Tanganyika, Lake Victoria, Lake Titicaca, Table 114); higher biomasses

Table 114 Biomass, chlorophyll and primary production values in some lakes of the Warm Belt

Lake (*Reservoir) (**River)	Biomass (mg wm m^{-3})	Chlorophyll a(mg m^{-3}) (mg m^{-2})	Productivity ^{b}(mg C m^{-3} d^{-1}) (mg C m^{-2} d^{-1})	Dominant algae	References
SOUTH AMERICA					
Maracaibo					
Valencia			2148	55% Cyano	Lewis & Weibezahn (1976)
Mucubaji	1800		168	75% Pyrro; 24% Chloro	Lewis & Weibezahn (1976)
Lagartijo*	2700		1252	89% Cyano; 5% Crypto	Lewis & Weibezahn (1976)
Guanapito*	1700		496	46% Bacillario; 27% Cyano; 14% Chloro	Lewis & Weibezahn (1976)
Guri*	450		65	84% Pyrro; 7% Crypto; 6% Chloro	Lewis & Weibezahn (1976)
Titicaca	130–280		1000 winter 2800 summer	Bacillario+Chloro Cyano+Chloro	Widmer et al. (1975) Richerson et al. (1977)
Rio Negro**		1–14	12–270^b 190	Chryso+Chloro	Rai (1978); Schmidt (1976) Fisher (1979)
Lower Rio Negro**	1000	61–162	440–2410	Bacillario; Cyano+Desmidiales	Schmidt & Uherkovich (1973)
Rio Tapajo**			63		Fisher (1979)
Amazon**			96^b		Marlier (1967)
Redondo			140		
Janauaca	15 000–24 000	52^a	2200	Cyano+Chloro	Fisher (1979)
Castanho		127	320–2150^b	Bacillario+Chloro+Cyano	Schmidt (1973)
Middle Parana water bodies		0.04–0.08 25^a	300–1000 50–1000 50–1000		Bonetto et al. (1969)

Table 114 (cont.)

Lake (*Reservoir) (**River)	Biomass (mg wm m^{-3})	Chlorophyll a(mg m^{-3}) (mg m^{-2})	Productivity h(mg C m^{-3} d^{-1}) (mg C m^{-2} d^{-1})	Dominant algae	References
CENTRAL AMERICA					
Gatun		2.5^a		Chloro (Desmidiales) + Bacillario	Gliwicz (1976)
Madden				Cyano + Pyrrophyta	
Izabal		5^a	430–2600	Cyano + Bacillario + Pyrro	Brinson et al. (1974)
Amatitlan			2000	Chloro (Desmidiales) + Pyrro	Deevey (1957)
AFRICA					
Victoria		1.2–5.5^a	1080–4200		Talling (1965)
George		150–300^a	5400	80% Cyano	Ganf (1972)
Roseires*			5500	1966–67 Microcystis	Hammerton (1976)
Sennar*			6600	1969–70 Chloro	Hammerton (1976)
Nile at Khartoum			6000	1969–70 Chloro	Hammerton (1976)
Mariut			5200	Bacillario + Cyano	Aleem & Samaan (1969)
Edku			604		Samaan (1974)
Manyara			2400		Melack & Kilham (1974)
Magadi			2800		Melack & Kilham (1974)
Elmenteita		97^a	2000		Melack & Kilham (1974); Melack (1979)
Nakuru		330^a	2370		Melack & Kilham (1974); Melack (1979)
Naivasha		10–17^a	1370–2300	Cyano	Melack (1979)
Simbi		200–650			Melack (1979)
Aranguadi		200–300 2000^a	1740–1960	Spirulina platensis	Talling et al. (1973)
Kivu	60–160	dry season	1000	Cyano + Chloro	Degens et al. (1973)
	300–900	rainy season	1100–290	Chloro + Cyano + Chryso	Hecky et al. (1978)
Tanganyika	25–1570	0.1–4.5^a			Hecky & Kling (1981)

Location				Dominant algae	Reference
Chad	30–2000		300–3700	Chloro+Cyano+Bacillario	Iltis (1977); Lemoalle (1979); Lévêque et al. (1972)
Volta (R)		18[a]	800–9300 121–478 mg m^{-3} h^{-1}	Crypto+Pyrro	Vinner (1969b); Biswas (1978)
MIDDLE EAST					
Kinneret	500–18 000[c] 50 000[d]	108–600 10–148[a] 275[a] [d]	900–4000 100–1460[b] 2590[b] [d]	Pyrrho+Chloro	Pollingher (1978b); Berman (1978)
Solar Lake			4960[b] 8015	Cyano	Cohen et al. (1977b)
ASIA					
Trigam			3000	Chloro+Cyano	
Dal			350–600	Eugleno+Cyano	
Naranbagh			36–905	Bacillario+Chloro+Pyrro	Khan & Zutschi (1980b)
Lanao	5000–6000		400–5000 Av. 1700	Bacillario+Chloro	Lewis (1974)
Mainit	18 000–28 000		1000–2000 700[b]		Lewis (1973a) Fish (1960)
Black waters South Malaysia					
Malaysian fish ponds			10 000	Cyano+Eugleno	Dunn (1967)
AUSTRALIA					
Corangamite		50–300[a]	1620 500–4000 383[b]	Cyano	Hammer et al. (1973) Hammer (1981)
Red Rock Tarn		125–1050[a]	1000–17 280 23–58 160[b]	Cyano	Hammer et al. (1973) Hammer (1981)
Werowrap			237–2900	*Gymnodinium* + *Chroococcus*	Walker (1973)

[c] = monthly averages; [d] = at optimal depth at the peak of dinoflagellate bloom
Cyano = Cyanophyta; Chryso = Chrysophyta; Chloro = Chlorophyta; Pyrro = Pyrrophyta; Bacillario = Bacillariophyta

are recorded in smaller lakes (Lake Janauaca, Lake Mariut, Lake Lanao, Lake George, Lake Kinneret) and in some very shallow and saline lakes (Lake Chad, Lake Corangamite, Lake Red Rock Tarn, Lake Nakuru, Lake Aranguadi).

The biomass values per unit area, computed for the euphotic zone, show large differences depending on each lake. The values are low in Lake Valencia: 7.4 g m^{-2} (Lewis & Weibezahn, 1976) and Lake Castanho: 19 g m^{-2} (Schmidt, 1973), and higher in Lake Mainit: 22.5 g m^{-2} (Lewis, 1973b) and Lake Lanao: 23.7 g m^{-2} (Lewis, 1978b). High standing crops are recorded in the shallow Lake George: 46.8 g m^{-2} (Burgis *et al.*, 1973) and Lake Chad: > 100 g m^{-2} (Iltis, 1974). In Lake Kinneret, due to the yearly dinoflagellate bloom, the mean annual phytoplankton biomass varies between 40 g m^{-2} y^{-1} (1975) and 80 g m^{-2} y^{-1} (1972) (Pollingher, 1981).

The phytoplankton biomass varies less than one order of magnitude over a year in Lake Lanao. No differences are found in Lake George, while large differences were observed in Lake Tanganyika. In Lake Kinneret, during the mixing period, the lowest phytoplankton biomass was recorded; 500 mg m^{-3} wm (monthly average). The biomass increases to 10 000 mg m^{-3} wm (monthly average) at the peak of the bloom, and 800 mg m^{-3} wm is the lowest monthly average recorded during the stratification period.

In spite of the small biomass, the photosynthetic activity in warm lakes is very high in comparison with lakes located in the temperate zone (Table 114). In Table 115, Likens (1975) gives a mean value of net primary production for aquatic water bodies at different latitudes, without taking into account the very high production recorded in soda and saline lakes. The mean annual averages of some warm lakes' productivity, followed throughout the year, were: 620 g C m^{-2} y^{-1} in Lake Lanao, 635 g C m^{-2} y^{-1} in Lake Kinneret, and 640 g C m^{-2} y^{-1} in Lake Victoria. Even the high-altitude Lake Titicaca has an annual mean of 511 g C m^{-2} y^{-1}, and the poor Great Lake Tanganyika produces 300 g C m^{-2} y^{-1}, which is twice the

Table 115 *Net production values for lakes and rivers located at different latitudes (from Likens, 1975)*

	mg C m^{-2} d^{-1}	g C m^{-2} y^{-1}
Tropical lakes	100–7600	30–2500
Temperate lakes	5–3600	2–950
Temperate rivers	< 1–3000	< 1–650
Tropical rivers	< 1–?	< 1–1000

Biological diversity 465

production of Lake Baikal, 122.5 g C m^{-2} y^{-1} (Moskalenko, 1972) and that of Lake Michigan, 140–150 g C m^{-2} y^{-1} (Vollenweider et al., 1974).

It is worth mentioning that many of the shallow warm lakes have large areas covered with macrophytes. Their mean net production varies between 500–2700 mg C m^{-2} d^{-1} for submerged plants and 4100–12 000 mg C m^{-2} d^{-1} for emergent plants (Likens, 1975). In Lake Redondo, the floating meadows of *Paspalum repens* and *Panicum* present a daily average production of 6300 mg C m^{-2} or 50 tons dm ha^{-1} y^{-1}. In Lake Edku, the *Potamogeton* and *Ceratophyllum* production is 1320 mg C m^{-2} d^{-1}; in Lake Mariut the *Potamogeton* area produces 2500 g C m^{-2} d^{-1}.

The few examples discussed indicate the high photosynthetic capacity of the phytoplankton and macrophyte communities in the lakes of the Warm Belt in comparison with their counterparts located in the temperate zones.

Protozoa

The study of the freshwater Protozoa, like other fields of limnology, was made in lakes located in temperate regions. Data on warm-lake Protozoa are scarce.

The first studies in African warm lakes were made by Daday, at the beginning of the century (1907, 1910) and in the 1920s by Van Oye (Rhizopoda). The study of ciliates started later and was limited to taxonomic work. Two new ciliates from an East-African soda lake were described by Dietz (1965). The ciliates from small water bodies in Cameroon have been recorded by Njine (1977). The most complete work concerning the African Protozoa is that of Dragesco (1980).

The thecamoebae are represented there by a high number of species and a low number of endemic forms. Among them Dragesco quoted *Difflugia corona*, *Cucurbitella dentata*, *Quadrulella tropica* and *Hoogenraadia africana*.

The Heliozoa have not yet been investigated. Dragesco has studied the African ciliates for seven years and has recorded about 150 species. In his opinion, a similar study in Europe would have recorded twice that number of species.

The diversity of ciliates in warm waters seems to be low and can be explained by temperature limitation. Among the 130 species described by Dragesco, 38 (26%) are new forms. No new genera were found.

Ruttner (1952) was the first to investigate the quantitative contribution of the Protozoa to the planktonic biomass in warm lakes of the Sunda Expedition. He has found *Difflugia hydrostatica* only in Lake Tjigombong (maximum at 3 m depth, 420 organisms l^{-1}). The organisms were smaller in body size than those from the Ploner See (Germany). *Acantocystis* sp. was recorded in Lakes Lamongan, Pakis, Toba and Singkarat (at 10 m depth, 50 organisms l^{-1}).

As in the lakes in the temperate regions, many ciliates live in the meta- and hypolimnions: *Uronema marina* in Pasir (at 9 m, 455 000 organisms l^{-1}), *Prorodon* sp. in Klindungan (850 l^{-1}), and *Metopus contortus* only in Pasir (38 l^{-1}). *Paradilepthus elephantinus* was present in three lakes (38 organisms l^{-1}), *Strobilidium* sp. appeared in three lakes and *Strombidium* cf. *viride* in only one – Tjigombong (800 l^{-1}). Ruttner noted the absence of tintinnids in the lakes of the Sunda Expedition, but Thomasson (1955) has mentioned their presence in Lake Victoria, Hecky et al. (1978) in Lake Tanganyika, and Pollingher & Kimor (1965) in Lake Kinneret. A preliminary list of the Protozoa found in the littoral zone of Lake Kinneret was given by Pollingher (1978c). Hecky et al. (1978), in their study of the planktonic ecology of Lake Tanganyika, have recorded 16 species of Protozoa and Hecky & Kling (1981) have estimated the biomass of *Strombidium* cf. *viride*, which nearly equalled or exceeded the phytoplankton biomass during the stratified period.

The Protozoa constitute a major portion of the total zooplankton biomass (15–50%) of Lake Mainit, Philippines (Lewis, 1973a).

Cairns (1966) gives a list of 22 ciliates recorded in the Amazon River; 21 species are cosmopolitan forms.

The ciliates may play an important role in the pelagic food chain of the warm lakes, and more importance should be given to their study.

Rotifera

According to Ruttner (1952), the rotifer plankton of warm lakes is similar to that of small water bodies in the temperate regions.

Most of the rotifers are cosmopolitan forms; however some endemics seem to occur – *Keratella keikoo* and *K. shieli* are possible endemic eurytropic forms in the Murray River (Australia) (Shiel, 1979).

Brachionus donneri and the two subspecies *Anuraeopsis fissa haueri* and *A. fissa lata* were recorded only in Kampuchea, India and Sri Lanka (Pejler, 1977; Fernando, 1980a).

Platyas leloupi, A. cristata and *A. fissa punctata* were described only from the Ethiopian region (Pejler, 1977).

Hauer (1965) has studied the rotifers from different types of water of the Amazon region. He gave a list of 100 rotifers, describing 10 new species and forms. Among them, some are American forms: *Brachionus dolabratus, B. zahniseri, Lecane habita, Platias patulus* var. *macracantus* and *Keratella americana*, which has a cosmopolitan distribution (Table 116).

Biological diversity

Table 116 *List of Rotifera common in lakes of the Warm Belt*

Anuraeopsis
  fissa — Mainit (As), Chimene R. (Af)
  navicula — Sunda lakes (As)
  racensi — Lakes of Venezuela (SA)
  sp. — Victoria (Af)
Ascomorpha
  saltans — Kinneret (ME)
  sp. — Sunda lakes (As)
Asplanchna
  brightwellii — George (Af), Nilnag (As), Kinneret (ME), Murray R. (A)
  priodonta — Khan Payao, Bum Borapet, Great Lake (As)
  sieboldii — Kinneret (ME), Sunda lakes (As)
  sp. — Titicaca (SA), Victoria (Af), Volta (Af), Lakes of Malaya (As)
Brachionus
  ahlstromi — Castanho (SA)
  angularis — Victoria (Af), Nilnag (As), Kinneret (ME), Sunda lakes (As)
  budapestinensis — Kinneret (ME), Murray R. (A)
  calyciflores — Valencia (SA), Castanho (SA), Redondo (SA), Naivasha (Af), Chilwa (Af), Gorewada (As), Khan Payao (As), Kinneret (ME), Titicaca (SA)
  caudatus — George (Af), Naivasha (Af), Great Lake (As)
  dimidiatus — Soda lakes (Af)
  diversicornis — Victoria (Af)
  dolabratus — Castanho (SA)
  donneri — Sri Lanka (As)
  falcatus — Castanho (SA), Jacaretinga (SA), Izabal (CA), Victoria (Af), Chilwa (Af), Khan Payao (As), Murray R. (A)
  forficula — Victoria (Af), Khan Payao (As)
  gessneri — Castanho (SA), Cristalino (SA)
  havanensis — Valencia (SA)
  keikoa — Murray R. (A)
  novae zealandia — Murray R. (A)
  patulus macrocanthus — Castanho (SA), Redondo (SA), Jacaretinga (SA)
  plicatilis — Soda lakes (Af), Lakes of Central South Australia
  quadrata — Titicaca (SA)
  quadridentatus — Jacaretinga (SA), Victoria (Af), Nilnag (As)
  urceolaris — Govindgarh (As), Murray R. (A)
  variabilis — Titicaca (SA)
  zahniseri reductus — Castanho (SA), Cristalino (SA), Taruma-Mirim (SA)
  sp. — Nasser (Af), Govindgarh (As)
Coelotheca
  sp. — Kinneret (ME)
Conochiloides
  dossuiarius — Castanho (SA), Redondo (SA), Jacaretinga (SA), Taruma-Mirim (SA), Toba (As), Lanao (As)
  sp. — Chad (Af)
Diurella
  dixon-nuttalli — Pakis, Singkarak (As)
  stylata — Tjigombong (As)
  sp. — Great lake (As), Malaya lakes (As)
Epiphanes
  macrourus — Chad (Af)
Euclanis
  dilatata — Murray R. (A), Bali (A)
  sp. — Lakes of Mt. Everest (As)

Table 116 (cont.)

Filinia	
longisetta	Nepal Valley lakes (As), Kinneret (ME), Murray R. (A)
mystacina	Chad (Af)
pejleri	Castanho (SA), Redondo (SA), Jacaretinga (SA), Cristalino (SA), Taruma-Mirim (SA)
pejleri var. *grandis*	Murray R. (A)
terminalis	Nilnag (As), Bangweulu (Af), Murray R. (A)
sp.	Victoria (Af), Volta (Af)
Gastropus	
stilifer	Cristalino (SA), Taruma-Mirim (SA)
Hexarthra	
brasilensis	Castanho (SA), Redondo (SA)
fennica	Central South Australia
insulana	Bali (As)
intermedia	Govindgarh (As), Murray R. (A), Chad (Af)
major	Nilnag (As)
mira	Kariba (Af)
sp.	Lakes Mt. Everest (As)
Horaella	
brichmi	Nilnag (As)
Kellicotia	
longispinna	Kinneret (ME)
Keratella	
americana	Valencia (SA)
australis	Murray R. (A)
cochlearis	Castanho (SA), Redondo (SA), Jacaretinga (SA), Cristalino (SA), Taruma-Mirim (SA), Nilnag (As), Nepal Valley Lakes (As), Gorewada (As), Kinneret (ME), Murray R. (A)
lenzi	Castanho (SA)
procurva	Lanao (As)
quadrata	Titicaca (SA)
serrulata	Murray R. (A)
shieli	Murray R. (A)
slacki	Murray R. (A)
tropica	George (Af), Chilwa (Af), Nilnag (As), Govindgarh (As), Kinneret (ME), Murray R. (A)
sp.	Victoria (Af), Nasser (Af)
Lecane	
curvicornis	Kariba (Af)
elsa	Taruma-Mirim (SA)
leontina	Castanho (SA), Redondo (SA), Jacaretinga (SA), Cristalino (SA), Taruma-Mirim (SA)
luna	Murray R. (A)
melini n.sp.	Rio Negro (SA)
rugosa	Jacaretinga (SA)
sp.	Nepal Valley Lakes (As)
Monostylla	
bulla	Castanho (SA), Redondo (SA), Jacaretinga (SA), Cristalino (SA)
Mytilina	
ventralis	Redondo (SA), Jacaretinga (SA)
Pedalia	
fennica	Kinneret (ME)
intermedia	Java (As)
mira	Toba (As)

Biological diversity 469

Table 116 (cont.)

Platyas	
leloupi	Albert (Af)
patulus	Castanho (SA), Kinneret (ME), Java (As)
quadricornis	Kariba (Af)
Polyarthra	
dolichoptera	Bali (As)
remota	
trigla	Toba (As)
vulgaris	Castanho (SA), Redondo (SA), Jacaretinga (SA), Govindgarh (As), Lanao (As), Kariba (Af)
sp.	Cristalino (SA), Taruma-Mirim (SA), Lakes Mt. Everest (As)
Pompolix	
complanatus	Nilnag (As), Murray R. (A)
sulcata	Govindgarh (As), Lamongan (As)
Rattulus	
chattoni	Lakes Sunda Exp. (As)
cylindricus	Lakes Sunda Exp. (As)
mus	Lakes Sunda Exp. (As)
Synchaeta	
oblonga	Kinneret (ME)
pectinata	Kinneret (ME), Murray R. (A)
Testudinella	
muc. hauriensis	Redondo (SA), Jacaretinga (SA)
patina	Khan Payao (As), Kariba (Af)
Tetramastix	
opoliensis	Lanao (As), Chad (Af)
Trichocerca	
birostris	Kariba (Af)
braziliensis	Lago Rotondo (SA)
chatoni	Castanho (SA), Cristalino (SA), Taruma-Mirim (SA)
opoliensis	Kariba (Af)
ruttneri	Lago Rotondo (SA)
similis	Castanho (SA), Redondo (SA), Jacaretinga (SA), Cristalina (SA), Taruma-Mirim (SA)
stylata	Kinneret (ME)
sp.	George (Af)
Trochosphaera	
equatorialis	Sri Lanka (As)

KEY: A = Australia, Af = Africa, As = Asia, CA = Central America, ME = Middle East, SA = South America

Other species are known only from the Amazon region: *Brachionus voigti, Dipleuchlanis macrodactyla, Eudactylota wulferti, Jtura claviger, Lecane murrayi, L. remanei* and *Chromogaster klementi*.

In the third group are species also known in South America and outside the Amazon region: *Brachionus gessneri, B. zahniseri* var. *reductus, Filinia longiseta* var. *saltator, Hexarthra intermedia* var. *brasiliensis, Keratella lenzi, Lecane melini, L. proiecta* and *Trichocerca braziliensis*.

The most abundant and frequent species were *Keratella americana, K. cochlearis* and *Hexarthra intermedia* var. *brasiliensis*.

The highest number of species (50) was found in Parana de Xiborena (a connecting canal between Solimoes and Rio Negro). In all other water bodies, the rotifer fauna was poor: 26 species in Itugui (a tributary of the Amazon River), 21 species in Lake Castanho (Hardy, 1980), 17 species in Lago Rotondo (Varzea lake), 16 species in Rio Negro, 13 species in Lago Tefe (brown water), nine species in Lago Janauaca (brown water) and only four species in the lakes Juanico (Solimoes water) and Cabeceira (Rio Negro and white water). In the opinion of Hauer, the low frequency rate in most of the species may be due to scarcity of food in these waters. Koste (1972, 1974), in his study of the rotifers from the water bodies of the Amazon drainage basin, gave a list of 206 species and has described 14 new species. In the clear-water biotope 124 species were found, in white water 105, in mixed water 61, and in blackwaters 46 species.

Pejler (1977), in his review on the global distribution of the Brachionidae, mentioned seven species which were recorded only from South America, among them *Notholca haueri* (probably endemic) and *Anuraeopsis racensi*, found only in one lake in Venezuela.

Fernando (1980*a*), in his study of the zooplankton from Sri Lanka, concluded that the rotifer fauna is typical of tropical regions: dominated by the genus *Brachionus*, accompanied by *Trichocerca, Testudinella* and *Lecane* species. Two species, *Keratella earlinae* (which according to Pejler occurs only in the Nearctic region) and *Kellicotia longispina* (considered by the same author as a temperate form), also occur in Sri Lankan water bodies. *Kellicotia* has also been found in Lake Victoria and Lake Kinneret.

The genus *Brachionus* is missing in the Arctic areas and increases in importance approaching the Equator (Pejler, 1977). East-African high-mountain lakes are poor in rotifers. Among the 40 lakes studied by Löffler (1964), only four contained planktonic rotifers: *Brachionus angularis* and *Trichocerca* sp. (at 3500 m altitude) and *Keratella valga tropica* (up to 4200 m).

In the warm lakes, 50–100% of the species belong to the genus *Brachionus*. *Anuraeopsis navicula, Keratella tropica* and *Lecane leontina* seem to be common forms in the warm-water bodies (Gillard, 1957). Conversely the genera *Notholca* and *Argonotholca* are found in Arctic and temperate zones (except for *N. squamula* in Lake Manzala, Egypt, and *N. haueri* in South America).

In Pejler's opinion (1977), the latitudinal distribution of the different rotifer genera seems to be correlated with the availability of optimal food. He correlates the infrequency of *Keratella cochlearis* and *Kellicotia* in warm lakes to the insufficiency of the Chrysophycean and Cryptophycean algae,

Biological diversity 471

which are their preferred food. Conversely, several of the phytoplanktonic algae eaten by *Brachionus* are present in tropical lakes.

Cladocera

The tropical zooplankton is characterized by the poverty of limnetic Cladocera and their small body size. The number of Cladocera in tropical regions (~ 60 species) is smaller than in temperate regions (90 species) (Fernando, 1980a,b).

The genus *Ceriodaphnia* is represented by fewer species in the tropics, whereas the opposite is true for the genus *Diaphanosoma*.

The genus *Daphnia* is represented by a small number of species in tropical limnetic plankton. Dumont (1980) explains this by the fact that the production of resting eggs in *Daphnia* is controlled by seasonal variations, which are not well defined in the warm zones. In Africa, *Daphnia* is absent from the non-seasonal equatorial forest and Guinea savanna. *D. longispina*, *D. lumholtzi* and *D. barbata* occur throughout the eastern half of the African continent. They have isolated populations in the Maghreb and westward from the Nile sources to Lake Chad, the Niger and Senegal River. These species occur in lakes flooded during monsoon (i.e. strong seasonal fluctuations). North of the Sahara, no *Daphnia* are found except in astatic and semipermanent lakelets (*D. similis* and *D. magna* (Dumont, 1980). *D. lumholtzi* is also common in Asia and was recorded in Australia (Timms, 1970). *D. gessneri* is common in South America (Table 117). The pigmented forms *D. peruviana* and *Pleuroxus caca* appear in shallow lakes in the Peruvian and Venezuelan Andes and *D. tibetana* in Himalayan lakes. The high ultraviolet radiation which occurs in tropical high mountain lakes is responsible for the existence of pigmentation in Cladocera (Löffler, 1964). The high mountain lakes of East Africa lack Cladocera, and if present no pigmented forms occur. *Daphnia laevis* and *D. galeata mendotae*, endemics of the American continent, were found in Lake Atitlan and Tecochotal (Central America), respectively (Van der Velde, Dumont & Grootaert, 1978).

The genus *Bosmina* is represented by fewer species than in the temperate zone. *Bosmina chilensis* is common in South America, *B. longirostris* in Africa and *Bosminopsis deitersii* is found in the equatorial and subequatorial zones of all continents. It was also found in Lake Peten Itza (Central America). In South-East Asia, two endemic species were recorded: *Alona macronyx* and *Indialona ganapati* (Fernando, 1980b).

The common limnetic Cladocera in South-East Asia are *Diaphanosoma sarsi*, *D. excisum*, *Moina micrura* and *Ceriodaphnia cornuta*. These species

472 Biological diversity

Table 117 *List of Cladocera common in lakes of the Warm Belt*

Alona	
affinis	L. Negra (SA), Peten (CA)
circumfimbriata	Peten (CA)
costata	Pokhara Valley (As)
gutata	Lakes of Malaya (As), Pokhara Valley (As)
macronix	Malaysia (As)
rectangularis	Nilnag (As)
verrucosa	Peten (CA)
sp.	Gorewada (As), Victoria (Af), Chilwa (Af), South Australia
Alonella	
excisa	Peten (CA), Pokhara Valley (As)
nana	Malaysia (As), Pokhara Valley (As)
Biapertura	
affinis	Pokhara Valley (As)
karua	Pokhara Valley (As)
Bosmina	
chilensis	Castanho (SA), Redondo (SA), Jacaretinga (SA), Cristalino (SA), Taruma-Mirim (SA)
hagmanii	Titicaca (SA), Rio Negro (SA)
longirostris	Izabal (CA), Victoria (Af), Tana (Af), Chad (Af), Kariba (Af), Bum Borapet (As), Great Lake (As), Kinneret (ME), Toba (As), Lanao (As), Indonesian lakes (As)
meridionalis	Toba (As)
sp.	Gorewada (As)
Bosminopsis	
deitersi	Rio Negro (SA), Castanho (SA), Redondo (SA), Jacaretinga (SA), Cristalino (SA), Taruma-Mirim (SA), Ambadi (Af), Bum Borapet (As), Great Lake (As), Lakes Malaya (As), Lakes Indonesia (As)
negrensis	Rio Negro (SA), Cristalino (SA), Taruma-Mirim (SA)
Camptocerus	
rectirostris	Lakes Malaya (As)
Ceriodaphnia	
bicuspidata	Kyoga (Af), Albert (Af), Tana (Af)
cornuta	Rio Negro (SA), Castanho (SA), Redondo (SA), Jacaretinga (SA), Cristalino (SA), Taruma-Mirim (SA), Izabal (CA), Victoria (Af), George (Af), Albert Nile (Af), Tana (Af), Chilwa (Af), Chad (Af), Kainji (Af), New South Wales (A), Indonesian Lakes (As), Lakes of South-East Asia, Nile (Af), Sokoto River (Af), Niger River (Af), Mekong (As), Salt Lakes in Bengal (As)
dubia	Albert Nile (Af), Gebel Aulia Dam (Af), Toba (As)
pulchella	Nilnag (As)
quadrangula	Titicaca (SA)
reticulata	Kinneret (ME)
rigaudii	Edward (Af), Gebel Aulia Dam (Af), Kivu (Af), Lakes of Pokhara and Katmandu valleys (As), Bum-Borapet (As), Great Lake (As), Kinneret (ME)
sp.	Victorian Lakes (A)
Chidorus	
barroisi	Kyoga (Af), Peten (CA), Mainit (As)
eurynotus	Nilnag (As), Pokhara Valley (As)
flaviformis	South America, Malaysia (As), Kashmir (As)
parvus	Pokhara Valley (As)

Biological diversity

Table 117 (cont.)

patagonicus	Laguna Negra (SA)
sphaericus	Victoria (Af), Khan Payao (As), Bum-Borapet (As), Lakes Malaya (As)
Dadaya	
macrops	Sri Lanka (As)
Daphnia	
barbata	George (Af), Albert Nile (Af), Gebel Aulia Dam (Af), Chilwa (Af), Chad (Af)
carinata	New South Wales (A), Ranau (As)
gessneri	Castanho (SA), Jacaretinga (SA)
longispina	Edward (Af), Tana (Af), Pokhara+Katmandu Valley Lakes (As), Chad (Af)
lumholtzi	George (Af), Albert (Af), Gebel Aulia Dam (Af), Tana (Af), New South Wales (A), Wisdom (As), Kinneret (ME), Chad (Af), Govindgarh (As)
magna	Kinneret (ME)
pulex	Titicaca (SA), Kivu (Af)
tibetana	Lakes Mt. Everest (As), Khan Payao (As)
sp.	Edku (Af), Victoria (Af)
Daphniopsis	
pusila	Wisdom (As)
Diaphanosoma	
brachyurum	Kinneret (ME), Amaravathy (As)
excisum	Albert Nile (Af), Gebel Aulia Dam (Af), Tana (Af), Edku (Af), Naivasha (Af), Chilwa (Af), Chad (Af), Kainji (Af), Gorewada (As), Khan Payao (As), Bum-Borapet (As), Great Lake (As), Lakes Malaya (As), New South Wales (A), Wisdom (As), Indonesian lakes (As)
fluviatile	Rio Negro (SA), Izabal (CA)
modighiani	Indonesian lakes (As)
paucispinosum	Govindgarh (As)
perarmatum	Danau di Atas (As)
sarsii	Castanho (Sa), Jacaretinga (SA), Taruma-Mirim (SA), Albert (Af), Toba (As), Indonesian lakes (As), Mainit (As)
unguicalatum	Murray R. (A)
Disparalona	
dadayi	Peten (CA)
Dunhevedia	
serrata	Sri Lanka (As)
Eubosmina	
tubicen	Peten (CA), Izabal (CA)
Euryalona	
orientalis	Tumba (Af)
Gladioferens	
spinosus	Victoria lakes (A)
Grinoldina	
brazzai	Sri Lanka (As)
Holopedium	
amazonicum	Rio Negro (SA), Cristalino (SA), Taruma-Mirim (SA)
Indialona	
ganapati	India (As), Malaysia (As)
Latanopsis	
australis	Wisdom (As), Maninjan (As)
fasciculata	Jacaretinga (SA)
mongolica	Central South Australia

474 Biological diversity

Table 117 (cont.)

Leydigia	
ciliata	Pokhara Valley (As)
Macrotrix	
laticornis	Pokhara Valley (As)
rosea	Peten (CA)
spinosa	Khan Payao (As), Lakes Malaya (As)
Moina	
dubia	Victoria (Af), Edward (Af), Albert (Af), Gebel Aulia Dam (Af), Tana (Af), Edku (Af), Kivu (Af), Khan Payao (Af), Great Lake (As), Lakes Malaya (As), Chad (Af)
latidens	Danau Bratan (As)
micrura	Izabal (CA), George (Af), Chilwa (Af), Chad (Af), Nilnag (As), New South Wales (A), Indonesian lakes (As)
minuta	Rio Negro (SA)
mongolica	Australia
rectirostris	Kinneret (ME)
reticulata	Castanho (SA), Redondo (SA), Jacaretinga (SA), Cristalino (SA), Taruma-Mirim (SA)
sp.	Victoria lakes (A), Gorewada (As)
Neobosmina	
chilensis	Laguna Negra (SA)
hagmanii	Joanicu (SA)
Oxyurella	
senegalensis	Pokhara Valley (As)
Pleuroxus	
similis	Pokhara Valley (As)
Simocephalus	
serrulatus	Manindjan (As), Peten (CA)
vetulus	Naivasha (Af), Khan Payao (As)

KEY: A = Australia; Af = Africa; As = Asia; CA = Central America; ME = Middle East; SA = South America

are also common in tropical Africa; in South America the *Diaphanosoma* species are different (Idris & Fernando, 1981).

As one moves from the tropical to subtropical regions, the larger zooplanktonic Cladocera become more numerous. Fernando (1980*b*) suggests that the lack of continuity of large freshwater lakes in the tropics could account for the lack of evolution of diverse zooplankton fauna. Food may be an important factor restricting species diversity. Temperature seems to play an important role in limiting the presence of large Cladocera in the tropical zone.

Copepoda

About 94 (47%) species and subspecies of the 200 world freshwater species of Calanoida were recorded in Africa (Dussart, 1980) and 32 species (16%) in South-East Asia (Lai & Fernando, 1980).

The most common species in Africa seem to belong to *Diaptomus* and

Biological diversity 475

Tropodiaptomus and in South-East Asia to *Tropodiaptomus* and *Neodiaptomus*.

The distribution of Calanoida in the Warm Belt is not homogeneous. There are no Calanoida at all in the pelagic fauna of Lake Tanganyika; only one calanoid was recorded in Lake Toba (Kiefer, 1933). *Tropodiaptomus* was not found in the watershed of the Black Volta River, but in the watershed of the White Volta, *T. senegambiae* and *T. malianus* were recorded. One species of *Arctodiaptomus* (boreal alpine form) was recorded in Ethiopia. The ranges of most Calanoida are limited. In the northern half of the African continent, there is a sharp separation between the tropical genera *Thermodiaptomus* and *Tropodiaptomus* and the Saharan representative *Metadiaptomus*. Both groups may co-occur in small ponds, but as soon as the pelagic space of a lake becomes large, *Metadiaptomus* is replaced by *Thermodiaptomus* and *Tropodiaptomus* (Dumont, 1980).

In African high mountain lakes, Calanoids are represented only by *Lovenula falcifera*, which is limited to the warmer lakes. In South America, in high mountain lakes, many tropical montane–subantarctic elements are present (Löffler, 1964). The Australasian family of Boeckellidae is represented by the genera *Boeckella* and *Pseudoboeckella*. *Boeckella titicacae* was recorded only in Lake Titicaca (Table 118). *B. gracilipes* and *B. gracilis* are present in lakes and small water bodies in South Chile and southern Argentina (Löffler, 1964). Many species of this family are found in Australia. *B. triarticulata* is an ubiquitous species there and in New Zealand. In Australian athalassic saline waters, two endemic species of the genus *Calamoecia* were recorded: *C. salina* and *C. clitellata* (Bayly & Williams, 1966a).

Lai & Fernando (1980), studying the composition and distribution of the Calanoids from the freshwater bodies in south-eastern Asia (there are no natural lakes), have observed a north–south diminishing of Calanoida diversity and southward qualitative changes. Fifty species of Calanoids were recorded in India, 25 species in China, 10 species in Sri Lanka, 12 species each in Thailand and Peninsular Malaysia, and about three and four species in Sulawesi and Java, respectively.

In Peninsular Malaysia, most neodiaptomids occurred in the northwest and tropodiaptomids in the south and south-east.

Fernando & Ponyi (1981), comparing the number of species of cyclopods in different latitudes, found that the tropics have relatively few species in comparison with subtropical and temperate regions, with the exception of Indonesia (where 30 species were recorded). These findings are contrary to the normal effects of latitude on species number. Fernando (1980a)

476 Biological diversity

Table 118 *List of Copepoda common in the lakes of the Warm Belt*

Acanthodiaptomus	
denticornis	Nilnag (As)
Arctodiaptomus	
jurisovichi	Lakes of Mt. Everest (As)
n.sp.	Bale mountain lakes (Af)
Aspinus	
acicularis	Rio Negro (SA)
sp.	Central American lakes
Boeckella	
fluvialis	New South Wales lakes (A), Southern Queensland lakes (A)
minuta	New South Wales lakes (A), Southern Queensland lakes (A), Victoria lakes (A)
montana	New South Wales lakes (A)
occidentalis	Titicaca (SA)
robusta	New South Wales lakes (A), Southern Queensland lakes (A)
symmetrica	New South Wales lakes (A), South Australia lakes (A), Victoria lakes (A)
titicacae	Titicaca (SA)
triarticulata	New South Wales lakes (A)
Calamoecia	
clitellata	Salt lakes of Australia
gibbosa	Lakes in South Australia
salina	Salt lakes of Australia
tasmanica	Victoria lakes (A), South-East Queensland (A)
Dactylodiaptomus	
pearsei	Rio Negro (SA)
Diaptomus	
banforanus	Chad (Af)
coronatus	Rio Negro (SA)
doriai	Toba (As)
dorsalis	Peten (CA), Izabal (CA)
melini n.sp.	Rio Negro (SA)
mixtus	Malawi (Af), endemic
negrensis	Rio Negro (SA)
salinus	Qarun (Af), 1936, now disappeared
simplex	Tanganyika (Af)
sp.	Gorewada (As), Amaravathy Res. (As)
Eudiaptomus	
gracilis	Kinneret (ME)
Heliodiaptomus	
kikuchii	Bum Borapet (As), Great Lake of Kampuchea (As)
viduus	Govindgarh (As)
Leptodiaptomus	
sp.	Guatemala lakes (CA)
Notodiaptomus	
amazonicus	Castanho (SA), Redondo (SA), Jacaretinga (SA)
coniferoides	Jacaretinga (SA)
venezolanus	Valencia (SA)
sp.	Rio Negro (SA)
Paradiaptomus	
africanus	Soda lakes (Af), East-African lakes
Phyllodiaptomus	
blanci	Govindgarh (As)
Pseudodiaptomus	
culebiensis	Izabal (CA)

Biological diversity

Table 118 (cont.)

Racodiaptomus	
calatus	Rio Negro (SA)
retroplexus	Rio Negro (SA)
Sinodiaptomus	
volcanoi	Bum-Borapet (As)
Thermodiaptomus	
galebi	Albert Nile (Af), Ambadi (Af), Gebel Aulia Dam (Af), Tana (Af), Nasser (Af), Edku (Af), Chad (Af)
Tropodiaptomus	
gigantoviger	Lanao (As)
incognitus	Chad (Af)
kraepelini	Chilwa (Af)
vicinus	Lanao (As)
sp.	Nasser (Af), Cabora Bassa (Af)
Acanthocyclops	
languidoides	Kinneret (ME)
Afrocyclops	
gibsoni	Tanganyika (Af)
Apocyclops	
dengizicus	Lakes of central South Australia (A)
Cryptocyclops	
hinjanticus	Tumba (Af)
tanganicae	Tanganyika (Af)
Cyclops	
hyalinus	Edward (Af), Kainji (Af), Bali (As)
ladakensis	Nilnag (As)
leuckartii	Edward (Af), Kainji (Af)
varicans rubellus	Peten (CA)
vicinus	Nilnag (As)
sp.	Bum Borapet (As), Great Lake of Kampuchea (As)
Ectocyclops	
fragilis	Peten (CA)
phaleratus	Central American lakes
rubescens	Tumba (Af), Tanganyika (Af)
sydniensis	Victoria lakes (A)
Eucyclops	
bondi	Central American lakes
caparti sp.nov.	Tanganyika (Af)
fragilis	Tumba (Af)
festivus	Central American lakes
gibsoni	Dariba Lake (Af)
nichollsi	South-East Queensland (A)
paucidenticulatus	
sp.nov.	Tanganyika (Af)
prasinus	Danau di Atas (As)
serrulatus	Kinneret (ME)
speratus	Central American lakes
sp.	Victorian lakes (A)
Halicyclops	
sp. nov	Lakes of Australia
sinensis	Yangtse-kiang River (A)
sp.	Central American lakes

478 Biological diversity

Table 118 (*cont.*)

Macrocyclops	
albidus	Tumba (Af), (CA)
ater	Central American lakes
fuscus	Central American lakes
principalis	L. Negra (SA)
Megacyclops	
viridis	Bale mountain lakes (Af), Central American lakes
Mesocyclops	
brasilianus	Valencia (SA)
crassus	Valencia (SA)
decipiens	Telaga Ngebel (As), Tjigombong (As)
edax	Izabal (CA)
ellipticus	Central American lakes
hyalinus	Dakataua (As), Singkarak (As), Indonesian lakes (As), Peten (CA)
inversus	Peten (CA)
lepotus	Andean lakes (SA)
leuckarti	Castanho (SA), George (Af), Albert (Af), Albert Nile (Af), Gebel Aulia Dam (Af), Tana (Af), Nasser (Af), Naivasha (Af), Chilwa (Af), Kivu (Af), Chad (Af), Kariba (Af), Nilnag (As), Great Lake (As), New South Wales (A), South Australia, Kinneret (ME), Toba (As), Lanao (As), Indonesian Lakes (As), Peten (CA)
longisetus	Central American lakes
meridianus	Central American lakes
nicaraguensis	Central American lakes
sp.	Edku (Af), Gorewada (As), Cabora Bassa (Af)
Microcyclops	
alius	Central American lakes
arnaudii	Saline lakes of Australia
ceibaensis	Central American lakes
dengizicus	Central American lakes
diversus	Central American lakes
hartmanni	Central American lakes
leptus	Titicaca (SA)
minutus	Kinneret (ME)
panamensis	Central American lakes
varicans	Tumba (Af), Tanganyika (Af), Central American lakes
Oithona	
amazonica	Cristalino (SA), Rio Negro (SA), Castanho (SA), Redondo (SA), Jacaretinga (SA)
amazonica continentalis subsp.nov.	Rio Negro
neotropica	Mucubaji (SA)
Orthocyclops	
modestus	Central American lakes
sp.	Gorewada (As), Bum Borapet (As), Great Lake (As)
Paracyclops	
affinis	Tumba (Af)
fimbriatus	Nilnag (As)
Thermocyclops	
crassus	Toba (As), Lanao (As), Great Lake of Kampuchea (As), Indonesian lakes (As), Nilnag (As), Mainit (As)

Table 118 (cont.)

decineus	New South Wales
decipiens	Central American lakes
dybowskii	Kinneret (ME)
hyalinus	Valencia (SA), George (Af)
incisus circusi	Chad (Af)
inversus	Izabal (CA)
minutus	Rio Negro (SA), Castanho (SA), Jacaretinga (SA), Taruma-Mirim (SA)
neglectus	Albert (Af), Gebel Aulia Dam (Af), Nasser (Af), Chad (Af)
schurmanii	Naivasha (Af)
tenuis	Central American lakes
sp.	Khan Payao (As), Bum Borapet (As), Great Lake of Kampuchea (As)
Tropocyclops	
confinis	Chilwa (Af)
extensus	Central American lakes
parvus	Central American lakes
prasinus	Bali (As), (CA)
prasinus mexicanus	Peten (CA)
sp.	Ambadi (Af)

KEY: A = Australia; Af = Africa; As = Asia; CA = Central America; ME = Middle East; SA = South America

recorded 11 cyclopids in Sri Lanka. All are cosmopolitan, tropical or widely distributed species. Only two species dominate in the limnetic zooplankton: the herbivorous *Thermocyclops crassus* and the carnivorous *Mesocyclops leuckarti* (Table 118). These two species are tropicopolitan and eurytropic and dominate in limnetic conditions. For many years, *M. leuckarti* has been considered as a widespread species; however, in 1981, Kiefer, revising the species of *Mesocyclops*, came to the conclusion that the species *M. leuckarti* sens. restr. occurs only in Europe and the western part of northern Asia. In contrast, the *Mesocyclops* described in Africa, southern and eastern Asia and Australia belong to other species.

In Lake George, a similar situation was reported by Burgis *et al.* (1973): *T. hyalinus* and *M. leuckarti* are dominant. In tropical South America, Brandorff (1977) found only three cyclopids in a lake. A different situation was described by Lindberg (1951) in Lake Tanganyika. Among the 30 cyclopids recorded by him, 19 species are known from other regions and 11 are endemic forms. Five species belong to *Eucyclops*, four to *Microcyclops* and *Cryptocyclops*, one to *Paracyclops* and another one to *Ectocyclops*. The 13 cyclopids recorded by Lindberg in Lake Kivu were all known from other African regions. In African high mountain lakes, the planktonic cyclopid genus *Thermocyclops* appears only downward of the Ericaceae belt. In South America, conversely, a typical cyclopid high mountain plankton is developed (*Metacyclops mendocinus – leptotus*) (Löffler, 1964). In Australian athalassic saline waters, a new species of *Halicyclops* and *Mesocyclops arnaudi* were recorded; the latter appears to be endemic to the Australasian region (Bayly & Williams, 1966a). *Mesocyclops leuckarti* is the most common species there. The distribution of two other species, *Thermocyclops decipiens* and *Microcyclops sydneyensis*, is described by Timms (1970).

Lindberg (1954a), studying the cyclopids from different small water bodies of Mexico, has recorded 19 species. Among them, two species were described only in South and Central America (*Thermocyclops inversus* and *T. tenuis*), and five species were found only in Central America: *Oithona alvarezi* sp. nov., *Eucyclops festivus*, *Tropocyclops prasinus aztequei*, *T. prasinus mexicanus* and *Apocyclops panamensis*. The first four species are known only from Mexico. Lindberg (1954b), studying the cyclopids from South America, concluded that the genus *Mesocyclops* is represented there by more species than in Africa; conversely, *Thermocyclops* comprises at least 12 endemic species in Africa and only two species in South America. The genus *Afrocyclops*, which is widely distributed in Africa and South-East Asia, is absent from South America.

Fish diversity in the Warm Belt

The great richness and diversity of flora and fauna is a distinctive feature of the Warm Belt. Dobzhansky (1950) contrasted the temperate forest, dominated by a few species, with the tropical forest, containing a remarkably high number of species without any being clearly dominant. Near Belem (Amazonia) Dobzhansky reported 60 species among 564 trees. In the investigated plot, 22 species were represented by only one individual, whereas 241 specimens of the commonest species were counted. The progressive increase in diversity from the poles to the Equator is well illustrated by the number of breeding species of birds at different latitudes reported by Dobzhansky: from 56 species in Greenland, this number increases gradually with latitude and culminates in 1395 species in Colombia. Similarly, the number of known species of snakes goes from 22 in Canada to 210 in Brazil. Fishes are no exception: the equatorial part of Brazil has 1383 species (so far determined) belonging to 46 families, Thailand 546 species belonging to 49 families, Central America 456 species, and Congo more than 408 species belonging to 24 families (Lowe-McConnell, 1969). In comparison, only 92 species are known south of the Zambezi, and the whole of Europe has no more than 192 fish species

Biological diversity 481

belonging to 25 families. Lakes Tanganyika, Victoria and Malawi have, respectively, 184, 208 and 234 fish species, whereas the Great Lakes of North America together have only 172 species. This well-established zonation of species diversity has produced much reflection and controversy. It is, however, generally admitted that a characteristic which affects so many groups of plants and animals must be caused by one or several large-scale phenomena. In the present state of knowledge, the main factors of fish diversity include: (1) the large-scale continental distribution of fish which allowed or prevented the access of a certain fish group to a given region; (2) the absence of Pleistocene glaciations which kept the tertiary fauna from destruction; (3) the capture of adjacent basins and their associated fish fauna by large rivers; and (4) speciation.

Geographical distribution
Large-scale geotectonic events, resulting from continental drift, have determined the general pattern of the world's freshwater-fish distribution. This has been described in detail by Lowe-McConnell (1975) and will be only briefly summarized here.

The various components of Gondwanaland remained in contact until the early Mesozoic. They have in common an old stock of fish families belonging to the Sarcopterygii and Brachopterygii, groups of bony fishes which appeared in the Devonian, and other families belonging to the non-ostariophysan teleosts which appeared in the Triassic and Jurassic.

Very early, probably in the early Mesozoic, Australia separated from Gondwanaland; it follows that the only true freshwater fishes of Australia descended from the old Gondwanalandian stock and are limited to the lungfish *Neoceratodus* (Sarcopterygii) and the osteoglossid *Scleropages leichardti*. In the Lower Jurassic, India and Madagascar left the main Gondwanaland mass and their presently basic freshwater fish fauna consists of dipnoi (Sarcopterygii) and osteoglossids. Certain species of the Gondwanalandian stock became extinct and have been found as fossils in Upper Triassic formations (*Ceratodus*). After India came in contact with Asia, many fish species from South China and Malaya invaded the peninsula. Similarly, many secondary freshwater groups, especially Cichlidae and Mugilidae, migrated from Africa to Madagascar, but the characoids which appeared in Gondwanaland after the Madagascar separation are totally absent from the island.

In the Upper Jurassic to Lower Cretaceous, the Ostariophysii made their appearance. The cyprinids which originated in Asia did not reach South America, and the characoids, so abundant in South America, did not reach

Asia. South America, by drifting away from Africa during the Cretaceous, interrupted the faunal exchange with the Old World while Africa and Asia remained in contact.

The present freshwater fish fauna of South America is incredibly limited in number of groups; for example, characids (and derived gymnotids) and the silurids form 85% of the Amazon fish fauna. The non-ostariophysan primary freshwater fishes are one genus of Lepidosirenidae, two genera of Osteoglossidae and two genera of Nandidae. All these groups are very ancient; some appeared in the Palaeozoic (Lepidosirenidae and Osteoglossidae), others in the early Mesozoic (characoids and siluroids). They were in South America before this continent drifted from Gondwanaland. The presently extremely diversified fauna of Amazonia stems from very few and ancient ancestors and is all the more interesting.

Characids and silurids represent only 48% of the freshwater fish fauna of the Zaire basin, which also includes 16% of cyprinids and a large variety of archaic forms (Protopteridae, Polypteridae, Mormyridae). The basic stock of fish fauna of Africa is then more diversified than that of South America; it includes a larger variety of archaic families but also a 'recently arrived' group, the cyprinids, which probably entered Africa in Plio–Pleistocene times. In contrast with Amazonia, the African continent has a relatively high altitude which limits considerably the penetration of marine fishes. This may have played an important role in the conservation of old species. However, it is among the secondary freshwater fishes (Cichlidae) of the Great Lakes that the most spectacular and recent developments occurred.

Tropical Asia is dominated by cyprinids and silurids representing 57% of the fish fauna in Thailand, for example. The diversity of fauna, very high west of the 'Wallace's line' (formed by the straits between Borneo and Celebes), is nearly non-existent east of this line. However, a few Bornean primary freshwater fishes have reached Mindanao and permitted the appearance of 20 species of the genus *Barbus* in Lake Lanao.

The preservation of old species in the Warm Belt during the Pleistocene

Although the forests of Northern Europe, Siberia and North America are somehow related in the minds of many people to notions of great antiquity, it is clear that these ecosystems appeared and developed after the Pleistocene glaciations. In comparison, the equatorial rain forest is assumed to have had a continuous existence from the Cretaceous until the present (Richards, 1966, 1973). Whereas the species of cold and temperate regions were partly destroyed and partly migrated towards low

Biological diversity 483

latitudes during the Quaternary glaciations, the old species of the Warm Belt were much better preserved and, in post-glacial times, represented a substantial portion of the earth's genetic reservoir. These well-known facts led to the concept that the extreme diversity of the tropics resulted from a 'maturing process' which 'must have taken place in the absence of any destructive natural catastrophes' (Schwabe, 1968). We shall see that this view has not been confirmed by recent work.

Capture of adjacent basins by large rivers

Roberts (1973) emphasizes that the Zaire and Amazon basins have incorporated adjacent basins by hydrological captures. The most famous example is the capture of the Qualuba River by the Zaire River. Similarly, the Tocantins is suspected to have been captured by the Amazon. These enlargements of the main hydrological basins of the world were accompanied by non-negligible gains of fish.

Speciation

Formation of new species is undoubtedly the main cause of diversity in the tropics. Although much has been written on the fish speciation of the African Great Lakes concerning the environmental conditions which may have favoured such a process (Brooks, 1950; Worthington, 1954; Greenwood, 1964, 1974; Lowe-McConnell, 1969, 1975; Fryer & Iles, 1972), the genetic aspect of fish speciation has been little discussed in the past due to lack of relevant data. What follows is only a modest attempt to raise the questions in a different perspective.

Two sexually reproducing animals are defined as belonging to the same species when they are able to produce fertile offspring. This implies the 'compatibility' of their genomes and a similitude of sexual behaviour which is necessary to a successful mating. Since the genetic material is sensitive to various environmental influences such as radiation and certain chemical products, modifications of the genome (mutations) do occur spontaneously and randomly. They form the basis of evolution. Mutations can modify one single gene of a chromosome (gene mutation) or affect the architecture of a whole chromosome (chromosome mutation or rearrangement). This latter process may cause the loss of a chromosome segment, its duplication, its inversion or its translocation (transfer to a non-homologous chromosome). Mutations may deeply modify the vital functions of the mutants and are frequently disadvantageous. Certain modifications, preventing the functional performance of the gene product, are lethal. In other cases, the mutation only alters the performance of the mutant which then can live but has little chance to leave descendants. This is the free game of Darwinian Natural Selection. It is clear that, with the exception of lethal mutations, there are no 'good' or 'bad' mutations; they are good or bad relative to the environment in which they appear. The blood of most fish is unable to exchange gases at low pH. A mutation which would modify the enzymatic system of the fish in such a way that the enzyme would optimally function at pH 4 would probably be lethal or deleterious to fishes living in alkaline water. Conversely, this would confer to fishes living in the acid Rio Negro a tremendous ecological advantage.

The frequency of mutations, their nature and their effect on the mutant depend very much on the philogenetic position and consequent structure of the genome of the considered group of animals. The genome of fish is characterized by two outstanding features. One is the great diversity of the genome size. Hinegardner & Rosen (1972) studied 275 species of teleostean fishes and reported chromosome diploid numbers varying from 22 to 120. In those fishes, DNA haploid content ranged from 0.4 to 4.4 pg and varied with the number of chromosomes. These and other observations suggest that in this group of fishes, modifications of the genome occurred through various modes of gene duplication. Certain species increased their genome by duplication of chromosomal segments. This, however, did not favour the evolution of these species since the redundancy of a gene without a parallel redundancy of all functionally related genes does not generally lead to the acquisition of new functional loci but simply to an increase in genetically inactive DNA. This is the case of the South American lungfish which, with its DNA value 35 times that of mammals, has not changed much from the Devonian. A second feature typical of the fish genome is the low differentiation of the sex chromosomes. This is possibly what allowed fish to practice polyploidy on a large scale. In animals with more differentiated sex chromosomes, polyploidy leads to intersexuality and sterility. The fish species which increase their genome by polyploidy could acquire new functional gene loci, since duplication of all the gene loci creates no problems with respect to functional gene loci (Ohno, 1970a,b).

Hinegardner & Rosen (1972) also found that more specialized fishes have less DNA than generalized fishes. This is true not only within the teleosts in general but also within each group. It is as if specialization was accompanied by a certain elimination of redundant DNA. Since specialization consists of developing one or several functions at the expense of other functions which become useless to the animal, there is a general trend of elimination from the genome of the extra DNA corresponding to the useless functions. As no fish has been found to have less than 0.4 pg haploid

DNA, it is probable that this value corresponds to the minimum amount of DNA required to ensure organized life in fish.

These findings make it theoretically possible to predict which groups of fishes are candidates for speciation. For example, Hinegardner & Rosen (1972), considering that the *Corydoras* catfishes have only recently acquired their abundant DNA 4.4 pg, think that this group 'would be ideally suited as a genetic storehouse that could provide the raw material for new groups of fishes: any nucleotide changes that were advantageous would be selected for'.

These lines of reasoning are supported by recent work done by Kornfield, Ritte, Richler & Wahrman, (1979) on six species of cichlids of Lake Kinneret (Israel). The species concerned were: *Haplochromis flavii-josephi, Tristramella sacra, T. simonis, Tilapia zillii, Sarotherodon aureus* and *S. galilaeus*. The diploid number of chromosomes was 44 in all species. Variation in chromosomal morphology included variation in length of chromosome no. 1, relative frequency of metacentrics and distribution of constitutive heterochromatin. The two *Tristramella* species had nearly similar karyotypes; the two *Sarotherodon* species had indistinguishable karyotypes. Constitutive heterochromatin was present in the centromeric region of all chromosomes in both *Sarotherodon* species but was lacking in 10 to 12 chromosomes of the related *Tilapia zillii*.

The measurement of electrophoretic similarity at 21 isozyme loci showed that interspecific similarities were very high within the *Tristramella* species and within the *Sarotherodon* species, indicating a recent divergence. The similarities between *Haplochromis flavii-josephi* and any other species were very small. The endemic genus *Tristramella* was equidistant between *Tilapia* and *Sarotherodon*. All these findings are in good agreement with the taxonomic classification.

The amount of DNA per haploid nucleus varied in the different species. A nearly identical value was found for the two species of *Tristramella* but *Sarotherodon aureus* had 15% more DNA than *S. galilaeus* which had the lowest concentration of DNA of all six species, closely followed by *H. flavii-josephi*. The variations of DNA in the different species were significant but were not related to the genetic distance between species. It is then clear in this particular case that speciation took place in the absence of gross chromosomal modifications. Differences in constitutive heterochromatin were detectable only in species separated by wide genetic distances. The hypothesis of King & Wilson (1975) that speciation results from 'a relatively small number of genetic changes in systems controlling the expression of genes' rather than from gross changes of structural genes may well apply to the cichlids. King & Wilson estimate that rearrangement of genes on chromosomes, modifying the neighbourhood of the concerned genes, may change the timing of gene expression which considerably affects the development of the animal, its phenotypic appearance and behavioural pattern.

Recent work by Kornfield (1978) on seven species of the 'Mbuna flock' of Lake Malawi indicates that, although the species have achieved reproductive isolation, they have very high values of allozymic similarity. Here, as in the case of the Kinneret cichlids, the incipient stages of speciation do not concern structural loci; new regulatory mechanisms exerted on the gene expression modify behavioural patterns in such a way that species become rapidly sexually incompatible. Then and only then, deeper genetic modifications start accumulating. Similar results are reported by Kornfield & Koehn (1975) in their study of speciation of cichlids from Texas and Mexico.

Until we have enough data on the cichlids of the Great African lakes, we can only hypothesize that speciation in these fish did not result from a sudden fundamental modification of the genome but from minor genetic changes which sexually isolated the mutants, and that more fundamental genetic changes occurred subsequently as a result of isolation. The combination of genetic differences, diversity of ecological niches, more permanent food supply than in temperate zones, constancy of temperature and Darwinian selection lead inevitably to narrow speciation. For example, a cichlid in a lake with a permanent temperature does not require a wide-range thermal regulating system. These 'steno-species' lose DNA and become irremediably dependent on the environment to which they have adapted. *Sarotherodon galilaeus* of Lake Kinneret, a specialized *Peridinium*-eater, has less DNA than the generalized *S. aureus* (Kornfield *et al.*, 1979). In Lake Kinneret, *S. galilaeus* grows rapidly during the *Peridinium* period. In ponds, it grows slowly, much more so than *S. aureus* (Ben Tuvia, 1981).

Nikolsky & Vasilev (1973) have shown that the number of fish chromosomes tends to decrease progressively with latitude from an average number of $2n = 72$ in the Arctic to $2n = 46.8$ in the tropics. These authors relate the diminution of the genome size in tropical species to trophic specialization in the Warm Belt.

The fixation of a genetic modification and its perpetuation require 'the reproductive isolation' of the new small population from the parent stock. Some argue that gene mutations cannot erect a strong enough reproductive barrier and that a new species would be irremediably 'diluted' by

Biological diversity 487

hybridization. Others claim that chromosomal rearrangements can be rapidly fixed through the appearance of homozygotes having equilibrated gametes. In any case, inbreeding seems a necessary step in the fixation of mutations. It follows that to a certain extent, social organization generally governed by feeding and reproductive habits considerably affects the probability of fixation of a mutation. Lower speciation rates are expected in lake sardines living in large schools, feeding on plankton and even breeding in pelagic areas, than in inshore species able to split into small communities; in the first case any genetic modification has all chances to remain in the heterozygous stage since the large genetic pool does not favour the formation of homozygotes. Allopatric speciation, involving the geographical isolation of a small group, is therefore the most common process through which mutations are fixed, although Bush (1975) describes examples of parapatric and sympatric speciation.

In many cases, natural events cause the permanent or temporary isolation of a given population. In Lake Nabagaboo, isolated 5000 years ago from Lake Victoria as a result of hydrogeological changes, the speciation of cichlids was a fairly rapid process. Large-scale climatic modifications were perhaps the most important factor of speciation in the Warm Belt during the Pleistocene period.

Recent work concerning speciation patterns of plants and animals in Africa, Australia and South America, and geological and palaeobotanical features of these areas (Hall & Moreau, 1970; Vuilleumier, 1971; Vanzolini, 1973; Tricart, 1974, 1975; Haffer, 1978; Simpson & Haffer, 1978) demonstrated that the notion of tropical stability had to be replaced by the idea of Pleistocene climatic changes in the Warm Belt. Geomorphological and palynological studies clearly indicate that areas presently covered with rain forest were, in the upper Quaternary, under semi-arid climatic conditions. During these periods, the wet forest was replaced by savanna vegetation. Moreover, in the Pliocene, vast lakes covered the Amazon and Congo valleys and separated the forest areas extending north and south of the water expanses. Climatological studies in South America have shown that areas now receiving high rainfall were also rainfall centres during all the Pleistocene, because topographic conditions and patterns of wind direction remained unchanged throughout this period. The permanently wet areas include the western coastal lowlands of Colombia and Ecuador, the eastern slopes of the Andes in Colombia and Ecuador, the coast of north-eastern Brazil and Guyana and various spots in central Brazil. It is highly probable that, during the dry episodes of the Quaternary, the rain forest persisted only in these areas which then became wet islands surrounded by dry savannas. As far as the trees and their associate fauna were concerned, these wet islands played the role of refuges or sanctuaries during dry phases. In these sanctuaries, the various species were preserved and/or modified. New humid episodes allowed the re-expansion of the preserved species in the area intermediate between two sanctuaries. If the speciation did not go too far during the process of isolation, the areas of redispersion were characterized by a high number of hybrids. The actual studies of distribution patterns of species and subspecies of numerous vertebrate populations in Amazonia indicate the existence of areas where groups of genetic characters are always found together with little variation and areas where these associations of characters are completely disorganized. Vanzolini & Williams (1970), who carried out statistical analysis of such character complexes in anoline lizards, concluded that the areas of little variation in population characters corresponded to refuges or core areas and the areas with large variations of these characters corresponded to areas of new colonization during humid episodes.

It is interesting to note that from the maps of distribution patterns of various groups of vertebrates, we can infer the presence of permanent refuges, the location of which coincides with the rainfall centres determined on the basis of climatological considerations. Simpson & Haffer (1978) conclude that a causal relation links both observations.

The overall picture which emerges from these multidisciplinary studies of South America is that the species diversity of tropical areas is a recent phenomenon which did not take place because of long-term evolution but rather because of relatively rapid climatic changes during the Quaternary.

The previous conclusions have been based on the study of terrestrial animals. How did the fish population react to the Pleistocene climatic events? Before answering this question, we have to remember that in South America, for example, the fish faunas of the upper and middle courses of the Amazon and Orinoco tributaries and of the Guianas show greater affinity among themselves than with the fauna of Amazonia itself. It is as if the upper watershed of the Amazon had preserved a primitive population, whereas the fish fauna of Central Amazonia had 'diversified only afterwards' (Gery, 1969). It is likely that the huge lake which covered Central Amazonia during the Pliocene offered a wide variety of biotopes which enhanced speciation and operated a severe selection which favoured the lacustrine species at the expense of riverine species. The drainage of the lake and the shaping of the várzea in the Pleistocene could only have accentuated these trends. Here again we find areas where faunal variation has been relatively slow and areas where it has been considerably faster. The former areas correspond essentially to the old stable cratonic areas and the latter to areas much more affected by geological and hydrological disturbances. Superimposed on the effect of the geological history, the climatic fluctuations of the Pleistocene have several times modified the availability of water; as a result isolated aquatic environments where old species were preserved and/or evolved into new species alternated with large lakes where genetic flow was restored. In any case, it seems that the present Amazonia fish species diversity is the result of Plio–Pleistocene geological and climatic events and is not due to a lengthy period of stability. Dynamic environmental processes of moderate magnitude lead to maximal diversification of species, since complete environmental stability allows little variation or may even lead to extinction, and extreme instability causes rapid destruction of the genetic reservoir (e.g. glaciations).

The data that we have reported concerning the variation of water levels in African lakes may indicate that Africa has been more affected by the Quaternary variations in climate than has South America. This was also assumed by Jackson (1963) to explain the discontinuous distribution of various fish species in Africa. Under these circumstances, in arid periods, fish species could be preserved only in permanent water bodies such as the Zaire basin and the Great Lakes of East Africa, which subsequently played the role of dispersion centres. The lack of large river systems and deep lakes north of the Zaire River and south of the Zambezi River is the reason for the impoverished fish fauna of these regions.

We may conclude that the present diversity of fish species in the Warm Belt originates basically from the absence of geo-climatological destructive events over a very long period of time, which allowed the preservation of genetic stocks. In the Pleistocene, relatively mild, non-destructive climatological fluctuations triggered increased speciation by successive contractions and expansions of the water bodies. It is probable that the phases of recolonization of new water bodies during humid periods were accompanied by a decreased selection pressure on the genome modifications. The survival of more mutants during colonization periods may have greatly contributed to diversity. Simultaneously, the low seasonality of the Warm Belt has made useless numerous mechanisms of thermal regulation or of resistance to food shortage which are so crucial for the survival of temperate species. The relative abundance of food allowed large communities to develop which enhanced competition for space. Fish were then selected for very narrow ranges of environmental conditions and gained in specialization what they lost in adaptive flexibility. The present man-made modifications of the tropical environment may well go beyond the adaptive capability of many tropical specialized fishes.

13

Food webs: case studies

Quantitative information concerning food webs of warm water bodies is still scanty and does not allow generalizations. Therefore in the following, we shall only give a few examples well documented by recent work.

A detrital, aquatico-terrestrial food web: the várzea

The food web of the lakes cannot be separated from the trophic history of their environment, and we have to consider a composite system including the flood plain with its stagnant water, the lakes with their considerable variations of water level, and the river with its running water (Junk, 1980). The flood of the river generates a chain of events and modifications in the different parts of the system as described in Chapter 4, pp. 57–83. The main characteristics of the food chain of this system concern the important role of the macrophytes, the migration and feeding habits of the fishes and the nutrient flux through the system.

Macrophytes

The extreme variations of water level favoured the floating plants: submerged species such as *Utricularia foliosa* and *Ceratophyllum demersum*, or emergent plants such as *Eichhornia crassipes*, *Pistia stratiotes*, *Ceratopteris pteridoides*, *Salvinia auriculata*, *S. minima*, *Limnobium stoloniferum*, *Neptunia oleracea* and *Azolla* sp. These plants are characterized by very high growth rates. Their ability to reproduce vegetatively allows them to colonize rapidly the newly available water surface at rising water. At low water, up to 90% of this vegetation dies and decays and the drying areas are colonized by swamp vegetation such as *Cyperus ferax*, *C. odoratus*, *Paspalum fasciculatum* and tree species able to resist inundation (*Salix*

Food webs: case studies   491

humboldtiana, *Eugenia inundata* and *Symmeria paniculata*). Such plants produce numerous seeds eaten and propagated by frugivorous fish.

The rapidly expanding floating plants skim the nutrients from the water and accumulate them in their biomass. These nutrients are subsequently released at receding water. The overall result is nutrient retention during the flood and nutrient release after the flood.

Fish migration and food habits

The fish have elaborated various strategies which make them adapted to their changing environment. Various species which live at high water in the inundated forest migrate towards the river at low water and then avoid being caught in the many isolated pools which appear at receding water. This migration means a drastic modification in the fish's food source and exposes them to a higher pressure of predators. In particular, the herbivorous and frugivorous species which enjoyed an abundant food in the flooded areas have to live on their fat reserve when they reach the river.

The spawning migrations are also triggered by changes in water level. In Rio Madeira, the characins migrate twice a year (Gouldin, 1981). The first one starts with the flood when the fishes descend the tributaries, join the Rio Madeira and spawn near the confluence of this river with the Amazon. Then the fishes migrate upstream again. During the second migration at low water, the fishes descend the tributaries but move upstream in the Rio Madeira. At the confluence of Rio Madeira with the Amazon, the flood plain is very large and, in the numerous lakes and lagoons rich in plankton and macrophytes, the new fingerlings find the food environment required for their rapid growth. In Rio Madeira, the silurids migrate upstream at low water and feed mostly on migrating characids. Since the adults of these latter fish feed most of the year on fruit, seeds and insects of the inundated forest, this illustrates the direct transfer of matter from the flooded forest to the river's predatory fish. Gouldin estimates that the food chain of about '75% of the total commercial fish of the Rio Madeira begins in the flooded forests'.

The nutrient flux

The flooding of the várzea provides nutrients to the lakes when plankton production reaches its annual peak. The growth of aquatic macrophytes is also considerably enhanced, and Junk (1980) reports an increase in biomass of 3000% for *Salvinia auriculata*. In certain species, the nutrients are taken up directly from the water, whereas many others find their nutritive substances in bottom sediments. A certain fraction of these plants is transported from the várzea to the river where it is consumed or decomposed. As these plants are rich in phosphorus (1 ha of aquatic plants in Middle Amazonia contains 63 times more phosphorus than 1 ha of lake water), this represents an efficient nutrient transfer from the flood-plain sediments to the river. The remaining part dries in the lake at low water, and the released nutrients are used by the developing terrestrial vegetation. Terrestrial plants are also extremely productive, and approximately half of the produced material is decomposed and returns to the aquatic phase. These considerable nutrient fluxes are made possible by the very high rates of decomposition. The herbaceous plants lose 50% dry mass after two weeks in the water; a varied fauna of decomposers, including insects, fungi and bacteria, prevents the accumulation of organic matter on the bottom and allows rapid recirculation and reutilization of the nutrients.

It is clear that separation between terrestrial and aquatic vegetation or distinction of lake fishes, flood-plain fishes and river fishes has no justification when trophic interactions are considered. In consequence, Junk (1980) presented a global food-web scheme including all the components of the várzea. The Solimoes–Amazon carries into the system dissolved and particulate nutrients, mainly inorganic, which are channelled towards production of aquatic macrophytes and phytoplankton as well as terrestrial vegetation by enrichment of the flood-plain sediments. During the flood, decomposition of past-season terrestrial plants enhances the plankton and macrophyte production which is partly consumed by numerous herbivorous fish (*Leporinus frederici, L. maculatus, Schizodon fasciatus, Colossoma bideus, Metynnis hypsauchen, Cichlosoma bimaculatum* and *C. festivum*). The remaining fraction of the produced vegetation is decomposed, partly exported to the river and partly sedimented. At low water, a new cycle begins where the aquatic plants are replaced by terrestrial vegetation and the herbivorous fishes by terrestrial herbivores. The final products exported from the várzea by the Amazon are rich in organic material.

The main characteristics of the várzea food web are as follows:

1. The plankton productivity plays a secondary role, however essential to young fishes.
2. The macrophytes of the lakes, flood plain and the trees of the inundated forest are the main carbon contributors of the system:
    a. they provide food to the adult fishes of the first trophic level in the form of leaves, epiphytic algae, fruit, seeds and associated insects and other invertebrates;

Food webs: case studies 493

b. a large fraction is decomposed and the released nutrients
  i) are immediately recycled in the várzea;
  ii) reach the Amazon. There they are utilized by the river seston and re-input into the food web or exported to the ocean.

According to Richey et al. (1980), the Amazon receives its maximal carbon input at rising water when the decomposition products of the organic matter oxidized at low water reach the river. An amount of 10^{14} g C is yearly conveyed by the Amazon to the ocean, which means that the river draining the vastest photosynthetic compound of the globe reaches the sea with an average concentration of only 16 mg l^{-1} organic carbon. It is interesting to note that the Jordan River, the main inlet of Lake Kinneret, Israel, reaches the lake with an organic carbon concentration of 15 mg l^{-1} after draining a rocky watershed without much vegetation. However, Richey (1981) found a $-7‰$ downriver decrease in the isotopic composition of the Amazon HCO_3^- which he attributed to within-river respiration. This suggests that considerable mineralization occurs in the river. As a result, the estimation of the total transport of organic carbon by the river cannot be based solely on the export of particulate and dissolved organic carbon but should also include part of the dissolved inorganic carbon of the river. These equilibria might be rapidly modified if, as reported by Herrera, Jordan, Medina & Klinge (1981), 200 000 km^2 of rain forest are annually destroyed and turned into agricultural land.

A planktonic food web based on grazing: Lake Tanganyika
  As far as trophic structure is concerned, Lake Tanganyika represents a kind of antithesis of the várzea. The living community of Lake Tanganyika is purely aquatic and the nutrient inputs and outputs from inflowing and outflowing waters are minor.

The living community of Tanganyika is spatially limited to the upper 200 m above the permanent thermocline, which corresponds to a volume of ~ 4000 km^3. The littoral zone has a surface area of ~ 1000 km^2.

The littoral areas have a more diversified food web than the pelagic zone. On rocky coasts, some 70 species of cichlids and the cyprinodont *Lamprichthys tanganikae* feed on aufwuchs, aquatic insects, crabs and shrimps. The sandy beaches are inhabited by detritivorous fishes and mollusc-eaters such as *Tylochromis polypepis*. In the deep benthic zone, the habitat is more specialized. Cichlids, all endemic, predominate, with *Trematocara* as the most abundant genus. They are prey to another cichlid, *Hemibates stenosoma*, which is itself eaten by *Lates mariae* and *L.*

angustifrons. The *Chrynsichthys* sp. are all benthic fishes feeding on benthic invertebrates and dead fish.

In the pelagic zone, the main food chain consists of the phytoplankton, a sometimes large population of the ciliate *Strombidium* associated with an algal symbiont, other species of ciliates, an abundant population of zooplankton dominated by *Diaptomus simplex*, and a large population of sardines (*Limnothrissa* and *Stolothrissa*). Both sardines feed on phytoplankton and zooplankton in the juvenile stages. The adult *Stolothrissa* is pelagic and feeds mainly on zooplankton, whereas the adult *Limnothrissa* eats zooplankton but also insect larvae and prawns. The sardines are small fishes (7 cm at maturity) and short-lived. They are important food items to *Lates mariae*, *L. microlepis* and *L. angustifrons* and to *Luciolates stappersii*.

Considering the respective sizes of the littoral area and the pelagic zone as well as the abundance of sardines, it seems that the pelagic food web is the main carbon pathway in the lake. Moreover, the limited extension of macrophytes and the reduced importance of allochthonous carbon seem to indicate that phytoplankton production is the main carbon source of the pelagic, littoral and benthic food webs. However, Hecky & Fee (1981) estimate the mean annual daily rate of algal productivity to be only 1 g C m^{-2} d^{-1} or even 0.8 g C m^{-2} d^{-1} after correction for cloudiness. This production is accomplished by a relatively small algal biomass (0.59 g C m^{-2} in comparison with 30 g C m^{-2} in Lake George) having an extremely high growth rate (0.9 d^{-1} in comparison with 0.14 d^{-1} for Lake George). This obviously points to a rapid consumption of the produced biomass, probably by zooplankton. This remarkably efficient growth of the phytoplankton does not account, however, for the high fish production of the lake. Hecky & Fee, applying linear regression equations relating fish yield and primary productivity, found that the measured primary production allows a fish yield of 9 to 20 kg ha y^{-1}. These values lie approximately one order of magnitude lower than the actual yield (125 kg ha y^{-1}). Since Hecky & Fee considered their measurements of productivity as net production, the obtained fish-yield values calculated on this basis are maximal. It is then possible to explain the top levels of the food web by such a narrow autotrophic basis. By adding the littoral production and the allochthonous carbon inputs to the plankton production, one can improve the total production by some 20%, which does not solve the problem. Not only does Lake Tanganyika have an unusually high fish yield relative to its algal production, but fish productivity, fish biomass and fish productivity per unit of biomass are also higher than in temperate lakes (Coulter, 1981). Hecky *et al*. (1981) believe that the long geological history of the lake and

Food webs: case studies 495

its high endemicity have created a pelagic community with high rates of trophic transfer. However, the high rates of heterotrophic production due to bacteria oxidizing reduced compounds might be 'the missing link' of the Lake Tanganyika food web. These authors also suggest that the hypolimnic reduced substances, presently oxidized, are the product of a formerly non-productive lake. This 'palaeohypolimnion' is presently injecting an additional source of energy into a relatively oligotrophic epilimnion. It is as if the fish production of Lake Tanganyika is partly developing on a 'fossil' source of energy.

The double food chain of Lake Kinneret

The main carbon inputs of Lake Kinneret originate from phytoplankton productivity. Two types of contributing algae should be distinguished: the nanoplanktonic algae giving a peak in winter after the turnover and reappearing, with different species, in summer; and the dinoflagellate *Peridinium cinctum* developing from February to June. The zooplankton biomass is stable and ranges from 25 to 50 g wm m^{-2}. From April to July, it is dominated by the herbivorous *Ceriodaphnia reticulata*, whereas in other months the predatory *Mesocyclops leuckarti* is most common. The presence of a large population of herbivorous zooplankton simultaneously with the bloom of the inedible large cells of *Peridinium* poses the problem of the food sources of the zooplankton at this period. Gut content examination showed that the small biomass of nanoplanktonic algae (15% of total algal biomass) which always accompanies the *Peridinium* bloom is the main food source of zooplankton and contributes up to 50% of the total plankton productivity in winter–spring. During this period, Lake Kinneret has two superimposed food chains: the active nanoplankton–zooplankton grazing pathway and the slow *Peridinium*–detritus pathway. In addition, approximately 12% of the carbon inputs are then sedimented on the lake bottom. Then, the algal plankton productivity is always larger than the requirements of the herbivorous zooplankton. This double pathway of carbon flow allows large amounts of nutrients to be stored in the *Peridinium* biomass and removed from recycling. This structure confers upon the system a great stability and slowdown of the overall rate of primary productivity (Serruya *et al.*, 1980).

In summer–autumn, the absence of rain and floods and the thermal stratification reduce the nutrient inputs and the losses to sedimentation. The algal biomass is far smaller than in winter–spring but is dominated by nanoplankton species with high turnover times.

In summer, nanoplanktonic algae, herbivorous zooplankton and bacteria are part of a steady-state system, where the algal biomass is controlled by zooplankton grazing and where the bacteria-mediated decomposition of zooplankton excretion permits algal growth. The conditions of 'zero net production' prevailing in summer are unfavourable to the growth of large animals; most of the fish species are then in suboptimal conditions and lose fat. In temperate zones, the reduction of algal production in winter corresponds to the period of minimal respiratory requirements; in subtropical stratified lakes, the summer minimal net production corresponds to the highest respiratory requirements of consumers. Consequently in late autumn, the planktonic algal productivity becomes smaller than the requirements of herbivorous animals. The biomass of algae is then depleted and reaches its annual minimum. Similarly, the concentration of dissolved organic carbon drops to its lowest annual value (2 mg l^{-1}) in December.

The overall carbon balance of Lake Kinneret is positive in winter–spring when the non-grazed *Peridinium* material and zooplankton excretion enrich the pool of organic carbon. This 'reserve' is used in summer to cover the increasing maintenance cost of the community and is nearly exhausted in December just before the turnover. In spite of its conspicuous presence, *Peridinium* provides only 27% of the annual carbon flux to the pool in comparison with 70% for the zooplankton excretion.

In certain years, because of special physico-chemical modifications of the environment, *Peridinium* is repressed and replaced by nano- and netplankton edible to herbivorous zooplankton. In such years, most of the produced carbon is channelled to the grazing pathway, causing an immediate increase of zooplankton fertility and biomass. The abundance of zooplankton determines a higher survival of the juveniles of *Mirogrex terrae-sanctae terrae-sanctae*. This cyprinid, improperly named 'sardine', is a zooplankton-eater during its whole life cycle. Consequently, the peaks of zooplankton were always followed by a rise of the sardine biomass and an increase of the predation pressure on zooplankton. In two instances, the overall result of this chain of events was a nearly complete collapse of the zooplankton population and a resulting summer bloom of nanoplanktonic algae without any addition of nutrients. The management scheme of Lake Kinneret should then include not only watershed control but also food-web control, since the previous observations demonstrate that substantial algal blooms can be produced by disturbing the higher trophic levels without any external addition of nutrients to the system.

At its higher level, the food web of Lake Kinneret is characterized by its abundance of zooplankton-eating fish and its nearly complete absence of

Food webs: case studies 497

piscivorous fish (with the exception of a small population of *Clarias lazera*). At the primary producers level, the parallel pathways of carbon flow and the presence of *Peridinium* seem to be the product of a long selective evolution. The life cycle of *Peridinium* fits precisely with the meteorological and physical cycle of the lake. The *Peridinium* cysts are resuspended from the sediments during the turnover period; the vegetative cells, mixed in the whole lake volume during the mixed period, have a very low division rate only $\sim 5\%$ of the population divides daily). In March, the switch of the winter easterly winds to the summer pattern of westerly winds determines a windless period during which the *Peridinium* cells concentrate at the top layer of the lake and reach a daily division rate of 40% (Pollingher & Serruya, 1976). The resulting bloom lasts until June when the high temperature and radiation lead to the death or encystation of the cells. This summer encystation of *Peridinium* is typical of Lake Kinneret. In the water bodies of the temperate zone, *Peridinium* blooms in summer and the cysts are overwintering forms; in Lake Kinneret the selective adaptation to the warm climate has led to the formation of 'oversummering' cysts. The overwhelming dominance of this dinoflagellate is also a product of a long-term adaptation: in a phosphorus-limited environment, it is a tremendous ecological advantage to store phosphorus from the mud–water interface, the only locus in the lake where this element is abundant. This is successfully achieved by the young vegetative cells emerging from the cysts at the sediment surface during the turnover period. These young cells with their rich internal phosphorus pool can easily develop and dominate any other planktonic alga which is bound to find its phosphorus requirements in the lake water.

The short food chain of a shallow, equatorial lake: Lake George

Lake George is essentially characterized by its shallowness and the constancy of its physical environment (see p. 151). These features cause the daily complete mixing of the lake, an event which has only a seasonal occurrence in other lakes. This daily mixing leads to the release of a nearly constant amount of phosphorus and ammonia from the upper layers of the sediments. In addition, the resulting oxygenation of the water enhances the bacterial mineralization of phytoplankton metabolites and zooplankton excreta.

This daily injection of nutrients allows a permanent and dense bloom of blue-green algae dominated by *Microcystis* spp. Lake George is characterized by its abundance of herbivorous cichlids feeding mostly on *Microcystis*. This lake has therefore a very short food chain leading to high

498   Food webs: case studies

yields. The secondary production (herbivorous fish) has, however, been found to represent only 3% of the net primary production (Burgis & Dunn, 1978), i.e. much less than the generally accepted figure of 10%. The authors relate this low conversion percentage to the scarcity of zooplankton of suitable size as fingerling food and the limited breeding areas. Even with this low conversion factor, the ratio of solar radiation to yield is 0.0014% in Lake George in comparison with only 0.00028% in Loch Leven, a small temperate lake in Scotland. This is in perfect agreement with the results of Ryder (1965), indicating that the yield of north temperate lakes is about one-tenth that of comparable tropical African waters.

Appendix

Recently, efforts have been made by Bäuerle in collaboration with Hollan (Bäuerle and Hollan, personal communication) to apply to Lake Tanganyika a new internal wave model developed by Bäuerle (1983).

His theory of a two-layer model takes into account the variable bathymetry of the basin and the latitudinal variation of the Coriolis force. Both conditions exert a strong influence on the long internal oscillations of the lake. The latter condition results from the geographical site of the lake between 3.35° S and 8.76° S, on which distance the Coriolis parameter increases with the latitude from 5.8% to 15.2% of its maximum value at the poles. In view of the large size of the lake, these values are too high to neglect the Coriolis force in the calculations of internal oscillations. The subdivision of the depth configuration has been based on the bathymetric chart published by Capart (1949). The numerical grid on the lake had been chosen with grid spacing of 0.05 meridian degrees (5.5 km). On the basis of Coulter's data (1968a), the mean stratification was parameterized for the model by the thickness of the epilimnion of 55 m and the density difference of 4.5×10^{-4} g cm^{-3} between the latter one and the deep layer.

According to Coulter (1968a), the period of the dominant internal wave during the season from September through April varies between 33 and 23 days. Shorter oscillations of a period of about 3 days are superimposed on the long one, as temperature records at the southern end of the lake revealed during March 1966.

The solutions which explain Coulter's findings are represented in Figs A and B by the distribution of the interface elevations from the equilibrium level at four selected phases of the wave cycle. The displacements are given in 10% contours of the maximum elevation of a mode encountered on the grid.

Fig. A shows the fundamental mode; its period of 31.3 days falls in the range of periods quoted by Coulter for the long-term oscillation. There is an amphidromic point in the centre of the lake, around which the wave propagates cyclonically (clockwise on the southern hemisphere). The first stage of the cycle is represented in Fig. A (A) and shows the nodal line in the longitudinal direction with upwelling and downwelling to the west and east, respectively.

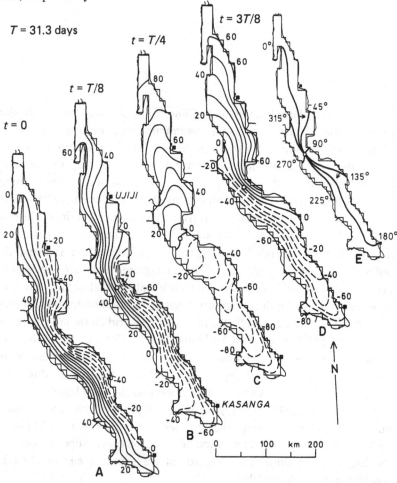

Fig. A First-order internal seiche of Lake Tanganyika. Period $T = 31.3$ days. The displacements of the interface are represented at the phases $t = 0$ (A), $t = T/8$ (B), $t = T/4$ (C) and $t = 3T/8$ (D) in percent of the maximum elevation encountered on the grid. The reduced diagram (E) on the right side displays the phase propagation by cotidal lines for the entire cycle

Appendix

One-eighth of the period later, i.e. after 4 days, Fig. A (B) illustrates an interim stage of positive and negative displacements which are in the process of occupying, respectively, the entire northern and southern halves of the lake. This latter configuration occurs at a quarter period and is depicted in Fig. A (C). At this phase, maximum elevation of the interface takes place at the southern end.

Another eighth of the period later, 12 days after the beginning, the nodal line arrives at a diagonal position from northwest to southeast, as recorded in Fig. A (D). Downward displacements have entered into the western part of the lake, as has upward motion in the eastern part.

After half a wave cycle, the same distribution is established as at the beginning except for an interchange of the direction of displacements. With that difference, the second half of the wave cycle proceeds in the same way as the first half.

The reduced diagram of Fig. A (E) displays the entire cycle of wave propagation at the same intervals by cotidal lines. In the previous four amplitude configurations these lines appear as nodal lines. The average phase velocity amounts to 1.7 km h^{-1}. With increasing order of the modes, more amphidromic points appear and bring about a more complicated structure with smaller horizontal scales. For example, the second mode with a period of 16.3 days has two cyclonic amphidromic systems. The lowest mode that also shows anticyclonic phase propagation is the 15th mode, the period of which is 2.6 days and less than the inertial period in the region of the lake, as expected.

The internal oscillation of about 3 days observed by Coulter is most likely explained by the 18th mode with the period of 2.5 days and high ranges in the southern half of the lake. Since the displacements are less than 10% elsewhere in the lake, only the southern part of the solution is represented in Fig. B. Corresponding to the smaller period, the four stages of the interface elevations during the first half of the cycle follow at intervals of 7.5 h. There are six amphidromic systems in the southern region, two of which are rotating anticlockwise (closed circles). For clear understanding of the more complicated structure, the sense of rotation around the amphidromic points is indicated by arrows and the regions of downward displacement are shown hatched. From the longitudinal position of the main nodal line in the first amplitude configuration shown in Fig. B (A) and the strong elevations on both sides of it, which remain confined more or less to the adjacent shores during the represented subsequent stages, the transverse character of the oscillation is apparent. The structure can be roughly described in terms of interchanging upwellings and downwellings,

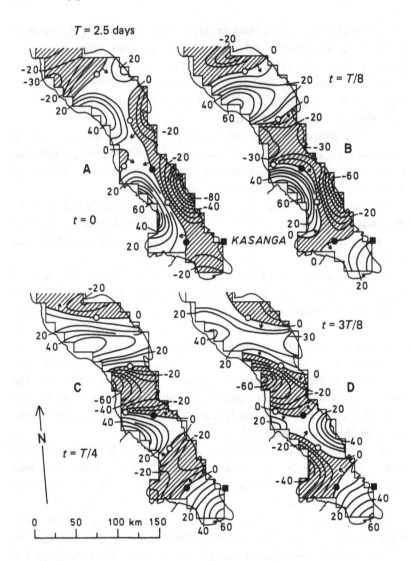

Fig. B Eighteenth-order internal seiche of Lake Tanganyika with period $T = 2.5$ days. The representation gives the interface elevations in the southern part of the lake for $t = 0$ (A), $t = T/8$ (B), $t = T/4$ (C) and $t = 3T/8$ (D), respectively. Units of elevation are the same as in Fig. A. Open circles denote clockwise and closed circles counterclockwise amphidromic systems. Regions of downward displacement are shown by hatching

Appendix 503

each extending about 50 km along the shore and lasting for about a day, though the outlines of those regions and the relative amount of displacements vary according to the phase propagation during that interval.

Both oscillations discussed above are excited most probably by the wind field over the lake. As to the dominant fundamental mode, the overall northward tilt of the thermocline during the season from April through September caused by strong southerly winds may be considered as an initial stage of this mode in September/October. The smaller horizontal scales of the mode with the period of 2.5 days imply a generation by local winds, which blow across the lake. Such wind fields are probably formed by the orography of the surroundings.

It would be most interesting to carry out similar work on other large and stratified lakes of the Warm Belt, especially those subjected to strong wind regimes.

Bibliography

Abdel Karim, A.G. & Saeed, D.M. (1978) Studies on the freshwater algae of the Sudan. 3. Vertical distribution of *Melosira granulata* (Ehren) Ralfs in the White Nile, with reference to certain environmental variables. *Hydrobiologia*, **57**: 73-9

Abdel Malek, S.A. & Ishak, M.M. (1980) Some ecological aspects of Lake Qarun, Fayoum, Egypt. Part II. Production of plankton and benthic organisms. *Hydrobiologia*, **75**: 201-8

Abdin, G. (1948a) Physical and chemical investigations relating to algal growth in the River Nile, Cairo. *Bulletin de l'Institut d'Egypte*, **29**: 19-44

Abdin, G. (1948b) Seasonal distribution of phytoplankton and sessile algae in the River Nile, Cairo. *Bulletin de l'Institut d'Egypte*, **29**: 369-82

Abdin, G. (1948c) The conditions of growth and periodicity of the algal flora of the Aswan Reservoir (Upper Egypt). *Bulletin of the Faculty of Science, Egyptian University*, **27**: 157-75

Abdin, G. (1949) Biological productivity of reservoirs with special reference to Aswan Reservoir (Egypt). *Hydrobiologia*, **1**: 469-75

Abu Gideiri, Y.B. (1967) Fishes of the Blue Nile between Khartoum and Roseires. *Revue de Zoologie et de Botanique Africaine*, **76**: 345-8

Abu Gideiri, Y.B. (1969) The development and distribution of plankton in the northern part of the White Nile. *Hydrobiologia*, **33**: 369-78

Abu Gideiri, Y.B. & Ali, M.T. (1975) A preliminary biological survey of Lake Nubia. *Hydrobiologia*, **46**: 535-41

Adeniji, H.A. (1973) Preliminary investigation into the composition and seasonal variation of the plankton in Kainji Lake, Nigeria. In *Man-made lakes: their problems and environmental effects*, eds W.C. Ackermann, G.F. White & E.B. Worthington, pp.617-19. Geophysical Monograph 17. Washington D.C.: American Geophysical Union

Akiyama, T., Kajunnulo, A.A. & Olsen, S. (1977) Seasonal variations of plankton and physicochemical conditions in Mwanza Gulf, Lake Victoria. *Bulletin of the Freshwater Fisheries Research Laboratory*, **27**: 49-60

Aleem, A.A. & Samaan, A.A. (1969a) Productivity of Lake Mariut, Egypt. Part I. Physical and chemical aspects. *Internationale Revue der gesamten Hydrobiologie*, **54**: 313-55

Aleem, A.A. & Samaan, A.A. (1969b) Productivity of Lake Mariut, Egypt. Part II. Primary production. *Internationale Revue der gesamten Hydrobiologie*, **54**: 491-529

Al-Kholy, A.A. & El-Wakeel, S.K. (1975) Fisheries of the south-eastern Mediterranean

sea along the Egyptian coast. Soviet–Egyptian Expedition 1970–71. *Bulletin of the Institute of Oceanography and Fisheries*, **5**: 279 pp.
Al-Saadi, H.A., Rattan, S.S., Muhsin, T.W. & Hameed, H.A. (1979) Possible relation between phytoplankton numbers and saprolegnioid fungi in Shatt-al-Arab near Basrah, Iraq. *Hydrobiologia*, **63**: 57–62
Al-Sahaf, M. (1975) Chemical composition of the water resources of Iraq. *Vodnie Resursi*, **4**: 173–85
Annandale, N. & Kemp, S. (1915) Introduction to the fauna of the Chilka Lake. *Memoirs of the Indian Museum*, **5**: 1–20
Anonymous (1972) Die Ionenfracht des Rio Negro, Staat Amazonas Brasilien nach Untersuchungen von Dr. Harald Ungemach. *Amazoniana*, **3**: 175–85
Ansari, Z.A. (1974) Macrobenthic production in Vembanad Lake. *Mahasagar*, **7**: 197–200
Apman, R.P. (1973) *Report on investigation of the effect of agricultural waste on the contamination of Lake Valencia, Venezuela.* (mimeograph, 19 pp.)
Arad, A. & Kafri, N. (1975) Geochemistry of groundwaters in the Chad basin. *Journal of Hydrology*, **25**: 105–27
Arad, A. & Morton, W. (1969) Mineral springs and saline lakes of the Western Rift Valley, Uganda. *Geochimica et Cosmochimica Acta*, **33**: 1169–81
Arakawa, H. (1969) (ed.) *Climates of Northern and Eastern Asia.* World Survey of Climatology **8**. Amsterdam: Elsevier
Arambourg, C. (1933) Contribution à l'étude géologique et paléontologique du Bassin de lac Rudolphe et de la basse Vallée de l'Omo. *Mission Scientifique de l'Omo*, Part I. Paris: Muséum de'l Histoire Naturelle
Ardiwinata, R.O. (1957) Fish culture on paddy-fields in Indonesia. *Proceedings Indo-Pacific Fisheries Council*, **7**: 119–54
Arumugam, P.T. & Furtado, J.I. (1980) Physico-chemistry, destratification and nutrient budget of a lowland eutrophicated Malaysian reservoir and its limnological implications. *Hydrobiologia*, **70**: 11–24
Assaf, G. (1976) The Dead Sea: a scheme for a solar lake. *Solar Energy*, **18**: 293–9
Assaf, G. & Nissenbaum, A. (1977) The evolution of the upper water mass of the Dead Sea. In *Desertic terminal lakes*, ed. D.C. Greer, pp. 61–72. Logan: Utah Water Research Laboratory, Utah State University
Atwell, R.G. (1970) Some effects of Lake Kariba on the ecology of a floodplain of the mid Zambezi Valley of Rhodesia. *Biological Conservation*, **2**: 189–96
Bacci, G. (1951–52). Elementi per una malacofauna del Abyssinia e delle Somalia. *Annali del Museo Civico di Storia Naturale Giacomo Doria*, **65**: 1–44.
Bachman, H. (1933) Phytoplankton von Victoria Nyanza, Albert Nyanza- und Kiogasee. *Berichte der Schweizerischen Botanischen Gessellschaft*, **42**: 705–17
Baker, B.H., Mohr, P.A. & Williams, L.A.J. (1972) Geology of the Eastern Rift System of Africa. *The Geological Society of America, Special paper* No.136
Baker, H.B. (1911) *The origin of the Moon.* Detroit: Detroit Free Press, 23 April, 1911
Balek, J. (1977) *Hydrology and water resources in tropical Africa.* 208 pp. Developments in water science **8**. Amsterdam: Elsevier
Ball, E. & Glucksman, J. (1978) Limnological studies of Lake Wisdom, a large New Guinea caldera lake with a simple fauna. *Freshwater Biology*, **8**: 455–68
Ball, E.E. & Glucksman, J. (1980) A limnological survey of Lake Dakatua, a large caldera lake on west New Britain, Papua New Guinea, with comparisons to Lake Wisdom a younger nearby caldera lake. *Freshwater Biology*, **10**: 73–84
Ball, E.E. & Johnson, R.W. (1976) Volcanic history of Long Island, Papua New Guinea. In *Volcanism in Australasia*, ed. R.W. Johnson, pp. 133–47. Amsterdam, Oxford & New York: Elsevier

Balon, E.K. (1973) Results of fish population size assessments in Lake Kariba coves (Zambia), a decade after their creation. In *Man-made lakes: their problems and environmental effects*, eds W.C. Ackermann, G.F. White & E.B. Worthington, pp.149–58. Geophysical Monograph 17. Washington D.C.: American Geophysical Union

Balon, E.K. (1974a) Fishes from the edge of Victoria Falls, Africa: demise of a physical border for downstream invasions. *Copeia*, 3: 643–60

Balon, E.K. (1974b) Fish production in a tropical ecosystem. In *Lake Kariba – a man-made tropical ecosystem in Central Africa*, eds E.K. Balon & A.G. Coche, pp. 249–573. Monographiae Biologicae 24. The Hague: Junk

Balon, E.K. & Coche A.G. (1974) *Lake Kariba – a man-made tropical ecosystem in Central Africa*. Monographiae Biologicae 24. 767 pp. The Hague: Junk

Banister, K.E. (1980) The fishes of the Euphrates and Tigris. In *Euphrates and Tigris*, ed. J. Rzoska, pp. 95–108, Monographiae Biologicae 38, The Hague: Junk

Bardach, J.E. (1959) *Etude sur la pêche au Cambodge soumis à la direction des eaux et fôrets*. 80 pp. Phnom-Penh, Kampuchea: Chasse et Pêche

Bardach, J.E. (1972) Some ecological implications of Mekong River development plans. In *The careless technology*, eds M.F. Farwar & J.P. Milton, pp. 236–44. Garden City, N.Y: Natural History Press

Bartholomew, G.A. & Pennycuick, C.J. (1973) The flamingo and pelican populations of the Rift Valley Lakes in 1968–69. *East African Wildlife Journal*, 2: 189–98

Battistini, R. & Richard-Vindard, G. (1972) (eds). *Biogeography and ecology in Madagascar*, 765 pp. Monographiae Biologicae 21. The Hague: Junk

Bäuerle, E. (1983) Baroclinic free oscillations of closed basins of non-uniform depths. *Journal of geophysical Research* (in press)

Baumgartner, A. and Reichel, E. (1975) *The world water balance: mean annual global, continental and run-off*. 179 pp. English translation by R. Lee, Amsterdam: Elsevier

Baxter, R.M., Prosser, M.V., Talling, J.F. & Wood, R.B. (1965) Stratification in tropical African lakes at moderate altitudes (1500 to 2000 m). *Limnology and Oceanography*, 10: 510–20

Bayly, I.A.E. (1964) Chemical and biological studies on some acidic lakes of East Australian sandy coastal lowlands. *Australian Journal of marine and freshwater Research*, 15: 56–72

Bayly, I.A.E. (1967) The general biological classification of aquatic environments with special reference to those of Australia. In *Australian inland waters and their fauna*, ed. A.H. Weatherley, pp. 78–104. Canberra: Australian National University Press

Bayly, I.A.E. (1969) The occurrence of calanoid copepods in athalassic saline waters in relation to salinity and anionic proportions. *Verhandlungen internationale Vereinigung für theoretische und angewandte Limnologie*, 17: 449–55

Bayly, I.A.E. (1970) Further studies on some saline lakes of South-East Australia. *Australian Journal of marine and freshwater Research*, 21: 117–29

Bayly, I.A.E. (1976) The plankton of Lake Eyre. *Australian Journal of marine and freshwater Research*, 27: 661–5

Bayly, I.A.E., Ebsworth, E.P. & Hang Tong Wan (1975) Studies on the lakes of Fraser Island, Queensland. *Australian Journal of marine and freshwater Research*, 26: 1–13

Bayly, I.A.E. & Williams, W.D. (1964) Chemical and biological observations on some volcanic lakes in the south-east of South Australia. *Australian Journal of marine and freshwater Research*, 15: 123–32

Bayly, I.A.E. & Williams, W.D. (1966a) Chemical and biological studies on saline lakes of south east Australia. *Australian Journal of marine and freshwater Research*, 17: 177–228

Bayly, I.A.E. & Williams, W.D. (1966b) Further observations on some volcanic lakes in the south east of South Australia. *Australian Journal of marine and freshwater*

Research, **17**: 229–37

Bayly, I.A.E. & Williams, W.D. (1972) The major ions of some lakes and other waters in Queensland, Australia. *Australian Journal of marine and freshwater Research*, **23**: 121–31

Beadle, L.C. (1932) Scientific results of the Cambridge Expedition to the East African Lakes 1930–31. 4. The waters of some East African lakes in relation to their flora and fauna. *Journal of the Linnean Society (Zoology)*, **38**: 157–211

Beadle, L.C. (1966) Prolonged stratification and deoxygenation in tropical lakes. I. Crater lake Nkugute, Uganda, compared with lakes Bunyonyi and Edwards. *Limnology and Oceanography*, **11**: 152–63

Beadle, L.C. (1974) *The inland waters of tropical Africa*. 365 pp. London: Longman

Beam, W. (1906) Report of the Chemical Section in Second Report, Wellcome Research Laboratories, Khartoum. *Report, Wellcome Tropical Research Laboratory*, **2**: 206–14

Beauchamp, P. (1939) Rotifères et Turbellariés. In Reports of the Percy Sladen Trust Expedition, ed. H.C. Gilson. *Transactions Linnean Society*, **1** (Ser. 3): 357 pp.

Beauchamp, R.S.A (1939) Hydrology of Lake Tanganyika. *Internationale Revue der gesamten Hydrobiologie*, **39**: 316–53

Beauchamp, R.S.A. (1946) Lake Tanganyika. *Nature*, **157**: 183–5

Beauchamp, R.S.A. (1953) Hydrological data from Lake Nyasa. *Journal of Ecology*, **41**: 226–39

Begg, G.W. (1970) Limnological observations on Lake Kariba during 1967, with emphasis on some special features. *Limnology and Oceanography*, **15**: 776–88

Begg, S.W. (1976) The relationship between the diurnal movements of some of the zooplankton and the sardine *Limnothrissa miodon* in L. Kariba. *Limnology and Oceanography*, **21**: 529–39

Bell-Cross, G. (1972) The fish fauna of the Zambezi River system. *Arnoldia (Rhodesia)*, **5**: 1–19

Ben Amotz, A. (1973). Photosynthetic and osmoregulation mechanisms in the halophilic alga *Dunaliella parva*. Unpublished Ph.D. Thesis (in Hebrew), Weizmann Institute of Science, Israel

Ben Amotz, A. & Avron, M. (1972) The role of glycerol in the osmotic regulation of the halophilic alga *Dunaliella parva*. *Plant Physiology*, **51**: 875–78

Ben Amotz, A. & Avron, M. (1980) Glycerol, β-carotene and dry algal mean production by commercial cultivation of *Dunaliella*. In *Algal Biomass*, eds G. Shelef & C.H. Soeder, pp. 603–10. Amsterdam: Elsevier/North Holland Biochemical Press

Ben Avraham, Z. (1981) The movement of continents. *American Scientist*, **69**: 291–9

Ben Avraham, Z., Nur, A., Jones, D. & Cox, A. (1981) Continental accretion: from oceanic plateaus to allochthonous terranes. *Science*, **213**: 47–54

Ben Avraham, Z., Shoham, Y., Klein, E., Michelson, H. & Serruya, C. (1980) Magnetic survey of Lake Kinneret Central Jordan Valley, Israel. *Marine Geophysical Research*, **4**: 257–76

Bennekom, Van, A.J., Berger, G.W., Helder, W. & deVries, R.T.P. (1978) Nutrient distribution in the Zaire estuary and river plume. *Netherlands Journal of Sea Research*, **12**: 296–323

Ben Tuvia, A. (1978) Fishes. In *Lake Kinneret*, ed. C. Serruya, pp. 407–30. Monographiae Biologicae **32**. The Hague: Junk

Ben Tuvia, A. (1981) Man-induced changes in the freshwater fish fauna of Israel. *Fisheries Management*, **12**: 139–48

Ben Tuvia, A. & Herman, T. (1973) Research on Bardawil Lagoon: biology of fish populations. *Internal report of Israel oceanographic and limnological Research* (in Hebrew) 30 pp.

Bergstein, T., Henis, Y. & Cavari, B.Z. (1978) Investigations on the photosynthetic sulphur bacterium *Chlorobium phaeobacteroides* causing seasonal blooms in Lake

Kinneret. *Canadian Journal of Microbiology*, **25**: 999–1007
Berman, T. (1976) Light penetrance in Lake Kinneret. *Hydrobiologia*, **49**: 41–8
Berman, T. (1978) Primary productivity and photosynthetic efficiency of light utilization by phytoplankton. In *Lake Kinneret*, ed. C. Serruya, pp. 253–8. Monographiae Biologicae **32**. The Hague: Junk
Berman, T. & Pollingher, U. (1974) Annual and seasonal variations of phytoplankton chlorophyll, and photosynthesis in Lake Kinneret. *Limnology and Oceanography*, **19**: 31–54
Berritt, G.R. (1964) Observations océanographiques cotières à Pointe Noire. *Cahiers ORSTOM (Oceanogr.)*, **2**: 13–55
Berry, L. & Whiteman, A.J. (1968) The Nile in the Sudan. *Geographical Journal*, **134**: 1–36
Bertram, C.K.R., Boley, H.J.H. & Trewavas, E.T. (1942) *Report on the fish and fisheries of L. Nyasa*. London: Crown Agents
Beyth, M. (1976) Chemical survey concerning the influence of the 'terminal brines' on the Dead Sea water in the area of the pumping station No.6. (in Hebrew) *Rep. IG/1/76 Geological Survey*. Jerusalem: Geological Survey of Israel
Beyth, M. (1978) Mixing of the end brines in the northern basin of the Dead Sea, April–Nov. 1977. *Geological Survey MG/2/78*. Jerusalem: Geological Survey of Israel
Bishai, H.M. (1962) The water characteristics of the Nile in the Sudan with a note on the effect of *Eichhornia crassipes* on the hydrobiology of the Nile. *Hydrobiologia*, **19**: 357–82
Bishop, J.A. (1967) The zoogeography of the Australian freshwater decapod crustacea. In *Australian inland waters and their fauna*, ed. A.H. Weatherley, pp. 107–22. Canberra: Australian National University Press
Bishop, J.E. (1973) *Limnology of a small Malayan river Sungai Gomback*, 485 pp. Monographiae Biologicae **22**. The Hague: Junk
Bishop, W.W. (1969) Pleistocene stratigraphy in Uganda. *Geological Survey Uganda Memoirs*, no.10: 128 pp.
Biswas, S. (1969a) Thermal changes in the Volta-lake in Ajena. In *Man-made lakes: the Accra Symposium*, ed. L.E. Obeng, pp. 103–9. Accra: Ghana University Press
Biswas, S. (1969b) The Volta lake: some ecological observations on the phytoplankton. *Verhandlungen internationale Vereinigung für theoretische und angewandte Limnologie*, **17**: 259–72
Biswas, S. (1973) Limnological observations during the early formation of Volta Lake in Ghana. In *Man-made lakes: their problems and environmental effects*, eds W.C. Ackermann, G.F. White & E.B. Worthington, pp. 121–8. Geophysical Monograph **17**. Washington D.C: American Geophysical Union
Biswas, S. (1975) Phytoplankton in Volta Lake, Ghana, during 1964–1973. *Verhandlungen internationale Vereinigung für theoretische und angewandte Limnologie*, **19**: 1928–34
Biswas, S. (1978) Observations on phytoplankton and primary productivity in Volta Lake, Ghana. *Verhandlungen internationale Vereinigung für theoretische und angewandete Limnologie*, **20**: 1672–6
Blache, J. (1951) Aperçu sur le plancton des eaux douces du Cambodge. *Cybium*, **6**: 62–94
Blache, J. (1964) Les poissons du bassin du Tchad et du bassin adjacent du Mayo Kebi. *Mémoires ORSTOM*, **4** (2): 483 pp.
Blanc, M. (1959) Mission hydrobiologique et océanographique au Cambodge (fevrier–mars 1959). *Cahiers du Pacifique*, **2**: 33–59
Blanc, M. & Daget, J. (1957) Les eaux et les poissons du Haute Volta. *Mémoires de l'I.F.A.N.*, **50**: 100–68
Bloomfield, K. (1966) A major ENE dislocation zone in central Malawi. *Nature*, **211**: 612–14

Bibliography

Bockh, A. (1956) *El Desecamiento del Lago de Valencia*. 246 pp. Caracas: Fundacion Eugenio Mendoza

Bockh, A. (1968) Consecuencias de las actividades humanas sincontrol eu la cuenca del Lago de Valencia. *Boletino del Instituto Conservacion del Lago de Valencia*, 51 pp.

Bonazzi, A. (1950) Speed of salt increase in the waters of Lake Tacarigua Venezuela. *Science*, **112**: 590–1

Bond, W.J. & Roberts, M.G. (1978) The colonization of Cabora Bassa Mozambique, a new man-made lake, by floating aquatic macrophytes. *Hydrobiologia*, **60**: 243–59

Bonetto, A. (1975) Hydrologic regime of the Parana River and its influence on ecosystems. In *Coupling of land and water systems*, ed. A.D. Hasler, pp. 175–97. Ecological Studies 10. New York: Springer Verlag

Bonetto, A., Dioni, W. & Pignalberi, C. (1969) Limnological investigations on biotic communities in the middle Parana River Valley. *Verhandlungen internationale Vereinigung für theoretische und angewandte Limnologie*, **17**: 1035–50

Boscan, L.A., Capote, F. & Farias, J. (1973) Contaminacion salina del lago de Maracaibo: efectos en la calidad y aplicacion de sus aguas. *Boletin del centro de Investigaciones biologicas*, **9**: 11–37

Bourne, D. (1973) The feeding of three commercially important fish species in Lake Chilwa, Malawi. *African Journal of tropical Hydrobiology and Fisheries*, **3**: 134–45

Bourrelly, P. (1957) Algues d'eau douce du Soudan Français, région du Macina (A.O.F.). *Bulletin de l'I.F.A.N.* (Ser. A.), **19**: 1047–123

Bourrelly, P. (1961) Algues d'eau douce de la République de Cote d'Ivoire. *Bulletin de l'I.F.A.N.*, **23**: 283–398

Bourrelly, P. (1964) Les algues des eaux courantes de Madagascar. *Verhandlungen internationale Vereinigung für theoretische und angewandte Limnologie*, **15**: 758–63

Bourrelly, P. (1966) *Les algues d'eau douce, algues vertes*. 511 pp. Paris: Editions N. Boubée et Cie

Bourrelly, P. (1973). Quelques algues d'eau douce recoltées lors du XVIIème Congrès International de limnologie en Israël. *Verhandlungen internationale Vereinigung für theoretische und angewandte Limnologie*, **18**: 1326–37

Bowen, H.J.M. (1966) *Trace elements in biochemistry*. London & New York: Academic Press

Bradbury, J.P. (1971) Paleolimnology of Lake Texcoco Mexico. Evidence from diatoms. *Limnology and Oceanography*, **16**: 180–200

Bradbury, J.P., Leyden, B., Salgado-Labourian, N., Lewis, W.M., Schubert, C., Binford, M.W., Frey, D.G., Whitehead, D.R. & Weibezahn, F.H. (1981) Late Quaternary environmental history of Lake Valencia, Venezuela. *Nature*, **214**: 1299–305

Brandorff, G.O. (1976*a*) A new species of *Bosminopsis* (Crustacea, Cladocera) from the Rio Negro. *Acta Amazonica*, **6**: 109–14

Brandorff, G.O. (1976*b*) The geographic distribution of the Diaptomidae in South America (Crustacea, Copepoda). *Revista Brasileira de Biologia*, **36**: 613–27

Brandorff, G.O. (1977) Untersuchungen zur Populations dynamik des Crustaceenplanktons in Lago Castanho (Amazona, Brazilien). Dissertation, Christian Albrechts Universität, Kiel, 108 pp.

Brandorff, G.O. (1978) Preliminary comparison of the Crustacean plankton of a white water and a black water lake in central Amazonia. *Verhandlungen internationale Vereinigung für theoretische und angewandte Limnologie*, **20**: 1198–202

Brandorff, G.O. & De Andrade, E.R. (1978) The relationship between the water level of the Amazon River and the fate of the zooplankton population in Lago Jacaretinga a Várzea lake in the Central Amazon. *Studies on Neotropical fauna and environment*, **13**: 63–70

Brehm, V. (1951) Cladocera and Copepoda. Calanoida von Cambodja. *Cybium*, **6**: 95–124

Brehm, V. (1956) Cladocera aus Venezuela. *Ergebnisse der Deutschen Limnologischen Venezuela Expedition 1952,* (West) Berlin: 217–32

Brezonik, P.L. & Fox, J.L. (1974) The limnology of selected Guatemalan lakes. *Hydrobiologia,* **45**: 467–87

Brinson, M.M., Brinson, L.G. & Lugo, A.E. (1974) The gradient of salinity, its seasonal movement and ecological implications for the Rio Dulce–Lake Izabal ecosystem, Guatemala. *Bulletin of marine Science,* **24**: 533–44

Brinson, M.M. & Nordlie, F.G. (1975) Lake Izabal, Guatemala. *Verhandlungen internationale Vereinigung für theoretische und angewandte Limnologie,* **19**: 1468–80

Brook, A. (1954) A systematic account of the phytoplankton of the Blue and White Niles. *Annals and Magazine of natural History,* Ser. 12, **7**: 648–56

Brook, A.J. & Rzoska, J. (1954) The influence of the Gebel Auliya dam on the plankton of the White Nile. *Journal of Animal Ecology,* **23**: 101–14

Brooks, H.K. (1970) A preliminary report on Lake Izabal geology–hydrology. *A report to the Organization of tropical studies.* University of Florida, Inc.

Brooks, J.L. (1950) Speciation in ancient lakes. *Quaternary Review of Biology,* **25**: 131–76

Brown, L.H. (1971) The flamingoes of Lake Nakuru. *New Scientist,* **51**: 97–101

Brown, L.H. & Root, A. (1971) The breeding of the lesser flamingo *Phoeniconaias minor. Ibis,* **113**(2): 147–72

Brunelli, G. & Cannicci, G. (1940) Le caratteristiche biologiche an Lago Tana. In *Missione di studio al Lago Tana. Richerche Limnologiche(B) Chimica e biologia,* **3**: 71–116. Reale Academia d'Italia XVIII

Burgis, M.J. (1969) A preliminary study of the ecology of zooplankton in Lake George, Uganda. *Verhandlungen internationale Vereinigung für theoretische und angewandte Limnologie,* **17**: 297–302

Burgis, M.J. (1970) The effect of temperature on the development time of eggs of *Thermocyclops* sp., a tropical cyclopid from Lake George, Uganda. *Limnology and Oceanography,* **15**: 742–7

Burgis, M.J. (1971) The ecology and production of copepods, particularly *Thermocyclops hyalinus,* in the tropical Lake George, Uganda. *Freshwater Biology,* **1**: 169–92

Burgis, M.J. (1973) Observations on the Cladocera of Lake George, Uganda. *Journal of Zoology,* (*London*), **170**: 339–49

Burgis, M.J. (1974) Revised estimates for the biomass and production of zooplankton in Lake George, Uganda. *Freshwater Biology,* **4**: 535–42

Burgis, M.J. (1978) Case studies of lake ecosystems at different latitudes: the tropics. The Lake George ecosystem. *Verhandlungen internationale Vereinigung für theoretische und angewandte Limnologie,* **20**: 1139–52

Burgis, M.J., Darlington, J.P.E.C., Dunn, I.G., Ganf, G.G., Gwahaba, J.J. & McGowan, L.M. (1973) The biomass and distribution of organisms in Lake George, Uganda. *Proceedings of the Royal Society,* (*London*), B, **184**: 271–98

Burgis, M.J. & Dunn, I.G. (1978) Production in three contrasting ecosystems. In *Ecology of freshwater fish production,* ed. S.D. Gerkin, pp. 137–58. Oxford: Blackwell Scientific Publications

Burgis, M.J. & Walker, A.F. (1971) A preliminary comparison of the zooplankton in a tropical and a temperate lake. (Lake George, Uganda and Loch Leven, Scotland). *Verhandlungen internationale Vereinigung für theoretische und angewandte Limnologie,* **18**: 647–55

Bush, G.L. (1975) Modes of animal speciation. *Annual Review of Ecology and Systematics,* **6**: 339–64

Butzer, K.W., Isaac, G.L., Richardson, J.L. & Washbourn-Kaman, C. (1972) Radiocarbon dating of East African lake levels. *Science,* **175**: 1069–76

Cadée, G.C. (1978) Primary production and chlorophyll in the Zaire river, estuary and plume. *Netherlands Journal of Sea Research,* **12**: 368–81

Bibliography

Cairns, J. (1966) III Protozoa. In *The Catherwood Foundation Peruvian Amazon Expedition: limnological and systematic studies*, eds R. Patrick & collaborators, pp. 53–61. Monographs of the Academy of Natural Sciences of Philadelphia No. 14. Philadelphia: G.W. Carpenter

Cano, G.J. (1978) Argentina, Brazil and the de la Plata River Basin. In *Water in a developing world*, eds. Ulton & Tecloff, pp. 127–206. Boulder, Colorado: Westview Press

Capart, A. (1949) Sondages et carte bathymetrique. *Exploration hydrobiologique du Lac Tanganika (1946–1947) Resultats scientifiques.* **2**: 1–16. Brussels: Institut Royal des Sciences Naturelles de Belgique

Capart, A. (1952) Le milieu géographique et géophysique. *Exploration hydrobiologique du Lac Tanganika (1946–1947). Resultats scientifiques*, **1**: 3–27. Brussels: Institut Royal des Sciences Naturelle de Belgique

Carlson, F.A. (1952) *Geography of Latin America*, 3rd edn, 569 pp. Englewood Cliffs, N.J.: Prentice Hall

Carmouze, J.P. (1979) The hydrochemical regulation of the Lake Tchad. In *Proceedings of the SIL–UNEP Workshop on African limnology*, ed. J.J. Symoens. Nairobi: SIL–UNEP

Carmouze, J.P., Arce, C. & Quintanilla, J. (1977) La regulation hydrique des Lacs Titicaca et Poopo. *Cahiers ORSTOM (Série Hydrobiologie)*, **11**: 269–83

Carmouze, J.P., DeJoux, C., Durand, J.R., Gras, R., Iltis, A., Lauzanne, L., Lemoalle, J., Lévêque, C., Loubens, G. & Saint Jean, L. (1972) Contribution à la connaissance du bassin Tchadien. *Cahiers ORSTOM (Série Hydrobiologie)*, **6**: 103–69

Carter, G.S. (1955) *The papyrus swamps of Uganda*, Cambridge: Heffer

Casey, D.J. (1978) Guyana, 'Land of waters', moves to harness hydropotential. *Water International*, **3**: 22–6

Cavari, B.Z. (1977) Denitrification in Lake Kinneret in the presence of oxygen. *Freshwater Biology*, **7**: 385–91

Chacko, P.J., Abraham, J.G. & Andal, R. (1953) Report on a survey of the flora, fauna and fisheries of Pulicat lake, Madras, India 1951–2. *Contribution, Freshwater Fisheries Biological Station, Madras*, No. **8**: 20 pp.

Chakraborty, K.D., Roy, P. & Singh, S.B. (1958) A quantitative study of the plankton and physico-chemical conditions of the river Jumuna at Allahabad in 1954–55. *Indian Journal of Fisheries*, **6**: 186–203

Chessman, B.C. & Williams, W.D. (1974) Distribution of fish in inland saline waters in Victoria, Australia. *Australian Journal of marine and freshwater Research*, **25**: 167–72

Chevey, P. (1940) Sur la periodicité sexuelle des poissons des eaux douces du Cambodge. *Bulletin de la Société centrale d'Aquiculture et de Pêche*, **46**: 77–81

Coche, A.G. (1968) Description of physico-chemical aspects of Lake Kariba, an impoundment in Zambia Rhodesia. *Fisheries Research Bulletin, Zambia*, **5**: 200–67

Coche, A.G. (1974) Limnological study of a tropical reservoir. Part I. In *Lake Kariba: a man-made tropical ecosystem in Central Africa*, eds E.K. Balon & A.G. Coche, pp. 1–248. Monographiae Biologicae 24. The Hague: Junk

Coe, M.J. (1966) The biology of *Tilapia grahami* Boulenger in Lake Magadi, Kenya. *Acta Tropica*, **23**: 146–77

Coe, M.J. (1969) Observations on *Tilapia alcalica* in Lake Natron. *Revue de Zoologie et de Botanique Africaine*, **80**: 1–14

Cohen, Y., Krumbein, W.E., Goldberg, M. & Shilo, M. (1977) Solar lake (Sinai) 1. Physical and chemical limnology. *Limnology and Oceanography*, **22**: 597–608

Cohen, Y., Krumbein, W. & Shilo, M. (1977*a*). Solar Lake (Sinai) 2. Distribution of photosynthetic microorganisms and primary production *Limnology and Oceanography*, **22**: 609–20

Cohen, Y., Krumbein, W.E. & Shilo, M. (1977*b*). Solar Lake (Sinai) 3. Bacterial

distribution and production. *Limnology and Oceanography,* **22**: 621-34
Cole, G.A. (1963) The American southwest and Middle America. In *Limnology in North America.* ed. D.G. Frey, pp. 393-434. Madison, Wisconsin: University of Wisconsin Press
Committee for the Coordination of Investigations of the Lower Mekong Basin (1970) *Report on indicative Basin E Plan CN 11/WRD/MGK/L.340* UNESCO
Committee for the Coordination of Investigations of the Lower Mekong Basin (1978) *Annual Report ST/ESCAP/79,* 123 pp. UNESCO
Compère, P. (1967) Algues du Sahara et de la région du Lac Tchad. *Bulletin du Jardin botanique national de Belgique,* **37**: 109-288
Compère, P. (1974) Algues de la région du Lac Tchad. II Cyanophycées. *Cahiers ORSTOM (Série Hydrobiologie),* **8**: 165-98
Compère, P. (1975a). Algues de la région du Lac Tchad. III Rhodophycées, Euglenophycées, Cryptophycées, Dinophycées, Chrysophycées, Xantophycées. *Cahiers ORSTOM (Série Hydrobiologie),* **9**: 167-92
Compère, P. (1975b) Algues de la région du Lac Tchad, IV Diatomophycées. *Cahiers ORSTOM (Série Hydrobiologie),* **9**: 203-90
Compère, P. (1976) Algues de la région du Lac Tchad V. VI Chlorophycophytes. *Cahiers ORSTOM (Série Hydrobiologie),* **10**: 77-118; 135-64
Compère, P. (1977) Algues de la région du Lac Tchad VII Chlorophycophytes (Desmidiées), *Cahiers ORSTOM (Série Hydrobiologie),* **11**: 77-177
Corbet, P.S.A. (1961) The food of non-cichlid fishes in the Lake Victoria Basin, with remarks on their evolution and adaptation to lacustrine conditions. *Proceedings Zoological Society London,* **136**: 1-101
Cordiviola de Yuan, E. & Pignalberi, C. (1981) Fish populations in the Parana River. 2. Santa Fe and Corrientes areas. *Hydrobiologia,* **77**: 261-72
Cott, H.B. (1963) Scientific results of an enquiry into the ecology and economic status of the Nile crocodile (*Crocodilus niloticus*) in Uganda and Northern Rhodesia. *Transactions of the Zoological Society of London,* **29**: 211-337
Coulter, G.W. (1962) Research results – Lake Tanganika. *Joint Fisheries Research Organization Annual Report,* no. 10, pp. 7-30. Lusaka: Government Printer
Coulter, G.W. (1963) Hydrological changes in relation to biological production in southern Lake Tanganyika. *Limnology and Oceanography,* **8**: 463-77
Coulter, G.W. (1968a) Hydrological processes and primary production in Lake Tanganyika. *Proceedings of the eleventh Conference on Great Lakes Research,* 609-26. International Association for Great Lakes Research
Coulter, G.W. (1968b) Thermal stratification in the deep hypolimnion of Lake Tanganyika. *Limnology and Oceanography,* **13**: 385-7
Coulter, G.W. (1970) Population changes within a group of fish species in Lake Tanganyika following their exploitation. *Journal of Fish Biology,* **2**: 329-53
Coulter, G.W. (1977) Approaches to estimating fish biomass and potential yield in Lake Tanganyika. *Journal of Fish Biology,* **11**: 393-408
Coulter, G.W. (1979) Diel migration of fish in Lake Tanganyika. In *Proceedings of the SIL-UNEP Workshop on African limnology,* ed. J.J. Symoens. Nairobi: SIL-UNEP
Coulter, G.W. (1981) Biomass, production and potential yield of the Lake Tanganyika pelagic fish community. *Transactions of the American Fisheries Society,* **110**: 325-35
Cowgill, U.M. & Hutchinson, G.E. (1966) La Aguada de Santa Ana Vieja: The history of a pond in Guatemala. *Archiv für Hydrobiologie,* **62**: 335-72
Cowgill, U.M., Hutchinson, G.E., Racek, A.A., Goulden, C.E., Patrick, R. & Tsukada, M. (1966) The history of Laguna de Petenxil, a small lake in northern Guatemala. *Memoirs of the Connecticut Academy of Arts and Sciences,* **17**: 1-26
Craig, H. & Craig, V. (1979) Geochemical studies of Lake Tanganyika. *Proceedings of*

Bibliography 513

the *SIL-UNEP Workshop on African limnology*, ed. J.J. Symoens. Nairobi: SIL-UNEP
Cronberg, G. (1977) The lago do Paranoa Restoration Project. Phytoplankton ecology and taxonomy. *Project PAHO/WHO 77/WT/BRA/2341/04* UNESCO
Daday, von, E. (1907) Plankton-Tiere aus dem Victoria Nyanza Sammelausbeute von A. Bogert 1904–1905. *Zoologische Jahrbucher* (syst.), **45**: 245–62
Daday, von, E. (1910) Beitrage zur Kenntnis der Mikroflora von Deutsch Ost-Africa. *Matematikai és természettudományi ertesito*, **25**: 402–20
Damas, H. (1937) La stratification thermique et chimique des Lacs Kivu, Edouard et Ndalaga. *Verhandlungen internationale Vereinigung für theoretische und angewandte Limnologie*, **8**: 51–67
Damas, H. (1955) Recherches limnologiques dans quelques lacs du Ruanda. *Verhandlungen internationale Vereinigung für theoretische und angewandte Limnologie*, **12**: 335–41
Davies, B.R., Hall, A. & Jackson, P.M.B. (1975). Some ecological aspects of the Cabora Bassa Dam. *Biological Conservation*, **8**: 189–201
De Buen, F. (1943) Los Lagos Michoacanos. I. Caracteres, Generales. El Lago de Zirahuen. *Revista de la Sociedad Mexicana de Historia natural*, **4**: 211–32
De Buen, F. (1944a) Limnobiologia de Patzcuaro. *Anales del Instituto de Biologia Universidad de México*, **15**: 261–312
De Buen, F. (1944b) Los Lagos michoacanos. II Patzcuaro. *Revista de la Sociedad Mexicana de Historia natural*, **5**: 99–125
De Buen, F. (1945) Resultados de una campana limnologica en Chapala y observaciones sobre otras aguas exploradas. *Revista de la Sociedad Mexicana de Historia natural*, **6**: 129–44
Deelstra, H. (1979) Review of the research efforts on the study of the pollution of the northern section of Lake Tanganyika. *Proceedings of the SIL-UNEP Workshop on African Limnology*, ed. J.J. Symoens. Nairobi: SIL-UNEP
Deevey, E.S. (1955) Limnological studies in Guatemala and Salvador. *Verhandlungen internationale Vereinigung für theoretische und angewandte Limnologie*, **12**: 278–82
Deevey, E.S. (1957) Limnologic studies in Middle America with a chapter on Aztec limnology. *Transactions of the Connecticut Academy of Arts and Sciences*, **39**: 213–328
Deevey, E.S., Jr., Brenner, M., Flannery, M.S. & Yezdani, G.H. (1980) Lakes Yaxha and Sacnab, Peten, Guatemala. Limnology and hydrology. *Archiv für Hydrobiologie, Supplement* **57**: 419–60
Deevey, E.S., Deevey, G.B. & Brenner, M. (1980) Structure of zooplankton communities in the Peten Lake district, Guatemala. In *Evolution and Ecology of zooplankton communities*, ed. W. Ch. Kerfoot, pp. 669–78. Special symposium volume 3, London, UK: ASLO University Press of New England
Deevey, E.S., Vaughan, H.H. & Deevey, G.B. (1977) Lakes Yaxha and Sacnab, Peten Guatemala: planktonic fossils and sediment focusing. In *Interactions between sediments and freshwater*, ed. H.L. Golterman, pp. 189–96. The Hague: Junk
Degens, E.T., Deuser, W.G., Von Herzen, R.P., Wong, H.K., Wooding, F.B., Jannasch, H.W. & Kanwisher, J.W. (1971) *Lake Kivu Expedition: geophysical hydrography and sedimentology (preliminary report)*. Woods Hole Oceanographic Institute
Degens E.T., Von Herzen, R.P. & How-Kin Wong (1971) Lake Tanganyika: water chemistry, sediments; geological structure. *Naturwissenschaften*, **58**: 229–41
Degens, E.T., Von Herzen, R.P., How-Kin Wong, Deuser, W.G. & Jannasch, H.W. (1973) Lake Kivu: structure, chemistry and biology of an East African Rift Lake. *Geologische Rundschau*, **62**: 245–77
De-Heer Amissah (1969) Some possible climatic changes that may be caused by the Volta Lake. In *Man-made lakes: the Accra Symposium*, ed. L.E. Obeng, pp. 73–82. Accra: Ghana University Press

514 Bibliography

Dejoux, C. (1969) Les insectes aquatiques du Lac Tchad. Aperçu systematique et bio-écologique. *Verhandlungen internationale Vereinigung für theoretische und angewandte Limnologie*, **17**: 900–6
Dejoux, C. (1976) Synécologie des Chironomides du Lac Tchad (Diptères, Nematocères). *Travaux et Documents ORSTOM*, no. **56**: 161 pp.
Delhaye, F. (1941) Les volcans au nord du Lac Kivu. *Bulletin Institut Royal Colonial Belge*, **12**: 409–59
Denny, P. (1972) Lakes of south western Uganda. I. Physical and chemical studies on Lake Bunyonyi. *Freshwater Biology*, **2**: 143–58
Denny, P. (1973) Lakes of south western Uganda. II. Vegetation studies on Lake Bunyonyi. *Freshwater Biology*, **3**: 123–35
Depetris. P.T. (1976) Hydrochemistry of the Parana River. *Limnology and Oceanography*, **21**: 736–9
Devasundaram, M.P. & Roy, J.C. (1954) A preliminary study of the plankton of the Chilka Lake for the years 1950–1951. In *Symposium on marine and freshwater plankton in the Indo-Pacific, Bangkok, Jan. 25–26, 1954*, pp. 48–54. Bangkok: FAO Indo-Pacific Fisheries Council and Djakarta: UNESCO
Dietrich, G., Kalle, K., Krauss, W. & Siedler, G. (1980) *General oceanography. An introduction*. New York & Toronto: Wiley Interscience
Dietz, G. (1965) Uber zwei neue Ciliaten aus einem Ostafrikanischen Sodasee, *Uroleptus natronophilus* n.sp. und *Spathidinium elmenteitanum*, n.sp. *Archiv für Protistenkunde*, **108**: 25–8
Dietz, R.S. (1961) Continent and ocean basin evolution by spreading of the sea floor. *Nature*, **190**: 854–7
Dietz, R.S. & Holden, J.C. (1970) Reconstruction of Pangea: breakup and dispersion of continents, Permian to present. *Scientific American*, **223**: 30–41
Dobzhansky, T. (1937) *Genetics and the origin of species*. 364 pp. New York: Columbia University Press
Dobzhansky, T. (1950) Evolution in the tropics. *American Scientist*, **38**: 209–21
Donselaar, J. van (1968) Water and marsh plants in the artificial Brokopondo Lake (Surinam) during the first three years of its existence. *Acta Botanica Neerlandica*, **17**: 183–96
Dragesco, J. (1980) Les Protozoaires. In *Flore et faune aquatique de l'Afrique Sahelo-Soudanienne*, eds J.R. Durand & C. Lévêque, pp. 160–92. Paris: T.I. ORSTOM
Dubois, J.T. (1959) Note sur la chimie des eaux du Lac Tumba. *Bulletin Séances Academie Royale Science Outremer*, **5**: 1321–34
Ducharme, A. (1975) Informe tecnico de biologia pesquera (limnologia). Projecto para el disarrollo de la pesea continental. *INDERENA-FAO Publ. No.4 DP/COL/71/552/4* Bogota, Colombia
Dudley, R.G. (1979) Changes in growth and size distribution of *Sarotherodon macrochir* and *S. andersoni* from the Kafue floodplain, Zambia Gorge Dam. *Journal of Fish Biology*, **14**: 205–23
Dufour, P. & Merle, J. (1972) Station cotière en Atlantique tropical: hydroclimat et production primaire. *Documents Scientifiques du Centre ORSTOM, Pointe Noire (n.s.)*, **25**: 1–48
Dumont, H.J. (1980) Zooplankton and the science of biogeography: the example of Africa. In *Evolution and ecology of zooplankton communities*, ed. W.Ch. Kerfoot, pp. 685–96. London, UK: University Press of New England
Dunn, J.G. (1967) Diurnal fluctuation of physico-chemical conditions in a shallow tropical pond. *Limnology and Oceanography*, **12**: 151–4
Dunn, I.G. (1973) The commercial fishery of Lake George, Uganda. *African Journal of tropical Hydrobiology and Fisheries*, **2**: 109–20
Dunn, I.G. (1975) Ecological notes on the *Haplochromis* (Pisces: Cichlidae) species flock

of Lake George, Uganda (East Africa). *Journal of Fish Biology,* **7**: 651–66
Dunn, I.G., Burgis, M.J., Ganf, G.G., McGowan, L.M. & Viner, A.B. (1969) Lake George, Uganda: a limnological survey. *Verhandlungen internationale Vereinigung für theoretische und angewandte Limnologie,* **17**: 284–8
Dussart, B. (1960) Problèmes sedimentologiques au Cambodge dans la région des Grands Lacs. *Cahiers du Pacifique,* **3**: 3–36
Dussart, B. & Gras, R. (1966) Faune planctonique du Lac Tchad. I. Crustacés, Copepodes, *Cahiers ORSTOM (Série Océanographique),* **4**: 77–91
Dussart, B.H. (1974) Biology of inland waters in humid tropical Asia. In *Natural resources of humid tropical Asia,* vol. XII, pp. 331–53. Paris: UNESCO
Dussart, B.H. (1980) Les Crustacés copépodes d'Afrique: catalogue et biogéographie. *Hydrobiologia,* **73**: 165–70
Du Toit, A.L. (1937) *Our wandering continents, an hypothesis of continental drifting.* 366 pp. Edinburgh & London: Oliver & Boyd
Dyer, T.G.J. (1976) Analysis of the temporal behaviour of the level of Lake Malawi. *South African Journal of Science,* **72**: 381–2
EAFRO (1954) *Annual Report for East African Fisheries Research Organization.* 40 pp. Jinja, Uganda: EAFRO
ECAFE (United Nations Economic Commission for Asia and the Far East) (1962) Factors affecting the selection of a dam site: selection of the first storage dam on the Indus. *Proceedings of regional symposium on dams and reservoirs, Tokyo. Flood control series,* No. **21**: 191–205
ECAFE (1966) *A compendium of major international rivers in the ECAFE region.* UN Water Resources Series No. 29
Eccles, D.H. (1962*a*) Research results, Lake Nyassa. 2 Hydrology. In *Joint Fisheries Research Organization Annual Report 11 (1961),* pp. 52–3. Lusaka: Government Printer
Eccles, D.H. (1962*b*). An internal wave in Lake Nyasa and its probable significance in the nutrient cycle. *Nature,* **194**: 832–3
Eccles, D.H. (1965) Hydrology. In *Annual Report of the Department of Agriculture and Fisheries for the year 1964.* (Fisheries Research) Part II, pp. 19–24. Zomba, Malawi: Government Printer
Eccles, D.H. (1974) An outline of the physical limnology of Lake Malawi (Lake Nyasa). *Limnology and Oceanography,* **19**: 730–43
Eckholm, E.P. (1975) The deterioration of mountain environments. *Science,* **189**: 764–70
Eckman, V.W. (1905) On the influence of the earth's rotation on ocean currents. *Arkiv for Matematik, Astronomi och Fysik,* **2**: 52 pp.
Eckstein, Y. (1970) Physico-chemical limnology and geology of a meromictic pond on the Red Sea shore. *Limnology and Oceanography,* **15**: 363–72
Edwards, A.M.C. & Thornes, J.B. (1970) Observations on the dissolved solids of the Casiquiare and Upper Orinoco, April–June 1968. *Amazoniana,* **2**: 245–56
Ehrlich, A. (1975) The diatoms from the surface sediments of the Bardawil Lagoon (northern Sinai) – Paleoecological significance. In *Third symposium on recent and fossil marine diatoms, Kiel.* ed. R. Simonsen, pp. 253–77. *Nova Hedwigia,* **53**
Eldredge, N. & Cracraft, J. (1980) *Phylogenetic patterns and the evolutionary process,* 350 pp. New York: Columbia University Press
El-Hawary, M.A. (1960) A preliminary study on the zooplankton of Lake Maryut and Lake Edku. *Notes and Memoirs, Hydrobiology and Fisheries Directorate, Cairo,* **52**: 1–12
El Moghraby, A.M., Hashem, M.T. & El Sedfy, H.M. (1973) Some biological characters of *Mugil capito* (cirv.) in Lake Borullus. *Bulletin of the Institute of Oceanography and Fisheries,* **3**: 55–82
El Moghraby, A.I. (1972) The zooplankton of the Blue Nile. Ph.D. Thesis, University of Khartoum

El Moghraby, A.I. (1977) A study on diapause of zooplankton in a tropical river. The Blue Nile. *Freshwater Biology*, **7**,(3): 207–12
Elster, J.H. & Jensen, K.W. (1960) Limnological and fishery investigations of the Nouzha Hydrodome near Alexandria. Notes and Memoirs **43**, 99 pp. Alexandria: Institute of Hydrology
Elster, H.J. & Vollenweider, R.A. (1961) Beitrage zur Limnologie Agyptens. *Archiv für Hydrobiologie*, **57**: 241–343
El Zarka Salah Eldin (1973) Kainji Lake, Nigeria. In *Man-made lakes: their problems and environmental effects*. eds W.C. Ackermann, G.F. White & E.B. Worthington, pp. 197–219. Geophysical Monograph **17**, Washington D.C.: American Geophysical Union
Entz, B. (1976) Lake Nasser and Lake Nubia. In *The Nile, biology of an ancient river*, ed. J. Rzoska, pp. 271–98. Monographiae Biologicae **29**. The Hague: Junk
Entz, B.A.G. (1969) Observations on limnochemical conditions of the Volta Lake. In *Man-made lakes: The Accra Symposium*, ed. L.E. Obeng, pp. 110–15. Accra: Ghana University Press
Eugster, H.P. (1970) Chemistry and origin of the brines of Lake Magado; Kenya. *Mineralogical Society of America, Special paper No.3*: 213–35
Eugster, H.P. & Hardie, L.A. (1978) Saline lakes. In *Lakes, chemistry, geology, physics*, ed. A. Lerman, pp. 237–93. New York, Heidelberg & Berlin: Springer Verlag
Evans, J.H. (1962) The distribution of phytoplankton in some Central East African waters. *Hydrobiologia*, **19**: 299–315
Evans, W.A. & Vanderpuye, J. (1973) Early development of the fish populations and fisheries of Volta Lake. In *Man-made lakes: their problems and environmental effects*, eds W.C. Ackermann, G.F. White & E.B. Worthington, pp. 114–20. Geophysical Monograph **17**. Washington D.C.: American Geophysical Union
Ewer, D.W. (1966) Biological investigations on the Volta Lake. May 1964 to May 1965. In *Man-made lakes*, ed. R.H. Lowe-McConnell, pp. 21–31. Symposia of the Institute of Biology No.**15**. London: Academic Press
Ezzat, A. (1972) The bottom fauna of Lake Edku (Egypt–UAR) *Rapport et Proces-verbaux des Réunions*, **20**(4): 503–5. Commission Internationale pour l'exploration scientifique de la Mer Mediterranée
FAO (1962) *Report to the Government of the Philippines on freshwater fisheries investigations based on the work of J.W. Parsons*. 17 pp. Report No.1565. Rome: FAO
FAO (1963) Project de pesca continental informi al gobierno de El Salvador. Report No. 1735. Rome: FAO
Farrell, T.P., Finlayson, C.M. & Griffiths, D.J. (1979) Studies of the hydrobiology of a tropical lake in North-Western Queensland. I. Seasonal changes in chemical characteristics. *Australian Journal of marine and Fisheries Research*, **30**: 579–95
Fernando, C.H. (1973) Man-made lakes of Ceylon: a biological resource. In *Man-made lakes: their problems and environmental effects*, eds W.C. Ackermann, G.F. White & E.B. Worthington, pp. 664–71. Geophysical Monograph **17**. Washington D.C.: American Geophysical Union
Fernando, C.H. (1980a) The freshwater zooplankton of Sri-Lanka, with a discussion of tropical freshwater zooplankton composition. *Internationale Revue der gesamten Hydrobiologie*, **65**: 85–125
Fernando, C.H. (1980b) The species and size composition of tropical freshwater zooplankton with special reference to the oriental region (South East Asia). *Internationale Revue der gesamten Hydrobiologie*, **65**: 411–25
Fernando, C.H. (1980c) Some important implications for tropical limnology. In *Proceedings of the first workshop on the promotion of limnology in the developing countries*, eds S. Mori and I. Kusima, pp. 103–7. (Organizing Committee XXI SIL Congress 29–30 Aug. 1980, Kyoto) Kyoto: SIL
Fernando, C.H. & Furtado, J.I. (1975) Reservoir fishery resources of South East Asia.

Bulletin of the Fisheries Research Station, Sri Lanka, **26**: 83–95
Fernando, C.H. & Ponyi, J.E. (1981) The freeliving freshwater cyclopid copepoda (Crustacea) of Malaysia and Singapore. *Hydrobiologia*, **78**: 113–23
Fink, W.L. & Fink, S.V. (1979) Central Amazonia and its fishes. *Comparative biochemical Physiology*, **62A**: 13–29
Fish, G.R. (1952) Appendix A. Chemical analyses Lake Albert. In *East African Fisheries Research Organization Annual Report for 1952*, p. 27. Jinja, Uganda: East African Commission
Fish, G.R. (1957) A seiche movement and its effect on the hydrology of Lake Victoria. *Fishery Publications, Colonial Office*, **10**: 68 pp.
Fish, G.R. (1960) The productivity of certain fish ponds in Malaya. In *Proceedings, Centenary, Bicentenary Congress of Biology, Singapore*, pp. 18–26. Singapore: University of Malaya Press
Fisher, R.T. (1979) Plankton and primary production in aquatic systems of the central Amazon basin. *Comparative biochemical Physiology*, **62A**: 31–8
Fisher, R.T. & Parsley, P.E. (1979) Amazon lakes: water storage and nutrient stripping by algae. *Limnology and Oceanography*, **24**: 547–53
Fittkau, E.J. (1970) Role of caymans in the nutrient regime of mouth lakes of Amazon affluents (A hypothesis). *Biotropica*, **2**: 138–42
Fittkau, E.J. (1973a) Crocodiles and the nutrient metabolism of Amazonian waters. *Amazoniana*, **4**: 103–33
Fittkau, E.J. (1973b) Artenmannigfaltigkeit amazonischer lebensräume aus ökologischer Sicht. *Amazoniana*, **4**: 321–40
Fittkau, E.J. (1974) Für ökologischen Gliederung Amazoniens. I. Die erdgeschichtliche Entwicklung Amazoniens. *Amazoniana*, **5**: 77–134
Fittkau, E.J., Illies, J., Klinge, H., Schwabe, G.H. & Sioli, H. (1968) (ed.) *Biogeography and ecology in South America*, vol.1, 447 pp. Monographiae Biologicae 18. The Hague: Junk
Fittkau, E.J., Illies, J., Klinge, H., Schwabe, G.H. & Sioli, H. (1969) (ed.) *Biogeography and ecology in South America*, vol.2, 946 pp. Monographiae Biologicae 19. The Hague: Junk
Fittkau, E.J., Irmler, U., Junk, W.J., Reiss, F. & Schmidt, G.W. (1975) Productivity, biomass and population dynamics in Amazonian water bodies. In *Tropical ecological systems. Trends in terrestrial and aquatic research*, eds F.B. Golley & E. Medina, pp. 289–311. New York & Berlin: Springer Verlag
Fittkau, E.J., Junk, W., Klinge, H. & Sioli, H. (1975) Substrate and vegetation in the Amazon region. *Berichte der Internationalen Symposien der Internationalen Vereinigung für Vegetationskunde*, 73–90
Flores, G. (1970) Suggested origin of the Mozambique Channel. *Transactions of the Geological Society of South Africa*, **73**: 1–16
Fontes, J.Ch., Florkowski, T., Pouchan, P. & Zuppi, G.M. (1979) Preliminary isotope study of Lake Asal System (republic of Djibouti). In *Isotopes in lake studies*, ed. R. Gonfiantini, pp. 163–74. Proceedings of an advisory group meeting, Vienna 29 Aug.–2 Sept. 1977. Vienna: International Atomic Energy Agency
Forel, F.A. (1892) *Le Léman*. Monographie limnologique. Lausanne: F. Rouge
Forster, K. (1963) Desmidiaceen aus Brasilien I. Nord-Brasilien. *Revue Algologique*. **7**: 38–91
Forster, K. (1964) Desmidiaceen aus Brasilien. 2 Teil: Bahia, Goyaz, Pianhyund Nord Brasilien. *Hydrobiologia*, **23**: 321–505
Forster, K. (1969) Amazonische Desmidiaceen. 1 Teil: Areal Santarem. *Amazoniana*, **2**: 5–232
Forster, K. (1972) The Desmids of the haloplankton of the Lake Valencia, Venezuela. *Internationale Revue der gesamten Hydrobiologie*, **57**: 409–29

Forster, K. (1974) Amazonische Desmidiaceen. 2 Areal maues. Abacaxis. *Amazoniana*, **5**: 135–242
Freson, R.E. (1972) Aspect de la limnochimie et de la production primaire au lac de la Lubumbashi. *Verhandlungen internationale Vereinigung für theoretische und angewandte Limnologie*, **18**: 661–5
Frey, D.G. (1969) A limnological reconnaissance of Lake Lanao. *Verhandlungen internationale Vereinigung für theoretische und angewandte Limnologie*, **17**: 1090–102
Frey, D.G. (1979) Paleolimnology of Lake Valencia, Venezuela. International Project on Paleolimnology and Late Cenozoic Climate, ed. S. Horie. *Newsletter*, No.2
Friedman, I., Norton, D.R., Carter, D.B. & Redfield, A.C. (1956) The deuterium balance of Lake Maracaibo. *Limnology and Oceanography*, **1**: 239–46
Fryer, G. (1957) Freeliving freshwater Crustacea from Lake Nyassa and adjoining waters. *Archiv für Hydrobiologie*, **53**: 527–36
Fryer, G. (1959) Some aspects of evolution in Lake Nyassa. *Evolution*, **13**: 440–51
Fryer, G. & Iles, T.D. (1972) *The cichlid fishes of the Great Lakes of Africa*. 641 pp. Edinburgh: Oliver & Boyd
Furch, K. (1976) Haupt- und spuren Metallgehalte zentralamazonischer Gewassertypen (erste Ergebnisse). *Biogeographica*, **7**: 27–43
Furch, K. & Klinge, H. (1978) Towards a regional characterization of the biogeochemistry of alkali and alkali-earth metals in northern South-America. *Acta Cientia Venezolana*, **29**: 434–44
Furse, M.T. (1972) A report of experimental trawling in Lake Chilwa, Malawi, 1971–2. In *Report to UNESCO/IBP (PF) conference Reading*, pp 28–34 (cyclostyled)
Ganapati, S.V. (1957) Limnological studies of two upland waters in the Madras State. *Archiv für Hydrobiologie*, **53**: 30–61
Ganapati, S.V. (1969) A major man-made lake in south India. In *Man-made lakes: The Accra Symposium*, ed. L.E. Obeng, pp. 57–72. Accra: Ghana Universities Press
Ganapati, S.V. (1973) Man-made lakes in South India. In *Man-made lakes: their problems and environmental effects*, eds W.C. Ackermann, G.F. White & E.B. Worthington, pp. 65–73. Geophysical Monograph 17. Washington, D.C.: American Geophysical Union
Ganapati, S.V. & Pathak, C.H. (1972) Photosynthetic productivity in the Ajwa Reservoir at Baroda, West India. In *Productivity problems of freshwaters, Proceedings of the IBP-UNESCO Symposium, Kazimierz Dolny, Poland*, eds Z. Kajak & A. Hillbricht-Ilkowska, pp. 685–92. Warsaw & Krakow: PWN Polish Scientific Publishers
Ganapati, S.V. & Sreenivasan, A. (1968) Aspects of limno-biology, primary production and fisheries in the Stanley Reservoir, Madras State. *Hydrobiologia*, **32**: 551–69
Ganapati, S.V. & Sreenivasan, A. (1970) Energy flow in natural aquatic ecosystems in India. *Archiv für Hydrobiologie*, **66**: 458–98
Ganf, G.G. (1972) The regulation of net primary production in Lake George, Uganda, East Africa. In *Productivity problems of freshwaters, Proceedings of IBP-UNESCO Symposium, Kazimierz Dolny, Poland*, eds Z. Kajak & A. Hillbricht-Ilkowska, pp. 693–708. Warsaw & Krakow: PWN Polish Scientific Publishers
Ganf, G.G. (1974a) Rates of oxygen uptake by the planktonic community of a shallow equatorial lake (Lake George, Uganda). *Oecologia*, **15**: 17–32
Ganf, G.G. (1974b) Diurnal mixing and the vertical distribution of phytoplankton in a shallow equatorial lake (Lake George, Uganda). *Journal of Ecology*, **62**: 611–29
Ganf, G.G. (1974c) Phytoplankton biomass and distribution in a shallow eutrophic lake (Lake George, Uganda). *Oecologia*, **16**: 9–29
Ganf, G.G. (1974d) Incident solar irradiance and underwater light penetration as factors controlling the chlorophyll a content of a shallow equatorial lake (Lake George, Uganda). *Journal of Ecology*, **62**: 593–609
Ganf, G.G. (1975) Photosynthetic production and irradiance–photosynthesis relationships

of the phytoplankton from a shallow equatorial lake (Lake George, Uganda). *Oecologia,* **18**: 165–83

Ganf, G.G., & Blazka, P. (1974) Oxygen uptake, ammonia and phosphate excretion by zooplankton of a shallow equatorial lake (Lake George, Uganda). *Limnology and Oceanography,* **19**: 313–25

Ganf, G.G. & Horne, A.J. (1975) Diurnal stratification, photosynthesis and nitrogen-fixation in a shallow equatorial lake (Lake George, Uganda). *Freshwater Biology,* **5**: 13–39

Ganf, G.G. & Milburn T.R. (1971) A conductimetric method for the determination of total inorganic and particulate organic carbon in freshwater. *Archiv für Hydrobiologie,* **69**: 1–13

Ganf, G.G. & Viner, A.B. (1973) Ecological stability in a shallow equatorial lake (Lake George, Uganda). *Proceedings Royal Society (B),* **184**: 321–46

Garson, M.S. (1960) The geology of the Lake Chilwa area. *Memoirs Nyasaland Geological Survey,* **12**: 1–67

Gasse, F. & Street, F.A. (1978) The main stages of the late quaternary evolution of the northern Rift Valley and Afar Lakes (Ethiopia and Djibouti). *Polish Archives of Hydrobiology,* **25**: 145–50

Gat, Y. & Stiller, M. (1981) La mer Morte. *Recherche,* **126**: 1084–93

Gaudet, J.J. (1977) Natural drawdown on Lake Naivasha (Kenya) and the origin of papyrus swamps. *Aquatic Botany,* **3**: 1–47

Gaudet, J.J. & Melack, J.M. (1981) Major ion chemistry and solute budget in a tropical African lake basin. *Freshwater Biology,* **11**: 309–33

Geisler, R. (1969) Untersuchungen über den Sauerstoffgehalt, den biochemischen Sauerstoff bedarf und der Sauerstoff Verbrauch von Fischer in einem tropischen Schwarzwasser (Rio Negro – Amazonien, Brazil). *Archiv für Hydrobiologie,* **66**: 307–25

Geisler, R., Knoppel, H.A. & Sioli, H. (1975) The ecology of freshwater fishes in Amazonia: present status and future tasks for research. *Animal Research and Development,* **1**: 102–19

Geisler, R., Schmidt, G.W. & Sookvibul, S. (1979) Diversity and biomass of fishes in three typical streams of Thailand. *Internationale Revue der gesamten Hydrobiologie,* **64**: 673–97

Geitler, L. & Ruttner, F. (1936) Die Cyanophyceen der Deutschen Limnologischen Sunda-Expedition, ihre Morphologie, Systematik und Okologie. *Archiv für Hydrobiologie,* Supplement **14**: 553–715

Gentilli, J. (1971) (ed.) *Climates of Australia and New Zealand.* World Survey of Climatology **12**. Amsterdam: Elsevier

George, M.G. (1966) Comparative plankton ecology of five fish tanks in Delhi, India. *Hydrobiologia,* **17**: 81–108

Gery, J. (1969) The freshwater fishes of South America. In *Biogeography and ecology in South America,* vol.2, eds E.J. Fittkau *et al.,* pp. 828–48. Monographiae Biologicae **19**. The Hague: Junk

Gessner, F. (1953) Auf den Spuren Alexander von Humboldts in Venezuela. *Natur und Wissenschaft,* **138**: 20

Gessner, F. (1955) Die limnologishe Verhaltnisse in den Seen und Flüssen von Venezuela. *Verhandlungen internationale Vereinigung für theoretische und angewandte Limnologie,* **12**: 284–95

Gessner, F. (1956) Das Plankton des Lago Maracaibo. *Ergebnisse der Deutschen Limnologischen Venezuela Expedition,* **1**: 67–92

Gessner, F. (1959) *Hydrobotanik II.* 701 pp. Berlin: VEB Deutscher Verlag der Wissenschaften

Gessner, F. (1960) Untersuchungen über den Phosphataushalt des Amazonas. *Internationale Revue der gesamten Hydrobiologie,* **45**: 339–45

Gessner, F. (1965) Zur Limnologie des unteren Orinoco. *Internationale Revue der gesamten Hydrobiologie*, **50**: 305-33
Gessner, F. (1968) Zur ökologischen Problematik der Uberschwemmungswälder des Amazonas. *Internationale Revue der gesamten Hydrobiologie*, **53**: 525-47
Gessner, F. & Hammer, L. (1967) Limnologische Untersuchungen an Seen der Venezuelanischen Hochanden. *Internationale Revue der gesamten Hydrobiologie*, **52**: 301-20
Gibbs, R.J. (1967) The geochemistry of the Amazon river system Part 1. The factors that control the salinity and the composition and concentration of the suspended solids. *Bulletin of the Geological Society of America*, **78**: 1203-32
Gillard, A. (1957) Rotifères. *Exploration hydrobiologique du Lac Tanganika (1946-47). Résultats scientifiques*, **3**: 1-42, Brussels: Institut Royal des Sciences Naturelles de Belgique
Gilson, H.C. (1964) Lake Titicaca. *Verhandlungen internationale Vereinigung für theoretische und angewandte Limnologie*, **15**: 112-27
Ginzburg, M., Sachs, L. & Ginzburg, B.Z. (1971) Ion metabolism in a halobacterium. 2: Ion concentrations in cells at different levels of metabolism. *Journal of Membrane Biology*, **5**: 78-101
Gladish, D.W. & Munawar, M. (1980) The phytoplankton biomass and species composition of two stations in Western Lake Erie, 1975/76. *Internationale Revue der gesamten Hydrobiologie*, **65**: 691-708
Gliwicz, Z.M. (1975) Effect of zooplankton grazing on photosynthetic activity and composition of phytoplankton. *Verhandlungen internationale Vereinigung für theoretische und angewandte Limnologie*, **19**: 1490-7
Gliwicz, Z.M. (1976) Plankton photosynthetic activity and its regulation in two neotropical man-made lakes. *Polish Archives of Hydrobiology*, **23**: 61-93
Gocke, K. (1981) Morphometric and basic limnological data of Laguna Grande de Chirripo, Costa Rica. *Revista de Biologia tropical*, **29** (in press).
Gonfiantini, R., Zuppi, G.M., Eccles, D.H. & Ferro, W. (1979) Isotope investigation of Lake Malawi. In *Isotopes in lake studies*, pp. 195-207. Proceedings of an advisory group meeting, Vienna, 29 Aug. - 2 Sept. 1977. Vienna: International Atomic Energy Agency
Goossens, J. (1951) Technique de pêche au Cambodge en region inondée. *Cybium*, **6**: 8-40
Gophen, M. (1972) Zooplankton distribution in Lake Kinneret (Israel), 1969-1970. *Israel Journal of Zoology*, **21**: 17-27
Gophen, M. (1978) Zooplankton. In *Lake Kinneret*, ed. C. Serruya, pp. 297-311. Monographiae Biologicae 32. The Hague: Junk
Gophen, M., Cavari, B.Z. & Berman, T. (1974) Zooplankton feeding on differentially labelled algae and bacteria. *Nature*, **247**: 390-4
Gosse, J.P. (1963) Le milieu aquatique et l'écologie des poissons dans la région de Yangambi. *Annales du Musée Royal de l'Afrique Central Sciences Zoologiques*, **116**: 113-270
Gouldin, M. (1981) *Man and fisheries on an Amazon frontier*. 137 pp. Developments in Hydrobiology 4. The Hague: Junk
Graham, M. (1929) *The Victoria Nyanza and its fisheries*. London: Crown Agents
Gras, R., Itlis, A. & Lévêque-Duwat, S. (1967) Le plancton du Bas Chari et de la partie Est du Lac Tchad. *Cahiers ORSTOM (Série Hydrobiologie)*, **1**: 25-96
Gras, R. & Saint Jean, L. (1969) Biologie des Crustacés du Lac Tchad. 1. Durée de développement embryonnaire et post embryonnaires: premiers résultats. *Cahiers ORSTOM (Série Hydrobiologie)*, **3**: 43-60
Gras, R. & Saint Jean, L. (1976) Durée de développement embryonnaire chez quelques

espèces de Cladocères et de Copépodes du Lac Tchad. *Cahiers ORSTOM (Série Hydrobiologie)*, **10**: 233–54

Gras, R. & Saint Jean, L. (1978a) Taux de natalité et relations entre les parametres d'accroissement et d'abondance dans une population à structure d'âge stable: cas d'une population de Cladocères à reproduction par parthénogenèse. *Cahiers ORSTOM (Série Hydrobiologie)*, **12**: 19–63

Gras, R. & Saint Jean, L. (1978b) Durée et caracteristiques du développement juvenile de quelques Cladocères du Lac Tchad. *Cahiers ORSTOM (Série Hydrobiologie)*, **12**: 119–36

Green, J. (1965) Zooplankton of lakes Mutanda, Bunyonyi and Muhele. *Proceedings of the Zoological Society of London*, **144**: 383–402

Green, J. (1971) Association of Cladocera in the zooplankton of the lake sources of the White Nile. *Journal of Zoology (London)*, **165**: 373–414

Green, J., Corbet, S.A., Watts, E. & Oey, B.L. (1976) Ecological studies on Indonesian lakes. Overturn ar restratification of Ranu Lamongan. *Journal of Zoology (London)*, **180**: 315–54

Green, J., Corbet, S.A., Watts, E. & Oey Biauw Lan (1978) Ecological studies on Indonesian lakes. The montane lakes of Bali. *Journal of Zoology (London)*, **186**: 15–38

Green, J.A.I., El Moghraby & Ali, O.M.M. (1979) Biological observations on the crater lakes of Jebel Marra, Sudan. *Journal of Zoology (London)*, **189**: 493–502

Greenwood, P.H. (1964) Explosive speciation in African lakes. *Proceedings of the Royal Institution of Great Britain*, **40**: 256–69

Greenwood, P.H. (1974) The cichlid fishes in L. Victoria, East Africa: the biology and evolution of a species flock. *Bulletin of British Museum, London, Zoological Supplement (Nat. Hist.)*, **6**: 1–134

Greenwood, P.H. (1976) Fish fauna of the Nile. In *The Nile, biology of an ancient river*, ed. J. Rzoska, pp. 127–42. Monographiae Biologicae 29. The Hague: Junk

Gresswell, R.K. & Huxley, A. (1965) *Standard encyclopedia of the world's rivers and lakes.* 384 pp. London: Weidenfeld & Nicholson

Griffiths, J.F. (1972) (ed.) *Climates of Africa.* World Survey of Climatology 10. Amsterdam: Elsevier

Grönblad, R. (1945) De algis brasiliensibus. *Acta Societatis scientiarum fennicae, Nov. Ser. B.*, **II**: 1–43

Grönblad, R. (1962) Sudanese Desmids II. *Acta Botanica Fennica*, **63**: 1–19

Grönblad, R., Prowse, G.A. & Scott, A.M. (1958) Sudanese desmids. *Acta Botanica Fennica*, **58**: 1–82

Grönblad, R., Scott, A.M. & Croasdale, H. (1964) Desmids from Uganda and Lake Victoria. *Acta Botanica Fennica*, **66**: 1–57

Grospietsch, Th. (1975) Beitrag zur Kenntnis der Testaceen-Fauna des Lago Valencia (Venezuela). *Verhandlungen internationale Vereinigung für theoretische und angewandte Limnologie*, **19**: 2778–84

Grove, A.T. & Gondie, A. (1971) Late quaternary lake levels in Rift Valley of Southern Ethiopia and elsewhere in tropical Africa. *Nature*, **234**: 403–5

Guest, N.J. & Stevens, J.A. (1951) Lake Natron: its springs, rivers, brines and visible saline reserves. *Geological Survey Tanganyika, Mineral Resources Pamphlet*, No. **58**: 21 pp.

Guilcher, A. (1965) *Précis d'Hydrologie marine et continentale.* 389 pp. Paris: Masson & Cie

Gunther, R.T. (1898) A further contribution to the anatomy of *Limnocnida tanganyikae*. *Quarterly Journal of the Microscopical Society*, **36**: 271

Gwahaba, J.J. (1973) Effects of fishing on the *Tilapia nilotica* (Linné) 1757 population in Lake George, Uganda over the past 20 years. *East African Wildlife Journal*, **11**: 317–28

Gwahaba, J.J. (1975) The distribution, population density and biomass of fish in an equatorial lake, Lake George, Uganda. *Proceedings of the Royal Society, London (B)*, **190**: 393–414

Gwahaba, J.J. (1978) The biology of cichlid fishes (Teleostei) in an equatorial lake (Lake George, Uganda). *Archiv für Hydrobiologie*, **83**: 538–51

Haas, F. (1955) Mollusca: Gastropoda. In *Report of the Percy Sladen Trust Expedition*, ed. H.C. Gilson *Transactions, Linnean Society, London* **1** (ser. 3): 275–308

Haffer, J. (1978) Distribution of Amazon forest birds. *Bonner zoologische Beiträge*, **29**: 38–79

Haines, A.K. (1979) *An ecological survey of fish of the lower Purari River System, Papua New Guinea. Purari River (Wabo) hydroelectrical scheme.* Environmental Studies, **6**. Waigani, Papua New Guinea. Office of Environment and Conservation, Central Government Office

Hall, A., Davies, B.R. & Valente, I. (1976) Cabora Bassa: some preliminary physicochemical and zooplankton preimpoundment results. *Hydrobiologia*, **50**: 17–25

Hall, B.P. & Moreau, R.E. (1970) *An atlas of speciation in African passerine birds*. London: British Museum

Hall, J.K. & Neev, D. (1975) *The Dead Sea Geophysical Survey Report 2/75*. Jerusalem: Geological Survey of Israel

Halstead, L.B. (1973) Evolution of shoreline features of Kainji Lake, Nigeria and Lake Kariba, Zambia and Southern Rhodesia. In *Man-made lakes: their problems and environmental effects*, eds W.C. Ackermann, G.F. White & E.B. Worthington, pp. 792–7. Geophysical Monograph 17. Washington D.C.: American Geophysical Union

Hamdan, G. (1961) Evolution de l'agriculture irriguée en Egypte. In *A history of land use in arid regions*, pp. 183–61. Paris: UNESCO

Hammer, L. (1965) Photosynthese und Primarproduktion im Rio Negro. *Internationale Revue der gesamten Hydrobiologie*, **50**: 335–9

Hammer, U.T. (1981) A comparative study of primary production and related factors in four saline lakes in Victoria, Australia. *Internationale Revue der gesamten Hydrobiologie*, **66**: 701–44

Hammer, U.T., Walker, R.F. & Williams, W.D. (1973) Derivation of daily phytoplankton production estimates from short-term experiments in some shallow, eutropic Australian saline lakes. *Australian Journal of marine and freshwater Research*, **24**: 259–66

Hammerton, D. (1972) The Nile River. A case history. In *River ecology and man*, eds R.T. Oglesby, C.A. Carlson & J.A. McCann, pp. 171–207. New York & London: Academic Press

Hammerton, D. (1976) The Blue Nile in the plains. In *The Nile, biology of an ancient river*, ed. J. Rzoska, pp. 243–57. Monographiae Biologicae 29. The Hague: Junk

Hancock, F.D. (1979) Diatom associations and succession in Lake Kariba, South Central Africa. *Hydrobiologia*, **67**: 33–50

Harding, D. (1964) Research on Lake Kariba, in *Joint Fisheries Research Organization Annual Report 1961*, pp. 25–50. Lusaka: Government Printer

Harding, D. (1966) Lake Kariba: the hydrology and development of fisheries. In *Man-made lakes*, ed. R.H. Lowe-McConnell, pp. 7–20. Symposia of the Institute of Biology No. **15**. London & New York: Academic Press

Harding, J.P. (1955a). Crustacea: Copepoda. In *Reports of the Percy Sladen Trust Expedition*, ed. H.C. Gilson. *Transactions Linnean Society London*, 1(Ser.3): 219–47

Harding, J.P. (1955b) Crustacea: Cladocera. In *Reports of the Percy Sladen Trust Expedition*, ed. H.C. Gilson. *Transactions Linnean Society London* 1(Ser.3): 329–54

Harding, J.P. (1957) Crustacea: Cladocera. *Exploration hydrobiologique du Lac Tanganika (1946–47). Résultats scientifiques*, 3,(6): 55–89. Brussels, Institut Royal de Sciences Naturelles de Belgique

Hardy, E.R. (1980) Composicão do zooplankton em cinco lagos da Amazonia Central. *Acta Amazonica*, **10**: 577–609
Harrison, A.D. & Rankin, J.J. (1976) Hydrobiological studies of Eastern Lesser Antillean Islands. II St. Vincent freshwater fauna, its distribution, tropical river zonation and biogeography. *Archiv für Hydrobiologie*, Supplement **50**: 275–311
Hart, B.T., McGregor, R.Y. & Perriman, W.S. (1976) Nutrient status of the sediments in Lake Mulwala. I Total phosphorus. *Australian Journal of marine and freshwater Research*, **27**, 129–35
Hartman, G. (1959) Beitrag zur Kenntnis des Nicaragua – Sees unter besonderer Berucksichtigung seiner Ostracoden. *Zoologischer Anzeiger*, **162**: 270–94
Hashem, M.T., El Maghraby, A.M. & El Sedfy, H.M. (1973) The grey mullet fishery of Lake Borullus. *Bulletin of the Institute of Oceanography and Fisheries, Cairo*, **3**: 29–54
Hauer, J. (1956) Rotatorien aus Venezuela und Kolumbien. *Ergebnisse der Deutschen Limnologischen Venezuela Expedition 1952, (West)* Berlin, 277–314
Hauer, J. (1965) Zur Rotatorienfauna des Amazonsgebietes. *Internationale Revue der gesamten Hydrobiologie*, **50**: 341–89
Haughton, S.H. (1963) *The stratification history of Africa, South of the Sahara*. 365 pp. Edinburgh: Oliver & Boyd
Haworth, E.Y. (1977) The sediments of Lake George (Uganda). V. The diatom assemblages in relation to the ecological history. *Archiv für Hydrobiologie*, **80**: 200–215
Hayes, C.W. (1899) Physiography and geology of the region adjacent to the Nicaragua Canal route. *Bulletin of the Geological Society of America*, **10**: 285–348
Hecky, R.E. (1978) The Kivu–Tanganyika basin: the last 14 000 years. *Polish Archives of Hydrobiology*, **25**: 159–65
Hecky, R.E. & Fee, E.J. (1981) Primary production and rates of algal growth in Lake Tanganyika. *Limnology and Oceanography*, **26**: 532–47
Hecky, R.E., Fee, E.J., Kling, H. & Rudd, J.M.W. (1978) Studies on the planktonic ecology of Lake Tanganyika. *Canada Fisheries and Marine Service Technical Report* No. **816**: 51 pp. Winnipeg, Manitoba: Western Region Department of Fisheries and the Environment
Hecky, R.E., Fee, E.J., Kling, H.J. & Rudd, J.W.M. (1981) Relationship between primary production and fish production in Lake Tanganyika. *Transactions of the American Fisheries Society*, **110**: 336–45
Hecky, R.E. & Kling, H.J. (1981) The phytoplankton and protozooplankton of the euphotic zone of Lake Tanganyika. Species composition, biomass, chlorophyll content, and spatiotemporal distribution. *Limnology and Oceanography*, **26**: 548–64
Hegewald, E., Aldave, A. & Hakulit, T. (1976) Investigations of the lakes of Peru and their phytoplankton 1. Review of literature, description of the investigated waters and chemical data. *Archiv für Hydrobiologie* **78**: 494–506
Hegewald, E., Aldave, A. & Schnepf, E. (1978) Investigations of the lakes of Peru and their phytoplankton. 2. The algae of pond La Laguna, Huanuco with special reference to *Scenedesmus intermedius* and *S. armatus*. *Archiv für Hydrobiologie*, **82**: 207–15
Heide, van der, J. (1972) Plankton development during the first year of inundation of the Brokopondo Reservoir in Suriname, South America. *Verhandlungen internationale Vereinigung für theoretische und angewandte Limnologie* **18**: 1784–91
Heide, van der, J. (1976) *Brokopondo Research Report, Suriname. Part II. Hydrobiology of the man-made Brokopondo Lake*. 95 pp. Utrecht: Publication of the Foundation for Scientific Research in Surinam and Netherlands Antilles No. **90**
Heide, van der, J. (1978) Stability of diurnal stratification in the forming Brokopondo Reservoir in Suriname, South America. *Verhandlungen internationale Vereinigung für theoretische und angewandte Limnologie*, **20**: 1702–9
Heintzelman, H. & Highsmith, R.M. (1973) *World regional geography*. 432 pp. Englewood Cliffs, N.J.: Prentice-Hall

Henderson, F. (1973) Stratification and circulation in Kainji Lake. In *Man-made lakes: their problems and environmental effects*. eds W.C. Ackermann, G.F. White & E.B. Worthington, pp. 489–94. Geophysical Monograph 17. Washington, D.C.: American Geophysical Union

Hernandez Devia, E.J. (1976) Contaminacion acuatica en Colombia. *Museo del Mar informe* No. 17. Fundacion Universidad de Bogota, Jorge Tadeo Lozano, Facultad de Ciencias del Mar

Herre, W.C.T. (1924) The Philippine Cyprinidae. *Philippine Journal of Sciences*, **24**: 249–307

Herre, W.C.T. (1933) The fishes of Lake Lanao: a problem in evolution. *American Naturalist*, **67**: 154–62

Herrara, R., Jordan, C.F., Klinge, H. & Medina, E. (1978) Amazon ecosystems. Their structure and functioning with particular emphasis on nutrients. *Interciencia*, **3**: 223–31

Herrera, R., Jordan, C.F., Medina, E. & Klinge, H. (1981) How human activities disturb the nutrient cycles of a tropical rainforest in Amazonia. *Ambio*, **10**: 109–14

Herrera, R., Merida, T., Stark, N. & Jordan, C.F. (1978) Direct phosphorus transfer from leaf litter to roots. *Naturwissenschaften*, **65**: 208–9

Herzen, R.P. von & Vacquier, V. (1967) Terrestrial heat flow in Lake Malawi, Africa. *Journal of geophysical Research*, **72**: 4221–6

Hess, H.H. (1962) History of ocean basins. In *Petrologic studies: a volume in honor of A.F. Buddington*, eds A.E.J. Engle, H.L. James & B.L. Leonard, pp. 599–620. New York: Geological Society of America

Hickel, B. (1973a) Phytoplankton in two ponds in Kathmandu Valley (Nepal). *Internationale Revue der gesamten Hydrobiologie*, **58**: 835–42

Hickel, B. (1973b) Limnological investigations in lakes of the Pokhara Valley, Nepal. *Internationale Revue der gesamten Hydrobiologie*, **58**: 659–72

Hinegardner, R.H. & Rosen, D.E. (1972) Cellular DNA content and the evolution of teleostean fishes. *American Naturalist*, **106**: 621–44

Hirsch, P. (1978) Microbial mats in a hypersaline solar lake; types composition. In *Abstracts of 3rd international symposium on environmental biogeochemistry*, ed. W.E. Krumbein, pp. 56–7. Occasional Publication, University of Oldenburg, West Germany

Holden, N.Y. & Green, J. (1960) The hydrology and plankton of the river Sokoto. *Journal of Animal Ecology*, **29**: 65–84

Holthuis, L.B. & Hassan, A.M. (1975) The introduction of *Palaemon elegans* Rathke, 1837. (Decapoda, Natantia) in Lake Abu-Dibic, Iraq. *Crustaceana*, **29**: 141–8

Horne, A.J. & Viner, A.B. (1971) Nitrogen fixation and its significance in tropical Lake George, Uganda. *Nature*, **232**: 417–18

Howard-Williams, C. (1974) Nutritional quality and calorific value of Amazonian forest litter. *Amazoniana*, **5**: 67–77

Howard-Williams, C. & Junk, W. (1976) The decomposition of aquatic macrophytes in the floating meadows of a central Amazonian varzea lake. *Biogeographica*, **7**: 115–23

Howard-Williams, C. & Junk, W. (1977) The chemical composition of Central Amazonian aquatic macrophytes with special reference to their role in the ecosystem. *Archiv für Hydrobiologie*, **79**: 446–64

Howard-Williams, C. & Lenton, G. (1975) The role of the littoral zone in the function of a shallow lake ecosystem. *Freshwater Biology*, **5**: 445–59

Hug, M.F., Al-Saadi, H.A. & Hameed, H.A. (1978) Phytoplankton ecology of Shatt al Arab River at Basrah, Irak. *Verhandlungen internationale Vereinigung für theoretische und angewandte Limnologie*, **20**: 1552–6

Hug, M.F., Al-Saadi, H.A. & Hameed, H.A. (1981) Studies on the primary production of the River Shatt al Arab at Basrah, Irak. *Hydrobiologia*, **77**: 25–9

Hundeshagen, F. (1909) Analyse einiger Ostafrikanischer Wässer. *Zeitschrift für öffentliche Chemie*, **15**: 202–5, 311–12

Hurst, H.E. (1952) *The Nile*, 2nd revised edn. London: Constable
Hurst, H.E., Black, R.P. & Simaika, Y.M. (1946) *The Nile basin. Vol.VII. The future conservation of the Nile.* Cairo: Government Press
Hurst, H.E. & Phillips, P. (1931) *The Nile basin. Vol.I. General description of the basin. Meteorology, topography of the White Nile basin.* Cairo: Government Press
Hussainy, S.U. & Abdulappa, M.K. (1973) A limnological reconnaissance of Lake Gorewada, Nagpur, India. In *Man-made lakes: their problems and environmental effects*, eds W.C. Ackermann, G.F. White & E.B. Worthington, pp. 500–6. Geophysical Monograph 17. Washington D.C.: American Geophysical Union
Hustedt, F. (1938) Systematische und ökologische Untersuchungen über die Diatomenflora von Iava, Bali und Sumatra nach dem material der Deutschen Limnologischen Sunda-Expedition. *Archive für Hydrobiologie*, Supplement **15**: 131–77, 187–295, 393–506
Hutchinson, G.E. (1957) *A treatise on limnology. I. Geography, physics and chemistry.* 1015 pp. New York & London: Wiley
Hutton, J.T. & Leslie, T.I. (1958) Accession of non-nitrogenous ions dissolved in rainwater to soils in Victoria. *Australian Journal of Agriculture Research*, **9**: 492–507
Idris, B.A.G. & Fernando, C.H. (1981) Cladocera of Malaysia and Singapore with new records redescriptions and remarks on some species. *Hydrobiologia*, **77**: 233–56
Iles, T.D. (1960) Activities of the Organisation in Nyassaland. D. Hydrology. In *Annual Report.* **9** (1959) pp. 11–12, 28–40. Lusaka: Joint Fisheries Research Organization, Government Printer
Iltis, A. (1971) Phytoplankton des eaux natronées du Kanem (Tchad). V. Les lacs mésohalins. *Cahiers ORSTOM (Série Hydrobiologie)*, **5**: 73–84
Iltis, A. (1974) Le phytoplancton des eaux natronées du Kanem (Tchad). Doctoral thesis, University of Paris
Iltis, A. (1977) Peuplements phytoplanctoniques du Lac Tchad. I State normal. II State petit Tchad. III Remarques générales. *Cahiers ORSTOM (Série Hydrobiologie)*, **11**: 35–52, 53–72, 189–99
Iltis, A. & Compère, P. (1974) Algues de la région du Tchad. I. Caracteristiques generales du milieu. *Cahiers ORSTOM (Série Hydrobiologie)*, **7**: 141–64
Iltis, A. & Lemoalle, J. (1979) La végétation aquatique du Lac Tchad. 17 pp. In *Proceedings of the SIL–UNEP Workshop on African limnology*, ed. J.J. Simoens. Nairobi: SIL–UNEP
Imevbore, A.M.A. & Adeniyi, F. (1977) Contribution on the role of suspended solids to the chemistry of Lake Kainji. In *Interactions between sediments and freshwater. Proceedings of international symposium, Amsterdam, Sept. 1976*, ed. H.L. Golterman, pp. 335–42. The Hague: Junk
Imevbore, A.M.A. & Bakare, O. (1974) A pre-impoundment study of swamps in the Kainji Lake Basin. *African Journal of tropical Hydrobiology and Fisheries*, **3**: 79–94
Indo-Pacific Fisheries Council (1952) *Report of Proceedings of the Indo-Pacific Fisheries Council, FAO, 3rd meeting, 1–16 Feb. 1951, Madras, India.* Rome: FAO
Infante, de A. (1978a) Zooplankton of Lake Valencia (Venezuela) species composition and abundance. *Verhandlungen internationale Vereinigung für theoretische und angewandte Limnologie*, **20**: 1186–91
Infante, de A. (1978b) Natural food of herbivorous zooplankton of Lake Valencia (Venezuela). *Archiv für Hydrobiologie*, **82**: 347–58
Infante, de A. (1981) Natural food of copepod larvae from Lake Valencia Venezuela. *Verhandlungen internationale Vereinigung für theoretische und angewandte Limnologie*, **21**: 709–14
Infante, de A., Infante, O., Marquez, M., Lewis, M.J. & Weibezahn, F.H. (1979) Conditions leading to mass mortality of fish and zooplankton in Lake Valencia, Venezuela. *Acta Cientifica Venezolana*, **30**: 67–73

Infante, O. (1981) Aspects of the feeding ecology of *Petenia kraussii* (Steindachner 1878) (Pisces, Perciformes) in Lake Valencia, Venezuela. *Verhandlungen internationale Vereinigung für theoretische und angewandte Limnologie*, **21**: 1326–33

Instituto de Fomento Nacional (1975) Departamento de Fomento y desarrollo, division de pesca. *Boletin Pesquero, Managua* **7**: 82 pp.

Irmler, V. (1979) Considerations on structure and function of the 'Central-Amazonian inundation forest ecosystem' with particular emphasis on selected soil animals. *Oecologia*, **43**: 1–18

Irion, G. (1976) Quaternary sediments of the upper Amazon Lowlands of Brazil. *Biogeographica*, **7**: 163–7

Ishak, M.M. & Abdel Malek, S.A. (1980) Some ecological aspects of Lake Qarum, Fayoum, Egypt. Part I. Physico-chemical environment. *Hydrobiologia*, **74**: 173–8

Islam, B.N. & Talbot, G.B. (1968) Fluvial migration, spawning, and fecundity of Indus River hilsa, *Hilsa ilisha*. *Transactions of American Fisheries Society*, **97**: 350–5

Jackson, P.B.N. (1961) Ichthyology, the fish of the Middle Zambezi. In *Kariba Studies*, pp. 1–36. Manchester: Manchester University Press

Jackson, P.B.N. (1963) Ecological factors affecting the distribution of freshwater fishes in tropical Africa. *Annual, Cape Province Museum*, **2**: 223–8

Jackson, P.B.N., Iles, T.D., Harding, D. & Fryer, G. (1963) *Report on a survey of northern Lake Nyasa by the Joint Fisheries Research Organization, 1953–55*. Zamba, Malawi: Government Printer

Jaeger, F. (1911) *Das Hochland des Riesenkrater und die umliegenden Hochländer Deutsch-Ostafrikas*. 133 pp. Mitteilungen aus dem Deutschen Schutzgebieten, Ergänzungsheft **4**, Berlin

Jannasch, H.W. (1975) Methane oxidation in Lake Kivu (Central Africa). *Limnology and Oceanography*, **20**: 860–4

Jayagoudar, I. (1980) Hydrobiological studies on the Ajwa Reservoir, the source of raw water supply to the Baroda water works. *Hydrobiologia*, **72**: 113–23

Jhingran, A.G. & Ghosh, K.K. (1978) The fisheries of the Ganga River system in the context of Indian aquaculture. *Aquaculture*, **14**(2): 141–2

Job, S.V. & Kannan, V. (1980) The detritus limnology of the Sathiar reservoir. *Hydrobiologia*, **72**: 81–4

Joeris, L.S. (1973) Lake Kariba; the UNDP program and North shore. In *Man-made lakes: their problems and environmental effect*, eds W.C. Ackermann, G.F. White & E.B. Worthington, pp. 143–7. Geophysical Monograph 17. Washington D.C.: American Geophysical Union

Johnson, D.S. (1967a) On the chemistry of freshwaters in Southern Malaya and Singapore. *Archiv für Hydrobiologie*, **63**: 477–96

Johnson, D.S. (1967b) Distributional patterns of Malayan freshwater fish. *Ecology*, **48**: 722–30

Johnson, D.S. (1968) Malayan blackwaters. In *Proceedings of the symposium on recent advances in tropical ecology, part I.*, eds R. Misra & B. Gopal, pp. 303–10. Faridabad, India: Publication of the International Society for Tropical Ecology, R.K. Jani, Today and Tomorrow's Printers & Publishers

JFRO (Joint Fisheries Research Organization) (1959) *Annual Report No.8 (1958)*. 58 pp. Lusaka: Government Printer

JFRO (Joint Fisheries Research Organization) (1964) *Annual Report No.11 (1961)*. Lusaka: Government Printer

Jones, B.F., Eugster, H.P. & Rettig, S.L. (1977) Hydrochemistry of the lake Magadi basin, Kenya. *Geochimica et Cosmochimica Acta*, **41**: 53–72.

Jothy, A.A. (1968) Preliminary observations of disused tin mining pools in Malaya and their potential for fish production. *Proceedings of Indo-Pacific Fisheries Council, 13th*

Session, Brisbane, Occasional Paper 69/11, 21 pp.
Jubb, R.A. (1964) The eels of South African rivers and observations on their ecology. In *Ecological studies in South Africa,* ed. D.H.S. Davis, pp. 186–205. Monographiae Biologicae **14**. The Hague: Junk
Juday, C. (1915) Limnological studies on some lakes in Central America. *Transactions of the Wisconsin Academy of Sciences, Arts and Letters,* **18**: 214–50
Junk, W.J. (1970) Investigations on the ecology and production biology of the 'floating meadows' (Paspalo-Echinochloetum) on the Middle Amazon. *Amazoniana,* **2**: 449–95
Junk, W.J. (1973*a*) Investigations on the ecology and production – biology of the 'floating meadows' (Paspalo-Echinochloetum) on the Middle Amazon. *Amazoniana,* **4**: 9–102
Junk, W.J. (1973*b*) Faunistisch ökologische untersuchungen als Moglichkeit der Definition von Lebensräumen dargestellt an Uberschwemmungsgebieten. *Amazoniana,* **4**: 263–71
Junk, W.J. (1975) The bottom fauna and its distribution in Bung Borapet, a reservoir in Central Thailand. *Verhandlungen internationale Vereinigung für theoretische und angewandte Limnologie,* **19**: 1935–46
Junk, W.J. (1976) Faunal ecological studies in inundated areas and the definition of habitats and ecological niches. *Animal Research and Development,* **4**: 47–54
Junk, W.J. (1980) Areas inundáveis – Um desafio para limnologia. *Acta Amazonica,* **10**(4): 775–95
Kafri, U. & Arad, A. (1979) Current subsurface intrusion of Mediterranean seawater a possible source of ground water salinity in the Rift Valley system, Israel. *Journal of Hydrology,* **44**: 267–87
Kalecsinski, A.V. (1901) Uber die Ungarischen Wärmen und Heissen Kochsalzseen als Naturliche Värmeaccumulatorem, sowie über die Herstellung von Wärmen Salzseen und Warmeaccumulatoren. *Zeitschrift für Gewässerkunde,* **4**: 226–48
Kaliyamurthy, M. (1974) Observations on the environmental characteristics of Pulicat Lake. *Journal of the Marine Biological Association of India,* **16**: 638–88
Kalk, M. (1971) The challenge of Lake Chilwa. *African Journal of tropical Hydrobiology and Fisheries,* **2**: 141–6
Kalk, M. (1979) Zooplankton in a quasi stable phase in an endorheic lake (Lake Chilwa, Malawi). *Hydrobiologia,* **66**: 7–15
Kalk, M., McLachlan, A.J. & Howard-Williams, C. (1979) *Lake Chilwa. Studies of change in a tropical ecosystem.* 462 pp. Monographiae Biologicae **35**. The Hague: Junk
Källqvist, T. (1979) Phytoplankton and primary production in lakes Naivasha and Baringo, Kenya. 32 pp. *Proceedings of the SIL-UNEP Workshop on African limnology,* ed. J.J. Simoens, Nairobi: SIL-UNEP
Kannan, V. & Job, S.V. (1980*a*) Diurnal depth-wise and seasonal changes of physico-chemical factors in Sathiar Reservoir. *Hydrobiologia,* **70**: 103–17
Kannan, V. & Job, S.V. (1980*b*) Diurnal seasonal and vertical study of primary production in Sathiar Reservoir. *Hydrobiologia,* **70**: 171–8
Kapestky, J.M., Escobar, J.J., Arias, P. & Zarate, M. (1978) *Algunos aspectos ecologicos de las cienagas del plano inundable del Magdalena.* 15 pp. Cartagena, Colombia: Investigaciones. INDEREA-FAO
Kaplan, I.R. (1963) *Mineralization problems in euxinic water.* Final Report NR 104-700 submitted to the Office of Naval Research
Kaplan, I.R. & Friedman, A. (1970) Biological productivity in the Dead Sea. Part I. Microorganisms in the water column. *Israel Journal of Chemistry,* **8**: 513–28
Kassas, M. (1972) Impact of river control schemes on the shoreline of the Nile Delta. In *The careless technology: ecology and international development.* eds M.T. Farvar & J.P. Milton, pp. 179–88. Garden City, N.Y.: Natural History Press

Katzer, F. (1903) *Grundzüge der Geologie des unteren Amazonas gebietes (des Staates Pará in Brasilien)*. 296 pp. Leipzig: Max Weg

Kell, V. & Saad, M.A.H. (1975) Investigations on the phytoplankton and some environmental parameters of the Shatt al Arab, Iraq. *Internationale Revue der gesamten Hydrobiologie*, **60**: 409–22

Kendall, R.L. (1969) An ecological history of the Lake Victoria basin. *Ecological Monographs*, **39**: 121–76

Kessler, A. (1970) Uber den Jahresgang der potentiellen Verdunstung in Titicaca Becken. *Archiv für Meteorologie, Geophysik und Bioklimatologie, Ser. B.* **18**: 239–52

Khan, M.A. & Zutshi, D.P. (1980a) Contribution to the high altitude limnology of the Himalayan system. I. Limnology and primary productivity of the plankton community of Nilnag Lake, Kashmir. *Hydrobiologia*, **75**: 103–12

Khan, M.A. & Zutshi, D.P. (1980b) Primary productivity and trophic status of a Kashmir Himalayan lake. *Hydrobiologia*, **68**: 3–8

Kiefer, F. (1933) Die freilebenden Copepoden der Binnenge wässer von Insulinde. *Archiv für Hydrobiologie*, Supplement **12**: 519–625

Kiefer, F. (1956) Freilebende Ruderfuss Krebse (Crustacea Copepoda) I. Calanoida und Cyclopoida. *Ergebnisse der deutschen limnologischen Venezuela Expedition*, **1**: 233–68

Kiefer, F. (1981) Contribution to the knowledge of morphology, taxonomy and geographical distribution of *Mesocyclops leuckarti* auctorum. *Archiv für Hydrobiologie*, Supplement **62**: 148–90

Kiener, A. & Richard-Vindard, G. (1972) Fishes of the continental waters of Madagascar. In *Biogeography and ecology in Madagascar*, eds R. Battestini & G. Richard-Vindard, pp. 477–99. Monographiae Biologicae 21. The Hague: Junk

Kilham, P. (1981) Pelagic bacteria: extreme abundances in African saline lakes. *Naturwissenschaften*, **68**: 380–1

Kilham, P. & Melack, J.M. (1972) Primary northupite deposition in Lake Mahega, Uganda? *Nature (Physical Science)*, **238**: 123

Kimor, B. & Berdugo, U. (1969) Preliminary report on the plankton of the Bardawil, a hypersaline lagoon in Northern Sinai. In *Biota of the Red Sea and Eastern Mediterranean. Interim Report*, pp. 90–95. (The Hebrew University–Smithsonian Institution Joint Project.) Jerusalem: Hebrew University

King, M.C. & Wilson, A.C. (1975) Evolution at two levels in humans and chimpanzees. *Science*, **188**: 107–16

Kirk, R.G. (1967) The fishes of Lake Chilwa. *Society of Malawi Journal*, **20**: 1–14

Kirk-Greene, A.H.M. (1966) *The Niger*. 32 pp. Oxford: Oxford University Press

Kitaka, G.E.B. (1971) An instance of cyclonic upwelling in the Southern offshore waters of L. Victoria. *African Journal of tropical Hydrobiology and Fisheries*, **2**: 85–92

Kittel, T. & Richerson, P.J. (1978) The heat budget of a large tropical lake, Lake Titicaca (Peru–Bolivia). *Verhandlungen internationale Vereinigung für theoretische und angewandte Limnologie*, **20**: 1203–9

Klein, C. (1961) On the fluctuations of the level of the Dead Sea since the beginning of the 19th century. *Min. Agr. Hydr. Ser. Israel, Hydrological Paper* No. 7: 1–73

Klinge, H. (1967) Podzol soils: a source of blackwater rivers in Amazonia. *Atas do Simposio sohe a biota Amazonica, Rio de Janeiro, (Limnologia)*, **3**: 117–25

Klinge, H. (1968a) Litter production in an area of Amazonian Terra Firme Forest. Part I. Litter fall, organic C and total N contents of litter. *Amazoniana*, **1**: 287–302

Klinge, H. (1968b) Litter production in an area of Amazonian Terra Firme Forest. Part II. Mineral content of the litter. *Amazoniana*, **1**: 303–10

Klinge, H. (1973) Struktur und Artenzeichtum des Zentral Amazonischen Regenwaldes. *Amazoniana*, **4**: 283–92

Klinge, H. (1976a) Balanzierung von Hauptnährstoffen in Okosystem tropischer

Regenwald (Manaus) – vorläufige Daten. *Biogeographica*, 7: 59–77
Klinge, H. (1976*b*) Nahrstoffen, Wasser und Durchwurzelung von Podsolen und Latosolen unter tropischem Regenwald bei Manaus (Amazonien). *Biogeographica*, 7: 45–58
Klinge, H. (1977) Fine litter production and nutrient return to the soil in three natural forest stands of eastern Amazonia. *Geo-Eco-Trop.*, 1: 159–67
Klinge, H. & Herrera, R. (1978) Biomass studies in Amazon caatinga forest in Southern Venezuela. 1. Standing crop of composite root-mass in selected stands. *Tropical Ecology*, 19: 93–110
Klinge, H., Medina, E. & Herrera, R. (1977) Studies on the ecology of Amazon caatinga forest in Southern Venezuela. 1. General features. *Acta Cientifica Venezolana*, 28: 270–6
Klinge, H. & Ohle, W. (1964) Chemical properties of rivers in the Amazonian area in relation to soil conditions. *Verhandlungen internationale Vereinigung für theoretische und angewandte Limnologie*, 15: 1067–76
Klinge, H. & Rodrigues, W.A. (1968) Litter production in an area of Amazonian terra firma forest I, II. *Amazoniana*, 1: 287–310
Knöppel, H.A. (1970). Food of Central Amazonian fishes. Contribution to the nutrient ecology of Amazonian rain-forest-streams. *Amazoniana*, 2: 257–353
Knorring, von, O. & DuBois, C.G.B. (1961) Carbonatitic lava from Fort Portal in Western Uganda. *Nature*, 192: 1064–5
Kobayashi, M. (1969) Chemical studies on river waters of South Eastern Asian countries. Quality of water in Thailand. *Nogaku Kengkyo*, 46: 63–112
Koechlin, J. (1972) Flora and vegetation of Madagascar. In *Biogeography and ecology in Madagascar*, eds R. Battistini & G. Richard-Vindard, pp. 145–90. Monographiae Biologicae 21. The Hague: Junk
Kornfield, I.L. (1978) Evidence for rapid speciation in African cichlid fishes. *Experientia*, 34: 335–36
Kornfield, I.L. & Koehn, R.K. (1975) Genetic variation and speciation in new world cichlids. *Evolution*, 29: 427–37
Kornfield, I.L., Ritte, U., Richler, C. & Wahrman, J. (1979) Biochemical and cytological differentiation among cichlid fishes of the Sea of Galilee. *Evolution*, 33: 1–14
Koste, W. (1972) Rotatorien aus Gewassern Amazoniens. *Amazoniana*, 3: 258–505
Koste, W. (1974) Rotatorien aus einem Ufersee des untere Rio Tapajos, dem Lago Paroni (Amazonien). *Gewässer und Abwässer* 53/54: 43–68
Koteswaram, P. (1974) Climate and meteorology of humid tropical Asia. In *Natural resources of humid tropical Asia*, vol. XII, pp. 27–85. Paris: UNESCO
Krieger, W. (1932) Die Desmidiaceen der Deutschen Limnologischen Sunda-Expedition. *Archiv für Hydrobiologie*, Supplementband, 11(3): 129–230
Krishman, M.S. (1975) III. Geology and biogeography in India. In *Monographiae Biologicae* 23, ed. M.S. Mani, pp. 60–98. The Hague: Junk
Krishnamoorthi, K.P., Gadkari, A. & Abdulappa, M.K. (1973) Limnological studies of Ambazari Reservoir, Nagpur, India, in relation to water quality. In *Man-made lakes: their problems and environmental effects*, eds W.C. Ackermann, G.T. White & E.B. Worthington, pp. 507–12 Geophysical Monograph 17. Washington D.C.: American Geophysical Union
Krishnamurthy, K.N. (1969) Observations on the food of the sandwhiting *Sillago sihama* (Forskål) from Pulicat Lake. *Journal of the Marine Biological Association of India*, 11: 295–303
Krishnamurthy, K.N. (1970) Preliminary studies on the bottom biota of Pulicat Lake. *Journal of the Marine Biological Association of India*, 13: 264–9
Krishnamurthy, K.V. & Ibrahim, A.M. (1973) *Hydrometeorological studies of Lakes*

Victoria, Kioga and Albert. Entebbe, Uganda: UNDP/WHO Hydrometeorological Survey

Kroebel, W. (1982) Die Hidrographie des Bodensees und ihre Interpretation nach quasisynoptischen Messungen durch Profilaufnahmen extrem höher Auflösung von zahreichen Parametern mit der Kieler Multisonde für die Zeit vom 24.07 bis 01.08.79 an 142 und für die Zeit vom 16.11 bis 04.12.79 an 222 über den See verteilten Mestationen. In *Report of the Landesanstalt für Umweltschutz Baden Wurtemberg Institut für Wasser und Abfallwirtschaft.* 450 pp. Karlsruhe

Krumbein, W.E., Cohen, Y. & Shilo, M. (1977) Stromatolitic cyanobacterial mats. *Limnology and Oceanography,* **22**: 635–56

Krumgalz, B.S., Hornung, H. & Oren, O.H. (1980) The study of a natural hypersaline lagoon in a desert area (the Bardawil Lagoon in Northern Sinai). *Estuarine and Coastal Marine Science,* **10**: 403–15

Kufferath, H. (1956a) Organismes trouvés dans les carottes de sondages et les vases prélevées au fond du Lac Tanganika. *Exploration hydrobiologique du Lac Tanganika, 1946–1947. Résultats scientifiques,* **4**: 1–73. Brussels: Institut Royal des Sciences Naturelles de Belgique

Kufferath, H. (1956b). Algues et protistes du fleuve Congo dans la Bas-Congo et de son estuaire. 1er. et 2ème partie. *Résultats scientifiques de l'Exploration océanographique Belge, Eaux Cotières Africaines de l'Atlantique Sud, (1948–49),* **5**: 1–75. Brussels: Institut Royal des Sciences Naturelles de Belgique

Kufferath, J. (1952) Le milieu biochimique. *Exploration hydrobiologique du Lac Tanganika, 1946–1947. Résultats scientifiques,* **1**: 31–47. Brussels: Institut Royal des Sciences Naturelles de Belgique

Kumano, S. (1978) Notes on freshwater algae from West Malaysia. *Botanical Magazine,* **91**: 97–107

Kuzoe, F.A.S. (1973) Entomological aspects of trypanosomiasis at Volta Lake. In *Man-made lakes: their problems and environmental effects,* eds W.C. Ackermann, G.F. White & E.B. Worthington, pp. 129–31. Geophysical Monograph 17. Washington D.C.: American Geophysical Union

Lafont, R. (1951) Les industries de la pêche au Cambodge. *Cybium,* **6**: 41–53

Lai, H.C. & Fernando, C.H. (1980) Zoogeographical distribution of southeast Asian freshwater Calanoida. *Hydrobiologia,* **74**: 53–66

Lake, J.S. (1967) Principal fishes of the Murray Darling River System. In *Australian inland waters and their fauna,* ed. A.H. Weatherley, pp. 192–213. Canberra: Australian National University Press

Lakshminarayana, S.S. (1965a) Studies on the phytoplankton of the River Ganges, Varanasi, India. Part I. The physico-chemical characteristics of the River Ganges. *Hydrobiologia,* **15**: 119–37

Lakshminarayana, S.S. (1965b) Studies on the phytoplankton of the River Ganges, Varanasi, India. Part II. The seasonal growth and succession of the plankton algae in the River Ganges. *Hydrobiologia,* **15**: 138–67

Lakshminarayana, S.S. (1965c) Studies on the phytoplankton of the River Ganges, Varanasi, India. Part III. Growth of certain phytoplankton in standard synthetic media. *Hydrobiologia,* **15**: 167–70

Lakshminarayana, S.S. (1965d) Studies on the phytoplankton of the River Ganges, Varanasi, India. Part IV. Phytoplankton in relation to fish population. *Hydrobiologia,* **15**: 171–75

Landsberg, H.E. (1961) Solar radiation at the Earth surface. *Solar Energy,* **3**: 95

Lartet, L. (1877) *Exploration géologique de la Mer Morte de la Palestine et de l'Idumée.* 326 pp. Paris: A. Bertrand Libraire de la Société géographique

Bibliography

Latif, A.F.A. (1976) Fishes and fisheries of Lake Nasser. In *The Nile, biology of an ancient river*, ed. J. Rzoska, pp. 299–306. Monographiae Biologicae **29**. The Hague: Junk

Lauzanne, L. (1968) Inventaire préliminaire des oligochetes du Lac Tchad. *Cahiers ORSTOM (Série Hydrobiologie)*, **2**: 83–110

Lauzanne, L. (1975) Regimes alimentaires et relations trophiques des poissons du Lac Tchad. *Cahiers ORSTOM (Série Hydrobiologie)*, **10**: 267–310

Le Van Dang (1970) Contribution to a biological study of the Lower Mekong. *Proceedings of the Regional Meeting of Inland Water Biologists in South East Asia*, pp. 65–90. Djakarta: UNESCO, Field Science Office

Leeden, van der, F. (1975) *Water resources of the World*. 568 pp. Port Washington, New York: Selected Statistics Water Information Center, Inc.

Leenheer, J.A. (1980) Origin and nature of humic substances in the waters of the Amazon River Basin. *Acta Amazonica*, **10**: 513–26

Leentvaar, P. (1966) The Brokopondo Research Project, Surinam. In *Man-made lakes*, ed. R.H. Lowe-McConnell, pp. 33–42. London: Academic Press

Leentvaar, P. (1973*a*) Lake Brokopondo. In *Man-made lakes: their problems and environmental effects*, eds W.C. Ackermann, G.F. White & E.B. Worthington, pp. 186–96. Geophysical Monograph 17. Washington D.C.: American Geophysical Union

Leentvaar, P. (1973*b*) Further developments in Lake Brokopondo, Surinam. *Amazoniana*, **11**: 1–8

Leentvaar, P. (1979) Additions and connections to the Brokopondo study (Surinam). *Amazoniana*, **6**: 521–8

Lelek, A. & El-Zarka, S. (1973) Ecological comparison of the preimpoundment and postimpoundment fish faunas of the River Niger and Kainji Lake, Nigeria. In *Man-made lakes: their problems and environmental effects*, eds W.C. Ackermann, G.F. White & E.B. Worthington, pp. 655–60. Geophysical Monograph 17. Washington D.C.: American Geophysical Union

Leloup, E. (1952) Les Invertébrés. *Exploration hydrobiologique du Lac Tanganika (1946–1947). Résultats scientifiques*, **1**: 71–100. Brussels: Institut Royal des Sciences Naturelles de Belgique

Lemoalle, J. (1979*a*) Activité photosynthétique du phytoplankton du Lac Tchad. In *Proceedings of the SIL-UNEP Workshop on African limnology*, ed J.J. Symoens, 30 pp. Nairobi: SIL-UNEP

Lemoalle, J. (1979*b*) Biomasse et production phytoplanktoniques du Lac Tchad (1968–1976). Relations avec les conditions du milieu. (Thesis) Paris: ORSTOM

Lenk, H. (1894) Uber Gesteine aus Deutsch Ost Afrika. In *Durch Massailand zur Nilquelle*, ed. O. Baumann, pp. 263–94. Berlin

Léonard, J. & Compère, P. (1967) *Spirulina platensis* (Gom.) Geitl, algue bleu de grande valeur alimentaire par sa richesse en proteines. *Bulletin du Jardin botanique national de Belgique*, **37** (supplement): 1–23

Lesaca, R.M. (1974) Coastal aquaculture and environment in the Philippines. In *Indo-Pacific Fisheries Council Proceedings 15, the session at Wellington, New Zealand, Oct. 1972. Section II. Coastal aquaculture and environment*, pp. 37–44. Bangkok: IPFC Secretariat FAO, Regional Office for Asia & the Far East

Lévêque, C. (1968) Mollusques aquatiques de la zone SE du Lac Tchad. *Bulletin de l'I.F.A.N.*, **29**: 1494–533

Lévêque, C. (1971) Prospection hydrobiologique du Lac de Léré et des mares avoisinantes. I. Milieu physique. *Cahiers ORSTOM (Série Hydrobiologie)*, **2**: 161–9

Lévêque, C. (1972) Mollusques benthiques du Lac Tchad: ecologie, étude des peuplements et estimation des biomasses. *Cahiers ORSTOM (Série Hydrobiologie)*, **6**: 3–45

Lévêque, C. (1979) Biological productivity of Lake Tchad. *Proceedings of the SIL–UNEP Workshop on African limnology*, ed. J.J. Symoens, 30 pp. Nairobi: SIL–UNEP

Lévêque, C., Carmouze, J.P., Dejoux, C., Durand, J.R., Gras, R., Iltis, A., Lemoalle, J., Loubens, G., Lausanne, L., Saint Jean, L. (1972) Recherches sur les biomasses et la productivité du Lac Tchad. In *Productivity problems in freshwaters*, eds Z. Kajak & A. Hillbricht-Ilkowska, pp. 165–81. Warsaw: Polish Scientific Publishers

Lévêque, C. & Saint Jean, L. (1979) Production secondaire (zooplankton–benthos) dans le Lac Tchad. *Proceedings of the SIL–UNEP workshop on African limnology*, ed. J.J. Symoens, 40 pp. Nairobi: SIL–UNEP

Lewis, W.M., Jr. (1973a) A limnological survey of Lake Mainit, Philippines. *Internationale Revue der gesamten Hydrobiologie*, **58**: 801–18

Lewis, W.M., Jr. (1973b) The thermal regime of Lake Lanao (Philippines) and its theoretical implications. *Limnology and Oceanography*, **18**: 200–17

Lewis, W.M., Jr. (1974) Primary production in the plankton community of a tropical lake. *Ecological Monographs*, **44**: 377–409

Lewis, W.M., Jr.(1977) Ecological significance of the shapes of abundance–frequency distributions for coexisting phytoplankton species. *Ecology*, **58**: 850–9

Lewis, W.M., Jr. (1978a) Spatial distribution of the phytoplankton in a tropical lake (Lake Lanao, Philippines). *Internationale Revue der gesamten Hydrobiologie*, **63**: 619–35

Lewis, W.M., Jr. (1978b). Dynamics and succession of the phytoplankton in a tropical lake: Lake Lanao, Philippines. *Journal of Ecology*, **66**: 849–80

Lewis, W.M., Jr. (1978c) A compositional, phytogeographical and elementary structural analysis of the phytoplankton in a tropical lake: Lake Lanao, Philippines. *Journal of Ecology*, **66**: 213–26

Lewis, W.M., Jr. & Weibezahn, F.H. (1976) Chemistry, energy flow and community structure in some Venezuelan freshwaters. *Archiv für Hydrobiologie*, Supplement **50**: 145–207

Likens, G.E. (1975) Primary production of inland aquatic systems. In *Primary production of the biosphere*. eds H. Lieth & R.H. Whittaker, pp. 185–215. Ecological Studies 14. Berlin: Springer-Verlag

Lim, R.P. (1974) Limnological studies on a Malaysian freshwater swamp, Tasek Bera, Pahang. 114 pp. M.Sc. Thesis, University of Malaya, Kuala-Lumpur, Malaysia (quoted in Arumugan & Furtado, 1980, *op cit*)

Lim, R.P. & Furtado, J.I. (1975) Population changes in the aquatic fauna inhabiting the bladderwort *Utricularia flexuosa* Vahl, in a tropical swamp, Tasek Bera, Malaysia. *Verhandlungen internationale Vereinigung für theoretische und angewandte Limnologie*, **19**: 1390–7

Limpadanai, D. & Brahamanonda, P. (1978) Salinity intrusion into Lake Songkla, a lagoonal lake of Southern Thailand. *Verhandlungen internationale Vereinigung für theoretische und angewandte Limnologie*, **20**: 1111–15

Lin, S.Y. (1961) Informe al Gobierno de Nicaragua. El desarrollo de un projecto de pesquerias continentales en dicho pars, 1959–60. *Informe FAO/ETAP*, No. **1347**

Lin, S.Y. (1963) Second Report to the Government of Guatemala on development of inland fisheries. *FAO/TA report*, No. 1719/E: 48 pp.

Lind, E.M. (1968) Notes on the distribution of phytoplankton in some Kenya waters. *British phycological Bulletin*, **3**: 481–93

Lind, E.M. & Visser, S.A. (1963) A study of a swamp at the north end of Lake Victoria. *Journal of Ecology*, **50**: 599–613

Lindberg, K. (1951) Cyclopides (Crustacés Copépodes). *Exploration hydrobiologique du Lac Tanganika (1946–47). Résultats scientifiques*, **3**: 47–78. Brussels: Institut Royal des Sciences Naturelles de Belgique

Lindberg, K. (1954a) Cyclopides (Crustacés Copépodes) du Mexique. *Arkiv för Zoology*, **7**: 459–89

Lindberg, K. *(1954b)* Cyclopides (Crustacés Copépodes) de l'Amerique du Sud. *Arkiv för Zoology,* **7**: 193–222
Lingen, Van der, M.J. (1973) Lake Kariba: early history and south shore. In *Man-made lakes: their problems and environmental effects,* eds W.C. Ackermann, G.F. White & E.B. Worthington, pp. 132–42. Geophysical Monograph 17. Washington D.C.: American Geophysical Union
Litterick, M.R., Gaudet, J.J., Kalff, J. & Melack, J.M. (1979) The limnology of an African lake, Lake Naivasha, Kenya. 74 pp. *Proceedings of the SIL–UNEP Workshop on African limnology,* ed. J.J. Symoens. Nairobi: SIL–UNEP
Livingstone, D.A. (1963) Chemical composition of rivers and lakes. In *Data of geochemistry,* ed. M. Fleischer, Chapter G. Geological survey professional paper 440-G, 64 pp. US Government Printing Office
Livingstone, D.A. (1965) Sedimentation and the history of water level change in Lake Tanganyika. *Limnology and Oceanography,* **10**: 607–9
Livingstone, D.A. (1975) Late quaternary climatic changes in Africa. *Annual Review of Ecology and Systematics,* **6**: 249–80
Livingstone, D.A. & Kendall, R.L. (1969) Stratigraphic studies of East African lakes. *Mitteilungen internationale Vereinigung für theoretische und angewandte Limnologie,* **17**: 147–53
Livingstone, D.A. & Melack, J.M. (1979) Lake Tanganyika. In *The lakes of subsaharan Africa. Proceedings of the SIL–UNEP Workshop on African limnology,* ed. J.J. Symoens. Nairobi: SIL–UNEP
Löffler, H. (1953) Limnologische Ergebnisse der Osterreichischen Iranexpedition 1949–50. *Naturwissenschaftliche Rundschau,* **6**: 64–8
Löffler, H. (1956) Limnologische Untersuchungen an Iranischen Binnengewässern. *Hydrobiologia,* **8**: 201–78
Löffler, H. (1959) Beiträge zur Kenntniss der Iranischen Binnengewasser 1–Der Nirizsee und sein Einzugsgebiet. *Internationale Revue der gesamten Hydrobiologie,* **44**: 227–76
Löffler, H. (1960) Limnologische Untersuchungen an Chilenischen und Peruanischen Binengewässern. 1. Die physikalisch-chemischen Verhaltnisse. *Arkiv för Geofysik,* **3**: 155–254
Löffler, H. (1961) Beiträger zur Kenntnis der Iranischen Binnengewasser. II Regional limnologische Studie mit besonderer Berücksichtigung der Crustaceenfauna. *Internationale Revue des gesamten Hydrobiologie,* **46**: 309–406
Löffler, H. (1964) The limnology of tropical high-mountain lakes. *Verhandlungen internationale Vereinigung für theoretische und angewandte Limnologie,* **15**: 176–93
Löffler, H. (1968) Geology of the mountainous regions of the tropical Americas. *Colloquium Geographicum,* **9**: 57–76
Löffler, H. (1969) High altitude lakes in Mt Everest region. *Verhandlungen internationale Vereinigung für theoretische und angewandte Limnologie,* **17**: 373–85
Löffler, H. (1972) Contribution to the limnology of high mountain lakes in Central America. *Internationale Revue der gesamten Hydrobiologie,* **57**: 397–408
Löffler, H. (1978) Limnological and paleolimnological data on the Bale mountain lakes (Ethiopia). *Verhandlungen internationale Vereinigung für theoretische und angewandte Limnologie,* **20**: 1131–8
Löffler, H. (1981) The winter condition of Lake Niriz in Southern Iran. *Verhandlungen internationale Vereinigung für theoretische und angewandte Limnologie,* **21**: 528–534
Loffredo, S. & Maldura, C.M. (1941) Risultadi generali delle ricerche di chimica limnologica sulle acquedei laghi dell'Africa orientale Italiana esplorati dalla missione ittologica. In *Esplorazione dei laghi della Fossa Galla,* ed. A. Piccioli, *Collezione Scientificae documentaria dell'Africa Italiana,* **III**(1): 181–200
Lortet, M.L. (1892) Researches on the pathogenic microbes of the mud of the Dead Sea. *Palestine Exploration Fund,* 48–50

Lowe, R.H. (1952) Report on the *Tilapia* and other fish and fisheries of Lake Nyasa. *Fishery Publications, Colonial Office*, **1**: 1–126

Lowe-McConnell, R.H. (1969) Species in tropical freshwater fishes. In *Speciation in tropical environments*, ed. R.H. Lowe-McConnell, pp. 51–75. *Biological Journal of the Linnean Society of London*, **1**

Lowe-McConnell, R.H. (1975) *Fish communities in tropical freshwaters*. 337 pp. London: Longman

Lucas, A. (1906) The salinity of Birket El-Qarum. *Survey notes*, **1**: 10–15

L'vovitch, M.T. (1975) World water balance (41–56). In *Selected works in water resources*, ed. A.K. Biswas, 382 pp. Champaign, Illinois: IWRA, AIRE

L'vovitch, M.I. (1977) World water resources present and future. *Ambio*, **6**: 13–21

L'vovitch, M.J. (1979) (ed.) *World water resources and their future*. English translation, ed. R.L. Nace, 415 pp. Washington D.C.: American Geophysical Union

Lynch, W.F. (1849) *Narrative of the United States' Expedition to the River Jordan and the Dead Sea*. 508 pp. Philadelphia: Lea & Blanchard

Lyra, F.H., Oliveira, L.A. & de Meno, F.M. (1976) Preliminary study of dams for hydroelectric developments of the Amazon left bank tributaries. In *XII International Congress on Large Dams, Mexico*, Vol. 3; pp. 603–14. Paris: International Commission on Large Dams

Magadza, C.H.D. (1980) The distribution of zooplankton in the Sanyati Bay, Lake Kariba; a multivariate analysis. *Hydrobiologia*, **70**: 57–67

Magis, N. (1962) Etude limnologique des lacs artificiels de la Lufira et du Lualaba (Haut Katanga). Le régime hydraulique, les variations saisonnières de la temperature. *Internationale Revue der gesamten Hydrobiologie*, **47**: 33–84

Maldonado-Koerdell, M. (1964) Geohistory and paleogeography of Middle America. In *Handbook of Middle American Indians. General*, ed. R. Wauchope, vol. I. *Natural environment and early cultures*, ed. R.C. West, pp. 1–32. Austin: University of Texas Press

Manacop, P.R. (1937) The fisheries of Lake Mainit and of northeastern Surigao, including the islands of Dinagat and Siargao. *Philippine Journal of Science*, **64**(4)

Mancy, K.H. (1981) Environmental and ecological impact of the Aswan high dam reservoir in Egypt. In *Development in arid zone ecology and environmental quality*. pp. 83–9. Rehovot & Philadelphia Pa: Balaban ISS

Mandahl-Barth, G. (1972) The freshwater Mollusca of Lake Malawi. *Revue de Zoologie et Botanique Africaine*, **86**: 257–89

Mani, M.S. (1974) Physical features. In *Ecology and biogeography in India*, ed. M.S. Mani, pp. 11–58. Monographiae Biologicae **23**. The Hague: Junk

Manohar, M. (1981) Coastal processes at the Nile delta coast. *Shore and Beach*, **49**: 8–15

Marcinek, I. (1964) *Abfluss von den Landflachen der Erde und seine Verteilung auf 5° zonen*. Berlin: Mitteilungen des Institutes für Wasserwirtschaft

Marlier, G. (1953) Etude biographique du Bassin de la Ruzizi, basée sur la distribution des poissons. *Annales Société Royal Zoologique de Belgique*, **1**(84): 175–224

Marlier, G. (1958) Recherches hydrobiologiques au Lac Tumba. *Hydrobiologia*, **10**: 382–5

Marlier, G. (1963) African fauna, freshwater biology. In *A review of the natural resources of the African continent*, pp. 341–94. Paris: UNESCO

Marlier, G. (1967) Ecological studies on some lakes of the Amazon Valley, *Amazoniana*, **1**: 91–115

Marlier, G. (1968) Etudes sur les lacs de l'Amazonie Centrale. II. Le plancton. III. Les poissons du Lac Redondo et leur regime alimentaire; les chaines trophiques du Lac Redondo. Les poissons du Rio Preto da Eva. *Cadernos de Amazonia* **11**: 21–57. Manaus: INPA

Marlier, G. (1973) Limnology of the Congo and Amazon rivers. In *Tropical forest ecosystems in Africa and South America: a comparative review*, eds B.J. Meggers, E.S. Ayensu & W.D. Duckworts, pp. 223–36. Washington D.C.: Smithsonian Institution Press

Mars, P. & Richard-Vindard, G. (1972) Contribution à l'étude écologique des peuplements du Lac Ihotry, region de Tuléar, Madagascar. *Verhandlungen internationale Vereinigung für theoretische und angewandte Limnologie*, **18**: 666–75

Mason, F. & Bryant, R.J. (1975) Production, mineral content and decomposition of *Phragmites communis* Trin and *Typha angustifolia* L. *Journal of Ecology*, **63**: 71–95

Mathew, P.M. (1977) Studies on the zooplankton of a tropical lake. In *Proceedings of the symposium on warm water zooplankton*, pp. 297–308. Dona Paula, Goa: Special Publication, National Institute of Oceanography

Matthes, H. (1964) Les poissons du Lac Tumba et de la region d'Ikela. *Annales du Musée Royal de l'Afrique Centrale*, **126**: 204 pp.

Matthes, H. (1968) The food and feeding habits of *Hydrocyon vittatus* (Cast., 1861), in Lake Kariba. *Beaufortia*, **15**: 143–53

McDougall, I., Morton, W.H. & Williams, M.A.J. (1975) Age and rates of denudation of trap series basalts at Blue Nile Gorge, Ethiopia. *Nature*, **254**: 207–9

McGowan, L.M. (1974) Ecological studies on *Chaoborus* (Diptera, Chaoboridae) in Lake George, Uganda. *Freshwater Biology*, **4**: 483–505

McKenzie, D.P. & Sclater, J.G. (1973) The evolution of the Indian Ocean. *Scientific American*, **228**: 62–72

McLachlan, A.J. (1974) Development of some lake ecosystems in tropical Africa with special reference to invertebrates. *Biological Reviews*, **49**: 365–97

McLachlan, A.J. & McLachlan, S.M. (1976) Development of the mud habitat during the filling of two new lakes. *Freshwater Biology*, **6**: 59–67

McLaren, I.A. (1963) Effects of temperature on growth of zooplankton and the adaptive value of vertical migration. *Canadian Journal of Fisheries and aquatic Sciences*, **20**: 685–727

Meek, S.E. (1908) The zoology of lakes Amatitlan and Atitlan, Guatemala with special reference to ichthyology. *Field Columbian Museum Publication*, **127**. *Zoology Series* **7**: 159–206

Meel, Van, L. (1953) Contribution à l'étude du Lac Upemba. A: Le milieu physico-chimique. *Exploration du Parc National de l'Upemba*, **1**: 1–190. Institut des Parcs Nationaux du Congo Belge

Meel, Van, L. (1954) Le phytoplancton. *Exploration hydrobiologique du Lac Tanganika 1946–1947, Résultats scientifiques*, **4**: 1–681. Brussels: Institut Royal des Sciences Naturelles de Belgique

Melack, J.M. (1978) Morphometric, physical and chemical features of the volcanic crater lakes of Western Uganda. *Archiv für Hydrobiologie*, **84**: 430–53

Melack, J.M. (1979) Photosynthetic rates in four tropical African freshwaters. *Freshwater Biology*, **9**: 555–71

Melack, J.M. (1979) Photosynthesis and growth of *Spirulina platensis* (Cyanophyta) in an equatorial lake (Lake Simbi, Kenya). *Limnology and Oceanography*, **24**: 753–60

Melack, J.M. (1980) An initial measurement of photosynthetic productivity in Lake Tanganyika. *Hydrobiologia*, **72**: 243–7

Melack, J.M. & Kilham, P. (1972) Lake Mahega: a mesothermic, sulphato-chloride lake in Western Uganda. *African Journal of tropical Hydrobiology and Fisheries*, **2**: 141–50

Melack, J.M. & Kilham, P. (1974) Photosynthetic rates of phytoplankton in East African alkaline, saline lakes. *Limnology and Oceanography*, **19**: 743–55

Mero, F. (1978) Hydrology. In *Lake Kinneret*, ed. C. Serruya, pp. 88–99. Monographiae Biologicae **32**. The Hague: Junk

Meshal, A.H. (1973) Water and salt budget of Lake Qarun, Fayoum, Egypt. 109 pp.

Ph.D. Thesis. Alexandria University
Meybeck, M. (1978) Note on dissolved elemental content of the Zaire River. *Netherlands Journal of Sea Research*, **12**: 293–5
Millbrink, G. (1977) On the limnology of two alkaline lakes (Nakuru and Naivasha) in the East Rift Valley system in Kenya. *Internationale Revue der gesamten Hydrobiologie*, **62**: 1–17
Miller, F.D., Hussein, M., Mancy, K.H. & Hilbert, M.S. (1978) Schistosomiasis in rural Egypt. *US Environmental Protection Agency Report EPA-600/1-78-070*
Miller, R.R. (1966) Geographical distribution of Central American freshwater fishes. *Copeia*, **4**: 773–802
Mills, M.L. (1977) A preliminary report on the planktonic microcrustacea of the Mivenda Bay area, Lake Kariba. *Transactions of the Rhodesia Scientific Association*, **58**: 28–37
Minty, A.E. (1973) Lake Burley Griffin, Australia. In *Man-made lakes: their problems and environmental effects*, eds W.C. Ackerman, G.F. White & E.B. Worthington, pp. 804–10. Geophysical Monograph 17. Washington D.C.: American Geophysical Union
Misra, J.N. & Singh, C.S. (1968) A preliminary study on periphyton growth in a temporary pond. In *Proceedings of the symposium on recent advances in tropical ecology*, Part 1, R. Misra & B. Gopal (eds), pp. 311–15. Varanasi: International Society for Tropical Ecology
Mitchell, D.S. (1973) Supply of plant nutrient chemicals in Lake Kariba. In *Man-made lakes: their problems and environmental effects*, eds W.C. Ackermann, G.F. White & E.B. Worthington, pp. 165–9. Washington D.C.: American Geophysical Union
Mizuno, T. & Mori, S. (1970) Preliminary hydrobiological survey of some southeast Asia inland waters. *Biological Journal of the Linnean Society*, **2**: 77–117
Mohr, P. (1961) The geology structure and origin of the Bishoftu explosion craters, Shoa, Ethiopia. *Bulletin of the Geophysical Observatory, University College, Addis Ababa*, **2**: 65–101
Mohr, P. (1962) *The geology of Ethiopia*. Addis Ababa: University College of Addis Ababa Press
Monakov, A.V. (1969) The zooplankton and zoobenthos of the White Nile and adjoining waters of the Republic of Sudan. *Hydrobiologia*, **33**: 161–85
Monheim, F. (1956) *Beiträge zur Klimatologie und Hydrologie des Titicacabeckens*. 152 pp. Heidelberg: Selbstverlag der Geographischen Institut der Universität Heidelberg
Morales, Ch. (1977) Rainfall variability – a natural phenomenon. *Ambio*, **6**: 30–3
Moreau, R.E. (1966) *The bird faunas of Africa and its islands*. 424 pp. New York: Academic Press
Morgan, P.R. (1971) The Lake Chilwa *Tilapia* and its fishery. *African Journal of tropical Hydrobiology and Fisheries*, **1**: 51–8
Morgan, A. & Kalk, M. (1970) Seasonal changes in the waters of Lake Chilwa (Malawi) in a drying phase, 1966–68. *Hydrobiologia*, **36**: 81–103
Moriarty, D.J.W. (1973) The physiology of digestion of blue-green algae in the cichlid fish, *Tilapia nilotica*. *Journal of Zoology*, **171**: 25–39
Moriarty, D.J.W., Darlington, J.P.E.C., Dunn, I.G., Moriarty, C.M. & Tevlin, M.P. (1973) Feeding and grazing in Lake George, Uganda. *Proceedings of the Royal Society, London, Series B*, **184**: 229–319
Moriarty, D.J.W. & Moriarty, C.M. (1973) The assimilation of carbon from phytoplankton by two herbivorous fishes: *Tilapia nilotica* and *Haplochromis nigripinnis*. *Journal of Zoology*, **171**: 41–55
Morris, P., Largen, M.J. & Yalden, D.W. (1976) Notes on the biogeography of the Blue Nile (Great Abbai) Gorge in Ethiopia. In *The Nile, biology of an ancient river*, ed. J. Rzoska, pp. 233–56. Monographiae Biologicae **29**. The Hague: Junk

Morrissy, N.M. (1979) Inland (non estuarine) halocline formation in a western Australian river. *Australian Journal of marine and freshwater Research*, **30**: 343–53

Moskalenko, B.K. (1972) Biological productivity system of Lake Baikal. *Verhandlungen internationale Vereinigung für theoretische und angewandte Limnologie*, **18**: 568–73

Moss, B. & Moss, J. (1969) Aspects of the limnology of an endorheic African lake (Lake Chilwa, Malawi). *Ecology*, **50**: 109–18

Mothersill, J.S., Freitag, R. & Barnes, B. (1980) Benthic macro invertebrates of north western Lake Victoria, East Africa: abundance, distribution, intra-phyletic relationships and relationships between taxa and selected element concentrations in the lake bottom sediments. *Hydrobiologia*, **74**: 215–24

Mukherjee, A.K. (1974) Some examples of recent faunal impoverishment and regression. In *Ecology and biogeography in India*, ed. M.S. Mani, pp. 330–68. Monographiae Biologicae **23**. The Hague: Junk

Munk, W.H. & Anderson, E.R. (1948) Notes on a theory of the thermocline. *Journal of marine Research*, **7**: 276–95

Mwanza, N.P. (1972) Some characteristics of a shallow endorheic lake in its drying phase and its recovery phase. Lake Chilwa (Malawi). In *Proceedings of the IBP–UNESCO symposium on productivity problems of freshwaters*, eds Z. Kajak & A. Hillbricht-Ilkowska, pp. 897–900. Warsaw: PWN Polish Scientific Publishers

Myers, G.S. (1960) The endemic fish fauna of Lake Lanao and the evolution of higher taxonomic categories. *Evolution*, **14**: 323–33

Myers, G.S. (1966) Derivation of the freshwater fish fauna of Central America. *Copeia*, **4**: 766–72

Naguib, M. (1958) Studies on the ecology of Lake Qarun. Part I. *Kieler Meeresforschungen*, **14**: 187–222

Naguib, M. (1961) Studies on the ecology of Lake Qarun (Faiyum – Egypt). Part II. *Kieler Meeresforschungen*, **17**: 94–131

Nazneen, S. (1980) Influence of hydrological factors on the seasonal abundance of phytoplankton in Kinjhar Lake, Pakistan. *Internationale Revue der gesamten Hydrobiologie*, **65**: 269–82

Neev, D. & Emery, K.O. (1967) The Dead Sea. *Geological Survey of Israel, Bulletin*, No. 41

Newell, B.S. (1960) The hydrology of Lake Victoria. *Hydrobiologia*, **15**: 363–83

Newell, N.D. (1949) Geology of the Lake Titicaca Region, Peru and Bolivia. *Memoirs, Geological Society of America*, **36**: 1–111

Newman, F.C. (1976) Temperature steps in Lake Kivu: a bottom heated saline lake. *Journal of physical Oceanography*, **6**: 157–63

Nijssen, H. (1969) Final remarks and tentative list of fish species. In *Biological Brokopondo Research Project, Surinam. Progress Report 4*, pp. 240–6. Utrecht: Foundation for Science Research in Surinam and the Netherlands Antilles

Nikolsky, G.V. & Vasilev, B.P. (1973) O nekotorykh zakonomernostyakh v respredelenii chisla khromosoma u ryb. *Voprosi Ikhtiologyi*, **13**: 2–22

Nissenbaum, A. (1970) Chemical analyses of Dead Sea and Jordan River Water, 1778–1830. *Israel Journal of Chemistry*, **8**: 281–7

Nissenbaum, A. (1975) The microbiology and biogeochemistry of the Dead Sea. *Microbial Ecology*, **2**: 139–61

Nissenbaum, A. (1978) Dead Sea asphalts. Historical aspects. *Bulletin of the American Association of Petroleum Geologists*, **62**: 837–44

Nissenbaum, A. (1979) Life in a Dead Sea – fables, allegories and scientific search. *Bio Science*, **29**(3): 153–7

Njine, T. (1977) Contribution à la connaissance des ciliés du Cameroun: Ecologie – Cytologie. *Annales de la Station biologique de Besse-en-Chandesse*, **11**: 1–55

Nordlie, F.G. (1970) *Final Report to the Organization for Tropical Studies, Inc.* University of Florida, July and August 1969 and March 1970
Nye, P.H. (1958) The mineral composition of some shrubs and trees in Ghana. *Journal of the West African Scientific Association,* **4**: 91–8
Obeng, L.E. (1969) The invertebrate fauna of aquatic plants of the Volta Lake in relation to the spread of helminth parasites. In *Man-made lakes: the Accra Symposium,* ed. L.E. Obeng, pp. 320–30. Accra: Ghana University Press
Obeng, L.E. (1973) Volta Lake: physical and biological aspects. In *Man-made lakes: their problems and environmental effects,* eds W.C. Ackermann, G.F. White & E.B. Worthington, pp. 87–97. Geophysical Monograph 17. Washington D.C.: American Geophysical Union
Obeng-Asamoa, E.K. (1977) A limnological study of the Afram arm of Volta Lake. *Hydrobiologia,* **55**: 257–64
Ohno, S. (1970*a*) The enormous diversity in genome sizes of fish as a reflection of nature's extensive experiments with gene duplication. *Transactions of the American Fisheries Society,* **99**: 120–30
Ohno, S. (1970*b*) *Evolution by gene duplication.* 160 pp. New York & Berlin: Springer-Verlag
Okemwa, E.N. (1981) Fish population studies on Lake Victoria (Kenya waters). *Abstracts of the 21st Congress of International Association of Limnology,* pp. 195. Kyoto
Osorio Tafall, B.F. (1944) Biodinamica del Lago de Patzcuaro. I. Ensayo de interpretation de sus relaciones troficas. *Revista de la Sociedad Mexicana de Historia Natural,* **5**: 197–227
Ostenfeld, C.H. (1908) Phytoplankton aus dem Victoria Nyanza. Sammelausbeute von A. Bogert, 1904–1905. *Botanisches Jahresblatt,* **41**: 330–50
Ostenfeld, C.H. (1909) Notes on the phytoplankton of Victoria Nyanza, East Africa. *Bulletin of the Museum of Comparative Zoology, Harvard,* **52**: 171–81
Oye, Van, P. (1922) Contribution a la connaissance de la flore et de la faune microscopiques des Indes Neerlandaises. *Annales de Biologie lacustre,* **11**: 130–51
Oye, Van, P. (1926) Potamoplankton de la Ruki au Congo Belge. *Revue d'Hydrobiologie,* **15**: 1–50
Oye, Van, P. (1927) Rhizopoden und Heliozoen von Belgisch Kongo. *Natuurwetenschappelijk Tijdschrift,* **9**: 4–18
Pais-Cuddou, I.C. (dos M.), Rawal, N.C. & Aggarwal, K.R. (1973) Quality of water of man-made lakes in India. In *Man-made lakes: their problems and environmental effects,* eds W.C. Ackermann, G.F. White & E.B. Worthington, pp. 513–16. Geophysical Monograph 17. Washington D.C.: American Geophysical Union
Pant, M.C., Sharma, A.P. & Sharma, P.C. (1980) Evidence for the increased eutrophication of Lake Nainital as a result of human interference. *Environmental Pollution, Series B,* 149–61
Pantulu, V.R. (1973) Fishery problems and opportunities in the Mekong. In *Man-made lakes: their problems and environmental effects,* eds W.C. Ackermann, G.F. White & E.B. Worthington, pp. 672–82. Geophysical Monograph 17. Washington D.C.: American Geophysical Union
Paperna, I. (1969) Snail vector of human schistosomiasis in the newly formed Volta Lake. In *Man-made lakes: the Accra symposium,* ed. L.E. Obeng, pp. 331–43. Accra: Ghana University Press
Paterson, C.G. & Walker, K.F. (1974) Seasonal dynamics and productivity of *Tanytarsus barbitarsis* Freeman (Diptera: Chironomidae) in the benthos of a shallow, saline lake. *Australian Journal of marine and freshwater Research* **25**: 151–65
Patnaik, S. (1971) Seasonal abundance and distribution of bottom fauna of the Chilka Lake. *Journal of the Marine Biological Association of India,* **13**: 106–25

Patnaik, S. & Sarkar, S.K. (1976) Observations on the distribution of phytoplankton in Chilka Lake. *Journal of the Inland Fisheries Society of India*, **8**: 38–48
Patrick, R. & collaborators (eds) (1966) *The Catherwood Foundation Peruvian Amazon Expedition: limnological and systematic studies*. Monographs of the Academy of Natural Sciences of Philadelphia, No. **14**: 495 pp. Philadelphia: G.W. Carpenter
Patterson, B. & Pascual, R. (1963) The extinct land mammals of South America. In *XVI Congress of Zoology, Program volume, Appendix*, J.A. Moore (ed.), pp. 138–48
Pavoni, M. (1963) Die Bedeutung des Nanoplanktons im Vergleich zum Netzplankton. *Schweizerische Zeitschrift für Hydrologie*, **25**: 219–341
Peeters, L. (1968) *Origin y evolucion de la cuenca del lago de Valencia*. 66 pp. Caracas: Venezuela Instituta para Conservacion del Lago (translated from the original German publication by the Geographical Institute of the Vzije University, Brussels)
Peeters, L. (1971) *Nuevos datos acerca de la evolucion de la cuenca del Lago de Valencia (Venezuelo) durante el Pleistoceno superior y el Holoceno*. 39 pp. Caracas: Instituta para Conservacion del Lago
Pejler, B. (1977) On the global distribution of the family Brachionidae (Rotatoria). *Archiv für Hydrobiologie*, Supplement **53**: 255–306
Pereira, H.C. (1973) *Land use and water resources in temperate and tropical climates*. Cambridge University Press
Petit, G. (1930) *L'Industrie des pêches à Madagascar*. 392 pp. Paris: Sociétié d'Editions Géographiques, Maritimes et Coloniales
Petr, T. (1975) *The Purari River – Wabo Scheme. Comments on the hydrobiology and fisheries development*. Internal report to the Office of Environment and Conservation, Central Government Office, Waigani, Papua New Guinea
Petr, T. (1976) Some chemical features of two Papuan fresh waters (Papua New Guinea). *Australian Journal of marine and freshwater Research*, **27**: 467–74
Pike, J.G. (1964) The hydrology of Lake Nyasa. *Journal of the Institution of Water Engineers*, **18**: 542–64
Pike, J.G. & Rimmington, G.T. (1965) *Malawi. A geographical study*. 229 pp. Oxford: Oxford University Press
Poll. M. (1942) Les poissons du Lac Tumba, Congo Belge. *Bulletin du Musée Royale d'Histoire Naturelle de Belgique*, **18**: 1–25
Poll, M. (1950) Histoire du peuplement et origine des espèces de la faune ichtyologique du Lac Tanganika. *Annales de la Societé Royale zoologique de Belgique*, **81**: 111–40
Poll, M. (1953) Poissons non cichlidae. *Exploration hydrobiologique du Lac Tanganika (1946–47), Résultats scientifiques*, **5**: 251 pp. Brussels: Institut Royal des Sciences Naturelles de Belgique
Poll, M. (1959) Aspects nouveaux de la faune ichtyologique du Congo Belge. *Bulletin de la Societé Zoologique de France*, **84**: 259–71
Pollingher, U. (1969) Fluctuations de la biomasse du phytoplankton du Lac Tiberiade. *Verhandlungen internationale Vereinigung für theoretische und angewandte Limnologie*, **17**: 352–7
Pollingher, U. (1978a) The algae of the River Jordan. In *Lake Kinneret*, ed. C. Serruya, pp. 223–8. Monographiae Biologicae **32**. The Hague: Junk
Pollingher, U. (1978b) The phytoplankton of Lake Kinneret. In *Lake Kinneret*, ed. C. Serruya, pp. 229–46. Monographiae Biologicae **32**. The Hague: Junk
Pollingher, U. (1978c) Protozoa. In *Lake Kinneret*, ed. C. Serruya, pp. 330. Monographiae Biologicae **32**. The Hague: Junk
Pollingher, U. (1981) The structure and dynamics of the phytoplankton assemblages in Lake Kinneret, Israel. *Journal of Plankton Research*, **3**: 93–105
Pollingher, U. & Berman, T. (1976) Autoradiographic screening for potential heterotrophs and estimation of relative phototrophic activity of natural phytoplankton

populations in Lake Kinneret. *Microbial Ecology*, **2**: 252–60
Pollingher, U. & Berman, T. (1977) Quantitative and qualitative changes in the phytoplankton of Lake Kinneret, Israel, 1972–1975. *Oikos*, **29**: 418–28
Pollingher, U & Kimor, B. (1965) The Tintinnid fauna of Lake Tiberias. *Lake Tiberias Investigations*, No. **2**: 17–21
Pollingher, U. & Serruya, C. (1976) Phased division of *Peridinium cinctum* fa *westii* (Dinophyceae) and the development of the blooms in Lake Kinneret (Israel). *Journal of Phycology*, **12**: 162–70
Pollingher, U. & Zemel, E. (1981) In situ and experimental evidence of the influence of turbulence on cell division processes of *Peridinium cinctum* forma *westii* (Lemm) Lefevre. *British phycological Journal*, **16**: 281–7
Por, F.D. (1968) The invertebrate zoobenthos of Lake Tiberias. I. Qualitative aspects. *Israel Journal of Zoology*, **13**: 78–88
Por, F.D. (1969a) The zoobenthos of the Sirbonian Lagoons. The Hebrew University of Jerusalem and the Smithsonian Institution, Washington D.C. Joint Research Project. In *Biota of the Red Sea and the Eastern Mediterranean. Interim report, March 1969*, pp. 96–8. Jerusalem: Hebrew University
Por, F.D. (1969b) Limnology of the heliothermal Solar Lake on the coast of Sinai (Gulf of Elat). *Verhandlungen internationale Vereinigung für theoretische und angewandte Limnologie*, **17**: 1031–4
Por, F.D. (1972) Hydrobiological notes on the high salinity waters of the Sinai Peninsula. *Marine Biology*, **14**: 111–19
Potts, M. (1979) Ethylene production in a hot brine environment. *Archiv für Hydrobiologie*, **87**: 359–73
Potts, M. (1980) Blue-green algae (Cyanophyta) in marine coastal environments of the Sinai Peninsula; distribution, zonation, stratification and taxonomic diversity. *Phycologia*, **19**: 60–73
Prasadam, R.D. (1977) Observations on zooplankton populations of some freshwater impoundments in Karnataka. In *Proceedings of the symposium on warm water zooplankton*. pp. 214–25. Dona Paula, Goa: Special publication, National Institute of Oceanography
Prosser, M.V., Wood, R.B. & Baxter, R.M. (1968) The Bishoftu crater lakes: a bathymetric and chemical study. *Archiv für Hydrobiologie*, **65**: 309–24
Proszynska, M. (1969) A preliminary report on the quantitative study of the Cladocera and Copepoda in the Volta Lake 1964–1965. In *Man-made lakes: the Accra symposium*, ed. L.E. Obeng, pp. 127–32. Accra: Ghana University Press
Prowse, G.A. & Talling, J.F. (1958) The seasonal growth and succession of plankton algae in the White Nile. *Limnology and Oceanography*, **3**: 222–38
Quensière. J. (1979) Synthèse des connaissances scientifiques sur la pêche et l'hydrologie du Tchad et les effets de la secheresse. *FAO Doc. Oceas. CPCA*, **8**: 18 pp.
Rai, H. (1974) Limnological studies on the River Yamona at Delhi, India, part II. The dynamics of potamoplankton in the River Yamuna. *Archiv für Hydrobiologie*, **73**: 492–517
Rai, H. (1978) Distribution of carbon, chlorophyll *a* pheo-pigments in the black water lake ecosystem of central Amazon region. *Archiv für Hydrobiologie*, **82**: 74–87
Rai, H. (1979a) Microbiology of Central Amazon lakes. *Amazoniana*, **6**: 583–99
Rai, H. (1979b) Glucose in freshwater of Central Amazon lakes: natural substrate concentrations determined by dilution bioassay. *Internationale Revue der gesamten Hydrobiologie*, **64**: 141–146
Rai, H. & Hill, G. (1980) Classification of Central Amazon lakes on the basis of their microbiological and physico-chemical characteristics. *Hydrobiologia*, **72**: 85–99
Rai, H. & Hill, G. (1981a) Bacterial biodynamics in Lago Tupe, a Central Amazonian black water 'Ria Lake'. *Archiv für Hydrobiologie*, Supplement **58**: 420–68

Bibliography

Rai, H. & Hill, G. (1981*b*) Physical and chemical studies of Lago Tupe; a Central Amazonian black water 'Ria Lake'. *Internationale Revue der gesamten Hydrobiologie*, **66**: 37–82

Rai, L.C. (1978) Ecological studies of algal communities of the Ganges River at Varanasi. *Indian Journal of Ecology*, **5**: 1–6

Raj, R.J.S. & Azariah, Y. (1967) Occurrence of the Cephalochordate *Branchiostoma lanceolatum* (Pallas) from the Pulicat lake, South India. *Journal of the Marine Biological Association of India*, **9**: 179–81

Rajagopal, J. (1969) Preliminary observations on the vertical distribution of plankton in different areas of the Volta Lake. In *Man-made lakes: the Accra symposium*, ed. L.E. Obeng, pp. 123–6. Accra: Ghana University Press

Rajan, S. (1971) Environmental studies of the Chilka lake. 2. Benthic animal communities. *Indian Journal of Fisheries*, **12**: 492–9

Ramanadham, R., Reddy, M.P.M. & Murty, A.V.S. (1964) Limnology of the Chilka lake. *Journal of the Marine Biological Association of India*, **6**: 183–201

Ramanankasina, E. (1969) Première contribution à l'étude faunistique de la rivière Andriandrano. *Verhandlungen internationale Vereinigung für theoretische und angewandte Limnologie*, **7**: 941–8

Ramanankasina, E. (1978) Recensement et analyses du peuplement planctonique de la rivière Andriandrano – Mandraka. *Verhandlungen internationale Vereinigung für theoretische und angewandte Limnologie*, **20**: 2743–9

Rao, A.V.P. (1970) Observations on the larval ingress of the milk fish, *Chanos chanos* (Forskål) into the Pulicat lake. *Journal of the Marine Biological Association of India*, **13**: 249–57

Redfield, A.C. (1958) Preludes to the entrapment of organic matter in the sediments of Lake Maracaibo, habitat of oil, *American Association of Petroleum Geologists*, 968–81

Redfield, A.C. (1961) The tidal system of Lake Maracaibo, Venezuela. *Limnology and Oceanography*, **6**: 1–12

Redfield, A.C. & Doe, L.A.E. (1965) Lake Maracaibo. *Verhandlungen internationale Vereinigung für theoretische und angewandte Limnologie*, **15**: 100–11

Reinson, G.E. (1976) Hydrogeochemistry of the Genoa River Basin, New South Wales–Victoria. *Australian Journal of marine and freshwater Research*, **27**: 165–86

Reiss, F. (1977*a*) Qualitative and quantitative investigations on the macrobenthic fauna of Central Amazon lakes I. Lago Tupé, a black water lake on the lower Rio Negro. *Amazoniana*, **6**: 203–35

Reiss, F. (1977*b*) The benthic zoocoenoses of central Amazon Varzea lakes and their adaptations to the annual water level fluctuations *Geo-Eco-Trop*. **1**: 65–75

Rey, J. & Saint Jean, L. (1968) Les Cladocères (Crustacés, Branchiopodes) du Lac Tchad. *Cahiers ORSTOM (Série hydrobiologique)*, **2**: 79–118

Rey, J. & Saint Jean, L. (1969) Les Cladocères (Crustacés, Branchiopodes) du Tchad. *Cahiers ORSTOM (Série hydrobiologique)*, **3**: 21–42

Reyes, E.F. (1972) *Estudio limnologico del Embalse de Lagartijo, Edo Miranda, Venezuela. Observaciones sobre el fitoplanctón*. Informe presentado al Instituto Nacional de Obras Sanitarias, Universidad Central de Venezuela, Caracas

Reyssac, J. & Dao, N.T. (1977) Sur quelques pêches de phytoplankton effectuées dans le Lac Titicaca (Bolivia – Perou) en decembre 1976. *Cahiers ORSTOM (Série hydrobiologique)*, **11**: 285–9

Ricardo, C.K. (1939) *Report on the fish and fisheries of Lake Rukwa in Tanganyika Territory and the Bangweulu region in Northern Rhodesia*. 78 pp. London: Crown Agents for the Colonies

Rich, F. (1932) Phytoplankton from the Rift Valley lakes in Kenya. Reports on the Percy Sladen expedition to some Rift Valley lakes in Kenya, 1929. *Annals and Magazine of Natural History*, Ser. **10**: 233–62

Richard, C. (1960) Importance des eaux de pluie à Saigon. *L'Eau*, **8**: 191–218
Richards, P.W. (1966) *The tropical rain forest: an ecological study*. 450 pp. Cambridge University Press
Richards, P.W. (1973) The tropical rain forest. *Scientific American*, **229**: 58–67
Richardson, J.L. (1966) Changes in level of Lake Naivasha Kenya, during postglacial times. *Nature*, **209**: 290–1
Richardson, J.L. (1969) Former lake level fluctuations – their recognition and interpretation. *Mitteilungen internationale Vereinigung für theoretische und angewandte Limnologie*, **17**: 78–93
Richardson, J.L. & Jin, L.T. (1975) Algal productivity of natural and artificially enriched freshwaters in Malaya. *Verhandlungen internationale Vereinigung für theoretische und angewandte Limnologie*, **19**: 1383–9
Richardson, J.L. & Richardson, A.E. (1972) History of an African rift lake and its climatic implications. *Ecological Monographs*, **42**: 499–534
Richerson, P.J., Widmer, C. & Kittel, T. (1977) *The limnology of Lake Titicaca (Peru – Bolivia) a large, high-altitude tropical lake*. 78 pp. Institute of Ecology Publication No. 14. Davis: University of California Press
Richey, J.E. (1981) Particulate and dissolved carbon in the Amazon River: a preliminary annual budget. *Verhandlungen internationale Vereinigung für theoretische und angewandte Limnologie*, **21**: 914–17
Richey, J.E., Brock, J.T., Naiman, R.J., Wissmar, R.C. & Stallard, R.F. (1980) Organic carbon: oxidation and transport in the Amazon River. *Science*, **207**: 1348–50
Robert, M. (1946) *Le Congo physique*. 3rd edn, 449 pp. Liège: Vaillant Carmanne
Roberts, T.R. (1973) Ecology of fishes in the Amazon and Congo Basins. In *Tropical forest ecosystems in Africa and South America*, eds B.J. Meggers, E.S. Ayensu & W.D. Duckworth, pp. 239–54. Washington D.C.: Smithsonian Institution Press
Robinson, A.H. & Robinson, P.K. (1971) Seasonal distribution of zooplankton in the northern basin of Lake Chad. *Journal of Zoology (London)*, **163**: 25–61
Rodhe, W. (1972) Evaluation of primary production and its conditions in Lake Kinneret (Israel). *Verhandlungen internationale Vereinigung für theoretische und angewandte Limnologie*, **18**: 93–104
Roest, F.C. (1978) *Stolothrissa tanganicae*: population dynamics, biomass evolution and life history in the Burundi waters of Lake Tanganyika. In *Symposium on river and flood plain fisheries in Africa*, ed. R.L. Welcomme. FAO ISBN 92-5-000674-8-: 42-64 CIFA Tech. Pap. No. 5
Rot, Y. (1972) Research on Bardawil Lagoon: chemical aspects. Internal Report of Israel Oceanographic and Limnological Research (in Hebrew), Haifa, Israel
Rubio, R.E. (1975) Reconocimiento limnologico del Embalse 5 de Noviembre. *Servicio de Recursos pesqueros informe tecnico*, **2**(11): 1–28
Rudd, J.W.M. (1980) Methane oxidation in Lake Tanganyika (East Africa). *Limnology and Oceanography*, **25**: 958–63
Ruello, N.V. (1976) Observations on some massive fish kills in Lake Eyre. *Australian Journal of marine and freshwater Research*, **27**: 667–72
Rufli, H. (1979) Pelagic biomass of Lake Tanganyika and Lake Malawi as estimated from hydroacoustic surveys. *Proceedings of the SIL–UNEP workshop on African limnology*, ed. J.J. Symoens. Nairobi: SIL–UNEP
Ruttner, F. (1931) Hydrographische und hydrochemische Beobachtungen auf Java, Sumatra und Bali. *Archiv für Hydrobiologie*, Supplement **8**: 197–454
Ruttner, F. (1937) Stabilität und Umschichtung in tropischen und temperierten Seen. *Archiv für Hydrobiologie*, Supplement **15**: 178–86
Ruttner, F. (1952) Planktonstudien der Deutschen Limnologischen Sunda Expedition. *Archiv für Hydrobiologie*. Supplement **21**: 1–274

Ruttner-Kolisko, A. (1966) The influence of climatic and edaphic factors on small astatic waters in the East Persian salt desert. *Verhandlungen internationale Vereinigung für theoretische und angewandte Limnologie*, **6**: 524–31

Ruttner, A.W. & Ruttner-Kolisko, A. (1973) The chemistry of springs in relation to the geology in an arid region of the Middle East (Khurasan Iran). *Verhandlungen internationale Vereinigung für theoretische und angewandte Limnologie*, **18**: 1751–2

Ryder, R.A. (1965) A method for estimating the potential fish production of north temperate lakes. *Transactions of the American Fisheries Society*, **94**: 214–18

Rzoska, J. (1957) Notes on the crustacean plankton of Lake Victoria. *Proceedings of the Linnean Society of London*, **168**: 116–25

Rzoska, J. (1974) The upper Nile swamps, a tropical wetland study. *Freshwater Biology*, **4**: 1–30

Rzoska, J. (1976) (ed.) *The Nile, biology of an ancient river*. 417 pp. Monographiae Biologicae **29**. The Hague: Junk

Rzoska, J. (1978) *On the nature of rivers with case stories of Nile, Zaire and Amazon*. 67 pp. The Hague & London: Junk

Rzoska, J. (1980) *Euphrates and Tigris. Mesopotamian ecology and destiny*. 122 pp. Monographiae Biologicae **38**. The Hague: Junk

Rzoska, J., Brook, A.J. & Prowse, G.A. (1955) Seasonal plankton development in the White and Blue Nile near Khartoum. *Verhandlungen internationale Vereinigung für theoretische und angewandte Limnologie*, **12**: 327–34

Saad, M.A. (1974) Calcareous deposits of the brackish water lakes in Egypt. *Hydrobiologia*, **44**: 381–7

Saad, M.A.H. (1978) Distribution of phosphate, nitrite and silicate in Lake Edku, Egypt. *Verhandlungen internationale Vereinigung für theoretische und angewandte Limnologie*, **20**: 1124–30

Saad, M.A.H. & Antoine, S.E. (1978*a*) Limnological studies on the River Tigris, Irak. Environmental conditions. *Internationale Revue der gesamten Hydrobiologie*, **63**: 685–704

Saad, M.A.H. & Antoine, S.E. (1978*b*) Limnological studies on the River Tigris, Irak. II. Seasonal variations of nutrients. *Internationale Revue der gesamten Hydrobiologie*, **63**: 705–19

Saad, M.A.H. & Antoine, S.E. (1978*c*) Limnological studies on the River Tigris, Irak, III. Phytoplankton. *Internationale Revue der gesamten Hydrobiologie*, **63**: 801–14

Saad, M.A.H. & Kelly V. (1975) Observations on some environmental conditions as well as phytoplankton blooms in the lower reaches of Tigris and Euphrates. *Wissenschaftliche Zeitschrift Universität Rostock*, **24**: 781–7

Salah, M.M. (1961) Biological productivity of Lake Mariut and Lake Edka. *Notes and Memoirs, Hydrobiology and Fisheries Directorate Cairo*, **63**: 1–32

Samaan, A.A. (1974) Primary production of Lake Edku. *Bulletin Institute of Oceanography and Fisheries*, **4**: 261–317

Samaan, A.A. & Aleem, A.A. (1972*a*) The ecology of zooplankton in Lake Mariut. *Bulletin of the Institute of Oceanography and Fisheries*, **2**: 339–73

Samaan, A.A. & Aleem, A.A. (1972*b*) Quantitative estimation of bottom fauna in Lake Mariut. *Bulletin of the Institute of Oceanography and Fisheries*, **2**: 375–97

Samman, J. & Thomas, M.P. (1978) Changes in zooplankton populations in the White Nile with particular reference to the effect of abate. *Journal of Environmental Studies*, **12**: 207–14

Sandon, H & Amin el Tayeb (1953) The food of some common Nile fish. *Sudan notes and records*, **30**: 245–51

Saul, W.G. (1975) An ecological study of fishes at a site in Upper Amazonian Ecuador. *Proceedings of the Academy of Natural Sciences, Philadelphia*, **127**: 93–134

Saunders, G.W., Jr., Coffman, W.P., Michael, R.G. & Krishnaswamy, S. (1975) Photosynthesis and extracellular release in ponds of South India. *Verhandlungen internationale Vereinigung für theoretische und angewandte Limnologie*, **19**: 2309–14

Saunders, J.F. (1980) Diel patterns of reproduction in rotifer populations from a tropical lake. *Freshwater Biology*, **10**: 35–9

Schindler, O. (1955) Limnologische Studien im Titicacasee. *Archiv für Hydrobiologie*, **51**: 118–24

Schmidt, G.W. (1970) Numbers of bacteria and algae and their interrelations in some Amazonia waters. *Amazoniana*, **2**: 393–400

Schmidt, G.W. (1972a) Seasonal changes in water chemistry of a tropical lake (Lago do Castanho, Amazonia South America). *Verhandlungen internationale Vereinigung für theoretische und angewandte Limnologie*, **18**: 613–21

Schmidt, G.W. (1972b) Amounts of suspended solids and dissolved substances in the middle reaches of the Amazon over the course of one year (August 1969–July, 1970). *Amazoniana*, **3**: 208–23

Schmidt, G.W. (1973) Primary production of phytoplankton in the three types of Amazonian waters. III. Primary productivity of phytoplankton in a tropical floodplain lake of central Amazonia, Lago do Castanho, Amazonas, Brazil. *Amazoniana*, **4**: 379–404

Schmidt, G.W. (1976) Primary production of phytoplankton in the three types of Amazonian waters. IV. On the primary productivity of phytoplankton in a bay of the lower Rio Negro (Amazonas, Brazil). *Amazoniana*, **5**: 517–27

Schmidt, G.W. & Uherkovich, G. (1973) Zur Artenfulle des Phytoplanktons in Amazonien. *Amazoniana*, **4**: 243–52

Schmitz, D.M. & Kufferath, J. (1955) Problèmes posés par la présence de gas dessous dans les eaux profondes du Lac Kivu. *Bulletin des Séances Académie des Sciences coloniales, N.S.*, **1**: 326–56

Schwabe, G.H. (1968) Towards an ecological characterisation of the South American Continent. In *Biogeography and ecology in South America*, eds E.J. Fittkau *et al.*, vol.1, pp. 113–36. Monographiae Biologicae **18**. The Hague: Junk

Schwerdtfegar, W. (1976) (ed.) Climates of Central and South America. *World Survey of Climatology*, **12**. Amsterdam: Elsevier

Scott, J.A. (1937) The incidence and distribution of human schistosomes in Egypt. *American Journal of Hygiene*, **25**: 566–614

Scrutton, R.A. (1976) Microcontinents and their significance. In *Geodynamics progress and prospects*, ed. C.L. Drake, pp. 177–82. Washington D.C.: American Geophysical Union

Scudder, T. (1972) Ecological bottlenecks, and the development of the Kariba Lake basin. In *The careless technology, ecology and international development*, ed. M.T. Farvar & J.P. Milton, pp. 206–35. Garden City, N.Y.: National History Press

Selby, J. (1981) A salty problem for the River Murray. *New Scientist*, 842–4

Sellers, W.D. (1965) *Physical climatology*. 212 pp. Chicago: University of Chicago Press

Sellers, W.D. (1977) Water circulation on the global scale: natural factors and manipulation by man. *Ambio*, **6**: 10–12

Serruya, C. (1975) Nitrogen and phosphorus balances and load–biomass relationship in Lake Kinneret (Israel). *Verhandlungen internationale Vereinigung für theoretische und angewandte Limnologie*, **19**: 1357–69

Serruya, C. (1978a) The benthic fauna. In *Lake Kinneret*, ed. C. Serruya, pp. 329–81. Monographiae Biologicae. **32**. The Hague: Junk

Serruya, C. (1978b) Vertical distribution of benthic fauna. In *Lake Kinneret*, ed. C. Serrruya, pp. 391–4. Monographiae Biologicae **32**. The Hague: Junk

Serruya, C. (1978c) *Lake Kinneret*. Monographiae Biologicae **32**. The Hague: Junk

Serruya, C. (1980) Water quality problems of Lake Kinneret. In *Water quality*

management under conditions of scarcity, ed. H. Shuval, pp. 167–87. New York: Academic Press Inc.
Serruya, C. & Berman, T. (1975) Phosphorus, nitrogen and the growth of algae in Lake Kinneret. *Journal of Phycology*, **11**: 155–62
Serruya, C., Gophen, M. & Pollingher, U. (1980) Lake Kinneret: carbon flow patterns and ecosystem management. *Archiv für Hydrobiologie*, **83**: 265–302
Serruya, C., Serruya, S. & Pollingher, U. (1978) Wind, phosphorus release and division rate of *Peridinium* in Lake Kinneret. *Verhandlungen internationale Vereinigung für theoretische und angewandte Limnologie*, **20**: 1096–102
Serruya, S. (1975) Wind water temperature and motions in Lake Kinneret. General pattern. *Verhandlungen internationale Vereinigung für theoretische und angewandte Limnologie*, **19**: 73–87
Servant, M. (1970) Données stratigraphiques sur le Quaternaire supérieur et récent au nord-est du Lac Tchad. *Cahiers ORSTOM, Série Géologie*, **2**(1): 95–114
Shaheen, A.H. & Yosef, S.F. (1978) The effect of the cessation of Nile flood on the hydrographic features of Lake Manzala, Egypt. *Archiv für Hydrobiologie*, **84**: 339–67
Shalie, Van der, H. (1972) World Health Organization Project Egypt 10: A case history of a schistosomiasis control project. In *The careless technology, ecology and international development*, eds M.T. Farvar & J.P. Milton, pp. 116–36. Garden City, N.Y.: Natural History Press
Sharaf El Din, S.H. (1977) Effect of the Aswan High Dam on the Nile flood and on the estuarine and coastal circulation pattern along the Mediterranean Egyptian coast. *Limnology and Oceanography*, **22**: 194–207
Shibl, Y. (1971) *The Aswan High Dam*. Beirut: Arab Institute for Research and Publishing
Shiel, R.J. (1979) Synecology of the Rotifera of the River Murray, South Australia. *Australian Journal of marine and freshwater Research*, **30**: 255–63
Sick, W.D. (1969) Geographical substance. In *Biogeography and ecology in South America*, eds E.J. Fittkau *et al.* pp. 449–74. Monographiae Biologicae 19. The Hague: Junk
Simpson, B.B. & Haffer, J. (1978) Speciation patterns in the Amazonian forest biota. *Annual Review of Ecology and Systematics*, **9**: 497–518
Sioli, H. (1950) Das Wasser in Amazonasgebiet. *Forschungen und Fortschritte*, **26**: 274–80
Sioli, H. (1954) Gewässerchemie und Vorgänge in den Boden im Amazonasgebiet. *Naturwissenschaften*, **41**: 456–7
Sioli, H. (1955) Beitrage zur regionalen Limnologie des Amazonasgebietes. III. Uber einige Gewässer des oberen Rio Negro-Gebietes. *Archiv für Hydrobiologie*, **50**: 1–32
Sioli, H. (1968a) Hydrochemistry and geology in the Brazilian Amazon region. *Amazoniana*, **1**: 267–77
Sioli, H. (1968b) Principal biotopes of primary production in waters of Amazonia. In *Proceedings of the symposium on recent advances in tropical ecology*, eds. R. Misra & B. Gopal, (Part II) pp. 591–600. Faridabad, India: R.K. Jain, Today & Tomorrow's Printers and Publishers
Sioli, H. (1975a) Tropical river: the Amazon. In *River ecology*, ed. B.A. Whitton, pp. 461–88. Oxford: Blackwell Scientific Publications
Sioli, H. (1975b) Tropical rivers: expressions of their terrestrial environments. In *Tropical ecological systems. Trends in terrestrial and aquatic research*, eds. F.B. Golley & E. Medina, pp. 275–88. Berlin: Springer-Verlag
Sioli, H. (1977) Amazonasgebiet – Zerstörung des ökologischen Gleichgewichtes? *Geologische Rundschau*, **66**(3): 782–95
Sioli, H. (1979) Principles and models as tools for ecosystem research, with examples from Amazon basin. *Biogeographica*, **16**: 145–58
Sitaramaiah, P. (1967) Community metabolism in a tropical freshwater pond.

Hydrobiologia, **29**: 93–112
Skuja, H. (1948) Taxonomie des Phytoplankton einiger Seen in Uppland, Schweden. *Symbolae Botanicae Upsalienses*, **9**(3): 400 pp. Uppsala.
Smith, H.M. (1945) The freshwater fishes of Siam or Thailand. *Bulletin United States National Museum Smithsonian Institution*, **188**: 622 pp.
Sorentino, C. (1979) Mercury in marine and freshwater fish of Papua New Guinea. *Australian Journal of marine and freshwater Research*, **30**: 617–24
Spataru, P. (1976) The feeding habits of *Tilapia galilea* (Artedi) in Lake Kinneret (Israel). *Aquaculture*, **9**: 47–59
Spataru, P. & Zorn, M. (1978) Food and feeding habits of *Tilapia aurea* (Steindachner) (Cichlidae) in Lake Kinneret (Israel). *Aquaculture*, **13**: 67–79
Sreenivasan, A. (1964a) A hydrological study of a tropical impoundment Bhavanisagar reservoir Madras State, India, for the years 1956–61. *Hydrobiologia*, **24**: 514–39
Sreenivasan, A. (1964b) Limnological studies and fish yield in three upland lakes of Madras State, India. *Limnology and Oceanography*, **9**: 564–75
Sreenivasan, A. (1964c) A hydrological study of some mountain lakes and ponds in Madras State with reference to their suitability for breeding and culture of German and English carps. *Archiv für Hydrobiologie*, **60**: 225–40
Sreenivasan, A. (1965) Limnology of tropical impoundments. III. Limnology and productivity of Amaravathy reservoir (Madras State) India. *Hydrobiologia*, **26**: 501–16
Sreenivasan, A. (1966) Limnology of tropical impoundments. I. Hydrological features and fish production in Stanley reservoir, Mettur Dam. *Internationale Revue der gesamten Hydrobiologie*, **51**: 295–306
Ssentongo, G.W. (1974) On the fishes and fisheries of Lake Baringo. *African Journal of tropical Hydrobiology and Fisheries*, **3**: 95–106
Stanhill, G. & Neumann, J. (1978) Energy balance and evaporation. In *Lake Kinneret*, ed. C. Serruya, pp. 173–82. Monographiae Biologicae 32. The Hague: Junk
Stark, N.M. (1969) Mycorrhiza and nutrient cycling in the tropics. Proceedings First North American Conference on Mycorrhizae, April 1969. *Miscellaneous Publication* **1189**, *USDA Forest Service Separate* No. **FS 303**: 228–9
Steenis, van, C.G.G. & Ruttner, F. (1932) Die Pteridophyten und Phanerogamen der Deutschen Limnologischen Sunda-Expedition. Mit Vegetationsskizzen nach Tagebuchaufzeichnungen. *Archiv für Hydrobiologie*, Supplement **9**: 231–387
Sternberg, H.O.H. (1968) Man and environmental change in South America. In *Biogeography and ecology in South America*, ed., E.J. Fittkau *et al.*, pp. 413–45. Monographiae Biologicae 18. The Hague: Junk
Stuart, L.C. (1964) Fauna of middle America. In *Handbook of Middle American Indians General*, ed. R. Wanchope, vol. I. *Natural environment and early cultures*, ed. R.C. West, pp. 316–62. Austin: University of Texas Press
Swain, F.M. (1964) Limnology and geology of Lake Nicaragua, Nicaragua. *Verhandlungen internationale Vereinigung für theoreische und angewandte Limnologie*, **15**: 149–50
Swain, F.M. (1966) Bottom sediments of Lake Nicaragua and Lake Managua, Western Nicaragua. *Journal of sedimentary Petrology*, **36**: 522–40
Swain, F.M. & Gilby, J.M. (1964) Ecology and taxonomy of Ostracoda and an alga from Lake Nicaragua. *Publicazioni della Stazione Zoologica di Napoli*, Supplemento, 361–86
Symoens, J.J. (1968) Exploration hydrobiologique du basin du Lac Bangweolo et de la Luapula. *Cercle hydrobiologique de Bruxelles*, **2**(1): 1–199
Tabor, H. (1963) Solar ponds. Large area solar collectors for power production. *Solar Energy*, **7**(4): 189–94

Bibliography

Tabor, H. (1980) The promise of solar ponds. *Kidma*, **19**: 8-11
Tabor, H. & Matz, R. (1965) A status report on a solar pond project. *Solar Energy*, **9**(4): 177-82
Takahashi, K. & Arakawa, H. (1981) (eds) *Climates of Southern and Western Asia.* World Survey of Climatology **9**. Amsterdam: Elsevier
Talling, J.F. (1957) The longitudinal succession of water characteristics in the White Nile. *Hydrobiologia*, **11**: 73-92
Talling, J.F. (1963) Origin of stratification in an African Rift Lake. *Limnology and Oceanography*, **8**: 68-78
Talling, J.F. (1965a) The photosynthetic activity of phytoplankton in East African lakes. *Internationale Revue der gesamten Hydrobiologie*, **50**: 1-32
Talling, J.F. (1965b) Comparative problems of phytoplankton production and photosynthetic productivity in a tropical and a temperate lake. *Memorie dell'Istituto Italiano di Idrobiologia*, **18** (supplemento): 399-424
Talling, J.F. (1966) The annual cycle of stratification and phytoplankton growth in Lake Victoria (East Africa). *Internationale Revue der gesamten Hydrobiologie*, **51**: 545-621
Talling, J.F. (1976) Water characteristics. In *The Nile, biology of an ancient river*, ed. J. Rzoska, pp. 357-84. Monographiae Biologicae 29. The Hague: Junk
Talling, J.F. & Rzoska, J. (1967) The development of plankton in relation to hydrological regime in the Blue Nile. *Journal of Ecology*, **55**: 637-62
Talling, J.F. & Talling, I.B. (1965) The chemical composition of African Lake waters. *Internationale Revue der gesamten Hydrobiologie*, **50**: 421-63
Talling, J.F., Wood, R.B., Prosser, M.V. & Baxter, R.M. (1973) The upper limit of photosynthetic productivity by phytoplankton: evidence from Ethiopian soda lakes. *Freshwater Biology*, **3**: 53-76
Tamayo, J.L. in collaboration with West, R.C. (1964) The hydrography of Middle America. In *Handbook of Middle American Indians General*, ed. R. Wauchope, vol.1. *Natural environment and early cultures*, ed. R.C. West, pp. 84-121. Austin: University of Texas Press
Taylor, F.B. (1910) Bearing of the Tertiary mountain belt on the origin of the Earth's plan. *Bulletin of the Geological Society of America*, **21**: 179-266
Tchernov, E. (1975) *The early Pleistocene molluscs of Erq el Ahmar.* 36 pp. Jerusalem: Publications of Israel Academy of Sciences & Humanities
Thienemann, A. (1957) Die Fische der Deutschen Limnologischen Sunda Expedition. *Archiv für Hydrobiologie*, Supplementband, **23**: 471-7
Thomas, A.J. (1970) Crab fishery of the Pulicat lake. *Journal of the Marine Biological Association of India*, **13**: 278-80
Thomasson, K. (1955) A plankton sample from Lake Victoria. *Svensk Botanisk Tidskrift*. **49**: 259-74
Thomasson, K. (1956) Reflections on arctic and alpine lakes. *Oikos*, **7**: 117-43
Thomasson, K. (1960) Notes on the plankton of Lake Bangweulu. *Nova Acta Regiae Societatis Scientiarum Upsaliensis Serie IV*, **17**: 43 pp.
Thomasson, K. (1965) Notes on algal vegetation of Lake Kariba. *Nova Acta Regiae Societatis Scientiarum Upsaliensis Serie IV*, **19**: 1-34
Thomasson, K. (1971) Amazonian algae. *Institut Royal des Sciences Naturelles de Belgique, Mémoires. Ser. 2*, No. **86**: 1-57
Thomasson, K. (1980) Plankton of Lake Kariba re-examined. *Acta Phitogeographica Sueccica* **68**, *Studies in Plant Ecology*, 157-62
Thompson, K. (1976) Swamp development in the head-waters of the White Nile. In *The Nile, biology of an ancient river*, ed. J. Rzoska, pp. 177-96. Monographiae Biologicae **29**. The Hague: Junk

Thornes, J.B. (1969) Variability in specific conductance and pH in the Casiquiare–Upper Orinoco. *Nature*, **221**: 461–2
Thorson, T.B., Cowan, C.M. & Watson, D.E. (1966) Sharks and sawfish in the Lake Izabal–Rio Dulce system, Guatemala. *Copeia*, **3**: 620–2
Tilzey, R.D.J. (1976) Observations on interactions between indigenous Galaxiidae and introduced Salmonidae in the Lake Eucumbene catchment, New South Wales. *Australian Journal of marine and freshwater Research*, **27**: 551–64
Timms, B.V. (1970) Chemical and zooplankton studies of lentic habitats in North Eastern New South Wales. *Australian Journal of marine and freshwater Research*, **21**: 11–33
Timms, B.V. (1972) A meromictic lake in Australia. *Limnology and Oceanography*, **17**: 918–21
Timms, B.V. (1974a) Aspects of the limnology of Lake Tali Karng, Victoria. *Australian Journal of marine and freshwater Research*, **25**: 273–9
Timms, B.V. (1974b) Morphology and benthos of three volcanic lakes in the Mt. Cambier district, South Australia. *Australian Journal of marine and freshwater Research*, **25**: 287–97
Timms, B.V. (1976) A comparative study of the limnology of three maar lakes in western Victoria. I. Physiography and physico-chemical features. *Australian Journal of marine and freshwater Research*, **27**: 35–60
Todd, D.K. (1970) (ed.) *The water encyclopedia: a compendium of useful information on water resources.* 559 pp. Port Washington, N.Y.: Water Information Center
Tricart, J. (1974) Existence de périodes sèches au Quaternaire en Amazonie et dans les régions voisines. *Revue de Géomorphologie dynamique*, **4**: 145–58
Tricart, J. (1975) Influence des oscillations climatiques récentes sur le modèle en Amazonie Orientale (Region de Santarém) d'après les images radar lateral. *Zeitschrift für Geomorphologie, NF.* **19**: 140–63
Tsurnamal, M. (1978) *Typhlocaris galilaea*. The blind prawn of Galilee. In *Lake Kinneret*, ed. C. Serruya, pp. 353–64. Monographiae Biologicae **32**. The Hague: Junk
Turner, J.L. (1977) Some effects of demersal trawling in Lake Malawi (Lake Nyasa) from 1968 to 1979. *Journal of Fish Biology*, **10**: 261–72
Turner, J.S. (1973) *Buoyancy effects in fluids.* 367 pp. Cambridge University Press
Tutin, T.G. (1940) The algae. In *Reports of the Percy Sladen Trust Expedition*, ed. H.C. Gilson, pp. 191–202. *Transactions Linnean Society London*, **1**
Tweddle, D., Lewis, D.S.C. & Willoughby, N.G. (1979) The nature of the barrier separating the Lake Malawi and Zambezi fish faunas. *Ichthyological Bulletin J.L.B. Smith Institute of Ichthyology Grahamstown*, **39**: 1–9
Ueno, M. (1976) Zooplankton of Lake Titicaca on the Bolivian side. *Hydrobiologia*, **29**: 547–68
Uherkovich, G. (1976) Algen aus den Flussen Rio Negro und Rio Tapajos. *Amazoniana*, **5**: 465–516
Uherkovich, G. & Rai, H. (1979) Algen aus dem Rio Negro und seinen Nebenflüssen. *Amazoniana*, **6**: 611–38
Uherkovich, G. & Schmidt, G.W. (1974) Phytoplanktontaxa in dem Zentralamazonischen Schwemmlandsee Lago do Castanho. *Amazoniana*, **5**: 243–83
UNESCO, (1969). *Discharge of selected rivers of the world*, vol.1. A contribution to the International Hydrological decade. General and regime characteristics of selected UNESCO stations. Paris: UNESCO
UNESCO (1971) *Studies and reports in hydrology. Discharge of selected rivers of the world*, vol. 2, 194 pp. Paris: UNESCO
UNESCO (1976) *Studies and reports in hydrology. World catalogue of very large floods.* 424 pp. Paris: UNESCO

US Department of Energy (1979) *Joint Egypt/US Report on Egypt.* US Cooperative Energy Assessment. **DOE/1A-0002/04.** Washington D.C.: US Government Printing Office.

Utermöhl, H. (1958) Zur Gewässer typenfrage tropischer Seen. *Verhandlungen internationale Vereinigung für theoretische und angewandte Limnologie,* **13**: 236–51

Vaas, K.F. (1954) On the nutritional relationships between plankton and fish in Indonesian freshwater ponds. In: *Symposium on marine and freshwater plankton in the Indo-Pacific.* pp. 90–7. Bangkok: FAO–UNESCO Indo-Pacific Fish Council

Vaas, K.F. & Sachlan, M. (1955) Limnological studies on diurnal fluctuations in shallow ponds in Indonesia. *Verhandlungen internationale Vereinigung für theoretische und angewandte Limnologie,* **12**: 309–19

Vaas, K.K. (1980) On the trophic status and conservation of Kashmir lakes. *Hydrobiologia,* **68**: 9–15

Vanzolini, P.E. (1973) Paleoclimates, relief and species multiplication in equatorial forest. In *Tropical forest ecosystems in Africa and South America: a comparative review,* eds B.J. Meggers, E.S. Ayensu & W.D. Duckworth, pp. 255–8. Washington, D.C.: Smithsonian Institution Press

Vanzolini, P.E. & Williams, E.E. (1970) South American anoles; the geographic differentiation and evolution of the *Anolis chrysolepis* species group (Sauria, Ignanidae). *Arquivos de Zoologia (Sao Paulo, Brazil),* **19**: 1–298

Vareschi, E. (1978) The ecology of Lake Nakuru (Kenya). I. Abundance and feeding of the lesser flamingo. *Oecologia (Berlin),* **32**: 11–35

Vareschi, E. (1979) The ecology of Lake Nakuru (Kenya). II. Biomass and spatial distribution of fish (*Tilapia grahami*) Boulenger *Sarotherodon alcalicum grahami* (Boulenger). *Oecologia (Berlin),* **37**: 321–35

Vareschi, V. (1963) Die Gabelteilung des Orinoco. *Petermanns Geographische Mitteilungen,* **1963**: 242–8

Velde, Van der, J., Dumont, H.J. & Grootaert, P. (1978) Report on a collection of Cladocera from Mexico and Guatemala. *Archiv für Hydrobiologie,* **83**: 391–404

Verbeke, J. (1957) Recherches écologiques sur la faune des grand lacs de l'Est Congo Belge. *Exploration Hydrobiologique des lacs Kivu, Edouard et Albert (1952–54). Résultats scientifiques,* **3**: 177 pp. Brussels: Institut Royal des Sciences Naturelles Belgique

Villwock, W. (1963) Die gattung *Orestias* (Pisces, Microcyprini) und die Frage der intralakustrischen Speziation im Titicaca – Seengebiet. *Verhandlungen der Deutschen Zoologischen Gesellschaft, Wien 1962,* **26**: 610–24

Villwock, W. (1972) Gefahren für die endemische Fisch fauna durch Einbürgerungsversuche und Akklimatisation von Fremdfischen am Beispiel des Titicaca-Sees (Peru–Bolivien) und des Lanao-Sees (Mindanao/Philippinen). *Verhandlungen internationale Vereinigung für theoretische und angewandte Limnologie,* **18**: 1227–34

Viner, A.B. (1969a) The chemistry of the water of Lake George, Uganda. *Verhandlungen internationale Vereinigung für theoretische und angewandte Limnologie,* **17**: 289–96

Viner, A.B. (1969b) Hydrobiological observations on the Volta Lake, North of Ajena, April 1965–April 1966. In *Man-made lakes: the Accra Symposium,* ed. L.E. Obeng, pp. 133–43. Accra: Ghana University Press

Viner, A.B. (1970) Hydrogeology of the Volta Lake, Ghana. 1. Stratification and circulation of water. *Hydrobiologia,* **35**: 209–29

Viner, A.B. (1972) Responses of a mixed phytoplankton population to nutrient enrichments of ammonia and phosphate and some associated ecological implications. *Proceedings of the Royal Society of London, B,* **183**: 351–70

Viner, A.B. (1975a) The supply of minerals to tropical rivers and lakes (Uganda). In *An introduction to land water interactions*, ed. J. Olson, pp. 227–61. Berlin: Springer-Verlag

Viner, A.B. (1975*b*) Non biological factors affecting phosphate recycling in the water of a tropical eutrophic lake. *Verhandlungen internationale Vereinigung für theoretische und angewandte Limnologie*, **19**: 1404–15

Viner, A.B. (1975*c*) The sediments of Lake George, Uganda, I: Redox potentials, oxygen consumption and carbon dioxide output. *Archiv für Hydrobiologie*, **76**: 181–97

Viner, A.B. (1975*d*) The sediments of Lake George Uganda, II: Release of ammonium and phosphate from an undisturbed mud surface. *Archiv für Hydrobiologie*, **76**: 368–78

Viner, A.B. (1975*e*) The sediments of Lake George Uganda, III: The uptake of phosphate. *Archiv für Hydrobiologie*, **76**: 393–410

Viner, A.B. (1977*a*) The sediments of Lake George, Uganda, IV: The vertical distribution of chemical characteristics. *Archiv für Hydrobiologie*, **80**: 40–69

Viner, A.B. (1977*b*) The influence of sediments upon nutrient exchanges in tropical lakes. In *Interactions between sediments and freshwater*, ed. H.L. Golterman, pp. 210–15. SIL–UNESCO Symposium. The Hague: Junk

Viner, A.B. (1979) Observations upon nutrient recycling in two contrasting African lakes. *Proceedings of the SIL–UNEP workshop on African limnology*, ed. J.J. Symoens. Nairobi: SIL–UNEP

Viner, A.B. & Smith I.R. (1973) Geographical, historical and physical aspects of Lake George. *Proceedings of the Royal Society of London, B*, **184**: 235–70

Visser, S.A. (1973) Preimpoundment features of the Kainji area and their possible influence on the ecology of the newly formed lake. In *Man-made lakes: their problems and environmental effects*, eds W.C. Ackermann, G.F. White & E.B. Worthington, pp. 590–5. Geophysical Monograph 17. Washington D.C.: American Geophysical Union

Visser, S.A. (1974) Composition of waters of lakes and rivers in East and West Africa. *African Journal of tropical Hydrobiology and Fisheries*, **3**: 43–60

Visser, S.A. & Villeneuve, J.P. (1975) Similarities and differences in the chemical composition of waters from West, Central and East Africa. *Verhandlungen internationale Vereinigung für theoretische und angewandte Limnologie*, **19**: 1416–26

Volcani, E.B. (1936) Life in the Dead Sea. *Nature*, **138**: 467

Volcani, E.B. (1940). Studies on the microflora of the Dead Sea. Unpublished Ph.D. Thesis (in Hebrew). Hebrew University, Jerusalem

Volcani, E.B. (1943*a*) Bacteria in the bottom sediments of the Dead Sea. *Nature*, **152**: 27

Volcani, E.B. (1943*b*) A dimastigoamoeba in the bed of the Dead Sea. *Nature*, **152**: 301

Vollenweider, R.A., Munawar, M. & Stadelmann, P. (1974) A comparative review of phytoplankton and primary production in the Laurentian Great Lakes. *Journal of the Fisheries Research Board of Canada*, **1**: 739–62

Vuilleumier, B.S. (1971) Pleistocene changes in the fauna and flora of South America. *Science*, **173**: 771–80

Vyas, L.N. (1968) Studies on phytoplankton ecology of Pichhola Lake, Udaipur. In *Proceedings of the symposium of recent advances in tropical ecology*, eds R. Misra & B. Gopal, Part I. pp. 334–47. Varanasi: International Society for Tropical Ecology

Vyas, L.N. & Kumar, H.D. (1968) Studies on the phytoplankton and other algae of Indrasagar tank, Udaipur, India. *Hydrobiologia*, **31**: 421–34

Wahby, S.D. (1961) Chemistry of Lake Mariut. *Notes and Memoirs, Hydrobiological Department Alexandria*, **65**: 25 pp.

Wahby, S.D. & Abd El Moneim, M.A. (1979) The problem of phosphorus in the eutrophic Lake Mariut. *Estuarine and coastal marine Science*, **9**: 615–22

Wahby, S.D., Kinawy, S.M., El Tabbagh, T.I. & Abdel Moneim, M.A. (1978) Chemical characteristics of Lake Mariut a polluted lake south of Alexandria, Egypt. *Estuarine and coastal marine Science*, **7**: 17–28

Wahby, S.D., Youssef, S.F. & Bishara, N.F. (1972) Further studies on the hydrography and chemistry of Lake Manzalah. *Bulletin of the Institute of Oceanography and Fisheries*, **2**: 399–422

Walker, K.F. (1973) Studies on a saline lake ecosystem. *Australian Journal of marine and freshwater Research*, **24**: 21–71

Walton, S. (1981) Aswan revisited: US Egypt Nile Project studies. High Dam's effects. *Bioscience*, **31**: 9–13

Ward, P.R.B. (1979) Seiches, tides and wind set up on Lake Kariba. *Limnology and Oceanography*, **24**: 151–7

Warmann, J. St. G. (1969) Onchocerciasis and the Volta dam construction. In *Man-made lakes: the Accra symposium*, ed. L.E. Obeng, pp. 352–60. Accra: Ghana University Press

Washbourn, C. (1967) Lake levels and Quaternary climates in the Eastern Rift Valley of Kenya. *Nature*, **216**: 672–3

Waterbury, J. (1978) *Egypt. Burdens of the past/option for the future.* 318 pp. (Balance of People, Land and Water, pp. 85–112) Bloomington & London: American University Field Staff, Indiana University Press

Weatherley A.H. & Lake, J.S. (1967) Introduced fish species. In *Australian inland waters and their fauna*, ed. A.H. Weatherley, pp. 217–39. Canberra: Australian National University Press

Weers, E.T. & Zaret, F.M. (1975) Grazing effects of nanoplankton in Gatun Lake, Panama. *Verhandlungen internationale Vereinigung für theoretische und angewandte Limnologie*, **19**: 1480–3

Wegener, A. (1912) Die Entstehung der Kontinente. *Petermann's Mitteilungen*, 185–95, 253–6, 305–9

Weir, J. (1959) Molluscicides. *WHO Chronicle*, **13**: 19

Welcomme, R.L. (1972) The inland waters of Africa. Les eaux continentales d'Afrique. *CIFA Technical Paper* **1**: 117 pp.

Welcomme, R.L. (1979) The inland fisheries of Africa. 68 pp. *FAO–CIFA Occasional Paper* No. 7. Rome: FAO

West, G.S. (1907) Report on the freshwater algae, including phytoplankton of the third Tanganyika Expedition conducted by Dr W.A. Cunnington, 1904–1905. *Journal Linnean Society (Botany)*, **38**: 81–197

West, R.C. & Angelli, J.P. (1976) *Middle America; its land and peoples*. 2nd edn, 494 pp. Englewood Cliffs, N.J.: Prentice-Hall

White, E. (1973) Zambia's Kafue hydroelectric scheme and its biological problems. In *Man-made lakes: their problems and environmental effects*, eds W.C. Ackermann, G.F. White & E.B. Worthington, pp. 620–8. Geophysical Monograph 17. Washington D.C.: American Geophysical Union

Widmer, C., Kittel, T. & Richerson, P.J. (1975) A survey of the biological limnology of Lake Titicaca. *Verhandlungen internationale Vereinigung für theoretische und angewandte Limnologie*, **19**: 1504–10

Wilbert, N. & Kahan, D. (1981) Ciliates of Solar Lake on the Red Sea shore. *Archiv für Protistenkunde*, **128**: 70–95

Willen, E. (1973) Phytoplankton in Lake Mälaren 1965 and 1965. *SNV PM 394 NLU Report*, **67**: 50 pp. Uppsala: National Swedish Environment Protection Board, Limnological Survey

Williams, M.A.J., Bishop, P.M., Dakin, F.M. & Gillespie, R. (1977) Late Quaternary lake levels in Southern Afar and the adjacent Ethiopian Rift. *Nature*, **267**: 690–3

Williams, W.D. (1964a) A contribution to lake typology in Victoria, Australia. *Verhandlungen internationale Vereinigung für theoretische und angewandte Limnologie*, **15**: 158–68

Williams, W.D. (1964b) Limnological observations on Lake Tali Karung. *Australian*

Society for Limnology, Newsletter, **3**: 14–15
Williams, W.D. (1966) Conductivity and the concentration of total dissolved solids in Australian lakes. *Australian Journal of marine and freshwater Research*, **17**: 167–176
Williams, W.D. (1967) The chemical characteristics of lentic surface waters in Australia. A review. In *Australian inland waters and their fauna*, ed. A.H. Weatherley, pp. 18–77. Canberra: Australian National University Press
Williams, W.D. (1978) Limnology of Victorian salt lakes, Australia. *Verhandlungen internationale Vereinigung für theoretische und angewandte Limnologie*, **20**: 1165–74
Williams, W.D. & Buckney, R.T. (1976a) Stability of ionic proportions in five salt lakes in Victoria, Australia. *Australian Journal of marine and freshwater Research*, **27**: 367–77
Williams, W.D. & Buckney, R.T. (1976b) Chemical composition of some inland surface waters in South Western and Northern Australia. *Australian Journal of marine and freshwater Research*, **27**: 379–97
Williams, W.D. & Siebert, B.D. (1963) The chemical composition of some surface waters in central Australia. *Australian Journal of marine and freshwater Research*, **14**: 166–75
Williams, W.D., Walker, K.F. & Brand, G.W. (1970) Chemical composition of some inland surface waters and lake deposits of New South Wales, Australia. *Australian Journal of marine and freshwater Research*, **21**: 103–6
Wimpenny, R.S. & Titterington, E. (1936) The tow-net plankton of Lake Qarûn, Egypt, December 1930 to December 1931. *Fisheries Research Directorate, Notes and Memoirs*, no. 14. Cairo
Wissmar, R.C., Richey, J.E., Stallard, R.F. & Edmond, J.M. (1980) Metabolismo do plancton e ciclo do carbono no rio Amazonas, Sus tributarias e aguas varzea, Peru–Brazil, maio junho. *Acta Amazonica*, **10**: 823–34
Woloszynska, J. (1914) Studien über das Phytoplankton des Victoriasees. *Hedwigia*, **55**: 184–223
Wood, R.B., Prosser, M.V. & Baxter, R.M. (1976) The seasonal pattern of thermal characteristics of four of the Bishoftu crater lakes, Ethiopia. *Freshwater Biology*, **6**: 519–30
Wood, W.E. (1924) Increase of salt in soil and streams following the destruction of the natural vegetation. *Journal and Proceedings of the Royal Society of Western Australia*, **10**: 35–47
Worthington, E.B. (1930) Observations on the temperature, hydrogen-ion concentration and other physical conditions of the Victoria and Albert Nyanzas. *Internationale Revue der gesamten Hydrobiologie*, **24**: 328–57
Worthington, E.B. (1932) A report on the fisheries of Uganda. *Uganda Protectorate Crown Agents Publications*, 47–59
Worthington, E.B. (1954) Speciation of fishes in African Lakes. *Nature*, **173**: 1064
Worthington, E.B. (1972) The Nile catchment technological change and aquatic biology. In *The careless technology, ecology and international development*, eds M.T. Farvar & J.P. Milton, pp. 189–205. Garden City, N.Y.: Natural History Press
Worthington, E.B. & Ricardo, C.K. (1936) Scientific results of the Cambridge Expedition to the East African lakes 1930–31. No. 17, The vertical distribution and movements of the plankton in lakes Rudolf, Naivasha, Edward and Bunyonyi. *Journal Linnean Society (Zool.)*, **11**: 56–69
Wright, C.A. (1976) Schistosomiasis in the Nile Basin. In *The Nile, biology of an ancient river*, ed. J. Rzoska, pp. 321–4. Monographiae Biologicae **29**. The Hague: Junk
Wynne, D. (1977) Alternations in activity of phosphatases during the *Peridinium* bloom in Lake Kinneret. *Physiologia Plantarum*, **40**: 219–24
Zalcman, D. & Por, F.D. (1975) The food web of Solar Lake (Sinai Coast – Gulf of Eilat). *Rapport et proces-verbaux des réunions comission internationale pour l'exploration scientifique de la mer Méditerranée*, **23**: 133–4

Zaret, T.M., Devol, A.H. & Santos, A. Dos (1979) Nutrient addition experiments in Lago Jacaretinga, Central Amazon, Brazil. *Proceedings of the SIL-UNEP Workshop on African limnology*, ed. J.J. Symoens. Nairobi: SIL-UNEP

Zaret, T.M., Devol, A.H. & Santos, A. Dos (1981) Nutrient addition in Lago Jacaretinga, Central Amazon basin, Brazil. *Verhandlungen internationale Vereinigung für theoretische und angewandte Limnologie*, **21**: 721-4

Ziesler, R. & Ardizzone, G.D. (1979) Las aguas continentales de América latina. *COPESCAL Doc. Tec.* **1**: 171 pp.

Zoppis, B.L. & Del Guidice, D. (1958) Geologia de la costa del Pacifico de Nicaragua. *Boletin Servicio Geologico Nacional de Nicaragua*, **2**: 33-68

Zutshi, D.P. (1975) Associations of macrophytic vegetation in Kashmir lakes. *Vegetatio*, **30**: 61-5

Zutshi, D.P., Kaul, V. & Vass, K.K. (1972) Limnology of high altitude Kashmir lakes. *Verhandlungen internationale Vereinigung für theoretische und angewandte Limnologie*, **18**: 599-604

Zutshi, D.P., Subla, B.A., Khan, M.A. & Wanganeo, A. (1980) Comparative limnology of nine lakes of Jammu and Kashmir Himalayas. *Hydrobiologia*, **72**: 101-12

Index

Abaya, L.,
 chemical features 206
 location 184
 morphometric features 205
 volume 27
Abhe, L., sediment cores 134
Abiata, L.,
 chemical features 206, 440
 location 184
 morphometric features 205
 volume 27
Abu-Dibbis, L.,
 chemical features 311
 location 311
Afar Lakes, physico-chemical features 209
Afram, R., 280
Africa, 131-287
 area and water balance 26
 aridity 132, 133
 climatic changes 132
 geological history 131, 132
 lake volume 27, 28
 volcanic crater lakes 155-7
 Warm Belt 131
Ajwa Reservoir
 physico-chemical features 349
 plankton 349
Alaotra, L., 285
Albert, L., 133
 algae 160, 448
 chemical composition 159, 439
 density differences 426, 427
 location 157
 morphometric features 158
 origin 157
 physical features 158-9
 salt springs 159
 temperature 158
 zooplankton 160

Albert Nile, 162-3
 discharge 162
 plankton 162, 163
Alipather, L.,
 algae 319
 physico-chemical features 318
allochthonous terraines 8, 9
Alpha-Helix Expedition, fish findings 79-81
Amaravathy Reservoir,
 algae 345
 area and volume 345
Amatitlan, L.,
 fish 120, 121
 heat budget 425
 location 120
 physico-chemical features 119, 439
 plankton 120
Amazon, R.,
 see also blackwaters, várzea
 affluents 58, 59
 bank colonization 70
 biotypes 64
 clear water 60, 68
 discharge and drainage area 25, 58
 flood period 59
 flora and nutrient reserve 63
 geology and vegetation 59, 60
 litter decomposition 63, 64
 length and altitude 57
 physico-chemical features 58
 várzea 58, 59, 64
 whitewaters 60, 64, 70
Amazonia, 55-83
 bacteria 444
 closed nutrient circulation 62
 fish species 79-83
 food chain 82
 geological development 55, 57, 441
 litter fall 63

Index

low salinity 441
map 56
mineral poorness 60-2
rainfall 57
rapid mineralization 445
soil composition 60, 61
throughfall richness 62
tree root features 62, 63
water types 60
Amazon Valley 55, 56
Ambadi, L. 168
  algae 168
Ambarasi Reservoir, location and features 342
Anavilhanas Archipelago 58
Anchar, L., features 320-2
Andes 84-96
Asal, L., sediment cores 134
Asia,
  area and water balance 26
  volume of lakes 27, 28
Aswan High Dam,
  algae 174, 449
  capacity 175
  cost-benefit 188
  ecological impact 187-92
  fisheries impact 191
  history 174-5
  land reclamation 187
  R.Nile shoreline 190
  schistosomiasis 190, 191
  seepage and water balance 188
  see also Nasser-Nubia, L.
Atelomixis 385
Atherton Tablelands, lakes and water 406
Atitlan, L.,
  heat budget 117, 425
  location and biological features 117-19
  physico-chemical features 119, 438
  volume 27
Atmosphere,
  composition 14
  heat transport 16, 17
  pressure and altitude 16
  radiation absorbed 14, 15
  thickness 14
Australia, 397
  area and water balance 26
  biology of waters 415, 418-20
  Gondwanaland 397-420
  lake map 399
  New South Wales waters 403-6
  Northern Territory waters 408
  Queensland waters 406-8
  rainfall 397, 398
  South Australia waters 409-13
  topography 397
  Victoria waters 398-403

volume of lakes 27
water chemistry 413-17
Western Australia waters 408, 409
Awasa, L.,
  features 204-6, 439
  location 184
  volume 27
Awash, R., 209

Bacteria,
  methane-oxidizing 445, 446
  sulphur 446
  Warm Belt 444-7
Bale Mountain Lakes,
  chemical characteristics 210
  fauna 211
  location 209
Bali
  morphological features of lakes 374
  plankton data, Sunda Expedition 376
Bangweulu, L., 27
  algae 234, 448
  chemical features 234
  location 184, 233
  morphometric features 233
  zooplankton and fish 234, 235
Bardawil lagoon,
  algae 289, 290
  features 288
  fish 290
  plankton 289
  plants 289
  water origin 289
Baringo, L., 184
  algae 201
  chemical characteristics 201
  fish 201
  location 184, 204
  origin 201
Barreiras series 59
Beachport-Robe lakes 412, 413
Beeac, L., 399-402
  chemical composition 401, 414, 417, 440
  morphological features 399
Besaka, L.,
  history 132
  sediment cores 134
Bhakra Reservoir,
  fish 324
  sedimentation 323, 324
  surface area 323
  volume 27
Bhawanisagar Reservoir, features 345
bilharzia see schistosomiasis
Bishoftu lakes
  chemical features 208
  crater age 207

morphological features 208
bitter lakes, features 290
blackwaters 56, 60, 237
  Amazon 65, 66
  bacteria 66
  carbon fixation and depth 65
  chlorophyll 66
  fish 81
  flora 65
  lake *see* L.Tupe
  Malaya 355
  zooplankton 65, 66
Blue Lake,
  algae 412
  chemical features 411, 417
  morphometric features 410
Blue Nile,
  flood regime 168
  flow and discharge 171
  origin 168
  reservoir impact 171–3
  watershed 168
Bori-Pat, R., 336, 371
Brahmaputra, R., (Tsang Po)
  rainfall 325
  watershed and origin 325
Brokopondo Reservoir, 32
  area drained 51
  chemical features 437
  chloride levels 55
  dead trees and flora 52, 53
  fish 54
  mixing 54
  plankton 52
  stagnation 52, 53
  temperature stratification 52, 53
  volume 27
Browne Lake, chemical features 411
Buchanan, L., 406
  chemical features 416, 440
Bullenmari, L., 399–402
  chemical composition 401
  morphological features 399
Bum Borapet Reservoir,
  algae 371
  chemical composition 439
  fish 371
  location 366, 370
  sections 370
  volume 27
Bunyoni, L.,
  algae 216, 217
  chemical features 216, 439
  location and origin 215
  mixing 217
  thermal profile 216
  volume 27
Burley Griffin, L., 404, 406, 439
Burullus, L., 178
  features 183
  fish farms 183
  volume 27

Cabora Bassa, L.,
  aquatic plants 259
  chemical composition of river water 258
  history 257
  morphometric features 258
  stratification 258
  volume 27
Calderas, L., physico-chemical features 119
Caribbean area 32–44
de Castanho, L.,
  Amazon connections 76
  chemical features 437
  fauna 78, 79
  floating meadows 76
  productivity and plankton 78, 449
  water features 77
  water mixing and effects 77, 78
Cauvery, R.,
  features 343
  man-made lakes 343–9
Central America,
  fish distribution 127–9
  geological history 104, 105
  high mountain lakes 126, 127
  volume of lakes 27
  rainfall 105
Ceratopteris spp., 53, 54
  control 54
Chad, L., 268–79
  algal productivity 274
  basins 269–71
  chemical composition 270–2, 437
  drainage area 269
  eastern archipelago, plankton biomass 275, 276
  ecosystem sensitivity 278
  energy transfer 279
  evaporation 272
  fauna productivity and biomass 276, 277
  fish 276, 278
  inlets 270
  insects 276, 277
  location 268
  molluscs 276, 277
  morphometric features 270, 271
  origin 269
  oxygen stratification 273
  sediment cores 134
  thermal profile 273, 431

Index

volume 27
water levels 432
worms 276, 277
zooplankton 274, 275
Chapala, L.,
 features 125
 volume 27
Chari, R., discharge and drainage areas 25
chemical features 436–43
 and salinity 437–40
 see also individual rivers and lakes
Chilka, L.,
 algae 338
 climate 338
 features 338
 fish 340
 map 339
 mollusc biomass 340
 volume 27
 zooplankton 340
Chilwa, L.,
 climate 259, 260
 drying effects 261, 262
 fish 260, 261
 fisheries 260
 inflow 259
 location 184, 259
 plankton 260, 262
 refilling effects 262, 263
 Tilapia stocking 263
 volume 27
chlorophyll, Warm Belt lakes 461–3
cichlids in man-made lakes 3
ciénagas,
 area 35
 fish biomass 36
 floating vegetation 36
 turbidity and productivity 35, 36
 volume 27
Cladocera, 471–4
 absent from L. Tanganyika 230, 231
 poverty in tropics 471
 Warm Belt lakes 472–4
Coatepeque, L., features 122
Colorado, R., discharge and drainage area 25
continental drift 8
continental movement 7–9
 Permian 9
 see also Gondwanaland
Copepoda, Warm Belt distribution 474–80
Corangamite, L., 399–402
 chemical composition 401, 417, 440
 fish 420
 morphological features 399
 volume 27

Coriolis force 16
crater lakes, circulation patterns 430
Cuba, lakes and lagoons 129
Cuitzeo, L., 125
Cyperis papyrus 137

Dakataua, L.,
 animals 396
 chemical features 395, 396, 439
 morphological features 394
 plankton 396
 volume 27
Dal, L., features 320–2
Danau Batur,
 algae 377, 381
 morphological features 374
 productivity 382
Danau Bratan,
 algae 377, 380
 chemical features 438
 morphological features 374
 zooplankton 381
Danau Buyan,
 morphological features 374
 phytoplankton 381
Danau Tamblingan,
 morphological features 374
 stratification 381
Dariba lakes, features and fauna 162
Dead Sea,
 algal and bacterial features 306, 307
 analysis 301
 asphalt emissions 307
 chemical features 303, 304
 glycerol production 308
 history 301, 302
 irrigation effects 302
 lower and upper water masses 305
 physical features and water level 302, 303
 salt origin 305
 salt use 307, 308
 sediments 305
 solar energy 308, 435
deforestation,
 India 315, 316, 333
 Indo-China 358
Descoberto, L., 103
DNA in fish and speciation 484

earthquakes 109
East African Rift Valley, 10
 lakes 184, 192, 193
 and water storage 10, 25, 27
Edku, L., 178
 area and salinity 182
 fish catch 183
 primary production 182, 448

558 Index

volume 27
Edward, L., (Africa) 133
  Chaoborus larvae 150–1
  chemical composition 148, 149, 160, 439
  fish 150, 151
  limnological features 148
  location 145, 147
  morphometric features 148
  origin 147
  phytoplankton 150, 160, 448
  temperature 147
  volume 27
  zooplankton 160
Edward L.,`(Australia),
  chemical composition 411, 417
  conifers 412
  morphometric features 410
Eichhornia crassipes (water hyacinth), 34, 53, 375
  Cabora, L., Bassa 259
  spread on Brokopondo Reservoir 54
Elementeita, L., 184
  features and flamingos 195
El Salvador, Lakes 121–4
Encantada, L., physico-chemical features 119
Enriquillo, L., 129, 130
epilimnion, factors affecting 428
Equator,
  barometric pressure 16
  Sun's rays angle 13
Ethiopian lakes 204
Euphrates, R., 308
  discharge and drainage area 25, 309
  features 309–12
  fish 311
  phytoplankton 311
  salinity 309
  zooplankton 311
Euramoo, L.,
  chemical features 414, 416
eutrophication 44
evaporation,
  continents 26
  and precipitation with latitude 19–21
  Warm Belt 22–4
Eyasi, L.,
  chemical features 192, 193
  volume 27
Eyre, L., features and plankton 413

Feia, L., and schistosomiasis 103
fish,
  acoustic estimations 231, 232
  Amazonia, 79–83
  air breathing 80

catfish 80
characoids 80
cichlids 81
  feeding 81, 82
  role 81
stingrays 80
breeding, South-East Asia 314
Central America 127–9
chromosomal number 486
diversity in Warm Belt 480, 481, 489
drying and refilling L.Chilwa 261, 262
farms in Malaya 356, 357
genetics and mutation 484
geographical distribution 481, 482
introduction into L.Titicaca 95
isolation 487
mineral content 372
ponds, Indonesia 382, 383
processing methods 368
self-maintaining 420
speciation 144
African Great Lakes 483–9
species, South America 480
underexploited, Lower Mekong 367–9
várzea 491
L.Victoria and falls 143
Wallace's line 482
flamingos,
  decline in African lakes 194, 195, 199
  filter feeding 199, 200
  migration patterns 200
  Spirulina food 200
Fly River system 393
food webs,
  George, L., 497, 498
  grazing 493–5
  Kinneret, L., 495–7
  várzea 490–3
Fraser Island Lakes, features and fish 407, 438
freshwater, use of term 21
fungal blooms 117

Galillee, L., (Australia), 406
  chemical features 416
Galla Lakes,
  chemical features 206
  features 204, 205
  location 184
  sediment cores 134
Ganges, R., (Ganga),
  algae 328, 329, 449
  chemical features 327–9
  map 327
  origin and tributaries 326
  physical features 327, 328
  watershed 327

Ganges basin, 326–37
  lakes, northern banks 330–4
  lakes, southern banks 334–7
Ganges–Brahmaputra–Meghna, R.,
  discharge and drainage area 25
Gatun, L., 27, 106
  characteristics 106, 437
  flooding (1914) 105
  mixing and thermal stratification 105
  phytoplankton and mixing 107, 449
  zooplankton grazing 107, 108
Gebel Aulia Dam and Reservoir,
  algae 161, 167
  chemical and biological features 161
  effect on Nile 167
  zooplankton 161, 167
Geneva, L., thermocline 427, 428
George, L., 133
  chemical composition 439
  energy input 152
  fish 154
  food chain 497, 498
  location 151
  origin 151
  physical and biological rhythm 154
  physical features 151, 497
  phytoplankton 153, 154
  thermal pattern 431, 432
  volume 27
  water budget 152
  zooplankton 94, 154, 479
Ghazal, R., 163
  chemical features 160
  plankton 160
Glen Helen 408
Godavari watershed 341–3
Gnotuk, L., 399–402
  chemical composition 401
  morphological features 399
Godavari, R., discharge and drainage
  areas 25
Gomback, R.,
  chemical features 351
  fish 352, 372
  location 350
  phytoplankton 351
Gondwanaland, 9
  fragmentation 10
  freshwater fauna 2
  Upper Permian composition 9
Gorewada, L., location and productivity
  342
Grande, R., discharge and drainage area
  25
Grande de Chirripo, L.,
  altitude 126
  flora and fauna 127
  location and features 126, 127
Gran Pantanal 98, 99
Great Lake of Kampuchea (Cambodia),
  chemical features 439
  features 362
  migration of man 363
  plankton 364, 365
  sources 362, 363
  thermal pattern 431
  variability 432
  volume 27
Guanapito Reservoir,
  features 49
  productivity and algae 49, 50, 449
Guatemala lakes 112–21
Guiana area,
  drainage area 51
  map 51
Guija, L.,
  heat budget 425
  volume 27, 123
  zooplankton 123
Guri Reservoir,
  algal flora 50
  area and drainage 50
  chemical features 437
  volume 27

Habbaniyah, L., 311
Hadley cells, water transport 18
Heat budget,
  temperate lakes 425
  Warm Belt lakes 425
high-altitude lakes, stratification 430, 431
Huacachina, L., 84
humidity transport 18
Hwang-ho, discharge and drainage area 25
hydrosphere,
  components 22
  freshwater 21, 22

Ihotry, L.,
  features 285
  fish 285, 286
  water chemistry 285, 286
Ilopango, L.,
  features 121, 122
  floating vegetation 121
  volcanic origin 121
India 314–49
  brackish water 338–41
  deforestation effects 315, 316
  flora and fauna development 315
  Himalayas and isolation 315
  hunting effects 315
Indo-China, 358–72
  climate 358

560 Index

geology 358
population 359
Indonesia, 372–83
  fish ponds 382, 383
  islands 373
  lake studies 373–5
  morphological lake features 374
  siesmic activity 372
Indrasagar Tank, algal fluctuations 336, 337
Indus, R.,
  discharge and drainage area 25
  features 317
  source 316
Indus basin, 316–25
  area 316
  climate 316, 317
  Siwalik lakes 322, 323
  valley lakes 320–2
Intertropical Convergence Belt 16
Iran, waters 312
Iraq, waters 308–12
Irrawadi, R., discharge and drainage areas 25
Itaipu Dam, 101
  area flooded 32
Izabel, L.,
  fish 117
  fungal blooms 117
  morphometric features 115
  orographic rainfall 115
  pH and chemistry 115, 116
  physico-chemical features 119, 438
  plankton 116, 117
  seawater 116
  volume 27
  watershed 113

Janauaca, L.,
  features 74, 75
  flora, fauna and mixing 75, 76
  nutrient levels 75
  zooplankton changes 75, 76
Java,
  morphological features of lakes 374
  plankton data, Sunda expedition 377–9
Jebel Auliya, capacity 175
Jhelum Valley Lakes
  features 320, 321
  phytoplankton 322
Jonglei project, 165, 166
  ecological effects 165
Jordan, R., 293
  algae 449
  chemical features 296, 297
  regulation of flow 293
  system, map 294

Jordan Valley, 293–308
  geological origin 293
Junin, L., chemical features 84

Kagera, R., 136
Kainji Reservoir and Dam,
  fish and dam closure 268
  morphometric features 265, 266
  R.Niger flow 266
  plankton fluctuations 267, 268
  suspended material 267
  thermal regime 266
  volume 27
Kariba, L.,
  algal succession 246, 448
  chemical features 244–6, 437
  density and stratification 426, 427
  ecological impact 250, 251
  fish origin and succession 248, 249
  fish production potential 249, 250
  heat budget 244
  hydrogen sulphide 245
  invertebrate succession 247, 248
  islands 242
  light transmission 243
  morphometric features 242
  origin 241
  Salvinia development 246, 247, 250
  seiches and tides 243
  seismic effects 242
  temperature 243, 244
  thermocline 244
  volume 27
  water budget 243
  zooplankton 247
Kariba Dam, 240
  purposes 242
Kashmir Himalaya lakes, high altitude 317–20
Kathmandu Valley Lakes, features and rainfall 331–3
Khan Payao, L.,
  location 366, 370
  plankton 370
Khasm Al-Girba, capacity 175
Kimberley waters 399
  features 409
Kinjar Reservoir,
  algae 324, 325
  features 324
  volume 27
Kinneret, L.,
  algae 299, 300, 448
  algal blooms 298
  bacteria 446
  benthic fauna 300
  carbon balance 496

Index

chemical features 296, 297
density differences 426, 427
double food chain 495–7
features 295
fish 300, 301,
   speciation 485
heat budget 296, 425
lake level 276
location 295
nutrient levels 298
Peridinium cinctum 298–300, 496
seiches 295
storms 429
vertical mixing 142
water budget 296
wind circulation 296
zooplankton 94, 108
Kitangiri, L., 184
  chemical features 192, 193, 439
  volume 27
Kivu, L., 217–23, 432
  algae 220, 222, 448
  chemical features 220, 221
  dead layer temperature 219
  dissolved carbon 220
  double diffusion convection 433
  fish 222, 223
  history 217
  hydrogen carbonates 220
  hydrothermal jets 218, 219
  location 184, 217
  magnetic survey 218
  morphometric features 218
  temperature profiles 219, 220
  thermocline 432–4
  volume 27
  zooplankton 222
Klindungan, L.,
  heat budget 425
  morphological features 374
  plankton 378
Kodai, L., features 347, 348
Korat plateau 366, 371
Kosciusko plateau lakes,
  chemical composition 404, 405, 438
Krishna, R.,
  dam 343
  discharge and drainage area 25
  plankton 343
  watershed 341–3
Krishnaraja Sagar, storage capacity 343
Kyoga, L., 133
  chemical composition 145, 147, 439
  fish 147
  location 145
  morphometric features 145

  origin 137, 144
  phytoplankton 145
  volume 27
Lagartijo Reservoir,
  conductivity 48
  density current 429, 459
  features 48
  plankton 49, 449
  productivity and turbulence 49
Laguna de Bay,
  duck rearing 390
  features 390
  pollution 391
  snails 390, 391
  volume 27
  zooplankton 390
lakes,
  crater, circulation patterns 430
  high-altitude, circulation pattern 430, 431
  large deep 432–4
  large shallow 431, 432
  seasonal 434
  solar 434
  Warm Belt capacities 24–7
Lanao, L.,
  chemical features 438
  circulation patterns 384
  endemic cyprinids 387
  energy budget 385
  heat budget 425
  light and productivity 386
  location and origin 383
  morphological features 384
  phytoplankton 385, 387, 449, 458
  rainfall 383, 384
  species diversity 386
  thermocline 427, 428
  volume 27
  zooplankton 387
Langano, L.,
  features 204–6
  location 184
  volume 27
Laurasia, 9
  Upper Permian composition 9
Leake, L.,
  chemical composition 411, 417
  ionic patterns 414
  morphometric features 410
Leg of Mutton Lake, 412
  chemical features 411
Lere, L., location and features 265
Litter decomposition, Amazonia 63, 64
Lower Mekong fish 367
Lower Mekong Basin Project 360–2

Lubumbashi, R.,
  chemical composition 236, 237
  location 236

Madagascar, 284-7
  flora and origin 285
  lakes 285
  phytoplankton 286, 287
Madden, L.,
  phytoplankton and mixing 107
  stratified 106
  volume 27, 106
  zooplankton grazing 107, 108
Mae Nam River,
  location 369, 370
  molluscs 370
Magadi, L., 184
  chemical features 193, 194, 440
  flamingo decline 194
  hot springs 193
  sediment cores 134
  volume 27
  water sources 194
Magdalena, R.,
  bimodal discharge 33, 34
  ciénagas 34-6
  copper content 34
  fish catch 36
  flora 34
  length and tributaries 33
  salt content 34
  sewage carried 34, 35
  watershed 33
Mainit, L.,
  algal production 389
  chemical features 388, 389
  climate 388
  fishes 389
  location 387
  morphological features 388
  volume 27
  zooplankton 389, 466
Maji Ndombe, L.,
  features 238
  volume 27
Makega, L., salinity 156, 440
Malacca, drought and pH 355
Malawi, L.,
  algae 257
  barrage 252
  chemical composition 254, 439
  climate 251, 252
  fish,
    distribution 256, 257
    origin 255, 256
    speciation 485
  gastropods 255
  history and origin 251, 255

  location 184, 251
  morphometric features 252
  plankton 254, 448
  stratification 253
  thermoclines 432-4
  vegetation 257
  volume 27
Malaysia,
  algae and pH 355, 356
  blackwaters 355
  chemical features of waters 356
  fish, 356
    ponds 356, 357
    species 357, 358
  waters of south 355-7
Malebo Pool 238
Managua, L.,
  chemical features 109, 110, 439
  eutrophication 111
  fauna 111, 112
  fish 110, 111
  geological history 108
  morphometric features 109
  plankton 110
  salt 110
  volume 27, 109
Manasbal, L., features 320-2
Mannavanur, L., features 347, 348
Mansar, L., features 322, 323
Manyara, L., 184
  chemical features 193, 440
  volume 27
Manzalah, L., 178
  features 183
  fish catch 184
  volume 27
Maracaibo, L.,
  algal blooms 40
  description 38
  deuterium balance 39
  fauna 39
  hydrography 38, 39
  lagunar 37
  location 37
  nutrients and productivity 40
  oil fields 37
  oil spills 41
  pollution 40, 41
  rainfall 38
  salinity 40
  satellite photograph, *ii*
  Strait of 38
  volume 27, 38
  water gains and losses 39
  water mixing 38, 39
  watershed 33
Maranon, R., 56
  source 57

Index

Mariut, L., 178
 chemical composition 179, 180
 fish 181
 phytoplankton 180
 size and history 179
 volume 27
 zooplankton 180, 181
Mary, L., chemical features 416, 439
Meghna, R., features 326
Mekong, R.,
 chemical features, Nong Khai 360
 course 359
 discharge and drainage area 25, 359, 360
 fisheries 367-9
 length 359
 map 361
 plankton 364, 365
 population 359
 suspended solids 363
 watershed and storage reservoirs 361
 water utilization 361
 see also Lower Mekong
Mexico,
 lakes 124-6
 Lerma system 124, 125
 Valley lakes 125, 126
microcontinents, properties 8, 9
Mississippi, R., discharge and drainage area 25
Monaro Peneplain Lakes, chemical composition 404, 405
monsoons 17
 mechanism in South-East Asia 313, 314
Moondara, L.,
 features 408
 man-made 407
Mucubaji, L.,
 algae 47
 location and features 46, 47, 437
 productivity 48
Mukerti Reservoir, features 346
Mulwala, L., 403
Murray, R., salinization 415
Murray-Darling, R., fish 419
mutation and fish speciation 484
Mwadingusha Reservoir,
 location 235
 volume 27
Mweru, L.,
 chemical features 234, 235
 fauna 235
 location 184, 235
 morphometric features 233
 volume 27

Nabagaboo, L., isolation and speciation 487
Nabugado, L., chemical composition 140, 141, 437

Naina, L., features and algae 335, 336
Nainital, L.,
 deforestation effects 333
 features 333
 plankton 333, 334
 uses 334
Naivasha, L., 184
 chemical features 196, 197, 439
 fish species 198
 hydrological balance 196, 197
 location 184, 195
 origin 196
 physical features 196
 phytoplankton 196, 197, 448
 sediment cores 134
 vegetation 195
 zooplankton 197
Nakuru, L.,
 alkaline springs 199
 chemical features 198, 199, 440
 cichlid introduction and predation 200
 flamingos 199
 location 184, 198
 morphometric features 198
 pelicans 200
 sediment cores 134
Nam Pong Reservoir,
 algae 365
 location 365, 366
 zooplankton 366
Napo, R., fish 83
Narambach, L., features 320-2
Nasser-Nubia, L.,
 Blue Nile water 177
 density differences 426, 427
 features 175
 fish 178
 hydrological features 177
 morphometric features 177
 phytoplankton 177
 vegetation 177-8
 volume 27, 176
 waterflow 175
Natron, L.,
 chemical features 193, 194, 440
 fish 195
 flamingos 194, 195
 volume 27
Nazdab, L., temporary 410
Negro, R.,
 blackwater 60
 mouthbay 58
 phytoplankton 65, 449
 productivity 59, 65
Nicaragua, lakes 109-11
Nicaragua, L.,
 chemical features 109, 110, 438
 fish 110, 111

geological history 109
morphometric features 109
plankton 110
sharks 108
volume 27
Niger, R.,
　chemical features 264, 265
　delta lakes 264, 265
　discharge and drainage area 25
　features 263
　hydraulic regime 264
　origin 263, 264
　watershed 263
Niger basin 263–8
Nile, R., *see also* Blue Nile, White Nile
　algal bloom 174, 449
　brackish water lakes 178–87
　chemical features 161
　coastal erosion 189, 190
　composition at mixing 173
　delta barrages 175, 189
　discharge and drainage area 25, 175, 188
　Mediterranean flood discharge 188
　phytoplankton 161
　pollution 191
　scouring 189
　sediment 178,
　　and Aswan Dam 189
　water storage schemes 174–5
　zooplankton 161
Nile basin 134–92
Nile system, 133–92
　chemical and biological features 160, 161
　historical drainage 136
Nilnag, L.,
　algae 320
　features and plants 319, 320
　zooplankton 320
Niriz, L., features and basins 312
Nkugute, L.,
　chemical composition 437
　stratification 430
Nong Raharn Reservoir,
　algae 366, 367
　location 366
North America, area and water balance 26
nutrient supply, Warm Belt 459
Nzilo Reservoir,
　features 236
　volume 27

oceans,
　currents and heat transport 17, 18
　external layers 9
　ocean spreading, 8
Okavango, R., 238

onchocerciasis 283
Ootacamund, L., features 346
Ooty, L.,
　features 347, 348
　pollution 348
Orinoco, R.,
　–Casiquiare waterway 45
　chemistry of upper 45, 46
　course 45
　delta area 46
　discharge and drainage area 25, 45
　fauna 47
　features of basin 45–7
　flora 47
　jets 45
　reservoirs and lakes 47–50
　water levels 46
　watershed 46
Orinoco system, 44–50
　location 44
Oti, R., 280
　chemical composition 279
Owen Falls Dam, capacity 175

Pahang, R., basin 354
Panama lakes 105–8
Papsalum repens 72
Papua New Guinea, 391–6
　geology 391
Parana basin, 96–105
　water bodies 99–101
Parana, R.,
　Amazonia similarities 97
　chemical features 437
　dams 101
　discharge and drainage area 25
　economic uses 101, 102
　fish abundance 101
　flooding 99
　flora 100
　geology 97
　map 98
　–Paraguay watershed 97
　rainfall 98
　vegetation 98, 99
　water chemistry 99
　zooplankton 100
Paranoa, L., eutrophication and Brasilia 102
Patzcuaro, L.,
　features and fauna 124, 125, 438
　volume 27
pelican breeding and feeding grounds 200
Peridinium sp.,
　high lakes 47, 48
　Kinneret, L., 297–9
Peru coastal plain 83, 84

Index

Peten Lakes,
  names and features 112, 113, 438
  plankton 113
  stratification 112
  water chemistry 112, 114
Peten Itza, L., volume 27
Phillipines, 383–91
  fishery production 383
  location and islands 383
phytoplankton,
  associations 458
  Bacillariophyta distribution 451, 452, 455
  biomass 461–5
  Chlorophyta distribution 453–5
  Chrysophyta distribution 457, 458
  Cyanophyta distribution 451, 454, 455
  Desmidiales distribution 454–6
  Dinophyta (Pyrrophyta) distribution 452, 457
  Euglenophyta distribution 452, 453, 457
  nutrient supply 459
  primary production 461–5
  sampling 447–9
  seasonal successions 458–60
  species found 448, 449
  Warm Belt distribution 451–4
Pichhola, L., features and pollution 336
Pistia stratiodes 34, 37
de la Plata, R., 96–105
  area of basin 96, 97
  geology 97
  hydraulic use 101
  map 98
  rainfall 98
  sewage 102
  vegetation 98, 99
plate tectonics 8, 10
plate translocation 9
  velocity 8
Pokhara Valley lakes
  features 331
  phytoplankton 332, 333
pollen analysis 132
pollution, 32
  Caribbean area 34, 35
Polochic, R., 115, 116
Poopó, L.,
  chemical features 91, 440
  chloride 92, 440
  hydrological balance 88, 89
  location 84, 85, 88
  map 85
  phytoplankton 92
  volume 27
El Prado reservoir, features 37
precipitation
  continents 26
  evaporation and latitude 19–21
  Warm Belt 22, 23
  Warm Belt Lakes 424, 425
Preto de Eva, L., fish 81
production values and lake latitude 464
Protozoa,
  African warm lakes 465
  ciliates 466
Pulicat, L.,
  features 340, 341
  fish yield 341
  map 339
  volume 27
Purari River,
  catchment area 391
  chemical features 392, 438
  dam impact 392
  System Project 391–3
Purrumbete, L., 399–402
  morphological features 399
Pykara, L., features 347, 348

Qarun, L.,
  basins 184
  plankton 185, 186
  fish 187
  location 178
  molluscs 186, 187
  salt accumulation 185
  underground income 185
Quaternary ice cover 1

radiation,
  atmosphere penetration 14
  balance of Warm Belt 15
  emitted by earth 15
  surplus and deficit areas 15
radiation flux, Warm Belt 12, 13
rain forest,
  Amazonia–Orinoco area 32
  interference effects 32
Ranu Klindungan, *see* Klindungan, L.
Ranu Lamongan,
  deforestation 380
  diatoms 375
  Eichhornia communities 375
  fish 380
  morphological features 374
  plankton 377
  turnover, stratification 375
Redondo, L.,
  fauna 74
  fish 74, 81
  food web 72
  location and features 70, 71
  meadow biomass 72–4
Reservoir 5th of November, 123
  features 123, 124

566 Index

Rhizophora mangle 47
rifting 10
rivers, runoff and drainage areas of Warm
  Belt 24, 25
Roots,
  features of Amazonia 62, 63
  and fungi 63
Roseires Dam,
  channel and biological effects 161, 171,
    172
  reservoir capacity 175
  silt desposition 173
Rotifera
  endemic Amazonian 469
  latitudinal distribution 470, 471
  Warm Belt lakes 466–71
Rudolf, L.,
  chemical characteristics 203, 439
  fish 204
  history 202, 203
  location 184, 202
  phytoplankton 204
  salinity 203
  sediment cores 134
  volume 27
Rukura, L., sediment cores 134
runoff,
  continents 26
  global 20
  Warm Belt 22–5
Ruppia maritima 289

Salween, R., discharge and drainage area
  25
Salvinia spp, L. Kariba 246
Santa Maria, L., 102
Sathiar Reservoir, formation and features
  348, 349
schistosomiasis, 60, 190, 191
  snail vectors 190
Schistosoma haematobium,
  L.Chad 284
  and irrigation 190, 191
Schistosoma mansoni, 60
  and irrigation 190, 191
Scirpus totora 96
sediment cores of African lakes 134
Senegal, R., discharge and drainage areas
  25
Sennar Dam,
  chemical and biological effects 161, 171,
    172
  reservoir capacity 175
sewage treatment, Sweden 3
Seychelles Islands, granite basement 9
Shala, L.,
  features 204–6, 440

location 184
volume 27
Shamo, L.,
  features 204–6
  location 184
  volume 27
Sinai Peninsula 288–93
Sirunumu impoundment, chemical features
  of water 392, 393
Sladen, Percy, expedition 89
snakes, Warm Belt 480
Sao Francisco, R., discharge and drainage
  area 25
Soil composition, 60, 61
  mineral-poor in Amazon 60–2
Solar Lake,
  algae and bacteria 291, 292
  algal history 292
  animals 292, 293
  biothermal processes 290
  temperature gradients 290, 291
solar ponds,
  anomalous stratification 434, 435
  energy production 308, 435
Solimoes, R., 68, 69 *see also* várzea
Songkhla, L.,
  features 372
  location 366
  volume 27
South America
  Andes east and west 31
  area and water balance 26
  hydrographic interference 32
  species diversity 488
  tectonic stability 31, 32
  topography 31
  volume of lakes 27, 28
  see also individual regions
South-East Asia, 313–96
  definition, UNESCO 313
  monsoon 313
  population density 314
  volume of lakes 27
  winter lacking 314
Stanley Reservoir,
  features 343, 344
  fish migration 344
  location 343
  volume 27
stelomixis 429
stratification, 426–35
  density differences 427
  vertical thermal gradient 426–8,
  factors affecting 428–30
Subang, L.,
  algae 353
  chemical features 352, 353, 438

Index

night mixing 429
physical features 352, 353
reservoir 352
subtropical belts of high pressure, 16, 17
  air motion 17
  deserts 16
  horse latitudes 17
Sudd swamps,
  canal bypass 166 (*see also* Jonglei project)
  fish 165
  large herbivores 165
  location 163
  and Nile water chemistry 163
Sumatra,
  morphological features of lakes 374
  plankton data, Sunda expedition 376, 377
Sunda expedition, 373, 374
  plankton data 376–9
Sun–Earth relationship, radiation flux 12
supralimital speciation 387
Suriname, R.,
  dam 51 (*see also* Brokopondo Reservoir)
  water features 51
Surinsar, L., features 322, 323
Syro-African Rift Valley 131

Tali Karng, L., features and zooplankton 402, 403
Tana, L.,
  and Blue Nile 168
  chemical features 169, 439
  fish 171
  molluscs 170
  morphometric features 168
  phytoplankton 169, 170
  vegetation 169
  volume 27
Tanganyika, L., 27, 184, 211, 223–33
  algal blooms 228, 229
  chemical features 225–7, 439
  clarity 227
  echo sounding profiles 223, 232
  fish yield 228, 230, 231
  grazing food web 493–5
  hydrology 223, 224
  location 184, 223
  morphometric features 224
  origin 232, 233
  physical features 223
  phytoplankton 227–9, 448
  Protozoa 228, 229
  salinity gradient 429
  sampling stations 229
  sardines 227, 231
  thermocline 224, 225, 432–4
  volume stability 225
  zooplankton 230, 231, 479

Taruma-Mirim, L., zooplankton 65, 66
Tasek Bera, location and features 354
Tasek Merah, features and desmids 354
Tasmania, features of lakes 414
temperature, Warm Belt lakes 423–5
Tertiary climate 12
Texcoco, L., 126
Thailand,
  climate 358
  fishery evolution 369
  hydrography 366
  southern waters 371, 372
  hermocline 429, 432–4
Tigris, R.,
  discharge and drainage area 25, 309
  features 309–12
  fish 311
  phytoplankton 311
  salinity 309
  zooplankton 311
Tiquina canal 86
Titicaca, L.,
  algal changes with time 92
  basin connections 86
  bird abundance 95, 96
  characteristics, morphometry 86
  charophytes 92, 93
  climate 89, 90
  density and stratification 426, 427
  evaporation 88, 90
  fauna 94–6
  flora 94–6
  heat budget 90, 425
  hydrological balance attempts 86–8
  L.Major and L.Pequeno 85, 88, 91, 92
  location 84, 85
  map 85
  mixing 431
  oxygen deficit 90
  phytoplankton 92, 93, 96
  productivity and algal biomass 93
  residence time 87
  species origin 96
  temperature 423
  trout introduction 95
  volume 27, 86
  water chemistry 91, 439
  watershed 87
  zooplankton 94
Titicaca depression 85
Tocantins, R., discharge and drainage area 25
Tonle Sap, 361, 362
  current 364
  fish 364
  flood plain 363
Tortugas, L., physico-chemical features 119

Tota, L.,
    features 36, 437
    oxygenation and productivity 36, 37
    volume 27
trade winds 17
Trigan, L., features 320–2
trypanosomiasis 250, 251, 284
tsetse fly 250, 251, 284
Tsola Tso, L., features 330, 331
Tumba, L.,
    blackwaters 237
    fauna 237
    features 237, 437
    fish 238
    location 237
    volume 27
Tupe, L.,
    bacteria 444
    blackwater lake 66, 67
    fauna 67
    pH 67
    stratified 67
    wind protection 426
Two Seas Project 308

Upemba, L.,
    chemical features 234, 235
    morphometric features 233
Uruguay, R., discharge and drainage area 25

Valencia, L.,
    chemical features 439
    disturbance and water level 42–4
    eutrophication 44
    features and age 41
    floating meadows 43
    location 41
    nutrient load 42
    productivity and fauna 42, 43, 449
    salinity 42
    stages I–III 41
    volume 27
    watershed 33
    zooplankton 43, 44
valley lakes,
    chemical features 411
    morphometric features 410
Várzea 58, 59, 64
    algal fluctuations 69
    fish migration 491
    flora 68, 69
    food web 490–3
    furo, channel 71
    lakes 70–9
    macrophytes 490, 491
    mineralization 74

nutrient flux 491, 492
productivity 69
see also individual lakes
Victoria, L., 137
    altitude 135
    chemical composition 140–1, 160, 437
    chemical stratification 139, 140
    drainage area 136
    Falls, 143
        barrier to fish 249
    fish 143–4
    geological history 137, 138
    invertebrates 143, 160
    morphometric parameters 139
    phytoplankton 141, 143, 160, 448
    radiocarbon dating 138
    rainfall pattern 137
    sediment cores 134
    seiche 140
    swamps 137
    vertical mixing 142
    volume 27, 139
    water budget 138, 139
    zooplankton 143, 160
Victoria amazonica 73
Volcanic crater lakes, W. Uganda 155–7
    see also crater lakes
Volta, L.,
    aquatic plants 283
    chemical composition 281, 282, 438
    clam fisheries 284
    disease 284
    fish groups 283, 284
    Gorge area 281
    inlets 280
    location 280
    morphometric features 280, 281
    phytoplankton 282, 283
    project origin 279
    thermal structure 280, 281
    volume 27, 280
    winds 280, 281
    zooplankton 283
Volta basin, 279–84
    rivers, chemical composition 279
    watershed 279

Wabo, L., 392
Warm Belt,
    atmospheric pressure and winds 16, 17
    circulation patterns 423–35
    climate 12–18
    continental water balance 24, 26
    drainage system 7
    energy balance 15
    geodynamic history 7–11
    Gondwanaland 9

Index

high-altitude lakes 10, 11
historic climate 1
humidity 2
lake distribution 25–7
lake temperature 1
oceanic water deficit 22
old species distribution 482, 483
Precambrian formations 10
radiation and insolation 12, 13
runoff and river discharge 24, 25
water balance 19–21, 23
water,
  active 21, 22
  depth 21
  exchange and oceans 19
  glacier 21
water balance,
  global 19, 20
  Warm Belt continents 24
water chemistry 436–43
  see also individual lakes
weathering and salinity 441, 442
Werowrap, L., 399–402
  chemical composition 401, 417
  ionic patterns 414
  morphological features 399
  phytoplankton 418
West Indies, 129, 130
  freshwater invertebrates 130
West Malaysia, 349–58
  climate 349, 350
  location 349
  rainfall 350
  rainforest 350
  water bodies 350
  (*see also* Malaysia)
White Nile, chemical and biological
  features 161, 166
winds,
  direction and hemisphere 16
  velocity, Warm Belt lakes 425, 426
  Warm Belt 16, 17
Wisdom, L.,
  chemical features 394, 395, 439

fauna 394
features and rainfall 393
morphometric features 394
plankton 394
temperature 393
volume 27

Yamuna, R., 334
  canal systems 334
  Delhi water supply 334
  phytoplankton 335
Yercand, L., features 347, 348

Zaire basin, 211–38
  features 211, 212
  lower, water bodies 237, 238
  surface 211
  upper, lakes 233–7
Zaire (Congo), R., 212–15
  chemical features 213, 214
  discharge and drainage area 25
  fish and Amazon 214, 215
  length 212
  level fluctuations 213
  phytoplankton 214
  volume 25
  Qualube River capture 483
Zambesi basin 238–63
Zambesi, R., 238–41
  chemical features 241, 257, 258
  climate 240
  discharge and drainage area 25, 240, 241
  flow and Victoria Falls 238, 240
  origin and map 238, 239
  silt 241
Zirahuen, L., 125
Ziway, L.,
  conductivity 206
  features 204–6
  location 184
  volume 27
Zooplankton, warm lake distribution
  466–80